Lecture Notes in Mathematics 1883

F. Marcellán · W. Van Assche (Eds.)

Orthogonal Polynomials and Special Functions

Computation and Applications

Springer

Editors

Francisco Marcellán
Departamento de Matemáticas
Universidad Carlos III de Madrid
Avenida de la Universidad 30
28911 Leganés
Spain
e-mail: pacomarc@ing.uc3m.es

Walter Van Assche
Department of Mathematics
Katholieke Universiteit Leuven
Celestijnenlaan 200B
3001 Leuven
Belgium
e-mail: walter@wis.kuleuven.be

Library of Congress Control Number: 2006923695

Mathematics Subject Classification (2000): 33C45, 33C50, 33E17, 33F05, 42C05, 41A55
65E05, 65F15, 65F25

ISSN print edition: 0075-8434
ISSN electronic edition: 1617-9692
ISBN-10 3-540-31062-2 Springer Berlin Heidelberg New York
ISBN-13 978-3-540-31062-4 Springer Berlin Heidelberg New York

DOI 10.1007/b128597

Springer is a part of Springer Science+Business Media
springer.com

Printed in The Netherlands

Typesetting: by the authors and SPI Publisher Services using a Springer LaTeX package
Cover design: *design & production* GmbH, Heidelberg

Printed on acid-free paper SPIN: 11605546 VA41/3100/SPI 5 4 3 2 1 0

Preface

These are the lecture notes of the fifth European summer school on *Orthogonal Polynomials and Special Functions*, which was held at the Universidad Carlos III de Madrid, Leganés, Spain from July 8 to July 18, 2004. Previous summer schools were in Laredo, Spain (2000) [1], Inzell, Germany (2001) [2], Leuven, Belgium (2002) [3] and Coimbra, Portugal (2003) [4]. These summer schools are intended for young researchers preparing a doctorate or Ph.D. and postdocs working in the area of special functions.

For this edition we were happy to have eight invited speakers who gave a series of lectures on a subject for which they are internationally known experts. Seven of these lectures are collected in this volume. The lecture of J. S. Geronimo on *WKB and turning point theory for second order difference equations* has been published elsewhere [5].

The lectures fall into two categories: on one hand we have lectures on *computational aspects* of orthogonal polynomials and special functions and on the other hand we have some modern *applications*. The computational aspects deal with algorithms for computing quantities related to orthogonal polynomials and quadrature (Walter Gautschi's contribution), but recently it was also found that computational aspects of numerical linear algebra are closely related to the asymptotic behavior of (discrete) orthogonal polynomials. The contributions of Andrei Martínez and Bernhard Beckermann deal with this interaction between numerical linear algebra, logarithmic potential theory and asymptotics of discrete orthogonal polynomials. The contribution of Adhemar Bultheel makes the transition between applications (linear prediction of discrete stationary time series) and computational aspects of orthogonal rational functions on the unit circle and their matrix analogues. Other applications in this volume are quantum integrability and separation of variables (Vadim Kuznetsov), the classification of orthogonal polynomials in terms of two linear transformations each tridiagonal with respect to an eigenbasis of the other (Paul Terwilliger), and the theory of nonlinear special functions arising from the Painlevé equations (Peter Clarkson).

Walter Gautschi gave a lecture about *Computational methods and software* for orthogonal polynomials, in particular related to quadrature and approximation. His lecture describes many algorithms which can be used in Matlab. The lecture of Andrei Martínez-Finkelshtein is about *Equilibrium problems of potential theory in the complex plane* and gives a brief introduction to the logarithmic potential in the complex plane and the corresponding equilibrium problems. Minimizing logarithmic energy is very close to best polynomial approximation. In his lecture the equilibrium problem is described in the classical sense, but also the extensions with external fields and with constraints, which are more recent, are considered. The lecture of Bernhard Beckermann on *Discrete orthogonal polynomials and superlinear convergence of Krylov subspace methods in numerical linear algebra* makes heavy use of the equilibrium problem with constraint and external field, which is a necessary ingredient for describing the asymptotics for discrete orthogonal polynomials. This asymptotic behavior gives important insight in the convergence behavior of several numerical methods in linear algebra, such as the conjugate gradient method, the Lanczos method, and in general many Krylov subspace methods.

The contribution of Adhemar Bultheel and his co-authors on *Orthogonal rational functions on the unit circle: from the scalar to the matrix case* extends on one hand the notion of orthogonal polynomials to orthogonal rational functions and on the other hand the typical situation with scalar coefficients to matrix coefficients. The motivation for using orthogonality on the unit circle lies in linear prediction for a discrete stationary time series. The motivation for using rational functions is the rational Krylov method (with shifts) and numerical quadrature of functions with singularities, thereby making the link with the lectures of Gautschi and Beckermann.

Vadim Kuznetsov's lecture on *Orthogonal polynomials and separation of variables* first deals with Chebyshev polynomials and Gegenbauer polynomials, which are important orthogonal polynomials of one variable for which he gives several well known properties. Then he considers polynomials in several variables and shows how they can be factorized and how this is relevant for quantum integrability and separability.

Paul Terwilliger describes *An algebraic approach to the Askey scheme of orthogonal polynomials.* The fundamental object in his contribution is a Leonard pair and a correspondence between Leonard pairs and a class of orthogonal polynomials is given. Even though the description is elementary and uses only linear algebra, it is sufficient to show how the three term recurrence relation, the difference equation, Askey-Wilson duality, and orthogonality can be expressed in a uniform and attractive way using Leonard pairs.

Finally, Peter Clarkson brings us to a very exciting topic: *Painlevé equations — Nonlinear special functions.* The six Painlevé equations, which are nonlinear second-order differential equations, are presented and many important mathematical properties are given: Bäcklund transformations, rational solutions, special function solutions, asymptotic expansions and connection formulae. Several applications of these Painlevé equations are described, such as partial differential equations, combinatorics, and orthogonal polynomials, which brings us back to the central notion in these lecture notes.

We believe that these lecture notes will be useful for all researchers in the field of special functions and orthogonal polynomials since all the contributions contain recent work of the invited speakers, most of which is not available in books or not easily accessible in the scientific literature. All contributions contain exercises so that the reader is encouraged to participate actively. Together with open problems and pointers to the available literature, young researchers looking for a topic for their Ph.D. or recent postdocs looking for new challenges have a useful source for contemporary research problems.

We would like to thank Guillermo López Lagomasino, Jorge Arvesú Carballo, Jorge Sánchez Ruiz, María Isabel Bueno Cachadiña and Roberto Costas Santos for their work in the local organizing committee of the summer school and for their help in hosting 50 participants from Austria, Belarus, Belgium, Denmark, England, France, Poland, Portugal, South Africa, Spain, Tunisia, and the U.S.A. This summer school and these lecture notes and some of the lecturers and participants were supported by INTAS Research Network on Constructive Complex Approximation (03-51-6637) and by the SIAM activity group on *Orthogonal Polynomials and Special Functions.*

References

1. R. Álvarez-Nodarse, F. Marcellán, W. Van Assche (Eds.), *Laredo Lectures on Orthogonal Polynomials and Special Functions*, Advances in the Theory of Special Functions and Orthogonal Polynomials, Nova Science Publishers, New York, 2004.
2. W. zu Castell, F. Filbir, B. Forster (Eds.), *Inzell Lectures on Orthogonal Polynomials*, Advances in the Theory of Special Functions and Orthogonal Polynomials, Nova Science Publishers, New York, 2005.
3. E. Koelink, W. Van Assche (Eds.), *Orthogonal Polynomials and Special Functions: Leuven 2002*, Lecture Notes in Mathematics **1817**, Springer-Verlag, Berlin, 2003.
4. A. Branquinho, A. Foulquié (Eds.), *Orthogonal Polynomials and Special Functions: Approximation and Iteration*, Advances in the Theory of Special Functions and Orthogonal Polynomials, Nova Science Publishers, New York (to appear).
5. J.S. Geronimo, O. Bruno, W. Van Assche, *WKB and turning point theory for second-order difference equations*, in 'Spectral Methods for Operators of Mathematical Physics' (J. Janas, P. Kurasov, S. Naboko, Eds.), Operator Theory: Advances and Applications **154**, Birkhäuser, Basel, 2004, pp. 101–138.

During the processing of this volume we received sad news of the sudden death on December 16, 2005 of Vadim Kuznetsov, one of the contributors. Vadim Kuznetsov enjoyed a very strong international reputation in the field of integrable systems and was responsible for a number of fundamental contributions to the development of separation of variables techniques by exploiting the methods of integrability, a topic on which he lectured during the summer school and which is the subject in his present contribution *Orthogonal polynomials and separation of variables.* We dedicate this volume in memory of Vadim Kuznetsov.

Vadim Kuznetsov 1963–2005

Leganés (Madrid) and Leuven,

Francisco Marcellán
Walter Van Assche

Contents

Orthogonal Polynomials, Quadrature, and Approximation: Computational Methods and Software (in Matlab)

Walter Gautschi

Department of Computer Sciences, Purdue University,
West Lafayette, IN 47907, USA
e-mail: wxg@cs.purdue.edu

Summary. One of the main problems in the constructive theory of orthogonal polynomials is the computation of the coefficients, if not known explicitly, in the three-term recurrence relation satisfied by orthogonal polynomials. Two classes of methods are discussed: those based on moment information, and those using discretization of the underlying inner product. Other computational problems considered are the computation of Cauchy integrals of orthogonal polynomials, and the problem of modification, i.e., of ascertaining the effect on the recurrence coefficients of multiplying the weight function by a (positive) rational function. Moment-based methods and discretization algorithms are also available for generating Sobolev orthogonal polynomials, i.e., polynomials orthogonal with respect to an inner product involving derivatives. Of particular interest here is the computation of their zeros.

Important applications of orthogonal polynomials are to the development of quadrature rules of maximum algebraic degree of exactness, most notably Gauss-type quadrature rules, but also Gauss-Kronrod and Gauss-Turán quadratures. Modification algorithms and discretization methods find application to constructing quadrature rules exact not only for polynomials, but also for rational functions with prescribed poles. Gauss-type quadrature rules are applicable also for computing Cauchy principal value integrals. Gaussian quadrature sums are expressible in terms of the related Jacobi matrix, which has interesting applications to generating orthogonal polynomials on several intervals and to the estimation of matrix functionals.

In the realm of approximation, the classical use of orthogonal polynomials, including Sobolev orthogonal polynomials, is to least squares approximation to which interpolatory constraints may be added. Among other uses considered are moment-preserving spline approximation and the summation of slowly convergent series.

All computational methods and applications considered are supported by a software package, called `OPQ`, of Matlab routines which are downloadable individually from the internet. Their use is illustrated throughout.

1 Introduction

Orthogonal polynomials, unless they are classical, require special techniques for their computation. One of the central problems is to generate the coefficients in the basic three-term recurrence relation they are known to satisfy. There are two general approaches for doing this: methods based on moment information, and discretization methods. In the former, one develops algorithms that take as input given moments, or modified moments, of the underlying measure and produce as output the desired recurrence coefficients. In theory, these algorithms yield exact answers. In practice, owing to rounding errors, the results are potentially inaccurate depending on the numerical condition of the mapping from the given moments (or modified moments) to the recurrence coefficients. A study of related condition numbers is therefore of practical interest. In contrast to moment-based algorithms, discretization methods are basically approximate methods: one approximates the underlying inner product by a discrete inner product and takes the recurrence coefficients of the corresponding discrete orthogonal polynomials to approximate those of the desired orthogonal polynomials. Finding discretizations that yield satisfactory rates of convergence requires a certain amount of skill and creativity on the part of the user, although general-purpose discretizations are available if all else fails.

Other interesting problems have as objective the computation of new orthogonal polynomials out of old ones. If the measure of the new orthogonal polynomials is the measure of the old ones multiplied by a rational function, one talks about modification of orthogonal polynomials and modification algorithms that carry out the transition from the old to the new orthogonal polynomials. This enters into a circle of ideas already investigated by Christoffel in the 1850s, but effective algorithms have been obtained only very recently. They require the computation of Cauchy integrals of orthogonal polynomials — another interesting computational problem.

In the 1960s, a new type of orthogonal polynomials emerged — the so-called Sobolev orthogonal polynomials — which are based on inner products involving derivatives. Although they present their own computational challenges, moment-based algorithms and discretization methods are still two of the main working tools. The computation of zeros of Sobolev orthogonal polynomials is of particular interest in practice.

An important application of orthogonal polynomials is to quadrature, specifically quadrature rules of the highest algebraic degree of exactness. Foremost among them is the Gaussian quadrature rule and its close relatives, the Gauss–Radau and Gauss–Lobatto rules. More recent extensions are due to Kronrod, who inserts $n+1$ new nodes into a given n-point Gauss formula, again optimally with respect to degree of exactness, and to Turán, who allows derivative terms to appear in the quadrature sum. When integrating functions having poles outside the interval of integration, quadrature rules of polynomial/rational degree of exactness are of interest. Poles inside the interval of integration give rise to Cauchy principal value integrals, which pose computational problems of their own. Interpreting Gaussian quadrature sums in terms of matrices allows interesting applications to orthogonal polynomials on several intervals, and to the computation of matrix functionals.

In the realm of approximation, orthogonal polynomials, especially discrete ones, find use in curve fitting, e.g. in the least squares approximation of discrete data. This indeed is the problem in which orthogonal polynomials (in substance if not in name) first appeared in the 1850s in work of Chebyshev. The presence of interpolatory constraints can be handled by a modification algorithm relative to special quadratic factors. Sobolev orthogonal polynomials also had their origin in least squares approximation, when one tries to fit simultaneously functions together with some of their derivatives. Physically motivated are approximations by spline functions that preserve as many moments as possible. Interestingly, these also are related to orthogonal polynomials via Gauss and generalized Gauss-type quadrature formulae. Slowly convergent series whose sum can be expressed as a definite integral naturally invite the application of Gauss-type quadratures to speed up their convergence. An example are series whose general term is expressible in terms of the Laplace transform or its derivative of a known function. Such series occur prominently in plate contact problems.

A comprehensive package, called `OPQ`, of Matlab routines is available that can be used to work with orthogonal polynomials. It resides at the web site

`http://www.cs.purdue.edu/archives/2002/wxg/codes/`

and all its routines are downloadable individually.

2 Orthogonal Polynomials

2.1 Recurrence Coefficients

Background and Notation

Orthogonality is defined with respect to an inner product, which in turn involves a measure of integration, $\mathrm{d}\lambda$. An *absolutely continuous* measure has the form

$$\mathrm{d}\lambda(t) = w(t)\mathrm{d}t \quad \text{on } [a,b], \qquad -\infty \le a < b \le \infty,$$

where w is referred to as a *weight function*. Usually, w is positive on (a,b), in which case $\mathrm{d}\lambda$ is said to be a *positive measure* and $[a,b]$ is called the *support* of $\mathrm{d}\lambda$. A *discrete measure* has the form

$$\mathrm{d}\lambda_N(t) = \sum_{k=1}^{N} w_k \delta(t - x_k)\mathrm{d}t, \qquad x_1 < x_2 < \cdots < x_N,$$

where δ is the Dirac delta function, and usually $w_k > 0$. The support of $\mathrm{d}\lambda_N$ consists of its N *support points* $x_1, x_2, \ldots, x_N$. For absolutely continuous measures, we make the standing assumption that all *moments*

$$\mu_r = \int_{\mathbb{R}} t^r \mathrm{d}\lambda(t), \qquad r = 0, 1, 2, \ldots,$$

exist and are finite. The *inner product* of two polynomials p and q relative to the measure $\mathrm{d}\lambda$ is then well defined by

$$(p,q)_{\mathrm{d}\lambda} = \int_{\mathbb{R}} p(t)q(t)\mathrm{d}\lambda(t),$$

and the *norm* of a polynomial p by

$$\|p\|_{\mathrm{d}\lambda} = \sqrt{(p,p)_{\mathrm{d}\lambda}}.$$

Orthogonal polynomials relative to the (positive) measure $\mathrm{d}\lambda$ are defined by

$$\pi_k(\,\cdot\,) = \pi_k(\,\cdot\,;\mathrm{d}\lambda) \text{ a polynomial of exact degree } k,\ \ k = 0, 1, 2, \ldots,$$

$$(\pi_k, \pi_\ell)_{\mathrm{d}\lambda} \begin{cases} = 0, \ k \neq \ell, \\ > 0, \ k = \ell. \end{cases}$$

There are infinitely many, if $\mathrm{d}\lambda$ is absolutely continuous, and they are uniquely defined up to the leading coefficient. If all leading coefficients are equal to 1, they are said to be *monic*. For a discrete measure $\mathrm{d}\lambda_N$, there are exactly N orthogonal polynomials $\pi_0, \pi_1, \ldots, \pi_{N-1}$. *Orthonormal polynomials* are defined and denoted by

$$\tilde{\pi}_k(\,\cdot\,;\mathrm{d}\lambda) = \frac{\pi_k(\,\cdot\,;\mathrm{d}\lambda)}{\|\pi_k\|_{\mathrm{d}\lambda}}, \qquad k = 0, 1, 2, \ldots .$$

They satisfy

$$(\tilde{\pi}_k, \tilde{\pi}_\ell)_{\mathrm{d}\lambda} = \delta_{k,\ell} = \begin{cases} 0, & k \neq \ell, \\ 1, & k = \ell. \end{cases}$$

Examples of measures resp. weight functions are shown in Tables 1 and 2. The former displays the most important "classical" weight functions, the latter the best-known discrete measures.

Three-Term Recurrence Relation

For any n ($< N-1$ if $\mathrm{d}\lambda = \mathrm{d}\lambda_N$), the first $n+1$ monic orthogonal polynomials satisfy a three-term recurrence relation

$$\begin{aligned} \pi_{k+1}(t) &= (t - \alpha_k)\pi_k(t) - \beta_k \pi_{k-1}(t), \quad k = 0, 1, \ldots, n-1, \\ \pi_{-1}(t) &= 0, \quad \pi_0(t) = 1, \end{aligned} \tag{2.1}$$

where the *recurrence coefficients* $\alpha_k = \alpha_k(\mathrm{d}\lambda)$, $\beta_k = \beta_k(\mathrm{d}\lambda)$ are real and positive, respectively. The coefficient β_0 in (2.1) multiplies $\pi_{-1} = 0$, and hence can be arbitrary. For later use, it is convenient to define

Table 1. "Classical" weight functions $\mathrm{d}\lambda(t) = w(t)\mathrm{d}t$

name	$w(t)$	support	comment
Jacobi	$(1-t)^\alpha(1+t)^\beta$	$[-1, 1]$	$\alpha > -1$, $\beta > -1$
Laguerre	$t^\alpha \mathrm{e}^{-t}$	$[0, \infty]$	$\alpha > -1$
Hermite	$\lvert t\rvert^{2\alpha}\mathrm{e}^{-t^2}$	$[-\infty, \infty]$	$\alpha > -\frac{1}{2}$
Meixner-Pollaczek	$\frac{1}{2\pi}\,\mathrm{e}^{(2\phi-\pi)t}\lvert\Gamma(\lambda+it)\rvert^2$	$[-\infty, \infty]$	$\lambda > 0$, $0 < \phi < \pi$

Table 2. "Classical" discrete measures $\mathrm{d}\lambda(t) = \sum_{k=0}^{M} w_k \delta(t-k)\mathrm{d}t$

name	M	w_k	comment
discrete Chebyshev	$N-1$	1	
Krawtchouk	N	$\binom{N}{k} p^k (1-p)^{N-k}$	$0<p<1$
Charlier	∞	$\mathrm{e}^{-a} a^k / k!$	$a>0$
Meixner	∞	$\frac{c^k}{\Gamma(\beta)} \frac{\Gamma(k+\beta)}{k!}$	$0<c<1, \beta>0$
Hahn	N	$\binom{\alpha+k}{k}\binom{\beta+N-k}{N-k}$	$\alpha>-1, \beta>-1$

$$\beta_0 = \beta_0(\mathrm{d}\lambda) = \int_{\mathbb{R}} \mathrm{d}\lambda(t). \tag{2.2}$$

The proof of (2.1) is rather simple if one expands $\pi_{k+1}(t) - t\pi_k(t) \in \mathbb{P}_k$ in orthogonal polynomials $\pi_0, \pi_1, \ldots, \pi_k$ and observes orthogonality and the obvious, but important, property $(tp, q)_{\mathrm{d}\lambda} = (p, tq)_{\mathrm{d}\lambda}$ of the inner product. As a by-product of the proof, one finds the formulae of Darboux,

$$\begin{aligned} \alpha_k(\mathrm{d}\lambda) &= \frac{(t\pi_k, \pi_k)_{\mathrm{d}\lambda}}{(\pi_k, \pi_k)_{\mathrm{d}\lambda}}, \quad k = 0, 1, 2, \ldots, \\ \beta_k(\mathrm{d}\lambda) &= \frac{(\pi_k, \pi_k)_{\mathrm{d}\lambda}}{(\pi_{k-1}, \pi_{k-1})_{\mathrm{d}\lambda}}, \quad k = 1, 2, \ldots. \end{aligned} \tag{2.3}$$

The second yields

$$\|\pi_k\|_{\mathrm{d}\lambda}^2 = \beta_0 \beta_1 \cdots \beta_k. \tag{2.4}$$

Placing the coefficients α_k on the diagonal, and $\sqrt{\beta_k}$ on the two side diagonals of a matrix produces what is called the *Jacobi matrix* of the measure $\mathrm{d}\lambda$,

$$\boldsymbol{J}(\mathrm{d}\lambda) = \begin{bmatrix} \alpha_0 & \sqrt{\beta_1} & & & \mathbf{0} \\ \sqrt{\beta_1} & \alpha_1 & \sqrt{\beta_2} & & \\ & \sqrt{\beta_2} & \alpha_2 & \ddots & \\ & & \ddots & \ddots & \\ \mathbf{0} & & & & \end{bmatrix}. \tag{2.5}$$

It is a real, symmetric, tridiagonal matrix of infinite order, in general. Its principal minor matrix of order n will be denoted by

$$\boldsymbol{J}_n(\mathrm{d}\lambda) = \boldsymbol{J}(\mathrm{d}\lambda)_{[1:n,1:n]}. \tag{2.6}$$

Noting that the three-term recurrence relation for the ortho*normal* polynomials is

$$\sqrt{\beta_{k+1}}\tilde{\pi}_{k+1}(t) = (t - \alpha_k)\tilde{\pi}_k(t) - \sqrt{\beta_k}\tilde{\pi}_{k-1}(t), \quad k = 0, 1, 2, \ldots,$$
$$\tilde{\pi}_{-1}(t) = 0, \quad \tilde{\pi}_0(t) = 1/\sqrt{\beta_0}, \tag{2.7}$$

or, in matrix form, with $\tilde{\boldsymbol{\pi}}(t) = [\tilde{\pi}_0(t), \tilde{\pi}_1(t), \ldots, \tilde{\pi}_{n-1}(t)]^{\mathrm{T}}$,

$$t\tilde{\boldsymbol{\pi}}(t) = \boldsymbol{J}_n(\mathrm{d}\lambda)\tilde{\boldsymbol{\pi}}(t) + \sqrt{\beta_n}\tilde{\pi}_n(t)\boldsymbol{e}_n, \tag{2.8}$$

one sees that *the zeros* τ_ν *of* $\tilde{\pi}_n(\,\cdot\,;\mathrm{d}\lambda)$ *are precisely the eigenvalues of* $\boldsymbol{J}_n(\mathrm{d}\lambda)$, *and* $\tilde{\boldsymbol{\pi}}(\tau_\nu)$ *corresponding eigenvectors.* This is only one of many reasons why knowledge of the Jacobi matrix, i.e. of the recurrence coefficients, is of great practical interest. For classical measures as the ones in Tables 1 and 2, all recurrence coefficients are explicitly known (cf. [10, Tables 1.1 and 1.2]). In most other cases, they must be computed numerically.

In the `OPQ` package, routines generating recurrence coefficients have the syntax `ab=r_`*name*`(N)`, where *name* identifies the name of the orthogonal polynomial and `N` is an input parameter specifying the number of α_k and of β_k desired. There may be additional input parameters. The αs and βs are stored in the `N`$\times$2 array `ab`:

α_0	β_0
α_1	β_1
$\vdots$	$\vdots$
α_{N-1}	β_{N-1}

$N \in \mathbb{N}$.

For example, `ab=r_jacobi(N,a,b)` generates the first `N` recurrence coefficients of the (monic) Jacobi polynomials with parameters α=`a`, β=`b`.

Demo#1 The first ten recurrence coefficients for the Jacobi polynomials with parameters $\alpha = -\frac{1}{2}$, $\beta = \frac{3}{2}$.

The Matlab command, followed by the output, is shown in the box below.

```
>> ab=r_jacobi(10,-.5,1.5)
ab =
      6.666666666666666e-01 4.712388980384690e+00
      1.333333333333333e-01 1.388888888888889e-01
      5.714285714285714e-02 2.100000000000000e-01
      3.174603174603174e-02 2.295918367346939e-01
      2.020202020202020e-02 2.376543209876543e-01
      1.398601398601399e-02 2.417355371900826e-01
      1.025641025641026e-02 2.440828402366864e-01
      7.843137254901961e-03 2.455555555555556e-01
      6.191950464396285e-03 2.465397923875433e-01
      5.012531328320802e-03 2.472299168975069e-01
```

2.2 Modified Chebyshev Algorithm

The first $2n$ moments $\mu_0, \mu_1, \ldots, \mu_{2n-1}$ of a measure $d\lambda$ uniquely determine the first n recurrence coefficients $\alpha_k(d\lambda)$ and $\beta_k(d\lambda)$, $k = 0, 1, \ldots, n-1$. However, the corresponding moment map $\mathbb{R}^{2n} \mapsto \mathbb{R}^{2n} : [\mu_k]_{k=0}^{2n-1} \mapsto [\alpha_k, \beta_k]_{k=0}^{n-1}$ is severely ill-conditioned when n is large. Therefore, other moment maps must be sought that are better conditioned. One that has been studied extensively in the literature is based on *modified moments*

$$m_k = \int_{\mathbb{R}} p_k(t) d\lambda(t), \quad k = 0, 1, 2, \ldots, \tag{2.9}$$

where $\{p_k\}$, $p_k \in \mathbb{P}_k$, is a given system of polynomials chosen to be close in some sense to the desired polynomials $\{\pi_k\}$. We assume that p_k, like π_k, satisfies a three-term recurrence relation

$$\begin{gathered} p_{k+1}(t) = (t - a_k)p_k(t) - b_k \pi_{k-1}(t), \quad k = 0, 1, 2, \ldots, \\ p_{-1}(t) = 0, \quad p_0(t) = 1, \end{gathered} \tag{2.10}$$

but with coefficients $a_k \in \mathbb{R}$, $b_k \geq 0$, that are known. The case $a_k = b_k = 0$ yields powers $p_k(t) = t^k$, hence ordinary moments μ_k, which however, as already mentioned, is not recommended.

The modified moment map

$$\mathbb{R}^{2n} \mapsto \mathbb{R}^{2n} : \quad [m_k]_{k=0}^{2n-1} \mapsto [\alpha_k, \beta_k]_{k=0}^{n-1} \tag{2.11}$$

and related maps have been well studied from the point of view of conditioning (cf. [10, §2.1.5 and 2.1.6]). The maps are often remarkably well-conditioned, especially for measures supported on a finite interval, but can still be ill-conditioned otherwise.

An algorithm that implements the map (2.11) is the *modified Chebyshev algorithm* (cf. [10, §2.1.7]), which improves on Chebyshev's original algorithm based on ordinary moments. To describe it, we need the *mixed moments*

$$\sigma_{k\ell} = \int_{\mathbb{R}} \pi_k(t; d\lambda) p_\ell(t) d\lambda(t), \quad k, \ell \geq -1, \tag{2.12}$$

which by orthogonality are clearly zero if $\ell < k$.

Algorithm 1 Modified Chebyshev algorithm

initialization:

$$\begin{aligned} &\alpha_0 = a_0 + m_1/m_0, \quad \beta_0 = m_0, \\ &\sigma_{-1,\ell} = 0, \quad \ell = 1, 2, \ldots, 2n-2, \\ &\sigma_{0,\ell} = m_\ell, \quad \ell = 0, 1, \ldots, 2n-1 \end{aligned}$$

continuation (if $n > 1$): for $k = 1, 2, \ldots, n-1$ do

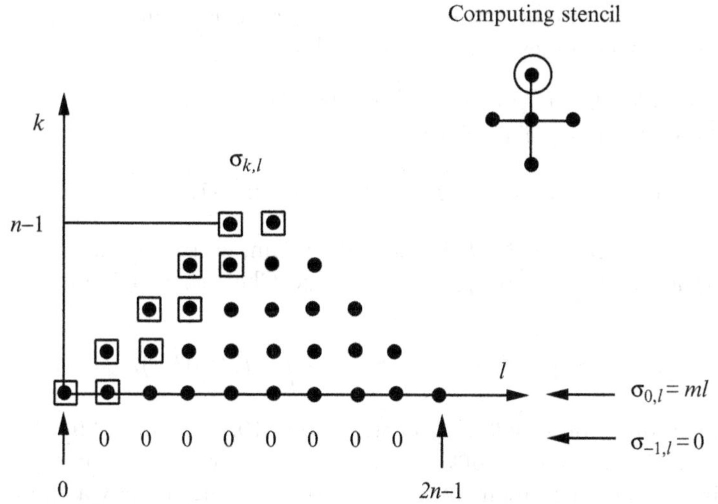

Fig. 1. Modified Chebyshev algorithm, schematically

$$\sigma_{k\ell} = \sigma_{k-1,\ell+1} - (\alpha_{k-1} - a_\ell)\sigma_{k-1,\ell} - \beta_{k-1}\sigma_{k-2,\ell}$$

$$+ b_\ell \sigma_{k-1,k-1}, \quad \ell = k, k+1, \ldots, 2n-k-1,$$

$$\alpha_k = a_k + \frac{\sigma_{k,k+1}}{\sigma_{kk}} - \frac{\sigma_{k-1,k}}{\sigma_{k-1,k-1}}, \quad \beta_k = \frac{\sigma_{kk}}{\sigma_{k-1,k-1}}.$$

If $a_k = b_k = 0$, Algorithm 1 reduces to Chebyshev's original algorithm.

Fig. 1 depicts the trapezoidal array of the mixed moments and the computing stencil indicating that the circled entry is computed in terms of the four entries below. The entries in boxes are used to compute the αs and βs.

The `OPQ` Matlab command that implements the modified Chebyshev algorithm has the form `ab=chebyshev(N,mom,abm)`, where `mom` is the 1×2`N` array of the modified moments, and `abm` the $(2$`N`$-1)\times 2$ array of the recurrence coefficients a_k, b_k from (2.10) needed in Algorithm 1:

m_0	m_1	m_2	$\cdots$	m_{2N-1}

`mom`

a_0	b_0
a_1	b_1
$\vdots$	$\vdots$
a_{2N-2}	b_{2N-2}

`abm`

If the input parameter abm is omitted, the routine assumes $a_k = b_k = 0$ and implements Chebyshev's original algorithm.

Demo#2 "Elliptic" orthogonal polynomials.

These are orthogonal relative to the measure

$$d\lambda(t) = [(1 - \omega^2 t^2)(1 - t^2)]^{-1/2} dt \quad \text{on } [-1, 1], \quad 0 \leq \omega < 1.$$

To apply the modified Chebyshev algorithm, it seems natural to employ Chebyshev moments (i.e. p_k = the monic Chebyshev polynomial of degree k)

$$m_0 = \int_{-1}^{1} d\lambda(t), \quad m_k = \frac{1}{2^{k-1}} \int_{-1}^{1} T_k(t) d\lambda(t), \quad k \geq 1.$$

Their computation is not entirely trivial (cf. [10, Example 2.29]), but a stable algorithm is available as OPQ routine mm_ell.m, which for given N generates the first 2N modified moments of $d\lambda$ with ω^2 being input via the parameter om2. The complete Matlab routine is as follows:

```
function ab=r_elliptic(N,om2)
   abm=r_jacobi(2*N-1,-1/2);
   mom=mm_ell(N,om2);
   ab=chebyshev(N,mom,abm)
```

For om2=.999 and N=40, results produced by the routine are partially shown in the box below.

```
ab =
         0 9.682265121100620e+00
         0 7.937821421385184e-01
         0 1.198676724605757e-01
         0 2.270401183698990e-01
         0 2.410608787266061e-01
         0 2.454285325203698e-01
         0 2.473016530297635e-01
         0 2.482587060199245e-01
         : :
         0 2.499915376529289e-01
         0 2.499924312667191e-01
         0 2.499932210069769e-01
```

Clearly, $\beta_k \to \frac{1}{4}$ as $k \to \infty$, which is consistent with the fact that $d\lambda$ belongs to the Szegő class (cf. [10, p. 12]). Convergence, in fact, is monotone for $k \geq 2$.

2.3 Discrete Stieltjes and Lanczos Algorithm

Computing the recurrence coefficients of a discrete measure is a prerequisite for discretization methods to be discussed in the next section. Given the measure

$$d\lambda_N(t) = \sum_{k=1}^{N} w_k \delta(t - x_k) dt, \tag{2.13}$$

the problem is to compute $\alpha_{\nu,N} = \alpha_\nu(d\lambda_N)$, $\beta_{\nu,N} = \beta_\nu(d\lambda_N)$ for all $\nu \le n-1$, $n \le N$, which will provide access to the discrete orthogonal polynomials of degrees up to n, or else, to determine the Jacobi matrix $\boldsymbol{J}_N(d\lambda_N)$, which will provide access to all discrete orthogonal polynomials. There are two methods in use, a discrete Stieltjes procedure and a Lanczos-type algorithm.

Discrete Stieltjes Procedure

Since the inner product for the measure (2.13) is a finite sum,

$$(p,q)_{d\lambda_N} = \sum_{k=1}^{N} w_k p(x_k) q(x_k), \tag{2.14}$$

Darboux's formulae (2.3) seem to offer attractive means of computing the desired recurrence coefficients, since all inner products appearing in these formulae are finite sums. The only problem is that we do not yet know the orthogonal polynomials $\pi_k = \pi_{k,N}$ involved. For this, however, we can make use of an idea already expressed by Stieltjes in 1884: combine Darboux's formulae with the basic three-term recurrence relation. Indeed, when $k = 0$ we know that $\pi_{0,N} = 1$, so that Darboux's formula for $\alpha_0(d\lambda_N)$ can be applied, and $\beta_0(d\lambda_N)$ is simply the sum of the weights w_k. Now that we know $\alpha_0(d\lambda_N)$, we can apply the recurrence relation (2.1) for $k = 0$ to compute $\pi_{1,N}(t)$ for $t = x_k$, $k = 1, 2, \ldots, N$. We then have all the information at hand to reapply Darboux's formulae for $\alpha_{1,N}$ and $\beta_{1,N}$, which in turn allows us to compute $\pi_{2,N}(t)$ for all $t = x_k$ from (2.1). In this manner we proceed until all $\alpha_{\nu,N}$, $\beta_{\nu,N}$, $\nu \le n-1$, are determined. If $n = N$, this will yield the Jacobi matrix $\boldsymbol{J}_N(d\lambda_N)$.

The procedure is quite effective, at least when $n \ll N$. As n approaches N, instabilities may develop, particularly if the support points x_k of $d\lambda_N$ are equally, or nearly equally, spaced.

The `OPQ` routine implementing Stieltjes's procedure is called by `ab=stieltjes(n,xw)`, where $\mathtt{n} \le N$, and `xw` is an $N \times 2$ array containing the support points and weights of the inner product,

x_1	w_1
x_2	w_2
$\vdots$	$\vdots$
x_N	w_N

`xw`

As usual, the recurrence coefficients $\alpha_{\nu,N}$, $\beta_{\nu,N}$, $0 \le \nu \le n-1$, are stored in the $\mathtt{n}\times 2$ array `ab`.

Lanczos-type Algorithm

Lanczos's algorithm is a general procedure to orthogonally tridiagonalize a given symmetric matrix $\boldsymbol{A}$. Thus, it finds an orthogonal matrix $\boldsymbol{Q}$ and a symmetric tridiagonal matrix $\boldsymbol{T}$ such that $\boldsymbol{Q}^{\mathrm{T}}\boldsymbol{A}\boldsymbol{Q} = \boldsymbol{T}$. Both $\boldsymbol{Q}$ and $\boldsymbol{T}$ are uniquely determined by the first column of $\boldsymbol{Q}$.

Given the measure (2.13), it is known that an orthogonal matrix $\boldsymbol{Q} \in \mathbb{R}^{(N+1)\times(N+1)}$ exists, with the first column being $\boldsymbol{e}_1 = [1, 0, \ldots, 0]^{\mathrm{T}} \in \mathbb{R}^{N+1}$, such that (see [10, Corollary to Theorem 3.1])

$$\boldsymbol{Q}^{\mathrm{T}} \begin{bmatrix} 1 & \sqrt{w_1} & \sqrt{w_2} & \cdots & \sqrt{w_N} \\ \sqrt{w_1} & x_1 & 0 & \cdots & 0 \\ \sqrt{w_2} & 0 & x_2 & \cdots & 0 \\ \vdots & \vdots & \vdots & \ddots & \vdots \\ \sqrt{w_N} & 0 & 0 & \cdots & x_N \end{bmatrix} \boldsymbol{Q} = \begin{bmatrix} 1 & \sqrt{\beta_0} & 0 & \cdots & 0 \\ \sqrt{\beta_0} & \alpha_0 & \sqrt{\beta_1} & \cdots & 0 \\ 0 & \sqrt{\beta_1} & \alpha_1 & \cdots & 0 \\ \vdots & \vdots & \vdots & \ddots & \vdots \\ 0 & 0 & 0 & \cdots & \alpha_{N-1} \end{bmatrix}, \qquad (2.15)$$

where $\alpha_k = \alpha_{k,N}$, $\beta_k = \beta_{k,N}$. We are thus in the situation described above, where $\boldsymbol{A}$ is the matrix displayed on the left and $\boldsymbol{T}$ the matrix on the right, the desired Jacobi matrix $\boldsymbol{J}_N(\mathrm{d}\lambda_N)$ bordered by a first column and a first row containing β_0. The computation can be arranged so that only the leading principal minor matrix of order $n+1$ is obtained.

Lanczos's algorithm in its original form (published in 1950) is numerically unstable, but can be stabilized using ideas of Rutishauser (1963). An algorithm and pseudocode of Gragg and Harrod [14], using a sequence of Givens rotations to construct the matrix $\boldsymbol{Q}$ in (2.15), forms the basis for the `OPQ` Matlab code `ab=lanczos(n,xw)`, where the input and output parameters have the same meaning as in the routine `stieltjes.m`.

This routine enjoys good stability properties but may be considerably slower than Stieltjes's procedure.

2.4 Discretization Methods

The basic idea is to discretize the given measure $\mathrm{d}\lambda$, i.e. approximate it by a discrete measure

$$\mathrm{d}\lambda(t) \approx \mathrm{d}\lambda_N(t), \qquad (2.16)$$

and then use the recurrence coefficients $\alpha_k(\mathrm{d}\lambda_N)$, $\beta_k(\mathrm{d}\lambda_N)$ of the discrete measure to approximate $\alpha_k(\mathrm{d}\lambda)$, $\beta_k(\mathrm{d}\lambda)$. The former are computed by either Stieltjes's procedure or a Lanczos-type algorithm. The effectiveness of the method is crucially tied to the quality of the discretization. We illustrate this by a simple, yet instructive, example.

Example 1. Chebyshev weight function plus a constant,

$$w(t) = (1-t^2)^{-1/2} + c \quad \text{on } [-1,1], \quad c > 0.$$

It suffices to approximate the inner product for the weight function w. This can always be done by using appropriate quadrature formulae. In the case at hand, it is natural to treat the two parts of the weight function separately, indeed to use Gauss–Chebyshev quadrature for the first part and Gauss–Legendre quadrature for the second,

$$\begin{aligned}(p,q)_w &= \int_{-1}^{1} p(t)q(t)(1-t^2)^{-1/2}\mathrm{d}t + c\int_{-1}^{1} p(t)q(t)\mathrm{d}t \\ &\approx \sum_{k=1}^{M} w_k^{\mathrm{Ch}} p(x_k^{\mathrm{Ch}})q(x_k^{\mathrm{Ch}}) + c\sum_{k=1}^{M} w_k^{\mathrm{L}} p(x_k^{\mathrm{L}})q(x_k^{\mathrm{L}}). \end{aligned} \tag{2.17}$$

Here, x_k^{Ch}, w_k^{Ch} are the nodes and weights of the M-point Gauss–Chebyshev quadrature rule, and x_k^{L}, w_k^{L} those of the Gauss–Legendre quadrature rule. The discrete measure implied by (2.17) is $\mathrm{d}\lambda_N$ with $N = 2M$ and

$$\mathrm{d}\lambda_N(t) = \sum_{k=1}^{M} w_k^{\mathrm{Ch}} \delta(t - x_k^{\mathrm{Ch}}) + c\sum_{k=1}^{M} w_k^{\mathrm{L}} \delta(t - x_k^{\mathrm{L}}). \tag{2.18}$$

What is attractive about this choice is the fact that the approximation in (2.17) is actually an equality whenever the product $p \cdot q$ is a polynomial of degree $\le 2M - 1$. Now if we are interested in computing $\alpha_k(w)$, $\beta_k(w)$ for $k \le n-1$, then the products $p \cdot q$ that occur in Darboux's formulae are all of degree $\le 2n - 1$. Therefore, we have equality in (2.17) if $n \le M$. It therefore suffices to take $M = n$ in (2.17) to obtain the first n recurrence coefficients exactly.

In general, the quadrature rules will not produce exact results, and M will have to be increased through a sequence of integers until convergence occurs.

Example 1 illustrates the case of a 2-component discretization. In a general *multiple-component discretization*, the support $[a, b]$ of $\mathrm{d}\lambda$ is decomposed into s intervals,

$$[a,b] = \bigcup_{j=1}^{s} [a_j, b_j], \tag{2.19}$$

where the intervals $[a_j, b_j]$ may or may not be disjoint. The measure $\mathrm{d}\lambda$ is then discretized on each interval $[a_j, b_j]$ using either a tailor-made M-point quadrature (as in Example 1), or a general-purpose quadrature. For the latter, a Fejér quadrature rule on $[-1, 1]$, suitably transformed to $[a_j, b_j]$, has been found useful. (The Fejér rule is the interpolatory quadrature formula based on Chebyshev points.) If the original measure $\mathrm{d}\lambda$ has also a discrete component, this component is simply added on. Rather than go into details (which are discussed in [10, §2.2.4]), we present the Matlab implementation, another illustrative example, and a demo.

The `OPQ` routine for the multiple-component discretization is `ab=mcdis (N,eps0,quad,Mmax)`, where in addition to the variables `ab` and `n`, which have the usual meaning, there are three other parameters, `eps0`: a prescribed accuracy tolerance, `quad`: the name of a quadrature routine carrying out the discretization on each subinterval if tailor-made (otherwise, **quadgp.m**, a general-purpose quadrature routine can be used), **Mmax**: a maximal allowable value for the discretization parameter M. The decomposition (2.19) is input via the mc×2 array

$$\texttt{AB}=\begin{array}{||c|c||}\hline a_1 & b_1 \\ a_2 & b_2 \\ \vdots & \vdots \\ a_{\rm mc} & b_{\rm mc} \\ \hline\end{array}$$

where mc is the number of components (the s in (2.19)). A discrete component which may possibly be present in $\mathrm{d}\lambda$ is input via the array

$$\texttt{DM}=\begin{array}{||c|c||}\hline x_1 & y_1 \\ x_2 & y_2 \\ \vdots & \vdots \\ x_{\rm mp} & y_{\rm mp} \\ \hline\end{array}$$

with the first column containing the support points, and the second column the associated weights. The number of support points is mp. Both mc and mp, as well as AB and DM, are global variables. Another global variable is iq, which has to be set equal to 1 if the user provides his or her own quadrature routine, and equal to 0 otherwise.

Example 2. The normalized Jacobi weight function plus a discrete measure.

This is the measure

$$\mathrm{d}\lambda(t) = (\beta_0^{\rm J})^{-1}(1-t)^\alpha(1+t)^\beta \mathrm{d}t + \sum_{j=1}^{p} w_j\delta(t-x_j)\mathrm{d}t \text{ on } [-1,1],$$

where

$$\beta_0^{\rm J} = \int_{-1}^{1}(1-t)^\alpha(1+t)^\beta \mathrm{d}t, \quad \alpha > -1,\ \beta > -1.$$

Here, one single component suffices to do the discretization, and the obvious choice of quadrature rule is the Gauss–Jacobi M-point quadrature formula to which the discrete component is added on. Similarly as in Example 1, taking $M = n$ yields the first n recurrence coefficients $\alpha_k(\mathrm{d}\lambda)$, $\beta_k(\mathrm{d}\lambda)$, $k \le n-1$, exactly. The global parameters in Matlab are here mc=1, mp=p, iq=1, and

$$\texttt{AB}=\begin{array}{||c|c||}\hline -1 & 1 \\ \hline\end{array} \qquad \texttt{DM}=\begin{array}{||c|c||}\hline x_1 & w_1 \\ x_2 & w_2 \\ \vdots & \vdots \\ x_p & w_p \\ \hline\end{array}$$

Demo#3 Logistic density function,

$$d\lambda(t) = \frac{e^{-t}}{(1+e^{-t})^2}\, dt, \quad t \in \mathbb{R}.$$

The discretization is conveniently effected by the quadrature rule

$$\begin{aligned}\int_{\mathbb{R}} p(t)d\lambda(t) &= \int_0^\infty \frac{p(-t)}{(1+e^{-t})^2}\, e^{-t}dt + \int_0^\infty \frac{p(t)}{(1+e^{-t})^2}\, e^{-t}dt \\ &\approx \sum_{k=1}^{M} \lambda_k^{\mathrm{L}} \frac{p(-\tau_k^{\mathrm{L}}) + p(\tau_k^{\mathrm{L}})}{(1+e^{-\tau_k^{\mathrm{L}}})^2},\end{aligned}$$

where τ_k^{L}, λ_k^{L} are the nodes and weights of the M-point Gauss–Laguerre quadrature formula. This no longer produces exact results for $M = n$, but converges rapidly as $M \to \infty$. The exact answers happen to be known,

$$\begin{aligned}&\alpha_k(d\lambda) = 0 \quad \text{by symmetry,} \\ &\beta_0(d\lambda) = 1, \quad \beta_k(d\lambda) = \frac{k^4\pi^2}{4k^2-1}, \quad k \geq 1.\end{aligned}$$

Numerical results produced by `mcdis.m` with N=40, `eps0`$=10^3\times$`eps`, along with errors (absolute errors for α_k, relative errors for β_k) are shown in the box below. The two entries in the bottom row are the maximum errors taken over $0 \leq n \leq 39$.

n	β_n	err α	err β
0	1.0000000000(0)	7.18(−17)	3.33(−16)
1	3.2898681337(0)	1.29(−16)	2.70(−16)
6	8.9447603523(1)	4.52(−16)	1.43(−15)
15	5.5578278399(2)	2.14(−14)	0.00(+00)
39	3.7535340252(3)	6.24(−14)	4.48(−15)
		6.24(−14)	8.75(−15)

2.5 Cauchy Integrals of Orthogonal Polynomials

The Jacobi Continued Fraction

The *Jacobi continued fraction* associated with the measure $d\lambda$ is

$$\mathcal{J} = \mathcal{J}(t; d\lambda) = \frac{\beta_0}{t-\alpha_0 -}\,\frac{\beta_1}{t-\alpha_1 -}\,\frac{\beta_2}{t-\alpha_2 -}\cdots, \tag{2.20}$$

where $\alpha_k = \alpha_k(d\lambda)$, $\beta_k = \beta_k(d\lambda)$. From the theory of continued fractions it is readily seen that the nth convergent of $\mathcal{J}$ is

$$\frac{\beta_0}{z-\alpha_0 -}\,\frac{\beta_1}{z-\alpha_1 -}\cdots\frac{\beta_{n-1}}{z-\alpha_{n-1}} = \frac{\sigma_n(z; d\lambda)}{\pi_n(z; d\lambda)}, \quad n = 1, 2, 3, \ldots, \tag{2.21}$$

where π_n is the monic orthogonal polynomial of degree n, and σ_n a polynomial of degree $n-1$ satisfying the same basic three-term recurrence relation as π_n, but with different starting values,

$$\sigma_{k+1}(z) = (z - \alpha_k)\sigma_k(z) - \beta_k\sigma_{k-1}(z), \quad k = 1, 2, 3, \ldots, \\ \sigma_0(z) = 0, \quad \sigma_1(z) = \beta_0. \tag{2.22}$$

Recall that $\beta_0 = \int_{\mathbb{R}} \mathrm{d}\lambda(t)$. If we define $\sigma_{-1} = -1$, then (2.22) holds also for $k = 0$. We have, moreover,

$$\sigma_n(z) = \int_{\mathbb{R}} \frac{\pi_n(z) - \pi_n(t)}{z - t} \mathrm{d}\lambda(t), \quad n = 0, 1, 2, \ldots\ , \tag{2.23}$$

as can be seen by showing that the integral on the right also satisfies (2.22). If we define

$$F(z) = F(z; \mathrm{d}\lambda) = \int_{\mathbb{R}} \frac{\mathrm{d}\lambda(t)}{z - t} \tag{2.24}$$

to be the *Cauchy transform* of the measure $\mathrm{d}\lambda$, and more generally consider

$$\rho_n(z) = \rho_n(z; \mathrm{d}\lambda) = \int_{\mathbb{R}} \frac{\pi_n(t)}{z - t} \mathrm{d}\lambda(t), \tag{2.25}$$

the *Cauchy integral* of the orthogonal polynomial π_n, we can give (2.23) the form

$$\sigma_n(z) = \pi_n(z)F(z) - \rho_n(z), \tag{2.26}$$

and hence

$$\frac{\sigma_n(z)}{\pi_n(z)} = F(z) - \frac{\rho_n(z)}{\pi_n(z)}. \tag{2.27}$$

An important result from the theory of the moment problem tells us that, whenever the moment problem for $\mathrm{d}\lambda$ is determined, then

$$\lim_{n\to\infty} \frac{\sigma_n(z)}{\pi_n(z)} = F(z) \quad \text{for } z \in \mathbb{C}\backslash[a, b], \tag{2.28}$$

where $[a, b]$ is the support of the measure $\mathrm{d}\lambda$. If $[a, b]$ is a finite interval, then the moment problem is always determined, and (2.28) is known as *Markov's theorem.*

Note from (2.26) that, since $\sigma_{-1} = -1$, we have

$$\rho_{-1}(z) = 1, \tag{2.29}$$

and the sequence $\{\rho_n\}_{n=-1}^{\infty}$ satisfies the same three-term recurrence relation as $\{\pi_n\}_{n=-1}^{\infty}$. As a consequence of (2.27) and (2.28), however, it behaves quite differently at infinity,

$$\lim_{n\to\infty} \frac{\rho_n(z)}{\pi_n(z)} = 0, \tag{2.30}$$

which implies that $\{\rho_n(z)\}$ is the *minimal solution* of the three-term recurrence relation having the initial value (2.29). It is well known that a minimal solution of a three-term recurrence relation is uniquely determined by its starting value, and, moreover, that

$$\frac{\rho_n(z)}{\rho_{n-1}(z)} = \frac{\beta_n}{z-\alpha_n-}\,\frac{\beta_{n+1}}{z-\alpha_{n+1}-}\,\frac{\beta_{n+2}}{z-\alpha_{n+2}-}\cdots, \tag{2.31}$$

i.e. the successive ratios of the minimal solution are the successive *tails* of the Jacobi continued fraction (Pincherle's theorem). In particular, by (2.31) for $n = 0$, and (2.21), (2.28) and (2.29),

$$\rho_0(z) = F(z), \tag{2.32}$$

i.e., ρ_0 is the Cauchy transform of the measure.

We remark that (2.25) is meaningful also for real $z = x$ in (a, b), if the integral is interpreted as a *Cauchy principal value integral* (cf. (4.36))

$$\rho_n(x) = ⨍_{\mathbb{R}} \frac{\pi_n(t; \mathrm{d}\lambda)}{x-t}\,\mathrm{d}\lambda(t), \quad x \in (a, b), \tag{2.33}$$

and the sequence $\{\rho_n(x)\}$ satisfies the basic three-term recurrence relation with initial values

$$\rho_{-1}(x) = 1, \quad \rho_0(x) = ⨍_{\mathbb{R}} \frac{\mathrm{d}\lambda(t)}{x-t}, \tag{2.34}$$

but is no longer minimal.

Continued Fraction Algorithm

This is an algorithm for computing the minimal solution $\rho_n(z)$, $z \in \mathbb{C}\backslash[a, b]$, of the basic three-term recurrence relation. Denote the ratio in (2.31) by

$$r_{n-1} = \frac{\rho_n(z)}{\rho_{n-1}(z)}. \tag{2.35}$$

Then, clearly,

$$r_{n-1} = \frac{\beta_n}{z-\alpha_n-r_n}. \tag{2.36}$$

If, for some $\nu \geq N$, we knew r_ν, we could apply (2.36) for $r = \nu, \nu-1, \ldots, 0$, and then obtain

$$\rho_n(z) = r_{n-1}\rho_{n-1}(z), \quad n = 0, 1, \ldots, N. \tag{2.37}$$

The *continued fraction algorithm* is precisely this algorithm, except that r_ν is replaced by 0. All quantities generated then depend on ν, which is indicated by a superscript.

Algorithm 2 Continued fraction algorithm
backward phase; $\nu \geq N$:

$$r_\nu^{[\nu]} = 0, \quad r_{n-1}^{[\nu]} = \frac{\beta_n}{z-\alpha_n-r_n^{[\nu]}}, \quad n = \nu, \nu-1, \ldots, 0.$$

forward phase:

$$\rho_{-1}^{[\nu]}(z) = 1, \quad \rho_n^{[\nu]}(z) = r_{n-1}^{[\nu]} \rho_{n-1}^{[\nu]}(z), \quad n = 0, 1, \ldots, N.$$

It can be shown that, as a consequence of the minimality of $\{\rho_n(z)\}$ (cf. [10, pp. 114–115]),

$$\lim_{\nu \to \infty} \rho_n^{[\nu]}(z) = \rho_n(z), \quad n = 0, 1, \ldots, N, \quad \text{if } z \in \mathbb{C} \backslash [a, b]. \tag{2.38}$$

Convergence is faster the larger $\operatorname{dist}(z, [a, b])$. To compute $\rho_n(z)$, it suffices to apply Algorithm 2 for a sequence of increasing values of ν until convergence is achieved to within the desired accuracy.

The OPQ command implementing this algorithm is

```
[rho,r,nu]=cauchy(N,ab,z,eps0,nu0,numax)
```

where the meanings of the output variables `rho`, `r` and input variable `ab` are as shown below.

rho
$\rho_0(z)$
$\rho_1(z)$
$\vdots$
$\rho_N(z)$

r
$r_0(z)$
$r_1(z)$
$\vdots$
$r_N(z)$

ab	
α_0	β_0
α_1	β_1
$\vdots$	$\vdots$
α_{numax}	β_{numax}

The input variable `eps0` is an error tolerance, the variable `nu0` a suitable starting value of ν in Algorithm 2, which is incremented in steps of, say 5, until the algorithm converges to the accuracy `eps0`. If convergence does not occur within $\nu \le$ `numax`, an error message is issued, otherwise the value of ν yielding convergence is output as `nu`.

2.6 Modification Algorithms

By "modification" of a measure $d\lambda$, we mean here multiplication of $d\lambda$ by a rational function r which is positive on the support $[a, b]$ of $d\lambda$. The *modified measure* thus is

$$d\hat{\lambda}(t) = r(t)d\lambda(t), \quad r \text{ rational and } r > 0 \text{ on } [a, b]. \tag{2.39}$$

We are interested in determining the recurrence coefficients $\hat{\alpha}_k$, $\hat{\beta}_k$ for $d\hat{\lambda}$ in terms of the recurrence coefficients α_k, β_k of $d\lambda$. An algorithm that carries out the transition from α_k, β_k to $\hat{\alpha}_k$, $\hat{\beta}_k$ is called a *modification algorithm.* While the passage from the orthogonal polynomials relative to $d\lambda$ to those relative to $d\hat{\lambda}$ is classical (at least in the case when r is a polynomial), the transition in terms of recurrence coefficients is more recent. It was first treated for linear factors in 1971 by Galant.

Example 3. Linear factor $r(t) = s(t - c)$, $c \in \mathbb{R}\backslash[a, b]$, $s = \pm 1$.

Here, s is a sign factor to make $r(t) > 0$ on (a, b). Galant's approach is to determine the Jacobi matrix of $\mathrm{d}\hat{\lambda}$ from the Jacobi matrix of $\mathrm{d}\lambda$ by means of one step of the symmetric, shifted LR algorithm: by the choice of s, the matrix $s[\boldsymbol{J}_{n+1}(\mathrm{d}\lambda) - c\boldsymbol{I}]$ is symmetric positive definite, hence admits a Cholesky decomposition

$$s[\boldsymbol{J}_{n+1}(\mathrm{d}\lambda) - c\boldsymbol{I}] = \boldsymbol{L}\boldsymbol{L}^{\mathrm{T}},$$

where $\boldsymbol{L}$ is lower triangular. The Jacobi matrix $\boldsymbol{J}_n(\mathrm{d}\hat{\lambda})$ is now obtained by reversing the order of the product on the right, adding back the shift c, and then discarding the last row and column,[1]

$$\boldsymbol{J}_n(\mathrm{d}\hat{\lambda}) = \left(\boldsymbol{L}^{\mathrm{T}}\boldsymbol{L} + c\boldsymbol{I}\right)_{[1:n,1:n]}.$$

Since the matrices involved are tridiagonal, the procedure can be implemented by simple nonlinear recurrence relations. These can also be obtained more systematically via Christoffel's theorem and its generalizations.

Generalized Christoffel's Theorem

We write

$$\mathrm{d}\hat{\lambda}(t) = \frac{u(t)}{v(t)}\,\mathrm{d}\lambda(t), \quad u(t) = \pm\prod_{\lambda=1}^{\ell}(t - u_\lambda), \;\; v(t) = \prod_{\mu=1}^{m}(t - v_\mu), \tag{2.40}$$

where u_λ and v_μ are real numbers outside the support of $\mathrm{d}\lambda$. The sign of $u(t)$ is chosen so that $\mathrm{d}\hat{\lambda}$ is a positive measure. Christoffel's original theorem (1858) relates to the case $v(t) = 1$, i.e. $m = 0$. The generalization to arbitrary v is due to Uvarov (1969). It has a different form depending on whether $m \le n$ or $m > n$. In the first case, it states that

$$u(t)\pi_n(t; \mathrm{d}\hat{\lambda}) = \text{const} \times \begin{vmatrix} \pi_{n-m}(t) & \cdots & \pi_{n-1}(t) & \pi_n(t) & \cdots & \pi_{n+\ell}(t) \\ \pi_{n-m}(u_1) & \cdots & \pi_{n-1}(u_1) & \pi_n(u_1) & \cdots & \pi_{n+\ell}(u_1) \\ \cdots & \cdots & \cdots & \cdots & \cdots & \cdots \\ \pi_{n-m}(u_\ell) & \cdots & \pi_{n-1}(u_\ell) & \pi_n(u_\ell) & \cdots & \pi_{n+\ell}(u_\ell) \\ \rho_{n-m}(v_1) & \cdots & \rho_{n-1}(v_1) & \rho_n(v_1) & \cdots & \rho_{n+\ell}(v_1) \\ \cdots & \cdots & \cdots & \cdots & \cdots & \cdots \\ \rho_{n-m}(v_m) & \cdots & \rho_{n-1}(v_m) & \rho_n(v_m) & \cdots & \rho_{n+\ell}(v_m) \end{vmatrix}, \tag{2.41}$$

where

$$\rho_k(z) = \int_{\mathbb{R}} \frac{\pi_k(t; \mathrm{d}\lambda)}{z - t}\,\mathrm{d}\lambda(t), \quad k = 0, 1, 2, \ldots,$$

[1] See, e.g. [9], where it is also shown how a quadratic factor $(t - c_1)(t - c_2)$ can be dealt with by one step of the QR algorithm; see in particular §3.2 and 3.3

are the Cauchy integrals of the orthogonal polynomials π_k. They occur only if $m > 0$. To get monic polynomials, the constant in (2.41) must be taken to be the reciprocal of the (signed) cofactor of the element $\pi_{n+\ell}(t)$.

If $m > n$, the generalized Christoffel theorem has the form

$$u(t)\pi_n(t;\mathrm{d}\hat{\lambda}) = \text{const} \times \begin{vmatrix} 0 & 0 & \cdots & 0 & \pi_0(t) & \cdots & \pi_{n+\ell}(t) \\ 0 & 0 & \cdots & 0 & \pi_0(u_1) & \cdots & \pi_{n+\ell}(u_1) \\ \cdots & \cdots & \cdots & \cdots & \cdots & \cdots & \cdots \\ 0 & 0 & \cdots & 0 & \pi_0(u_\ell) & \cdots & \pi_{n+\ell}(u_\ell) \\ 1 & v_1 & \cdots & v_1^{m-n-1} & \rho_0(v_1) & \cdots & \rho_{n+\ell}(v_1) \\ \cdots & \cdots & \cdots & \cdots & \cdots & \cdots & \cdots \\ 1 & v_m & \cdots & v_m^{m-n-1} & \rho_0(v_m) & \cdots & \rho_{n+\ell}(v_m) \end{vmatrix} . \tag{2.42}$$

Both versions of the theorem remain valid for complex u_λ, v_μ if orthogonality is understood in the sense of *formal orthogonality*.

Linear Factor

Generalizing Example 3 to arbitrary complex shifts, we let

$$\mathrm{d}\hat{\lambda}(t) = (t-z)\mathrm{d}\lambda(t), \quad z \in \mathbb{C}\backslash[a,b]. \tag{2.43}$$

Using Christoffel's theorem, letting $\hat{\pi}_n(\,\cdot\,) = \pi_n(\,\cdot\,;\mathrm{d}\hat{\lambda})$, we have

$$(t-z)\hat{\pi}_n(t) = \frac{\begin{vmatrix} \pi_n(t) & \pi_{n+1}(t) \\ \pi_n(z) & \pi_{n+1}(z) \end{vmatrix}}{-\pi_n(z)} = \pi_{n+1}(t) - r_n\pi_n(t), \tag{2.44}$$

where

$$r_n = \frac{\pi_{n+1}(z)}{\pi_n(z)} . \tag{2.45}$$

Following Verlinden [17], we write $(t-z)t\hat{\pi}_k(t)$ in two different ways: in the first, we use the three-term recurrence relation for π_k to obtain

$$\begin{aligned}(t-z)t\hat{\pi}_k(t) &= t\pi_{k+1}(t) - r_k \cdot t\pi_k(t) \\ &= \pi_{k+2}(t) + (\alpha_{k+1} - r_k)\pi_{k+1}(t) + (\beta_{k+1} - r_k\alpha_k)\pi_k(t) - r_k\beta_k\pi_{k-1}(t);\end{aligned}$$

in the second, we use the three-term recurrence relation directly on $\hat{\pi}_k$, and then apply (2.44), to write

$$\begin{aligned}(t-z)t\hat{\pi}_k(t) &= (t-z)[\hat{\pi}_{k+1} + \hat{\alpha}_k\hat{\pi}_k(t) + \hat{\beta}_k\hat{\pi}_{k-1}(t)] \\ &= \pi_{k+2}(t) + (\hat{\alpha}_k - r_{k+1})\pi_{k+1}(t) + (\hat{\beta}_k - r_k\hat{\alpha}_k)\pi_k(t) - r_{k-1}\hat{\beta}_k\pi_{k-1}(t).\end{aligned}$$

Since orthogonal polynomials are linearly independent, the coefficients in the two expressions obtained must be the same. This yields

$$\hat{\alpha}_k - r_{k+1} = \alpha_{k+1} - r_k, \quad r_{k-1}\hat{\beta}_k = r_k \beta_k,$$

hence the following algorithm.

Algorithm 3 Modification by a linear factor $t - z$

initialization:

$$r_0 = z - \alpha_0, \quad r_1 = z - \alpha_1 - \beta_1/r_0,$$
$$\hat{\alpha}_0 = \alpha_1 + r_1 - r_0, \quad \hat{\beta}_0 = -r_0\,\beta_0.$$

continuation (if $n > 1$): for $k = 1, 2, \ldots, n-1$ do

$$r_{k+1} = z - \alpha_{k+1} - \beta_{k+1}/r_k,$$
$$\hat{\alpha}_k = \alpha_{k+1} + r_{k+1} - r_k,$$
$$\hat{\beta}_k = \beta_k r_k / r_{k-1}.$$

Note that this requires α_n, β_n in addition to the usual n recurrence coefficients α_k, β_k for $k \leq n-1$. Algorithm 3 has been found to be numerically stable.

The `OPQ` Matlab command implementing Algorithm 3 is

```
ab=chri1(N,ab0,z)
```

where `ab0` is an (N+1)×2 array containing the recurrence coefficients α_k, β_k, $k = 0, 1, \ldots,$ N.

Quadratic Factor

We consider (real) quadratic factors $(t-x)^2 + y^2 = (t-z)(t-\overline{z})$, $z = x + \mathrm{i}y$, $y > 0$. Christoffel's theorem is now applied with $u_1 = z$, $u_2 = \overline{z}$ to express $(t-z)(t-\overline{z})\hat{\pi}_n(t)$ as a linear combination of π_n, π_{n+1}, and π_{n+2},

$$(t-z)(t-\overline{z})\hat{\pi}_n(t) = \pi_{n+2}(t) + s_n\pi_{n+1}(t) + t_n\pi_n(t), \tag{2.46}$$

where

$$s_n = -\left(r'_{n+1} + \frac{r''_{n+1}}{r''_n}\, r'_n\right), \quad t_n = \frac{r''_{n+1}}{r''_n}\,|r_n|^2. \tag{2.47}$$

Here we use the notation

$$r'_n = \operatorname{Re} r_n(z),\ r''_n = \operatorname{Im} r_n(z),\ |r_n|^2 = |r_n(z)|^2, \quad n = 0, 1, 2, \ldots, \tag{2.48}$$

where $r_n(z)$ continues to be the quantity defined in (2.45). The same technique used before can be applied to (2.46): express $(t-z)(t-\overline{z})t\hat{\pi}_k(t)$ in two different ways as a linear combination of $\pi_{k+3}, \pi_{k+2}, \ldots, \pi_{k-1}$ and compare the respective coefficients. The result gives rise to the following algorithm.

Algorithm 4 Modification by a quadratic factor $(t-z)(t-\overline{z})$, $z = x + \mathrm{i}y$
initialization:

$$r_0 = z - \alpha_0,\ r_1 = z - \alpha_1 - \beta_1/r_0,\ r_2 = z - \alpha_2 - \beta_2/r_1,$$
$$\hat{\alpha}_0 = \alpha_2 + r_2' + \frac{r_2''}{r_1''} r_1' - \left(r_1' + \frac{r_1''}{r_0''} r_0' \right),$$
$$\hat{\beta}_0 = \beta_0(\beta_1 + |r_0|^2).$$

continuation (if $n > 1$): for $k = 1, 2, \ldots, n-1$ do

$$r_{k+2} = z - \alpha_{k+2} - \beta_{k+2}/r_{k+1},$$
$$\hat{\alpha}_k = \alpha_{k+2} + r_{k+2}' + \frac{r_{k+2}''}{r_{k+1}''} r_{k+1}' - \left(r_{k+1}' + \frac{r_{k+1}''}{r_k''} r_k' \right),$$
$$\hat{\beta}_k = \beta_k \frac{r_{k+1}'' r_{k-1}''}{[r_k'']^2} \left| \frac{r_k}{r_{k-1}} \right|^2.$$

Note that this requires α_k, β_k for k up to $n+1$. Algorithm 4 is also quite stable, numerically.

The `OPQ` routine for Algorithm 4 is

```
ab=chri2(N,ab0,x,y)
```

with obvious meanings of the variables involved.

Since any real polynomial can be factored into a product of real linear and quadratic factors of the type considered, Algorithms 3 and 4 can be applied repeatedly to deal with modification by an arbitrary polynomial which is positive on the support $[a, b]$.

Linear Divisor

In analogy to (2.43), we consider

$$\mathrm{d}\hat{\lambda}(t) = \frac{\mathrm{d}\lambda(t)}{t-z}, \quad z \in \mathbb{C}\backslash[a,b]. \tag{2.49}$$

Now the *generalized* Christoffel theorem (with $\ell = 0$, $m = 1$) comes into play, giving

$$\hat{\pi}_n(t) = \frac{\begin{vmatrix} \pi_{n-1}(t) & \pi_n(t) \\ \rho_{n-1}(z) & \rho_n(z) \end{vmatrix}}{-\rho_{n-1}(z)} = \pi_n(t) - r_{n-1}\pi_{n-1}(t), \tag{2.50}$$

where now

$$r_n = \frac{\rho_{n+1}(z)}{\rho_n(z)}. \tag{2.51}$$

Similarly as before, we express $t\hat{\pi}_k(t)$ in two different ways as a linear combination of $\pi_{k+1}, \pi_k, \ldots, \pi_{k-2}$ and compare coefficients. By convention,

$$\hat{\beta}_0 = \int_{\mathbb{R}} \mathrm{d}\hat{\lambda}(t) = \int_{\mathbb{R}} \frac{\mathrm{d}\lambda(t)}{t-z} = -\rho_0(z).$$

The result is:

Algorithm 5 Modification by a linear divisor

initialization:

$$\hat{\alpha}_0 = \alpha_0 + r_0, \quad \hat{\beta}_0 = -\rho_0(z).$$

continuation (if $n > 1$): for $k = 1, 2, \ldots, n-1$ do

$$\hat{\alpha}_k = \alpha_k + r_k - r_{k-1},$$
$$\hat{\beta}_k = \beta_{k-1} r_{k-1} / r_{k-2}.$$

Note that here no coefficient α_k, β_k beyond $k \leq n-1$ is needed, not even β_{n-1}.

The ratios r_k of Cauchy integrals that appear in Algorithm 5 can be precomputed by Algorithm 2, where only the backward phase is relevant, convergence being tested on the $r_k^{[\nu]}$. Once converged, the algorithm also provides $\rho_0(z) = r_{-1}^{[\infty]}$.

As z approaches the support interval $[a, b]$, the strength of minimality of the Cauchy integrals $\{\rho_k(z)\}$ weakens and ceases altogether when $z = x \in [a, b]$. For z very close to $[a, b]$, Algorithm 2 therefore converges very slowly. On the other hand, since minimality is very weak, one can generate ρ_k with impunity, if n is not too large, by forward application of the basic three-term recurrence relation, using the initial values $\rho_{-1}(z) = 1$ and $\rho_0(z)$.

All of this is implemented in the `OPQ` routine

```
[ab,nu]=chri4(N,ab0,z,eps0,nu0,numax,rho0,iopt)
```

where all variables except `rho` and `iopt` have the same meaning as before. The parameter `rho` is $\rho_0(z)$, whereas `iopt` controls the method of computation for r_k: Algorithm 2 if `iopt`=1, and forward recursion otherwise.

Quadratic Divisor

We now consider

$$\mathrm{d}\hat{\lambda}(t) = \frac{\mathrm{d}\lambda(t)}{(t-z)(t-\overline{z})} = \frac{\mathrm{d}\lambda(t)}{(t-x)^2 + y^2}, \quad z = x + \mathrm{i}y, \ x \in \mathbb{R}, \ y > 0. \quad (2.52)$$

Here we have

$$\hat{\alpha}_0 = \frac{\displaystyle\int_{\mathbb{R}} t\mathrm{d}\lambda(t)/|t-z|^2}{\displaystyle\int_{\mathbb{R}} \mathrm{d}\lambda(t)/|t-z|^2} = x + y\,\frac{\operatorname{Re}\rho_0(z)}{\operatorname{Im}\rho_0(z)}, \quad \hat{\beta}_0 = -\frac{1}{y}\operatorname{Im}\rho_0(z). \tag{2.53}$$

We are in the case $\ell = 0$, $m = 2$ of the generalized Christoffel theorems (2.41) and (2.42), which give respectively

$$\hat{\pi}_n(t) = \frac{\begin{vmatrix} \pi_{n-2}(t) & \pi_{n-1}(t) & \pi_n(t) \\ \rho_{n-2}(z) & \rho_{n-1}(z) & \rho_n(z) \\ \rho_{n-2}(\overline{z}) & \rho_{n-1}(\overline{z}) & \rho_n(\overline{z}) \end{vmatrix}}{\begin{vmatrix} \rho_{n-2}(z) & \rho_{n-1}(z) \\ \rho_{n-2}(\overline{z}) & \rho_{n-1}(\overline{z}) \end{vmatrix}}, \; n \geq 2; \quad \hat{\pi}_1(t) = \frac{\begin{vmatrix} 0 & \pi_0(t) & \pi_1(t) \\ 1 & \rho_0(z) & \rho_1(z) \\ 1 & \rho_0(\overline{z}) & \rho_1(\overline{z}) \end{vmatrix}}{\begin{vmatrix} 1 & \rho_0(z) \\ 1 & \rho_0(\overline{z}) \end{vmatrix}}. \tag{2.54}$$

This becomes

$$\hat{\pi}_n(t) = \pi_n(t) + s_n\pi_{n-1}(t) + t_n\pi_{n-2}(t), \quad n \geq 1, \tag{2.55}$$

where

$$s_n = -\left(r'_{n-1} + \frac{r''_{n-1}}{r''_{n-2}}\,r'_{n-2}\right), \; n \geq 1; \quad t_n = \frac{r''_{n-1}}{r''_{n-2}}\,|r_{n-2}|^2, \; n \geq 2, \tag{2.56}$$

with r_n as defined in (2.51) and notation as in (2.48). Exactly the same procedure used to obtain Algorithm 5 yields

Algorithm 6 Modification by a quadratic divisor
initialization:

$$\hat{\alpha}_0 = x + \rho'_0 y/\rho''_0, \quad \hat{\beta}_0 = -\rho''_0/y,$$
$$\hat{\alpha}_1 = \alpha_1 - s_2 + s_1, \quad \hat{\beta}_1 = \beta_1 + s_1(\alpha_0 - \hat{\alpha}_1) - t_2,$$
$$\hat{\alpha}_2 = \alpha_2 - s_3 + s_2, \quad \hat{\beta}_2 = \beta_2 + s_2(\alpha_1 - \hat{\alpha}_2) - t_3 + t_2.$$

continuation (if $n > 3$): for $k = 3, 4, \ldots, n-1$ do

$$\hat{\alpha}_k = \alpha_k - s_{k+1} + s_k, \quad \hat{\beta}_k = \beta_{k-2}\,t_k/t_{k-1}.$$

The `OPQ` routine for Algorithm 6 is

```
[ab,nu]=chri5(N,ab0,z,eps0,nu0,numax,rho0,iopt)
```

where the input and output variables have the same meaning as in the routine `chri4.m`.

Just like Algorithms 3 and 4, also Algorithms 5 and 6 can be applied repeatedly to deal with more general polynomial divisors.

Exercises to §2 (Stars indicate more advanced exercises.)

1. Explain why, under the assumptions made about the measure $\mathrm{d}\lambda$, the inner product $(p,q)_{\mathrm{d}\lambda}$ of two polynomials p, q is well defined.
2. Show that monic orthogonal polynomials relative to an absolutely continuous measure are uniquely defined. {*Hint*: Use Gram-Schmidt orthogonalization.} Discuss the uniqueness in the case of discrete measures.
3. Supply the details of the proof of (2.1). In particular, derive (2.3) and (2.4).
4. Derive the three-term recurrence relation (2.7) for the orthonormal polynomials.
5. (a) With $\tilde{\pi}_k$ denoting the orthonormal polynomials relative to a measure $\mathrm{d}\lambda$, show that

$$\int_{\mathbb{R}} t\tilde{\pi}_k(t)\tilde{\pi}_\ell(t)\mathrm{d}\lambda(t) = \begin{cases} 0 & \text{if } |k-\ell|>1, \\ \sqrt{\beta_{k+1}} & \text{if } |k-\ell|=1, \\ \alpha_k & \text{if } k=\ell, \end{cases}$$

where $\alpha_k = \alpha_k(\mathrm{d}\lambda)$, $\beta_k = \beta_k(\mathrm{d}\lambda)$.
(b) Use (a) to prove

$$\boldsymbol{J} = \boldsymbol{J}_n(\mathrm{d}\lambda) = \int_{\mathbb{R}} t\boldsymbol{p}(t)\boldsymbol{p}^{\mathrm{T}}(t)\mathrm{d}\lambda(t),$$

where $\boldsymbol{p}^{\mathrm{T}}(t) = [\tilde{\pi}_0(t), \tilde{\pi}_1(t), \dots, \tilde{\pi}_{n-1}(t)]$.
(c) With notation as in (b), prove

$$t\boldsymbol{p}(t) = \boldsymbol{J}\boldsymbol{p}(t) + \sqrt{\beta_n}\tilde{\pi}_n(t)\boldsymbol{e}_n,$$

where $\boldsymbol{e}_n = [0,0,\dots,1]^{\mathrm{T}} \in \mathbb{R}^n$.
6. Let $\mathrm{d}\lambda(t) = w(t)\mathrm{d}t$ be symmetric on $[-a,a]$, $a>0$, that is, $w(-t)=w(t)$ on $[-a,a]$. Show that $\alpha_k(\mathrm{d}\lambda)=0$, all $k \geq 0$.

7*. Symmetry of orthogonal polynomials.
Let $\mathrm{d}\lambda(t) = w(t)\mathrm{d}t$ be symmetric in the sense of Exercise 6.
(a) Show that

$$\pi_{2k}(t;\mathrm{d}\lambda) = \pi_k^+(t^2), \quad \pi_{2k+1}(t;\mathrm{d}\lambda) = t\pi_k^-(t^2),$$

where $\pi_k^\pm$ are the monic polynomials orthogonal on $[0,a^2]$ with respect to $\mathrm{d}\lambda^\pm(t) = t^{\mp 1/2}w(t^{1/2})\mathrm{d}t$.
(b) Let (cf. Exercise 6)

$$\pi_{k+1}(t) = t\pi_k(t) - \beta_k\pi_{k-1}(t), \quad k=0,1,2,\dots,$$
$$\pi_{-1}(t)=0, \quad \pi_0(t)=1$$

be the recurrence relation for $\{\pi_k(\,\cdot\,;\mathrm{d}\lambda)\}$, and let $\alpha_k^\pm$, $\beta_k^\pm$ be the recurrence coefficients for $\{\pi_k^\pm\}$. Show that

$$\left.\begin{aligned} \beta_1 &= \alpha_0^+ \\ \beta_{2k} &= \beta_k^+/\beta_{2k-1} \\ \beta_{2k+1} &= \alpha_k^+ - \beta_{2k} \end{aligned}\right\} \quad k = 1, 2, 3, \ldots .$$

(c) Derive relations similar to those in (b) which involve α_0^+ and α_k^-, β_k^-.

(d) Write a Matlab program that checks the numerical stability of the nonlinear recursions in (b) and (c) when $\{\pi_k\}$ are the monic Legendre polynomials.

8. The recurrence relation, in Matlab, of the Chebyshev polynomials of the second kind.

   (a) Using Matlab, compute $U_k(x)$ for $1 \le k \le N$ either by means of the three-term recurrence relation $U_{n+1}(x) = 2xU_n(x) - U_{n-1}(x)$ for $n = 0, 1, \ldots, N-1$ (where $U_{-1}(x) = 0$, $U_0(x) = 1$), or else by putting $n = 1 : N$ in the explicit formula $U_n(\cos\theta) = \sin(n+1)\theta/\sin\theta$, where $x = \cos\theta$. For selected values of x and N, determine which of the two methods, by timing each, is more efficient.

   (b) Using Matlab, compute the single value $U_N(x)$ either by use of the three-term recurrence relation, or by direct computation based on the trigonometric formula for $U_N(\cos\theta)$. For selected values of x and N, determine which of the two methods, by timing each, is more efficient.

9*. Orthogonality on two separate (symmetric) intervals.
   Let $0 < \xi < 1$ and consider orthogonal polynomials π_k relative to the weight function

$$w(t) = \begin{cases} |t|^\gamma (1-t^2)^\alpha (t^2-\xi^2)^\beta, & t \in [-1, \xi] \cup [\xi, 1], \\ 0, & \text{otherwise.} \end{cases}$$

   Here, $\gamma \in \mathbb{R}$ and $\alpha > -1$, $\beta > -1$. Evidently, w is a symmetric weight function (in the sense of Exercise 6). Define $\pi_k^\pm$ as in Exercise 7(a).

   (a) Transform the polynomials $\pi_k^\pm$ orthogonal on $[\xi^2, 1]$ to orthogonal polynomials $\mathring{\pi}_k^\pm$ on the interval $[-1, 1]$ and obtain the respective weight function $\mathring{w}^\pm$.

   (b) Express β_{2k} and β_{2k+1} in terms of $\mathring{\gamma}_r^\pm$, the leading coefficient of the orthonormal polynomial of degree r relative to the weight function $\mathring{w}^\pm$ on $[-1, 1]$. {*Hint:* Use $\beta_r = \|\pi_r\|^2/\|\pi_{r-1}\|^2$ (cf. eqn (2.4)) and relate this to the leading coefficients γ_k, $\gamma_k^\pm$, and $\mathring{\gamma}_k^\pm$, with obvious notations.}

   (c) Prove that

$$\lim_{k\to\infty} \beta_{2k} = \frac{1}{4}(1-\xi)^2, \quad \lim_{k\to\infty} \beta_{2k+1} = \frac{1}{4}(1+\xi)^2.$$

   {*Hint:* Use the result of (b) in combination with the asymptotic equivalence

$$\mathring{\gamma}_k^{\pm} \sim 2^k \mathring{\gamma}^{\pm}, \quad \mathring{\gamma}^{\pm} = \pi^{-1/2} \exp\left\{ -\frac{1}{2\pi} \int_{-1}^{1} \ln \mathring{w}^{\pm}(x)(1-x^2)^{-1/2} \mathrm{d}x \right\},$$
$$\text{as } k \to \infty$$

(cf. [16, eqn (12.7.2)]). You may also want to use

$$\int_0^1 \ln(1-a^2x^2)(1-x^2)^{-1/2}\mathrm{d}x = \pi \ln \frac{1+(1-a^2)^{1/2}}{2}, \qquad a^2 < 1$$

(see [13, eqn 4.295.29]).}

(d) Prove that

$$\lim_{k\to\infty} \alpha_k^{\pm} = \frac{1+\xi^2}{2}, \qquad \lim_{k\to\infty} \beta_k^{\pm} = \left(\frac{1-\xi^2}{4}\right)^2.$$

{*Hint*: Express $\alpha_k^{\pm}$, $\beta_k^{\pm}$ in terms of $\mathring{\alpha}_k^{\pm}$, $\mathring{\beta}_k^{\pm}$, and use the fact that the weight function $\mathring{w}^{\pm}$ is in the Szegő class.}

(e) The recurrence coefficients $\{\beta_k\}$ must satisfy the two nonlinear recursions of Exercise 7(b),(c). Each of them can be interpreted as a pair of fixed-point iterations for the even-indexed and for the odd-indexed subsequence, the fixed points being respectively the limits in (c). Show that, asymptotically, both fixed points are "attractive" for the recursion in 7(b), and "repelling" for the one in 7(c). Also show that in the latter, the fixed points become attractive if they are switched. What are the numerical implications of all this?

(f) Consider the special case $\gamma = \pm 1$ and $\alpha = \beta = -\frac{1}{2}$. In the case $\gamma = 1$, use Matlab to run the nonlinear recursion of Exercise 7(b) and compare the results with the known answers

$$\beta_{2k} = \frac{1}{4}(1-\xi)^2 \frac{1+\eta^{2k-2}}{1+\eta^{2k}}, \qquad k = 1,2,3,\ldots, \qquad 0 \le t \le 1$$

and

$$\beta_1 = \frac{1}{2}(1+\xi^2), \quad \beta_{2k+1} = \frac{1}{4}(1+\xi)^2 \frac{1+\eta^{2k+2}}{1+\eta^{2k}}, \qquad k = 1,2,3,\ldots,$$

where $\eta = (1-\xi)/(1+\xi)$ (see [10, Example 2.30]). Likewise, in the case $\gamma = -1$, run the nonlinear recursion of Exercise 7(c) and compare the results with the exact answers

$$\beta_2 = \tfrac{1}{2}(1-\xi)^2, \quad \beta_{2k} = \tfrac{1}{4}(1-\xi)^2, \quad k = 2,3,\ldots,$$

and

$$\beta_1 = \xi, \quad \beta_{2k+1} = \tfrac{1}{4}(1+\xi)^2, \quad k = 1,2,3,\ldots.$$

Comment on what you observe.

10. Prove the validity of Algorithm 1.
    (a) Verify the initialization part.
    (b) Combine $\sigma_{k+1,k-1} = 0$ with the three-term recurrence relation for π_k to prove the formula for β_k in the continuation part.
    (c) Combine $\sigma_{k+1,k} = 0$ with the three-term recurrence relation for both, π_k and p_k, and use the result of (b), to prove the formula for α_k in the continuation part.
11*. Orthogonal polynomials $\{\pi_k(\,\cdot\,; w)\}$ relative to the weight function ("hat function")

$$w(t) = \begin{cases} 1+t & \text{if } -1 \leq t \leq 0, \\ 1-t & \text{if } 0 \leq t \leq 1, \\ 0 & \text{otherwise.} \end{cases}$$

    (a) Develop a modified Chebyshev algorithm for generating the first n recurrence coefficients $\beta_k(w)$, $k = 0, 1, \ldots, n-1$ (all $\alpha_k(w) = 0$; why?). Define modified moments with respect to a suitable system of (monic) orthogonal polynomials.
    (b) What changes in the routine are required if one wants $\{\pi_k(\,\cdot\,; 1-w)\}$, or $\{\pi_k(\,\cdot\,; w(1-w))\}$, or $\{\pi_k(\,\cdot\,; w^p)\}$ where $p > -1$?
    (c) Download from `OPQ` the routine `chebyshev.m`, write a routine `mom.m` for the modified moments to be used in conjunction with `chebyshev.m` to implement (a), and write a Matlab driver to produce results for selected values of n.
    (d) Devise a 2-component discretization scheme for computing the first n recurrence coefficients $\beta_k(w)$, $k = 0, 1, 2, \ldots, n-1$, which uses an n-point discretization of the inner product on each component interval and is to yield exact answers (in the absence of rounding errors).
    (e) Same as (b).
    (f) Download from `OPQ` the routine `mcdis.m`, write a quadrature routine `qhatf.m` necessary to implement (d), and append a script to the driver of (c) that produces results of the discretization procedure for selected values of n. Download whatever additional routines you need. Run the procedure with `irout` $= 1$ and `irout` $\neq 1$ and observe the respective timings and the maximum discrepancy between the two sets of answers. Verify that the routine "converges" after one iteration if `idelta` is properly set. Compare the results with those of (a).
    (g) Use the routines `acondG.m` and `rcondG.m` to print the absolute and relative condition numbers of the relevant map G_n. Do any of these correlate well with the numerical results obtained in (c)? If not, why not?
12*. Orthogonal polynomials $\{\pi_k(\,\cdot\,; w)\}$ relative to the weight function ("exponential integral")

$$w(t) = E_1(t), \quad E_1(t) = \int_1^\infty \frac{e^{-ts}}{s}\, ds \quad \text{on } [0, \infty].$$

    These are of interest in the theory of radiative transfer (Chandrasekhar [2, Chapter II, §23]).

(a) Develop and run a multiple-component discretization routine for generating the first n recurrence coefficients $\alpha_k(w)$, $\beta_k(w)$, $k = 0, 1, \dots, n-1$. Check your results for $n = 20$ against [3, Table 3]. {*Hint*: Decompose the interval $[0, \infty]$ into two subintervals $[0, 2]$ and $[2, \infty]$ (additional subdivisions may be necessary to implement the developments that follow) and incorporate the behavior of $E_1(t)$ near $t = 0$ and $t = \infty$ to come up with appropriate discretizations. For $0 \le t \le 2$, use the power series

$$E_1(t) - \ln(1/t) = -\gamma - \sum_{k=1}^{\infty} \frac{(-1)^k t^k}{k k!},$$

where $\gamma = .57721566490153286\dots$ is Euler's constant, and for $t > 2$ the continued fraction (cf. [1, eqn 5.1.22])

$$te^t E_1(t) = \frac{1}{1+}\,\frac{a_1}{1+}\,\frac{a_2}{1+}\,\frac{a_3}{1+}\,\frac{a_4}{1+}\cdots, \qquad a_k = \lceil k/2 \rceil / t.$$

Evaluate the continued fraction recursively by (cf. [7, §2])

$$\frac{1}{1+}\,\frac{a_1}{1+}\,\frac{a_2}{1+}\cdots = \sum_{k=0}^{\infty} t_k,$$

where

$$t_0 = 1, \quad t_k = \rho_1 \rho_2 \cdots \rho_k, \qquad k = 1, 2, 3, \dots,$$

$$\rho_0 = 0, \quad \rho_k = \frac{-a_k(1 + \rho_{k-1})}{1 + a_k(1 + \rho_{k-1})}, \qquad k = 1, 2, 3, \dots .$$

Download the array `abjaclog(101:200,:)` to obtain the recurrence coefficients `ab` for the logarithmic weight function $\ln(1/t)$.}

(b) Do the same for

$$w(t) = E_2(t), \quad E_2(t) = \int_1^{\infty} \frac{e^{-ts}}{s^2}\, ds \;\text{ on } [0, \infty].$$

Check your results against the respective two- and three-point Gauss quadrature formulae in Chandrasekhar [2, Table VI].

(c) Do the same for

$$w(t) = E_m(t) \;\text{ on } [0, c], \quad 0 < c < \infty, \; m = 1, 2.$$

Check your results against the respective two-point Gauss quadrature formulae in Chandrasekhar [2, Table VII].

13. Let $C = b_0 + \frac{a_1}{b_1+} \frac{a_2}{b_2+} \frac{a_3}{b_3+} \cdots$ be an infinite continued fraction, and $C_n = b_0 + \frac{a_1}{b_1+} \cdots \frac{a_n}{b_n} = \frac{A_n}{B_n}$ its nth convergent. From the theory of continued fractions, it is known that

$$\left.\begin{aligned} A_n &= b_n A_{n-1} + a_n A_{n-2} \\ B_n &= b_n B_{n-1} + a_n B_{n-2} \end{aligned}\right\} \quad n = 1, 2, 3, \ldots,$$

where

$$A_{-1} = 1,\ A_0 = b_0; \quad B_{-1} = 0,\ B_0 = 1.$$

Use this to prove (2.21) and (2.22).

14. Prove (2.23).
15. Show that (2.30) implies $\lim_{n\to\infty} \frac{\rho_n}{y_n} = 0$, where y_n is *any* solution of the three-term recurrence relation (satisfied by ρ_n and π_n) which is linearly independent of ρ_n. Thus, $\{\rho_n\}$ is indeed a minimal solution.
16. Show that the minimal solutions of a three-term recurrence relation form a one-dimensional manifold.
17. (a) Derive (2.47).
    (b) Supply the details for deriving Algorithm 4.
18. Supply the details for deriving Algorithm 5.
19. (a) Prove (2.53).
    (b) Prove (2.55), (2.56).
    (c) Supply the details for deriving Algorithm 6.

3 Sobolev Orthogonal Polynomials

3.1 Sobolev Inner Product and Recurrence Relation

In contrast to the orthogonal polynomials considered so far, the inner product here involves not only function values, but also successive derivative values, all being endowed with their own measures. Thus,

$$\begin{aligned} (p, q)_S = &\int_{\mathbb{R}} p(t)q(t)\mathrm{d}\lambda_0(t) + \int_{\mathbb{R}} p'(t)q'(t)\mathrm{d}\lambda_1(t) \\ &+ \cdots + \int_{\mathbb{R}} p^{(s)}(t)q^{(s)}(t)\mathrm{d}\lambda_s(t), \quad s \geq 1. \end{aligned} \tag{3.1}$$

If all the measures $\mathrm{d}\lambda_\sigma$ are positive, the inner product (3.1) has associated with it a sequence of (monic) polynomials $\pi_k(\,\cdot\,; S)$, $k = 0, 1, 2, \ldots$, orthogonal in the sense

$$(\pi_k, \pi_\ell)_S \begin{cases} = 0, \ k \neq \ell, \\ > 0, \ k = \ell. \end{cases} \tag{3.2}$$

These are called *Sobolev orthogonal polynomials.* We cannot expect them to satisfy a three-term recurrence relation, since the inner product no longer has the shift property $(tp, q) = (p, tq)$. However, like any sequence of monic polynomials of degrees $0, 1, 2, \ldots$, orthogonal or not, they must satisfy an extended recurrence relation of the type

$$\pi_{k+1}(t) = t\pi_k(t) - \sum_{j=0}^{k} \beta_j^k \pi_{k-j}(t), \quad k = 0, 1, 2, \ldots. \tag{3.3}$$

Associated with it is the upper Hessenberg *matrix of recurrence coefficients*

$$\boldsymbol{H}_n = \begin{bmatrix} \beta_0^0 & \beta_1^1 & \beta_2^2 & \cdots & \beta_{n-2}^{n-2} & \beta_{n-1}^{n-1} \\ 1 & \beta_0^1 & \beta_1^2 & \cdots & \beta_{n-3}^{n-2} & \beta_{n-2}^{n-1} \\ 0 & 1 & \beta_0^2 & \cdots & \beta_{n-4}^{n-2} & \beta_{n-3}^{n-1} \\ \cdots & \cdots & \cdots & \cdots & \cdots & \cdots \\ 0 & 0 & 0 & \cdots & \beta_0^{n-2} & \beta_1^{n-1} \\ 0 & 0 & 0 & \cdots & 1 & \beta_0^{n-1} \end{bmatrix}. \tag{3.4}$$

In the case $s = 0$ (of ordinary orthogonal polynomials) there holds $\beta_j^k = 0$ for $j > 1$, and the matrix $\boldsymbol{H}_n$ is tridiagonal. If symmetrized by a (real) diagonal similarity transformation, it becomes the Jacobi matrix $\boldsymbol{J}_n(\mathrm{d}\lambda_0)$. When $s > 0$, however, symmetrization of $\boldsymbol{H}_n$ is no longer possible, since $\boldsymbol{H}_n$ may well have complex eigenvalues (see Example 6).

3.2 Moment-Based Algorithm

There are now $s + 1$ sets of modified moments, one set for each measure $\mathrm{d}\lambda_\sigma$,

$$m_k^{(\sigma)} = \int_{\mathbb{R}} p_k(t)\mathrm{d}\lambda_\sigma, \quad k = 0, 1, 2, \ldots; \quad \sigma = 0, 1, \ldots, s. \tag{3.5}$$

The first $2n$ modified moments of all the sets will uniquely determine the matrix $\boldsymbol{H}_n$ in (3.4), i.e. there is a well-determined map

$$[m_k^{(\sigma)}]_{k=0}^{2n-1}, \ \sigma = 0, 1, \ldots, s \ \mapsto \boldsymbol{H}_n, \tag{3.6}$$

called *modified moment map* for Sobolev orthogonal polynomials. In the case where the polynomials p_k in (3.5) satisfy a three-term recurrence relation with known coefficients, and for $s = 1$, an algorithm has been developed that implements the map (3.6). It very much resembles the modified Chebyshev algorithm for ordinary orthogonal polynomials, but is technically much more elaborate (see [12]). The algorithm, however, is implemented in the `OPQ` routine

```
B=chebyshev_sob(N,mom,abm)
```

which produces the `N`×`N` upper triangular matrix `B` of recurrence coefficients, with β_j^k, $0 \le j \le k$, $0 \le k \le$N–1, occupying the position $(j + 1, k + 1)$ in the matrix. The input parameter `mom` is the $2 \times (2\mathrm{N})$ array of modified moments $m_k^{(\sigma)}$, $k = 0, 1, \ldots,$2N–1; $\sigma = 0, 1$, of the two measures $\mathrm{d}\lambda_0$ and $\mathrm{d}\lambda_1$, and `abm` the (2N−1) × 2 array of coefficients a_k, b_k, $k = 0, 1, \ldots, 2N-2$, defining the polynomials p_k.

Example 4. Althammer's polynomials (1962).
These are the Sobolev polynomials relative to the measures $d\lambda_0(t) = dt$, $d\lambda_1(t) = \gamma dt$ on $[-1, 1]$, $\gamma > 0$.

A natural choice of modified moments are the Legendre moments, i.e. $p_k(t)$ is the monic Legendre polynomial of degree k. By orthogonality of the Legendre polynomials, all modified moments $m_k^{(0)}$ and $m_k^{(1)}$ are zero for $k > 0$, while $m_0^{(0)} = 2$ and $m_0^{(1)} = 2\gamma$. The following Matlab routine, therefore, can be used to generate the Althammer polynomials.

```
mom=zeros(2,2*N);
mom(1,1)=2; mom(2,1)=2*g;
abm=r_jacobi(2*N-1);
B=chebyshev_sob(N,mom,abm);
```

3.3 Discretization Algorithm

Taking the inner product of both sides of (3.3) with π_{k-j} gives

$$0 = (\pi_{k+1}, \pi_{k-j})_S = (t\pi_k, \pi_{k-j})_S - \beta_j^k (\pi_{k-j}, \pi_{k-j})_S, \quad j = 0, 1, \ldots, k,$$

hence

$$\beta_j^k = \frac{(t\pi_k, \pi_{k-j})_S}{(\pi_{k-j}, \pi_{k-j})_S}, \quad j = 0, 1, \ldots, k; \; k = 0, 1, \ldots, n-1. \tag{3.7}$$

These are the analogues of Darboux's formulae for ordinary orthogonal polynomials, and like these, can be combined with the recurrence relation (3.3) to successively build up the recurrence coefficients β_j^k in the manner of Stieltjes's procedure. The technical details, of course, are more involved, since we must generate not only the polynomials π_k, but also their derivatives, in order to be able to compute the Sobolev inner products in (3.7). This all is implemented, for arbitrary $s \geq 1$, in the Matlab routine `stieltjes_sob.m`. The basic assumption in the design of this routine is the availability, for each measure $d\lambda_\sigma$, of an n_σ-point quadrature rule

$$\int_{\mathbb{R}} p(t)\, d\lambda_\sigma(t) = \sum_{k=1}^{n_\sigma} w_k^{(\sigma)} p(x_k^{(\sigma)}), \quad p \in \mathbb{P}_{2(n-\sigma)-1}, \quad \sigma = 0, 1, \ldots, s, \tag{3.8}$$

that is exact for polynomials p of degree $\leq 2(n-\sigma) - 1$. These are typically Gaussian quadrature rules, possibly with discrete components (present in $d\lambda_\sigma$) added on. The information is supplied to the routine via the $1 \times (s+1)$ array

$$\texttt{nd}= [n_0, n_1, \ldots, n_s]$$

and the md×$(2s+2)$ array

$$\texttt{xw=}\left[\begin{array}{ccc|ccc} x_1^{(0)} & \cdots & x_1^{(s)} & w_1^{(0)} & \cdots & w_1^{(s)} \\ x_2^{(0)} & \cdots & x_2^{(s)} & w_2^{(0)} & \cdots & w_2^{(s)} \\ \vdots & & \vdots & \vdots & & \vdots \\ x_{\text{md}}^{(0)} & \cdots & x_{\text{md}}^{(s)} & w_{\text{md}}^{(0)} & \cdots & w_{\text{md}}^{(s)} \end{array}\right]$$

where `md=max(nd)`. In each column of `xw` the entries after $x_{n_\sigma}^{(\sigma)}$ resp. $w_{n_\sigma}^{(\sigma)}$ (if any) are not used by the routine. Two more input parameters are needed; the first is `a0`, the coefficient $\alpha_0(\mathrm{d}\lambda_0)$, which allows us to initialize the matrix of recurrence coefficients,

$$\beta_0^0 = \frac{(t,1)_S}{(1,1)_S} = \frac{(t,1)_{\mathrm{d}\lambda_0}}{(1,1)_{\mathrm{d}\lambda_0}} = \alpha_0(\mathrm{d}\lambda_0).$$

The other, `same`, is a logical variable set equal to 1 if all quadrature rules have the same set of nodes, and equal to 0 otherwise. The role of this parameter is to switch to a simplified, and thus faster, procedure if `same`=1. A call to the routine, therefore, has the form

```
B=stieltjes_sob(N,s,nd,xw,a0,same)
```

Example 5. Althammer's polynomials, revisited.

Here, the obvious choice of the quadrature rule for $\mathrm{d}\lambda_0$ and $\mathrm{d}\lambda_1$ is the n-point Gauss–Legendre rule. This gives rise to the following routine:

```
s=1; nd=[N N];
a0=0; same=1;
ab=r_jacobi(N);
zw=gauss(N,ab);
xw=[zw(:,1) zw(:,1) ...
     zw(:,2) g*zw(:,2)];
B=stieltjes_sob(N,s,nd,xw,a0,same);
```

The results are identical with those obtained in Example 4.

3.4 Zeros

If we let $\boldsymbol{\pi}^{\mathrm{T}}(t) = [\pi_0(t), \pi_1(t), \ldots, \pi_{n-1}(t)]$, where π_k are the Sobolev orthogonal polynomials, then the recurrence relation (3.3) can be written in matrix form as

$$t\boldsymbol{\pi}^{\mathrm{T}}(t) = \boldsymbol{\pi}^{\mathrm{T}}(t)\boldsymbol{H}_n + \pi_n(t)\boldsymbol{e}_n^{\mathrm{T}} \tag{3.9}$$

in terms of the matrix $\boldsymbol{H}_n$ in (3.4). This immediately shows that the zeros τ_ν of π_n are the eigenvalues of $\boldsymbol{H}_n$ and $\boldsymbol{\pi}^{\mathrm{T}}(\tau_\nu)$ corresponding left eigenvectors. Naturally, there is no guarantee that the eigenvalues are real; some may well be complex. Also, if n is large, there is a good chance that some of the eigenvalues are ill-conditioned.

The OPQ routine for the zeros of π_n is

```
z=sobzeros(n,N,B)
```

where B is the N×N matrix returned by chebyshev_sob.m or stieltjes_sob.m, and z the n-vector of the zeros of π_n, $1 \le n \le N$.

Example 6. Sobolev orthogonal polynomials with only a few real zeros (Meijer, 1994).

The Sobolev inner product in question is

$$(u,v)_S = \int_{-1}^{3} u(t)v(t)\,\mathrm{d}t + \gamma \int_{-1}^{1} u'(t)v'(t)\,\mathrm{d}t + \int_{1}^{3} u'(t)v'(t)\,\mathrm{d}t, \quad \gamma > 0. \tag{3.10}$$

Meijer proved that for $n(\text{even}) \ge 2$ and γ sufficiently large, the polynomial $\pi_n(\,\cdot\,;S)$ has exactly two real zeros, one in $[-3,-1]$ and the other in $[1,3]$. If $n(\text{odd}) \ge 3$, there is exactly one real zero, located in $[1,3]$, if γ is sufficiently large. We use the routine stieltjes_sob.m and sobzeros.m to illustrate this for $n = 6$ and $\gamma = 44{,}000$. (The critical value of γ above which Meijer's theorem takes hold is about $\gamma = 43{,}646.2$; see [10, Table 2.30].)

The inner product corresponds to the case $s = 1$ and

$$\mathrm{d}\lambda_0(t) = dt \text{ on } [-1,3], \quad \mathrm{d}\lambda_1(t) = \begin{cases} \gamma \mathrm{d}t \text{ if } t \in [-1,1], \\ dt \ \text{ if } t \in (1,3]. \end{cases}$$

Thus, we can write, with suitable transformations of variables,

$$\int_{-1}^{3} p(t)\,\mathrm{d}\lambda_0(t) = 2\int_{-1}^{1} p(2x+1)\,\mathrm{d}x,$$
$$\int_{-1}^{3} p(t)\,\mathrm{d}\lambda_1(t) = \int_{-1}^{1} [\gamma p(x) + p(x+2)]\,\mathrm{d}x$$

and apply n-point Gauss–Legendre quadrature to the integrals on the right. This will produce the matrix $\boldsymbol{H}_n$ exactly. The parameters in the routine stieltjes_sob.m have to be chosen as follows:

$$\mathtt{nd} = [n, 2n], \quad \mathtt{xw} = \begin{bmatrix} 2\tau_1^G + 1 & \tau_1^G & 2\lambda_1^G & \gamma\lambda_1^G \\ \vdots & \vdots & \vdots & \vdots \\ 2\tau_n^G + 1 & \tau_n^G & 2\lambda_n^G & \gamma\lambda_n^G \\ & \tau_1^G + 2 & & \lambda_1^G \\ & \vdots & & \vdots \\ & \tau_n^G + 2 & & \lambda_n^G \end{bmatrix} \in \mathbb{R}^{2n\times 4},$$

where τ_ν^G, λ_ν^G are the nodes and weight of the n-point Gauss–Legendre quadrature rule. Furthermore, a0=1 and same=0. The complete program, therefore, is as follows:

```
N=6; s=1; a0=1; same=0; g=44000; nd=[N 2*N];
ab=r_jacobi(N); zw=gauss(N,ab);
xw=zeros(2*N,2*(s+1));
xw(1:N,1)=2*zw(:,1)+1; xw(1:N,2)=zw(:,1);
xw(1:N,3)=2*zw(:,2); xw(1:N,4)=g*zw(:,2);
xw(N+1:2*N,2)=zw(:,1)+2; xw(N+1:2*N,4)=zw(:,2);
B=stieltjes_sob(N,s,nd,xw,a0,same);
z=sobzeros(N,N,B)
```

It produces the output

```
z =

-4.176763898909848e-01 - 1.703657992747233e-01i
-4.176763898909848e-01 + 1.703657992747233e-01i
 8.453761089539369e-01 - 1.538233952529940e-01i
 8.453761089539369e-01 + 1.538233952529940e-01i
-1.070135059563751e+00
 2.598402134930250e+00
```

confirming Meijer's theorem for $n = 6$. A more detailed numerical study, also in the case of odd values of n, has been made in [10, Table 2.30].

Exercises to §3

1. Show that a Sobolev inner product does not satisfy the shift property $(tp, q) = (p, tq)$.
2. Prove (3.3).
3. The Sobolev inner product (3.1) is called *symmetric* if each measure $d\lambda_\sigma$ is symmetric in the sense of Problem 6, §2. For symmetric Sobolev inner products,
   (a) show that $\pi_k(-t; S) = (-1)^k \pi_k(t; S)$;
   (b) show that $\beta_{2r}^k = 0$ for $r = 0, 1, \ldots, \lfloor k/2 \rfloor$.
4. Consider a Sobolev inner product with $s = 1$ and

$$d\lambda_0 = d\lambda, \quad d\lambda_1 = \gamma d\lambda \ (\gamma > 0),$$

   and $d\lambda$ a symmetric measure. Use the routines chebyshev_sob.m and sobzeros.m to check numerically whether or not the positive zeros of the Sobolev orthogonal polynomials are monotonically increasing with γ. Experiment in turn with $d\lambda(t) = dt$ on $[-1, 1]$ (Althammer polynomials), $d\lambda(t) = (1-t^2)^\alpha$ on $[-1, 1]$, $\alpha > -1$, and $d\lambda(t) = \exp(-t^2)$ on $\mathbb{R}$. Identify any computational problems and how to deal with them.
5. Special Sobolev orthogonal polynomials.

(a) Consider the special Sobolev orthogonal polynomials that have an absolutely continuous (or, possibly, discrete) ground measure $\mathrm{d}\lambda_0$ and all $\mathrm{d}\lambda_\sigma$, $1 \leq \sigma \leq s$, identically zero except for $\mathrm{d}\lambda_{r_k}$, $k = 1, 2, \ldots, K$, where $1 \leq r_1 < r_2 < \cdots < r_K \leq s$, which are atomic measures located at the points c_k and having masses m_k. Assuming that the ground measure is given by the array `ab` of recurrence coefficients, write a Matlab routine `specsob.m` that uses the `OPQ` routine `stieltjes_sob.m` to compute the recurrence matrix `B` of the special Sobolev orthogonal polynomials.
(b) Use your routine together with the `OPQ` routine `sobzeros.m` to check Tables 2–4 in [8], relating to the Hermite measure $\mathrm{d}\lambda_0(t) = \exp(-t^2)\mathrm{d}t$ and a single atomic measure involving the rth derivative. In the cited reference, the results were obtained by a different method.
(c) In the case of the Laguerre measure $\mathrm{d}\lambda_0(t) = \exp(-t)\mathrm{d}t$ on $\mathbb{R}_+$ and $r_k = k$, $c_k = 0$, $m_k = 1$, it may be conjectured that any complex zero that occurs has negative real part. Use your routine and `sobzeros.m` to check out this conjecture.
(d) For $\mathrm{d}\lambda_0$, r_k, c_k, m_k as in (c), determine the pattern of occurrence of complex zeros. Cover the range $1 \leq s \leq 10$, $1 \leq n \leq 40$.
(e) Repeat (c) and (d) with $\mathrm{d}\lambda_0$ the Laguerre measure plus an atomic measure with mass 1 at the origin.

4 Quadrature

4.1 Gauss-Type Quadrature Formulae

Gauss Formula

Given a positive measure $\mathrm{d}\lambda$, the n-point *Gaussian quadrature formula* associated with the measure $\mathrm{d}\lambda$ is

$$\int_{\mathbb{R}} f(t)\mathrm{d}\lambda(t) = \sum_{\nu=1}^{n} \lambda_\nu^G f(\tau_\nu^G) + R_n^G(f), \tag{4.1}$$

which has maximum algebraic degree of exactness $2n-1$,

$$R_n^G(f) = 0 \quad \text{if } f \in \mathbb{P}_{2n-1}. \tag{4.2}$$

It is well known that the nodes τ_ν^G are the zeros of $\pi_n(\,\cdot\,;\mathrm{d}\lambda)$, and hence the eigenvalues of the Jacobi matrix $\boldsymbol{J}_n(\mathrm{d}\lambda)$; cf. §2.1. Interestingly, the weights λ_ν^G, too, can be expressed in terms of spectral data of $\boldsymbol{J}_n(\mathrm{d}\lambda)$; indeed, they are (Golub and Welsch, 1969)

$$\lambda_\nu^G = \beta_0 \boldsymbol{v}_{\nu,1}^2, \tag{4.3}$$

where $\boldsymbol{v}_{\nu,1}$ is the first component of the normalized eigenvector $\boldsymbol{v}_\nu$ corresponding to the eigenvalue τ_ν^G,

$$\boldsymbol{J}_n(\mathrm{d}\lambda)\boldsymbol{v}_\nu = \tau_\nu^G \boldsymbol{v}_\nu, \quad \boldsymbol{v}_\nu^{\mathrm{T}}\boldsymbol{v}_\nu = 1, \tag{4.4}$$

and, as usual, $\beta_0 = \int_{\mathbb{R}} \mathrm{d}\lambda(t)$. This is implemented in the OPQ Matlab routine

```
xw=gauss(N,ab)
```

where ab, as in all previous routines, is the N×2 array of recurrence coefficients for $\mathrm{d}\lambda$, and xw the N×2 array containing the nodes τ_ν^G in the first column, and the weights λ_ν^G in the second.

We remark, for later purposes, that the Gauss quadrature sum, for f sufficiently regular, can be expressed in matrix form as

$$\sum_{\nu=1}^{n} \lambda_\nu^G f(\tau_\nu^G) = \beta_0 \boldsymbol{e}_1^{\mathrm{T}} f(\boldsymbol{J}_n(\mathrm{d}\lambda))\boldsymbol{e}_1, \quad \boldsymbol{e}_1 = [1, 0, \ldots, 0]^{\mathrm{T}}. \tag{4.5}$$

This is an easy consequence of (4.3) and the spectral decomposition of $\boldsymbol{J}_n$,

$$\boldsymbol{J}_n(\mathrm{d}\lambda)\boldsymbol{V} = \boldsymbol{V}\boldsymbol{D}_\tau, \quad \boldsymbol{D}_\tau = \mathrm{diag}(\tau_1^G, \tau_2^G, \ldots, \tau_n^G),$$

where $\boldsymbol{V} = [\boldsymbol{v}_1, \boldsymbol{v}_2, \ldots, \boldsymbol{v}_n]$.

Example 7. Zeros of Sobolev orthogonal polynomials of Gegenbauer type (Groenevelt, 2002).

The polynomials in question are those orthogonal with respect to the Sobolev inner product

$$(u, v)_S = \int_{-1}^{1} u(t)v(t)(1-t^2)^{\alpha-1}\mathrm{d}t + \gamma \int_{-1}^{1} u'(t)v'(t)\frac{(1-t^2)^\alpha}{t^2+y^2}\,\mathrm{d}t.$$

Groenevelt proved that in the case $\gamma \to \infty$ the Sobolev orthogonal polynomials of even degrees $n \geq 4$ have complex zeros if y is sufficiently small. By symmetry, they must in fact be purely imaginary, and by the reality of the Sobolev polynomials, must occur in conjugate complex pairs. As we illustrate this theorem, we have an opportunity to apply not only the routine gauss.m, but also a number of other routines, specifically the modification algorithm embodied in the routine chri6.m, dealing with the special quadratic divisor $t^2 + y^2$ in the second integral, and the routine stieltjes_sob.m generating the recurrence matrix of the Sobolev orthogonal polynomials:

```
s=1; same=0; eps0=1e-14; numax=250; nd=[N N];
ab0=r_jacobi(numax,alpha);
z=complex(0,y);
nu0=nu0jac(N,z,eps0); rho0=0; iopt=1;
ab1=chri6(N,ab0,y,eps0,nu0,numax,rho0,iopt);
zw1=gauss(N,ab1);
ab=r_jacobi(N,alpha-1); zw=gauss(N,ab);
xw=[zw(:,1) zw1(:,1) zw(:,2) gamma*zw1(:,2)];
a0=ab(1,1); B=stieltjes_sob(N,s,nd,xw,a0,same);
z=sobzeros(N,N,B)
```

Demo#4 The case N=12, $\alpha = \frac{1}{2}$, and $\gamma = 1$ of Example 7.

Applying the above routine for $y = .1$ and $y = .09$ yields the following zeros (with positive imaginary parts; the other six zeros are the same with opposite signs):

y	zeros	y	zeros
.1	.027543282225	.09	.011086169153 i
	.284410786673		.281480077515
	.541878443180		.540697645595
	.756375307278		.755863108617
	.909868274113		.909697039063
	.989848649239		.989830182743

The numerical results (and additional tests) suggest that Groenevelt's theorem also holds for finite, not necessarily large, values of γ, and, when $\gamma = 1$, that the critical value of y below which there are complex zeros must be betweeen .09 and .1.

Gauss–Radau Formula

If there is an interval $[a, \infty]$, $-\infty < a$, containing the support of $\mathrm{d}\lambda$, it may be desirable to have an $(n+1)$-point quadrature rule of maximum degree of exactness that has a as a prescribed node,

$$\int_{\mathbb{R}} f(t)\mathrm{d}\lambda(t) = \lambda_0^a f(a) + \sum_{\nu=1}^{n} \lambda_\nu^a f(\tau_\nu^a) + R_n^a(f). \tag{4.6}$$

Here, $R_n^a(f) = 0$ for all $f \in \mathbb{P}_{2n}$, and τ_ν^a are the zeros of $\pi_n(\,\cdot\,; \mathrm{d}\lambda_a)$, $\mathrm{d}\lambda_a(t) = (t-a)\mathrm{d}\lambda(t)$. This is called the *Gauss–Radau formula.* There is again a symmetric, tridiagonal matrix, the *Jacobi–Radau matrix*

$$\boldsymbol{J}_{n+1}^{R,a}(\mathrm{d}\lambda) = \begin{bmatrix} \boldsymbol{J}_n(\mathrm{d}\lambda) & \sqrt{\beta_n}\boldsymbol{e}_n \\ \sqrt{\beta_n}\boldsymbol{e}_n^T & \alpha_n^R \end{bmatrix}, \quad \alpha_n^R = a - \beta_n \frac{\pi_{n-1}(a)}{\pi_n(a)}, \tag{4.7}$$

where $\boldsymbol{e}_n = [0, 0, \ldots, 1]^{\mathrm{T}} \in \mathbb{R}^n$, $\beta_n = \beta_n(\mathrm{d}\lambda)$, and $\pi_k(\,\cdot\,) = \pi_k(\,\cdot\,; \mathrm{d}\lambda)$, which allows the Gauss–Radau formula to be characterized in terms of eigenvalues and eigenvectors: all nodes of (4.6), including the node a, are the eigenvalues of (4.7), and the weights λ_ν^a expressible as in (4.3) in terms of the corresponding normalized eigenvectors $\boldsymbol{v}_\nu$ of (4.7),

$$\lambda_\nu^a = \beta_0 \boldsymbol{v}_{\nu,1}^2, \quad \nu = 0, 1, 2, \ldots, n. \tag{4.8}$$

As in (4.5), this implies that the Gauss–Radau quadrature sum, for smooth f, can be expressed as $\beta_0 \boldsymbol{e}_1^{\mathrm{T}} f(\boldsymbol{J}_{n+1}^{R,a})\boldsymbol{e}_1$, where $\boldsymbol{e}_1 = [1, 0, \ldots, 0] \in \mathbb{R}^{n+1}$.

Naturally, if the support of $\mathrm{d}\lambda$ is contained in an interval $[-\infty, b]$, $b < \infty$, there is a companion formula to (4.6) which has the prescribed node b,

$$\int_{\mathbb{R}} f(t)\,\mathrm{d}\lambda(t) = \sum_{\nu=1}^{n} \lambda_\nu^b f(\tau_\nu^b) + \lambda_{n+1}^b f(b) + R_n^b(f). \tag{4.9}$$

The eigenvalue/vector characterization also holds for (4.9) if in the formula for α_n^R in (4.7), the variable a, at every occurrence, is replaced by b.

The remainder terms of (4.6) and (4.9), if $f \in C^{2n+1}[a,b]$, have the useful property

$$R_n^a(f) > 0, \quad R_n^b(f) < 0 \quad \text{if } \operatorname{sgn} f^{(2n+1)} = 1 \text{ on } [a,b], \tag{4.10}$$

with the inequalities reversed if $\operatorname{sgn} f^{(2n+1)} = -1$.

For Jacobi resp. generalized Laguerre measures with parameters α, β resp. α, the quantity α_n^R is explicitly known (cf. [10, Examples 3.4 and 3.5]). For example, if $a = -1$ (in the case of Jacobi measures),

$$\alpha_n^R = -1 + \frac{2n(n+\alpha)}{(2n+\alpha+\beta)(2n+\alpha+\beta+1)}, \qquad \alpha_n^R = n, \tag{4.11}$$

whereas for $a = 1$, the sign of α_n^R must be changed and α and β interchanged.

The respective OPQ Matlab routines are

```
xw=radau(N,ab,end0)
xw=radau_jacobi(N,iopt,a,b)
xw=radau_laguerre(N,a)
```

In the first, ab is the (N+1)×2 array of recurrence coefficients for $\mathrm{d}\lambda$, and end0 either a (for (4.6)) or b (for (4.9)). The last two routines make use of the explicit formulae for α_n^R in the case of Jacobi resp. Laguerre measures, the parameters being α=a, β=b. The parameter iopt chooses between the two Gauss–Radau formulae: the left-handed, if iopt=1, the right-handed otherwise.

Gauss–Lobatto Formula

If the support of $\mathrm{d}\lambda$ is contained in the finite interval $[a,b]$, we may wish to prescribe two nodes, the points a and b. Maximizing the degree of exactness subject to these constraints yields the *Gauss–Lobatto formula*

$$\int_a^b f(t)\mathrm{d}\lambda(t) = \lambda_0^L f(a) + \sum_{\nu=1}^{n} \lambda_\nu^L f(\tau_\nu^L) + \lambda_{n+1}^L f(b) + R_n^{a,b}(f), \tag{4.12}$$

which we write as an $(n+2)$-point formula; we have $R_n^{a,b}(f) = 0$ for $f \in \mathbb{P}_{2n+1}$. The internal nodes τ_ν^L are the zeros of the polynomial $\pi_n(\,\cdot\,; \mathrm{d}\lambda_{a,b})$, $\mathrm{d}\lambda_{a,b}(t) = (t-a)(b-t)\mathrm{d}\lambda(t)$. All nodes and weights can be expressed in terms of eigenvalues and eigenvectors exactly as in the two preceding subsections, except that the matrix involved is the *Jacobi–Lobatto matrix*

$$\boldsymbol{J}_{n+2}^{L}(\mathrm{d}\lambda) = \begin{bmatrix} \boldsymbol{J}_{n+1}(\mathrm{d}\lambda) & \sqrt{\beta_{n+1}^{L}}\boldsymbol{e}_{n+1} \\ \sqrt{\beta_{n+1}^{L}}\boldsymbol{e}_{n+1}^{\mathrm{T}} & \alpha_{n+1}^{L} \end{bmatrix}, \tag{4.13}$$

where α_{n+1}^{L} and β_{n+1}^{L} are the solution of the 2×2 system of linear equations

$$\begin{bmatrix} \pi_{n+1}(a) & \pi_n(a) \\ \pi_{n+1}(b) & \pi_n(b) \end{bmatrix} \begin{bmatrix} \alpha_{n+1}^{L} \\ \beta_{n+1}^{L} \end{bmatrix} = \begin{bmatrix} a\pi_{n+1}(a) \\ b\pi_{n+1}(b) \end{bmatrix}. \tag{4.14}$$

For smooth f, the quadrature sum is expressible as $\beta_0 \boldsymbol{e}_1^{\mathrm{T}} f(\boldsymbol{J}_{n+2}^{L})\boldsymbol{e}_1$. For $f \in C^{2n+2}[a,b]$ with constant sign on $[a,b]$, the remainder $R_n^{a,b}(f)$ satisfies

$$R_n^{a,b}(f) < 0 \quad \text{if } \operatorname{sgn} f^{(2n+2)} = 1 \text{ on } [a,b], \tag{4.15}$$

with the inequality reversed if $\operatorname{sgn} f^{(2n+2)} = -1$.

The quantities α_{n+1}^{L}, β_{n+1}^{L} for Jacobi measures on $[-1,1]$ with parameters α, β and $a = -b = -1$ are explicitly known (cf. [10, Example 3.8]),

$$\begin{aligned} \alpha_{n+1}^{L} &= \frac{\alpha-\beta}{2n+\alpha+\beta+2}, \\ \beta_{n+1}^{L} &= 4\frac{(n+\alpha+1)(n+\beta+1)(n+\alpha+\beta+1)}{(2n+\alpha+\beta+1)(2n+\alpha+\beta+2)}. \end{aligned} \tag{4.16}$$

The `OPQ` Matlab routines are

```
xw=lobatto(N,ab,endl,endr)
xw=lobatto_jacobi(N,a,b)
```

with the meaning of `ab`, `a`, `b` the same as in the Gauss–Radau routines, and `endl`=a, `endr`=b.

We remark that both Gauss–Radau and Gauss–Lobatto formulae can be generalized to include boundary points of multiplicity $r > 1$. The internal (simple) nodes and weights are still related to orthogonal polynomials, but the boundary weights require new techniques for their computation; see Exercises 12–13.

4.2 Gauss–Kronrod Quadrature

In an attempt to estimate the error of the n-point Gauss quadrature rule, Kronrod in 1964 had the idea of inserting $n+1$ additional nodes and choosing them, along with all $2n+1$ weights, in such a way as to achieve maximum degree of exactness. The resulting quadrature rule can be expected to yield much higher accuracy than the Gauss formula, so that the difference of the two provides an estimate of the error in the Gauss formula. The extended formula thus can be written in the form

$$\int_{\mathbb{R}} f(t)\mathrm{d}\lambda(t) = \sum_{\nu=1}^{n} \lambda_\nu^K f(\tau_\nu^G) + \sum_{\mu=1}^{n+1} \lambda_\mu^{*K} f(\tau_\mu^K) + R_n^{GK}(f), \tag{4.17}$$

and having $3n+2$ free parameters λ_ν^K, λ_μ^{*K}, τ_μ^K at disposal, one ought to be able to achieve degree of exactness $3n+1$,

$$R_n^{GK}(f) = 0 \quad \text{for } f \in \mathbb{P}_{3n+1}. \tag{4.18}$$

A quadrature formula (4.17) that satisfies (4.18) is called a *Gauss–Kronrod formula.* The nodes τ_μ^K, called *Kronrod nodes*, are the zeros of the polynomial π_{n+1}^K of degree $n+1$ which is orthogonal to all polynomials of lower degree in the sense

$$\int_{\mathbb{R}} \pi_{n+1}^K(t) p(t) \pi_n(t; \mathrm{d}\lambda)\, \mathrm{d}\lambda(t) = 0 \quad \text{for all } p \in \mathbb{P}_n. \tag{4.19}$$

Note that the measure of orthogonality here is $\pi_n(t; \mathrm{d}\lambda)\mathrm{d}\lambda(t)$ and thus oscillates on the support of $\mathrm{d}\lambda$. Stieltjes (1894) was the first to consider polynomials π_{n+1}^K of this kind (for $\mathrm{d}\lambda(t) = \mathrm{d}t$); a polynomial π_{n+1}^K satisfying (4.19) is therefore called a *Stieltjes polynomial.* Stieltjes conjectured (in the case $\mathrm{d}\lambda(t) = \mathrm{d}t$) that all zeros of π_{n+1}^K are real and interlace with the n Gauss nodes— a highly desirable configuration! This has been proved only later by Szegő (1935) not only for Legendre measures, but also for a class of Gegenbauer measures. The study of the reality of the zeros for more general measures is an interesting and ongoing activity.

The computation of Gauss–Kronrod formulae is a challenging problem. An elegant solution has been given recently by Laurie (1997), at least in the case when a Gauss–Kronrod formula exists with real nodes and positive weights. It can be computed again in terms of eigenvalues and eigenvectors of a symmetric tridiagonal matrix, just like the previous Gauss-type formulae. The relevant matrix, however, is the *Jacobi–Kronrod matrix*

$$\boldsymbol{J}_{2n+1}^K(\mathrm{d}\lambda) = \begin{bmatrix} \boldsymbol{J}_n(\mathrm{d}\lambda) & \sqrt{\beta_n}\boldsymbol{e}_n & \boldsymbol{0} \\ \sqrt{\beta_n}\boldsymbol{e}_n^{\mathrm{T}} & \alpha_n & \sqrt{\beta_{n+1}}\boldsymbol{e}_1^{\mathrm{T}} \\ \boldsymbol{0} & \sqrt{\beta_{n+1}}\boldsymbol{e}_1 & \boldsymbol{J}_n^* \end{bmatrix}. \tag{4.20}$$

Here, $\alpha_n = \alpha_n(\mathrm{d}\lambda)$, $\beta_n = \beta_n(\mathrm{d}\lambda)$, etc, and $\boldsymbol{J}_n^*$ (which is partially known) can be computed by *Laurie's algorithm* (cf. [10, §3.1.2.2]). Should some of the eigenvalues of (4.20) turn out to be complex, this would be an indication that a Gauss–Kronrod formula (with real nodes) does not exist.

There are two routines in `OPQ`,

```
ab=r_kronrod(N,ab0)
xw=kronrod(n,ab)
```

that serve to compute Gauss–Kronrod formulae. The first generates the Jacobi-Kronrod matrix of order 2N+1, the other the nodes and weights of the quadrature formula, stored respectively in the first and second column of the (2N+1)×2 array `xw`. The recurrence coefficients of the given measure $d\lambda$ are input via the $\lceil$3N/2+1$\rceil$×2 array `ab0`.

4.3 Gauss–Turán Quadrature

The idea of allowing derivatives to appear in a Gauss-type quadrature formula is due to Turán (1950). He considered the case where each node has the same multiplicity $r \geq 1$, that is,

$$\int_{\mathbb{R}} f(t)\,d\lambda(t) = \sum_{\nu=1}^{n} [\lambda_\nu f(\tau_\nu) + \lambda'_\nu f'(\tau_\nu) + \cdots + \lambda_\nu^{(r-1)} f^{(r-1)}(\tau_\nu)] + R_n(f). \tag{4.21}$$

This is clearly related to Hermite interpolation. Indeed, if all nodes were prescribed and distinct, one could use Hermite interpolation to obtain a formula with degree of exactness $rn-1$ (there are rn free parameters). Turán asked, like Gauss before him, whether one can do better by choosing the nodes τ_ν judiciously. The answer is yes; more precisely, we can get degree of exactness $rn-1+k$, $k>0$, if and only if

$$\int_{\mathbb{R}} \omega_n^r(t) p(t)\,d\lambda(t) = 0 \quad \text{for all } p \in \mathbb{P}_{k-1}, \tag{4.22}$$

where $\omega_n(t) = \prod_{\nu=1}^{n}(t-\tau_\nu)$ is the *node polynomial* of (4.21). We have here a new type of orthogonality: the rth power of ω_n, not ω_n, must be orthogonal to all polynomials of degree $k-1$. This is called *power orthogonality.* It is easily seen that r must be odd,

$$r = 2s+1, \quad s \geq 0, \tag{4.23}$$

so that (4.21) becomes

$$\int_{\mathbb{R}} f(t)\,d\lambda(t) = \sum_{\nu=1}^{n}\sum_{\sigma=0}^{2s} \lambda_\nu^{(\sigma)} f^{(\sigma)}(\tau_\nu) + R_{n,s}(f). \tag{4.24}$$

Then in (4.22), necessarily $k \leq n$, and $k=n$ is optimal. The maximum possible degree of exactness, therefore, is $(2s+2)n-1$, and is achieved if

$$\int_{\mathbb{R}} [\omega_n(t)]^{2s+1} p(t)\,d\lambda(t) = 0 \quad \text{for all } p \in \mathbb{P}_{n-1}. \tag{4.25}$$

The polynomial $\omega_n = \pi_{n,s}$ satisfying (4.25) is called *s-orthogonal.* It exists uniquely and has distinct simple zeros contained in the support interval of $d\lambda$. The formula (4.24) is the *Gauss-Turán formula* if its node polynomial ω_n satisfies (4.25) and the weights $\lambda_\nu^{(\sigma)}$ are obtained by Hermite interpolation.

The computation of Gauss-Turán formulae is not as simple as in the case of ordinary Gauss-type formulae. The basic idea, however, is to consider the positive measure $\mathrm{d}\lambda_{n,s}(t) = [\pi_{n,s}(t)]^{2s}\mathrm{d}\lambda(t)$ and to note that $\pi_{n,s}$ is the nth-degree polynomial orthogonal relative to $\mathrm{d}\lambda_{n,s}$. The difficulty is that this defines $\pi_{n,s}$ implicitly, since $\pi_{n,s}$ already occurs in the measure $\mathrm{d}\lambda_{n,s}$. Nevertheless, the difficulty can be surmounted, but at the expense of having to solve a system of nonlinear equations; for details, see [10, §3.1.3.2]. The procedure is embodied in the OPQ routine

```
xw=turan(n,s,eps0,ab0,hom)
```

where the nodes are stored in the first column of the n×(2s+2) array xw, and the successive weights in the remaining 2s+1 columns. The input parameter eps0 is an error tolerance used in the iterative solution of the nonliner system of equations, and the measure $\mathrm{d}\lambda$ is specified by the ((s+1)n)×2 input array ab0 of its recurrence coefficients. Finally, hom=1 or hom≠1 depending on whether or not a certain homotopy in the variable s is used to facilitate convergence of Newton's method for solving the system of nonlinear equations.

4.4 Quadrature Formulae Based on Rational Functions

All quadrature formulae considered so far were based on polynomial degree of exactness. This is meaningful if the integrand is indeed polynomial-like. Not infrequently, however, it happens that the integrand has poles outside the interval of integration. In this case, exactness for appropriate rational functions, in addition to polynomials, is more natural. We discuss this for the simplest type of quadrature rule,

$$\int_{\mathbb{R}} g(t)\mathrm{d}\lambda(t) = \sum_{\nu=1}^{n} \lambda_\nu g(\tau_\nu) + R_n(g). \tag{4.26}$$

The problem, more precisely, is to determine λ_ν, τ_ν such that $R_n(g) = 0$ if $g \in \mathbb{S}_{2n}$, where $\mathbb{S}_{2n}$ is a space of dimension $2n$ consisting of rational functions and polynomials,

$$\begin{aligned} \mathbb{S}_{2n} &= \mathbb{Q}_m \oplus \mathbb{P}_{2n-m-1}, \quad 0 \le m \le 2n, \\ \mathbb{P}_{2n-m-1} &= \text{polynomials of degree } \le 2n-m-1, \\ \mathbb{Q}_m &= \text{rational functions with prescribed poles.} \end{aligned} \tag{4.27}$$

Here, m is an integer of our choosing, and

$$\mathbb{Q}_m = \text{span}\left\{ r(t) = \frac{1}{1+\zeta_\mu t},\quad \mu = 1,2,\ldots,m \right\}, \tag{4.28}$$

where

$$\zeta_\mu \in \mathbb{C}, \quad \zeta_\mu \neq 0, \quad 1 + \zeta_\mu t \neq 0 \text{ on } \mathrm{supp}(\mathrm{d}\lambda). \tag{4.29}$$

The idea is to select the poles $-1/\zeta_\mu$ of the rational functions in $\mathbb{Q}_m$ to match the pole(s) of g closest to the support interval of $\mathrm{d}\lambda$.

In principle, the solution of the problem is rather simple: put $\omega_m(t) = \prod_{\mu=1}^m (1 + \zeta_\mu t)$ and construct, if possible, the n-point (poynomial) Gauss formula

$$\int_{\mathbb{R}} g(t) \frac{\mathrm{d}\lambda(t)}{\omega_m(t)} = \sum_{\nu=1}^n \lambda_\nu^G g(\tau_\nu^G), \quad g \in \mathbb{P}_{2n-1}, \tag{4.30}$$

for the modified measure $\mathrm{d}\hat{\lambda}(t) = \mathrm{d}\lambda(t)/\omega_m(t)$. Then

$$\tau_\nu = \tau_\nu^G, \quad \lambda_\nu = \omega_m(\tau_\nu^G)\lambda_\nu^G, \quad \nu = 1, 2, \ldots, n, \tag{4.31}$$

are the desired nodes and weights in (4.26).

We said "if possible", since in general ω_m is complex-valued, and the existence of a Gauss formula for $\mathrm{d}\hat{\lambda}$ is not guaranteed. There is no problem, however, if $\omega_m \geq 0$ on the support of $\mathrm{d}\lambda$. Fortunately, in many instances of practical interest, this is indeed the case.

There are a number of ways the formula (4.30) can be constructed: a discretization method using Gauss quadrature relative to $\mathrm{d}\lambda$ to do the discretization; repeated application of modification algorithms involving linear or quadratic divisors; special techniques to handle "difficult" poles, that is, poles very close to the support interval of $\mathrm{d}\lambda$. Rather than going into details (which can be found in [10, §3.1.4]), we present an example taken from solid state physics.

Example 8. Generalized Fermi–Dirac integral.

This is the integral

$$F_k(\eta, \theta) = \int_0^\infty \frac{t^k \sqrt{1 + \theta t/2}}{\mathrm{e}^{-\eta + t} + 1} \mathrm{d}t,$$

where $\eta \in \mathbb{R}$, $\theta \geq 0$, and k is the Boltzmann constant ($= \frac{1}{2}, \frac{3}{2}$, or $\frac{5}{2}$). The ordinary Fermi–Dirac integral corresponds to $\theta = 0$.

The integral is conveniently rewritten as

$$F_k(\eta, \theta) = \int_0^\infty \frac{\sqrt{1 + \theta t/2}}{\mathrm{e}^{-\eta} + e^{-t}} \mathrm{d}\lambda^{[k]}(t), \quad \mathrm{d}\lambda^{[k]}(t) = t^k \mathrm{e}^{-t} \mathrm{d}t, \tag{4.32}$$

which is of the form (4.26) with $g(t) = \sqrt{1 + \theta t/2}/(\mathrm{e}^{-\eta} + \mathrm{e}^{-t})$ and $\mathrm{d}\lambda = \mathrm{d}\lambda^{[k]}$ a generalized Laguerre measure. The poles of g evidently are $\eta + \mu \mathrm{i}\pi$, $\mu = \pm 1, \pm 3, \pm 5, \ldots$, and all are "easy", that is, at a comfortable distance from the interval $[0, \infty]$. It is natural to take m even, and to incorporate the first $m/2$ pairs of conjugate complex poles. An easy computation then yields

$$\omega_m(t) = \prod_{\nu=1}^{m/2} [(1 + \xi_\nu t)^2 + \eta_\nu t^2], \quad 2 \leq m(\text{even}) \leq 2n, \tag{4.33}$$

where

$$\xi_\nu = \frac{-\eta}{\eta^2 + (2\nu - 1)^2\pi^2}, \quad \eta_\nu = \frac{(2\nu - 1)\pi}{\eta^2 + (2\nu - 1)^2\pi^2}. \tag{4.34}$$

Once the nodes and weights τ_ν, λ_ν have been obtained according to (4.31), the rational/polynomial quadrature approximation is given by

$$F_k(\eta, \theta) \approx \sum_{n=1}^{N} \lambda_n \frac{\sqrt{1 + \theta\tau_n/2}}{\mathrm{e}^{-\eta} + \mathrm{e}^{-\tau_n}}. \tag{4.35}$$

It is computed in the OPQ routine

```
xw=fermi_dirac(N,m,eta,theta,k,eps0,Nmax)
```

where eps0 is an error tolerance, Nmax a limit on the discretization parameter, and the other variables having obvious meanings.

4.5 Cauchy Principal Value Integrals

When there is a (simple) pole inside the support interval $[a, b]$ of the measure $\mathrm{d}\lambda$, the integral must be taken in the sense of a *Cauchy principal value integral*

$$(\mathcal{C}f)(x; \mathrm{d}\lambda) := \fint_a^b \frac{f(t)}{x - t}\,\mathrm{d}\lambda(t) = \lim_{\varepsilon \downarrow 0} \left(\int_a^{x-\varepsilon} + \int_{x+\varepsilon}^b \right) \frac{f(t)}{x - t}\,\mathrm{d}\lambda(t), \quad x \in (a, b). \tag{4.36}$$

There are two types of quadrature rules for Cauchy principal value integrals: one in which x occurs as a node, and one in which it does not. They have essentially different character and will be considered separately.

Modified Quadrature Rule

This is a quadrature rule of the form

$$(\mathcal{C}f)(x; \mathrm{d}\lambda) = c_0(x) f(x) + \sum_{\nu=1}^{n} c_\nu(x) f(\tau_\nu) + R_n(f; x). \tag{4.37}$$

It can be made "Gaussian", that is, $R_n(f; x) = 0$ for $f \in \mathbb{P}_{2n}$, by rewriting the integral in (4.36) as

$$(\mathcal{C}f)(x; \mathrm{d}\lambda) = f(x) \fint_{\mathbb{R}} \frac{\mathrm{d}\lambda(t)}{x - t} - \int_{\mathbb{R}} \frac{f(x) - f(t)}{x - t}\,\mathrm{d}\lambda(t) \tag{4.38}$$

and applying the n-point Gauss formula for $\mathrm{d}\lambda$ to the second integral. The result is

$$(\mathcal{C}f)(x; \mathrm{d}\lambda) = \frac{\rho_n(x)}{\pi_n(x)} f(x) + \sum_{\nu=1}^{n} \lambda_\nu^G \frac{f(\tau_\nu^G)}{x - \tau_\nu^G} + R_n(f; x), \tag{4.39}$$

where $\rho_n(x)$ is the Cauchy principal value integral (2.33) and τ_ν^G, λ_ν^G are the Gauss nodes and weights for $\mathrm{d}\lambda$.

Formula (4.39) is not without numerical difficulties. The major one occurs when x approaches one of the Gauss nodes τ_ν^G, in which case two terms on the right go to infinity, but with opposite signs. In effect, this means that for x near a Gaussian node severe cancellation must occur.

The problem can be avoided by expanding the integral (4.36) in Cauchy integrals $\rho_k(x)$. Let $p_n(f;\,\cdot\,)$ be the polynomial of degree n interpolating f at the n Gauss nodes τ_ν^G and at x. The quadrature sum in (4.39) is then precisely the Cauchy integral of p_n,

$$(\mathcal{C}f)(x;\mathrm{d}\lambda) = ⨍_a^b \frac{p_n(f;t)}{x-t}\,\mathrm{d}\lambda(t) + R_n(f;x). \tag{4.40}$$

Expanding p_n in the orthogonal polynomials π_k,

$$p_n(f;t) = \sum_{k=0}^{n} a_k \pi_k(t), \quad a_k = \frac{1}{\|\pi_k\|^2}\int_a^b p_n(f;t)\pi_k(t)\mathrm{d}\lambda(t), \tag{4.41}$$

and integrating, one finds

$$(\mathcal{C}f)(x;\mathrm{d}\lambda) = \sum_{k=0}^{n} a_k \rho_k(x) + R_n(f;x), \tag{4.42}$$

where

$$a_k = \frac{1}{\|\pi_k\|^2}\sum_{\nu=1}^{n} \lambda_\nu^G f(\tau_\nu^G)\pi_k(\tau_\nu^G),\ k<n; \quad a_n = \sum_{\nu=1}^{n} \frac{f(x)-f(\tau_\nu^G)}{(x-\tau_\nu^G)\pi_n'(\tau_\nu^G)}\,. \tag{4.43}$$

The Cauchy integrals $\rho_k(x)$ in (4.42) can be computed in a stable manner by forward recursion; cf. the paragraph surrounding (2.33) and (2.34). This requires $\rho_0(x)$, which is either explicitly known or can be computed by the continued fraction algorithm. Some care must be exercised in computing the divided difference of f in the formula for a_n.

The procedure is inplemented in the `OPQ` routine

```
cpvi=cauchyPVI(N,x,f,ddf,iopt,ab,rho0)
```

with `iopt`$\neq$1, which produces the (`N+1`)-term approximation (4.42) where $R_n(f;x)$ is neglected. The input parameter `ddf` is a routine for computing the divided difference of f in a stable manner. It is used only if `iopt`$\neq$1. The meaning of the other parameters is obvious.

Quadrature Rule in the Strict Sense

This rule, in which the node $t=x$ is absent, is obtained by interpolating f at the n Gauss nodes τ_ν^G by a polynomial $p_{n-1}(f;\,\cdot\,)$ of degree $n-1$,

$$f(t) = p_{n-1}(f;t) + E_{n-1}(f;t), \quad p_{n-1}(f;t) = \sum_{\nu=1}^{n} \frac{\pi_n(t)}{(t-\tau_\nu^G)\pi_n'(\tau_\nu^G)} f(\tau_\nu^G),$$

where E_{n-1} is the interpolation error, which vanishes identically if $f \in \mathbb{P}_{n-1}$. The formula to be derived, therefore, will have degree of exactness $n-1$ (which can be shown to be maximum possible). Integrating in the sense of (4.36) yields

$$(\mathcal{C}f)(x; \mathrm{d}\lambda) = \sum_{\nu=1}^{n} \frac{\rho_n(x) - \rho_n(\tau_\nu^G)}{(x-\tau_\nu^G)\pi_n'(\tau_\nu^G)} f(\tau_\nu^G) + R_n^*(f;x), \tag{4.44}$$

where $R_n^*(f;x) = \int\!\!\!\!-_a^b E_{n-1}(f;t)\mathrm{d}\lambda(t)/(x-t)$.

This formula, too, suffers from severe cancellation errors when x is near a Gauss node. The resolution of this problem is similar (in fact, simpler) as in the previous subsection: expand $p_{n-1}(f;\,\cdot\,)$ in the orthogonal polynomials π_k to obtain

$$\begin{aligned}(\mathcal{C}f)(x; \mathrm{d}\lambda) &= \sum_{k=0}^{n-1} a_k' \rho_k(x) + R_n^*(f;x), \\ a_k' &= \frac{1}{\|\pi_k\|^2} \int_a^b p_{n-1}(f;t)\pi_k(t)\mathrm{d}\lambda(t).\end{aligned} \tag{4.45}$$

It turns out that

$$a_k' = a_k, \quad k = 0, 1, \dots, n-1, \tag{4.46}$$

where a_k, $k < n$, is given by (4.43). This is implemented in the OPQ routine `cauchyPVI.m` with `iopt=1`.

4.6 Polynomials Orthogonal on Several Intervals

We are given a finite set of intervals $[c_j, d_j]$, which may be disjoint or not, and on each interval a positive measure $\mathrm{d}\lambda_j$. Let $\mathrm{d}\lambda$ be the "composite" measure

$$\mathrm{d}\lambda(t) = \sum_j \chi_{[c_j,d_j]}(t)\mathrm{d}\lambda_j(t), \tag{4.47}$$

where $\chi_{[c_j,d_j]}$ is the characteristic function of the interval $[c_j, d_j]$. Assuming known the Jacobi matrices $\boldsymbol{J}^{(j)} = \boldsymbol{J}_n(\mathrm{d}\lambda_j)$ of the component measures $\mathrm{d}\lambda_j$, we now consider the problem of determining the Jacobi matrix $\boldsymbol{J} = \boldsymbol{J}_n(\mathrm{d}\lambda)$ of the composite measure $\mathrm{d}\lambda$. We provide two solutions, one based on Stieltjes's procedure, and one based on the modified Chebyshev algorithm.

Solution by Stieltjes's Procedure

The main problem in applying Stieltjes's procedure is to compute the inner products $(t\pi_k, \pi_k)_{\mathrm{d}\lambda}$ and $(\pi_k, \pi_k)_{\mathrm{d}\lambda}$ for $k = 0, 1, 2, \dots, n-1$. This can be done by using n-point Gaussian quadrature on each component interval,

$$\int_{c_j}^{d_j} p(t)\mathrm{d}\lambda_j(t) = \sum_{\nu=1}^{n} \lambda_\nu^{(j)} p(\tau_\nu^{(j)}), \quad p \in \mathbb{P}_{2n-1}. \tag{4.48}$$

Here we use (4.5) to express the quadrature sum in terms of the Jacobi matrix $\boldsymbol{J}^{(j)}$,

$$\int_{c_j}^{d_j} p(t)\mathrm{d}\lambda_j(t) = \beta_0^{(j)} \boldsymbol{e}_1^{\mathrm{T}} p(\boldsymbol{J}^{(j)})\boldsymbol{e}_1, \quad \beta_0^{(j)} = \int_{c_j}^{d_j} \mathrm{d}\lambda_j(t). \tag{4.49}$$

Then, for the inner products $(t\pi_k, \pi_k)_{\mathrm{d}\lambda}$, $k \le n-1$, we get

$$\begin{aligned}
(t\pi_k, \pi_k)_{\mathrm{d}\lambda} &= \int_{\mathbb{R}} t\pi_k^2(t)\mathrm{d}\lambda(t) = \sum_j \int_{c_j}^{d_j} t\pi_k^2(t)\mathrm{d}\lambda_j(t) \\
&= \sum_j \beta_0^{(j)} \boldsymbol{e}_1^{\mathrm{T}} \boldsymbol{J}^{(j)} [\pi_k(\boldsymbol{J}^{(j)})]^2 \boldsymbol{e}_1 \\
&= \sum_j \beta_0^{(j)} \boldsymbol{e}_1^{\mathrm{T}} [\pi_k(\boldsymbol{J}^{(j)})]^{\mathrm{T}} \boldsymbol{J}^{(j)} \pi_k(\boldsymbol{J}^{(j)}) \boldsymbol{e}_1
\end{aligned}$$

and for $(\pi_k, \pi_k)_{\mathrm{d}\lambda}$ similarly (in fact, simpler)

$$(\pi_k, \pi_k)_{\mathrm{d}\lambda} = \sum_j \beta_0^{(j)} \boldsymbol{e}_1^{\mathrm{T}} [\pi_k(\boldsymbol{J}^{(j)})]^{\mathrm{T}} \pi_k(\boldsymbol{J}^{(j)}) \boldsymbol{e}_1.$$

This can be conveniently expressed in terms of the vectors

$$\boldsymbol{\zeta}_k^{(j)} := \pi_k(\boldsymbol{J}^{(j)})\boldsymbol{e}_1, \quad \boldsymbol{e}_1 = [1, 0, \ldots, 0]^T,$$

which, as required in Stieltjes's procedure, can be updated by means of the basic three-term recurrence relation. This leads to the following algorithm.

Algorithm 7 Stieltjes procedure for polynomials orthogonal on several intervals

initialization:

$$\begin{aligned}
&\boldsymbol{\zeta}_0^{(j)} = \boldsymbol{e}_1, \quad \boldsymbol{\zeta}_{-1}^{(j)} = \boldsymbol{0} \quad (\text{all } j), \\
&\alpha_0 = \frac{\sum_j \beta_0^{(j)} \boldsymbol{e}_1^{\mathrm{T}} \boldsymbol{J}^{(j)} \boldsymbol{e}_1}{\sum_j \beta_0^{(j)}}, \quad \beta_0 = \sum_j \beta_0^{(j)}.
\end{aligned}$$

continuation (if $n > 1$): for $k = 0, 1, \ldots, n-2$ do

$$\begin{aligned}
&\boldsymbol{\zeta}_{k+1}^{(j)} = (\boldsymbol{J}^{(j)} - \alpha_k \boldsymbol{I})\boldsymbol{\zeta}_k^{(j)} - \beta_k \boldsymbol{\zeta}_{k-1}^{(j)} \quad (\text{all } j), \\
&\alpha_{k+1} = \frac{\sum_j \beta_0^{(j)} \boldsymbol{\zeta}_{k+1}^{(j)\mathrm{T}} \boldsymbol{J}^{(j)} \boldsymbol{\zeta}_{k+1}^{(j)}}{\sum_j \beta_0^{(j)} \boldsymbol{\zeta}_{k+1}^{(j)\mathrm{T}} \boldsymbol{\zeta}_{k+1}^{(j)}}, \quad \beta_{k+1} = \frac{\sum_j \beta_0^{(j)} \boldsymbol{\zeta}_{k+1}^{(j)\mathrm{T}} \boldsymbol{\zeta}_{k+1}^{(j)}}{\sum_j \beta_0^{(j)} \boldsymbol{\zeta}_k^{(j)\mathrm{T}} \boldsymbol{\zeta}_k^{(j)}}.
\end{aligned}$$

In Matlab, this is implemented in the `OPQ` routine

```
ab=r_multidomain_sti(N,abmd)
```

where abmd is the array containing the (α, β)-coefficients of the measures $d\lambda^{(j)}$.

Example 9. Example 1, revisited.

This is the case of two identical intervals $[-1, 1]$ and two measures $d\lambda^{(j)}$ on $[-1, 1]$, one a multiple c of the Legendre measure, the other the Chebyshev measure. This was solved in Example 1 by a 2-component discretization method. The solution by the 2-domain algorithm of this subsection, in Matlab, looks as follows:

```
ab1=r_jacobi(N); ab1(1,2)=2*c;
ab2=r_jacobi(N,-.5);
abmd=[ab1 ab2];
ab=r_multidomain_sti(N,abmd)
```

It produces results identical with those produced by the method of Example 1.

Solution by the Modified Chebyshev Algorithm

The quadrature procedure used in the previous subsection to compute inner products can equally be applied to compute the first $2n$ modified moments of $d\lambda$,

$$m_k = \sum_j \int_{c_j}^{d_j} p_k(t) d\lambda_j(t) = \sum_j \beta_0^{(j)} \boldsymbol{e}_1^{\mathrm{T}} p_k(\boldsymbol{J}^{(j)}) \boldsymbol{e}_1. \tag{4.50}$$

The relevant vectors are now

$$\boldsymbol{z}_k^{(j)} := p_k(\boldsymbol{J}^{(j)}) \boldsymbol{e}_1, \quad \boldsymbol{e}_1 = [1, 0, \ldots, 0]^{\mathrm{T}},$$

and the computation proceeds as in

Algorithm 8 Modified moments for polynomials orthogonal on several intervals

initialization

$$\boldsymbol{z}_0^{(j)} = \boldsymbol{e}_1, \; \boldsymbol{z}_{-1}^{(j)} = \boldsymbol{0} \text{ (all } j), \quad m_0 = \sum_j \beta_0^{(j)}.$$

continuation: for $k = 0, 1, \ldots, 2n-2$ do

$$\begin{aligned} \boldsymbol{z}_{k+1}^{(j)} &= (\boldsymbol{J}^{(j)} - a_k \boldsymbol{I}) \boldsymbol{z}_k^{(j)} - b_k \boldsymbol{z}_{k-1}^{(j)} \quad \text{(all } j), \\ m_{k+1} &= \sum_j \beta_0^{(j)} \boldsymbol{z}_{k+1}^{(j)\mathrm{T}} \boldsymbol{e}_1. \end{aligned}$$

With these moments at hand, we can apply Algorithm 1 to obtain the desired recurrence coefficients. This is done in the OPQ routine

```
ab=r_multidomain_cheb(N,abmd,abmm)
```

The array abmd has the same meaning as in the routine r_multidomain_sti.m, and abmm is a $((2N-1)\times 2)$ array of the recurrence coefficients a_k, b_k generating the polynomials p_k.

Applied to Example 9, the Matlab program, using Legendre moments (p_k the monic Legendre polynomials), is as follows:

```
abm=r_jacobi(2*N-1);
ab1=abm(1:N,:); ab1(1,2)=2*c;
ab2=r_jacobi(N,-.5);
abmd=[ab1 ab2];
ab=r_multidomain_cheb(N,abmd,abm)
```

It produces results identical with those obtained previously, but takes about three times as long to run.

4.7 Quadrature Estimates of Matrix Functionals

The problem to be considered here is to find lower and upper bounds for the quadratic form

$$\boldsymbol{u}^{\mathrm{T}} f(\boldsymbol{A})\boldsymbol{u}, \quad \boldsymbol{u} \in \mathbb{R}^N, \ \|\boldsymbol{u}\| = 1, \tag{4.51}$$

where $\boldsymbol{A} \in \mathbb{R}^{N\times N}$ is a symmetric, positive definite matrix, f a smooth function (for which $f(\boldsymbol{A})$ makes sense), and $\boldsymbol{u}$ a given vector. While this looks more like a linear algebra problem, it can actually be solved, for functions f with derivatives of constant sign, by applying Gauss-type quadrature rules. The connecting link is provided by the *spectral resolution* of $\boldsymbol{A}$,

$$\boldsymbol{AV} = \boldsymbol{V\Lambda}, \ \boldsymbol{\Lambda} = \mathrm{diag}(\lambda_1, \lambda_2, \ldots, \lambda_N), \ \boldsymbol{V} = [\boldsymbol{v}_1, \boldsymbol{v}_2, \ldots, \boldsymbol{v}_N], \tag{4.52}$$

where λ_k are the eigenvalues of $\boldsymbol{A}$ (which for simplicity are assumed distinct), and $\boldsymbol{v}_k$ the normalized eigenvectors of $\boldsymbol{A}$. If we put

$$\boldsymbol{u} = \sum_{k=1}^{N} \rho_k \boldsymbol{v}_k = \boldsymbol{V\rho}, \ \boldsymbol{\rho} = [\rho_1, \rho_2, \ldots, \rho_N]^{\mathrm{T}}, \tag{4.53}$$

and again for simplicity assume $\rho_k \neq 0$, all k, then

$$\begin{aligned} \boldsymbol{u}^{\mathrm{T}} f(\boldsymbol{A})\boldsymbol{u} &= \boldsymbol{\rho}^{\mathrm{T}}\boldsymbol{V}^{\mathrm{T}}\boldsymbol{V} f(\boldsymbol{\Lambda})\boldsymbol{V}^{\mathrm{T}}\boldsymbol{V\rho} = \boldsymbol{\rho}^{\mathrm{T}} f(\boldsymbol{\Lambda})\boldsymbol{\rho}, \\ &= \sum_{k=1}^{N} \rho_k^2 f(\lambda_k) =: \int_{\mathbb{R}_+} f(t)\mathrm{d}\rho_N(t). \end{aligned} \tag{4.54}$$

This shows how the matrix functional is related to an integral relative to a discrete positive measure. Now we know from (4.10) and (4.15) how Gauss–Radau or Gauss–Lobatto rules (and for that matter also ordinary Gauss rules, in view of $R_n^G = [f^{(2n)}(\tau)/(2n)!] \int_a^b [\pi_n(t; \mathrm{d}\lambda)]^2 \mathrm{d}\lambda(t)$, $a < \tau < b$) can be applied to obtain two-sided bounds for (4.54) when some derivative of f has constant sign. To generate these quadrature rules, we need the orthogonal polynomials for the measure $\mathrm{d}\rho_N$, and for these the Jacobi matrix $\boldsymbol{J}_N(\mathrm{d}\rho_N)$. The latter, in principle, could be computed by the Lanczos-type algorithm of §2.3. However, in the present application this would require knowledge of the eigenvalues λ_k and expansion coefficients ρ_k, which are too expensive to compute. Fortunately, there is an alternative way to implement Lanczos's algorithm that works directly with the matrix $\boldsymbol{A}$ and requires only multiplications of $\boldsymbol{A}$ into vectors and the computation of inner products.

Lanczos Algorithm

Let ρ_k be as in (4.54) and $h_0 = \sum_{k=1}^N \rho_k \boldsymbol{v}_k$ $(= \boldsymbol{u})$, $\|h_0\| = 1$, as in (4.53).

Algorithm 9 Lanczos algorithm
initialization:

$$\boldsymbol{h}_0 \text{ prescribed with } \|\boldsymbol{h}_0\| = 1, \quad \boldsymbol{h}_{-1} = \boldsymbol{0}.$$

continuation: for $j = 0, 1, \ldots, N-1$ do

$$\begin{aligned}
\alpha_j &= \boldsymbol{h}_j^{\mathrm{T}} \boldsymbol{A} \boldsymbol{h}_j, \\
\tilde{\boldsymbol{h}}_{j+1} &= (\boldsymbol{A} - \alpha_j \boldsymbol{I})\boldsymbol{h}_j - \gamma_j \boldsymbol{h}_{j-1}, \\
\gamma_{j+1} &= \|\tilde{\boldsymbol{h}}_{j+1}\|, \\
\boldsymbol{h}_{j+1} &= \tilde{\boldsymbol{h}}_{j+1}/\gamma_{j+1}.
\end{aligned}$$

While γ_0 in Algorithm 9 can be arbitrary (it multiplies $\boldsymbol{h}_{-1} = \boldsymbol{0}$), it is convenient to define $\gamma_0 = 1$. The vectors $\boldsymbol{h}_0, \boldsymbol{h}_1, \ldots, \boldsymbol{h}_N$ generated by Algorithm 9 are called *Lanczos vectors*. It can be shown that α_k generated by the Lanczos algorithm is precisely $\alpha_k(\mathrm{d}\rho_N)$, and $\gamma_k = \sqrt{\beta_k(\mathrm{d}\rho_N)}$, for $k = 0, 1, 2, \ldots, N-1$. This provides us with the Jacobi matrix $\boldsymbol{J}_N(\mathrm{d}\rho_N)$. It is true that the algorithm becomes unstable as j approaches N, but in the applications of interest here, only small values of j are needed.

Examples

Example 10. Error bounds for linear algebraic systems.
Consider the system

$$\boldsymbol{A}\boldsymbol{x} = \boldsymbol{b}, \quad \boldsymbol{A} \text{ symmetric, positive definite.} \tag{4.55}$$

Given an approximation $\boldsymbol{x}^* \approx \boldsymbol{x} = \boldsymbol{A}^{-1}\boldsymbol{b}$ to the exact solution $\boldsymbol{x}$, and the residual vector $\boldsymbol{r} = \boldsymbol{b} - \boldsymbol{A}\boldsymbol{x}^*$, we have $\boldsymbol{x} - \boldsymbol{x}^* = \boldsymbol{A}^{-1}\boldsymbol{b} + \boldsymbol{A}^{-1}(\boldsymbol{r} - \boldsymbol{b}) = \boldsymbol{A}^{-1}\boldsymbol{r}$, thus

$$\|\boldsymbol{x}-\boldsymbol{x}^*\|^2 = (\boldsymbol{A}^{-1}\boldsymbol{r})^{\mathrm{T}}\boldsymbol{A}^{-1}\boldsymbol{r} = \boldsymbol{r}^{\mathrm{T}}\boldsymbol{A}^{-2}\boldsymbol{r},$$

and therefore

$$\|\boldsymbol{x}-\boldsymbol{x}^*\|^2 = \|\boldsymbol{r}\|^2 \cdot \boldsymbol{u}^{\mathrm{T}} f(\boldsymbol{A})\boldsymbol{u}, \tag{4.56}$$

where $\boldsymbol{u} = \boldsymbol{r}/\|\boldsymbol{r}\|$ and $f(t) = t^{-2}$. All derivatives of f are here of constant sign on $\mathbb{R}_+$,

$$f^{(2n)}(t) > 0, \quad f^{(2n+1)}(t) < 0 \quad \text{for } t \in \mathbb{R}_+. \tag{4.57}$$

By (4.54), we now have

$$\|\boldsymbol{x}-\boldsymbol{x}^*\| = \|\boldsymbol{r}\|^2 \int_{\mathbb{R}_+} t^{-2} \mathrm{d}\rho_N(t). \tag{4.58}$$

The n-point Gauss quadrature rule applied to the integral on the right of (4.58), by the first inequality in (4.57), yields a *lower bound* of $\|\boldsymbol{x}-\boldsymbol{x}^*\|$, without having to know the exact support interval of $\mathrm{d}\rho_N$. If, on the other hand, we know that the support of $\mathrm{d}\rho_N$ is contained in some interval $[a,b]$, $0 < a < b$, we can get a lower bound also from the right-handed $(n+1)$-point Gauss–Radau formula, and *upper bounds* from the left-handed $(n+1)$-point Gauss–Radau formula on $[a,b]$, or from the $(n+2)$-point Gauss–Lobatto formula on $[a,b]$.

Example 11. Diagonal elements of $\boldsymbol{A}^{-1}$.

Here, trivially

$$\boldsymbol{u}^{\mathrm{T}} f(\boldsymbol{A})\boldsymbol{u} = \boldsymbol{e}_i^{\mathrm{T}}\boldsymbol{A}^{-1}\boldsymbol{e}_i, \tag{4.59}$$

where $f(t) = t^{-1}$ and $\boldsymbol{e}_i$ is the ith coordinate vector. Using n-point Gauss quadrature in (4.54), with $n < N$, yields

$$(\boldsymbol{A}^{-1})_{ii} = \int_{\mathbb{R}_+} t^{-1}\,\mathrm{d}\rho_N(t) > \mathrm{e}_1^{\mathrm{T}}\boldsymbol{J}_n^{-1}\mathrm{e}_1, \quad \mathrm{e}_1^{\mathrm{T}} = [1,0,\ldots,0] \in \mathbb{R}^n. \tag{4.60}$$

Suppose we take $n = 2$ steps of Algorithm 9 to compute

$$\boldsymbol{J}_2 = \begin{bmatrix} \alpha_0 & \gamma_1 \\ \gamma_1 & \alpha_1 \end{bmatrix}.$$

We get

$$\begin{aligned}
\alpha_0 &= a_{ii}, \\
\tilde{\boldsymbol{h}}_1 &= (\boldsymbol{A} - \alpha_0 \boldsymbol{I})\boldsymbol{e}_i = [a_{1i}, \ldots, a_{i-1,i}, 0, a_{i+1,i}, \ldots, a_{Ni}]^{\mathrm{T}}, \\
\gamma_1 &= \sqrt{\sum_{k \neq i} a_{ki}^2} =: s_i, \\
\boldsymbol{h}_1 &= \tilde{\boldsymbol{h}}_1 / s_i, \\
\alpha_1 &= \frac{1}{s_i^2}\tilde{\boldsymbol{h}}_1^{\mathrm{T}}\boldsymbol{A}\tilde{\boldsymbol{h}}_1 = \frac{1}{s_i^2}\sum_{k\neq i}\sum_{\ell \neq i} a_{k\ell}a_{ki}a_{\ell i}.
\end{aligned} \tag{4.61}$$

But

$$J_2^{-1} = \frac{1}{\alpha_0\alpha_1 - \gamma_1^2}\begin{bmatrix} \alpha_1 & -\gamma_1 \\ -\gamma_1 & \alpha_0 \end{bmatrix}, \quad e_1^{\mathrm{T}} J_2^{-1} e_1 = \frac{\alpha_1}{\alpha_0\alpha_1 - \gamma_1^2},$$

so that by (4.60) with $n = 2$, and (4.61),

$$(A^{-1})_{ii} > \frac{\sum_{k\neq i}\sum_{\ell\neq i} a_{k\ell}a_{ki}a_{\ell i}}{a_{ii}\sum_{k\neq i}\sum_{\ell\neq i} a_{k\ell}a_{ki}a_{\ell i} - \left(\sum_{k\neq i} a_{ki}^2\right)^2}. \tag{4.62}$$

Simpler bounds, both lower and upper, can be obtained by the 2-point Gauss-Radau and Gauss-Lobatto formulae, which however require knowledge of an interval $[a, b]$, $0 < a < b$, containing the spectrum of A.

Exercises to §4 (Stars indicate more advanced exercises.)

1. Prove (4.5).
2. Prove that complex zeros of the Sobolev orthogonal polynomials of Example 7 must be purely imaginary.

3*. Circle theorems for quadrature weights (cf. [4]).

(a) Gauss–Jacobi quadrature
Let $w(t) = (1-t)^\alpha(1+t)^\beta$ be the Jacobi weight function. It is known [16, eqn (15.3.10)] that the nodes τ_ν and weights λ_ν of the n-point Gauss–Jacobi quadrature formula satisfy

$$\lambda_\nu \sim \frac{\pi}{n} w(\tau_\nu)\sqrt{1-\tau_\nu^2}, \quad n \to \infty,$$

for τ_ν on any compact interval contained in $(-1, 1)$. Thus, suitably normalized weights, plotted against the nodes, lie asymptotically on the unit circle. Use Matlab to demonstrate this graphically.

(b) Gauss quadrature for the logarithmic weight $w(t) = t^\alpha \ln(1/t)$ on $[0, 1]$ (cf. [10, Example 2.27]).
Try, numerically, to find a circle theorem in this case also, and experiment with different values of the parameter $\alpha > -1$. (Use the `OPQ` routine `r_jaclog.m` to generate the recurrence coefficients of the orthogonal polynomials for the weight function w.)

(c) Gauss–Kronrod quadrature.
With w as in (a), the analogous result for the $2n+1$ nodes τ_ν and weights λ_ν of the $(2n+1)$-point Gauss–Kronrod formula is expected to be

$$\lambda_\nu \sim \frac{\pi}{2n} w(\tau_\nu)\sqrt{1-\tau_\nu^2}, \quad n \to \infty.$$

That this indeed is the case, when $\alpha, \beta \in [0, \frac{5}{2})$, follows from Theorem 2 in [15]. Use Matlab to illustrate this graphically.

(d) Experiment with the Gauss–Kronrod formula for the logarithmic weight function of (b), when $\alpha = 0$.

4. Discrete orthogonality.
Let $\pi_k(\,\cdot\,;\mathrm{d}\lambda)$, $k = 0, 1, 2, \ldots$, be the orthogonal polynomials relative to an absolutely continuous measure. Show that for each $N \geq 2$, the first N of them are orthogonal with respect to the discrete inner product

$$(p, q)_N = \sum_{\nu=1}^{N} \lambda_\nu^G p(\tau_\nu^G) q(\tau_\nu^G),$$

where τ_ν^G, λ_ν^G are the nodes and weights of the N-point Gauss formula for $\mathrm{d}\lambda$. Moreover, $\|\pi_k\|_N^2 = \|\pi_k\|_{\mathrm{d}\lambda}^2$ for $k \leq N-1$.

5. (a) Consider the Cauchy integral

$$\rho_n(z) = \rho_n(z; \mathrm{d}\lambda) = \int_a^b \frac{\pi_n(t; \mathrm{d}\lambda)}{z-t}\, \mathrm{d}\lambda(t),$$

where $[a, b]$ is the support of $\mathrm{d}\lambda$. Show that

$$\rho_n(z) = O(z^{-n-1}) \quad \text{as } z \to \infty.$$

{*Hint*: Expand the integral defining $\rho_n(z)$ in descending powers of z.}

(b) Show that

$$\int_a^b \frac{\mathrm{d}\lambda(t)}{z-t} - \frac{\sigma_n(z)}{\pi_n(z)} = \frac{\rho_n(z)}{\pi_n(z)} = O(z^{-2n-1}) \quad \text{as } z \to \infty.$$

{*Hint*: Use (2.27).}

(c) Consider the partial fraction decomposition

$$\frac{\sigma_n(z)}{\pi_n(z)} = \sum_{\nu=1}^{n} \frac{\lambda_\nu}{z - \tau_\nu^G}$$

of $\sigma_n(z)/\pi_n(z)$ in (2.27). Use (b) to show that $\lambda_\nu = \lambda_\nu^G$ are the weights of the n-point Gaussian quadrature formula for $\mathrm{d}\lambda$. In particular, show that

$$\lambda_\nu^G = \frac{\sigma_n(\tau_\nu^G)}{\pi_n'(\tau_\nu^G)}.$$

(d) Discuss what happens if $z \to x$, $x \in (a, b)$.

6. Characterize the nodes τ_ν^b in (4.9) as zeros of an orthogonal polynomial of degree n, and identify the appropriate Jacobi–Radau matrix for (4.9).
7. Prove (4.10). {*Hint*: Use the fact that both formulae (4.6) and (4.9) are interpolatory.}
8. (a) Prove the first formula in (4.11). {*Hint*: Use the relation between the Jacobi polynomials $P_k = P_k^{(\alpha,\beta)}$ customarily defined and the monic Jacobi polynomials $\pi_k = \pi_k^{(\alpha,\beta)}$, expressed by $P_k(t) = 2^{-k}\binom{2k+\alpha+\beta}{k}\pi_k(t)$. You also need $P_k(-1) = (-1)^k\binom{k+\beta}{k}$ and the β-coefficient for Jacobi polynomials, $\beta_n^J = 4n(n+\alpha)(n+\beta)(n+\alpha+\beta)/(2n+\alpha+\beta)^2(2n+\alpha+\beta+1)(2n+\alpha+\beta-1)$.}

(b) Prove the second formula in (4.11). {*Hint*: With $\pi_k^{(\alpha)}$ and $L_k^{(\alpha)}$ denoting the monic resp. conventional generalized Laguerre polynomials, use $L_k^{(\alpha)}(t) = ((-1)^k/k!)\,\pi_k^{(\alpha)}(t)$. You also need $L_k^{(\alpha)}(0) = \binom{k+\alpha}{k}$, and $\beta_n^{\mathrm{L}} = n(n+\alpha)$.}

9. Prove (4.16). {*Hint*: With notation as in the hint to Exercise 8(a), use $P_k(1) = \binom{k+\alpha}{k}$ in addition to the information provided there.}

10. The (left-handed) *generalized Gauss–Radau formula* is

$$\int_a^\infty f(t)\,\mathrm{d}\lambda(t) = \sum_{\rho=0}^{r-1} \lambda_0^{(\rho)} f^{(\rho)}(a) + \sum_{\nu=1}^{n} \lambda_\nu^R f(\tau_\nu^R) + R_{n,r}^R(f),$$

where $r > 1$ is the multiplicity of the end point $\tau_0 = a$, and $R_{n,r}^R(f) = 0$ for $f \in \mathbb{P}_{2n-1+r}$. Let $\mathrm{d}\lambda^{[r]}(t) = (t-a)^r \mathrm{d}\lambda(t)$ and $\tau_\nu^{[r]}$, $\lambda_\nu^{[r]}$, $\nu = 1, 2, \ldots, n$, be the nodes and weights of the n-point Gauss formula for $\mathrm{d}\lambda^{[r]}$.

(a) Show that

$$\tau_\nu^R = \tau_\nu^{[r]}, \quad \lambda_\nu^R = \frac{\lambda_\nu^{[r]}}{(\tau_\nu^R - a)^r}, \qquad \nu = 1, 2, \ldots, n.$$

(b) Show that not only the internal weights λ_ν^R are all positive (why?), but also the boundary weights λ_0, λ_0' if $r = 2$.

11. The *generalized Gauss–Lobatto formula* is

$$\int_a^b f(t)\,\mathrm{d}\lambda(t) = \sum_{\rho=0}^{r-1} \lambda_0^{(\rho)} f^{(\rho)}(a) + \sum_{\nu=1}^{n} \lambda_\nu^L f(\tau_\nu^L) + \sum_{\rho=0}^{r-1} (-1)^\rho \lambda_{n+1}^{(\rho)} f^{(\rho)}(b) + R_{n,r}^L(f),$$

where $r > 1$ is the multiplicity of the end points $\tau_0 = a$, $\tau_{n+1} = b$, and $R_{n,r}^L(f) = 0$ for $f \in \mathbb{P}_{2n-1+2r}$. Let $\mathrm{d}\lambda^{[r]}(t) = [(t-a)(b-t)]^r \mathrm{d}\lambda(t)$ and $\tau_\nu^{[r]}$, $\lambda_\nu^{[r]}$, $\nu = 1, 2, \ldots, n$, be the nodes and weights of the n-point Gauss formula for $\mathrm{d}\lambda^{[r]}$.

(a) Show that

$$\tau_\nu^L = \tau_\nu^{[r]}, \quad \lambda_\nu^L = \frac{\lambda_\nu^{[r]}}{[(\tau_\nu^L - a)(b - \tau_\nu^L)]^r}, \qquad \nu = 1, 2, \ldots, n.$$

(b) Show that not only the internal weights λ_ν^L are all positive (why?), but also the boundary weights λ_0, λ_0' and λ_{n+1}, λ_{n+1}' if $r = 2$.

(c) Show that $\lambda_0^{(\rho)} = \lambda_{n+1}^{(\rho)}$, $\rho = 0, 1, \ldots, r-1$, if the measure $\mathrm{d}\lambda$ is symmetric.

12*. Generalized Gauss-Radau quadrature.

(a) Write a Matlab routine `gradau.m` for generating the generalized Gauss-Radau quadrature rule of Exercise 10 for a measure $d\lambda$ on $[a, \infty]$, having a fixed node a of multiplicity r, $r > 1$. {*Hint*: To compute the boundary weights, set up an (upper triangular) system of linear equations by applying the formula in turn with $\pi_n^2(t)$, $(t-a)\pi_n^2(t), \ldots, (t-a)^{r-1}\pi_n^2(t)$, where $\pi_n(t) = \prod_{\nu=1}^{n}(t - \tau_\nu^R)$.}

(b) Check your routine against the known formulae with $r = 2$ for the Legendre and Chebyshev measures (see [10, Examples 3.10 and 3.11]). Devise and implement a check that works for arbitrary $r \geq 1$ and another, in particular, for $r = 1$.

(c) Use your routine to explore positivity of the boundary weights and see whether you can come up with any conjectures.

13*. Generalized Gauss-Lobatto quadrature.

(a) Write a Matlab routine `globatto.m` for generating the generalized Gauss-Lobatto rule of Exercise 11 for a measure $d\lambda$ on $[a, b]$, having fixed nodes at a and b of multiplicity r, $r > 1$. For simplicity, start with the case $r \geq 2$ even; then indicate the changes necessary to deal with odd values of r. {*Hint*: Similar to the hint in Exercise 12(a).}

(b) Check your routine against the known formulae with $r = 2$ for the Legendre and Chebyshev measures (see [10, Examples 3.13 and 3.14]). Devise and implement a check that works for arbitrary $r \geq 1$ and another, in particular, for $r = 1$.

(c) Explore the positivity of the boundary weights $\lambda_0^{(\rho)}$ and the quantities $\lambda_{n+1}^{(\rho)}$ in the quadrature formula.

14. Show that the monic Stieltjes polynomial π_{n+1}^K in (4.19) exists uniquely.

15. (a) Let $d\lambda$ be a positive measure. Use approximation theory to show that the minimum of $\int_{\mathbb{R}} |\pi(t)|^p d\lambda(t)$, $1 < p < \infty$, extended over all monic polynomials π of degree n is uniquely determined.

(b) Show that the minimizer of the extremal problem in (a), when $p = 2s+2$, $s \geq 0$ an integer, is the s-orthogonal polynomial $\pi = \pi_{n,s}$. {*Hint*: Differentiate the integral partially with respect to the variable coefficients of π.}

16. (a) Show that r in (4.22) has to be odd.

(b) Show that in (4.22) with r as in (4.23), one cannot have $k > n$.

17. Derive (4.33) and (4.34).

18. Derive (4.39) from (4.38) and explain the meaning of $R_n(f; x)$. {*Hint*: Use Exercise 5(c) and (2.27).}

19. Show that $p_n(f; t)$ in (4.40) is

$$p_n(f;t) = \frac{\pi_n(t)}{\pi_n(x)} f(x) + \sum_{\nu=1}^{n} \frac{(t-x)\pi_n(t)}{(t-\tau_\nu^G)(\tau_\nu^G - x)\pi_n'(\tau_\nu^G)} f(\tau_\nu^G),$$

and thus prove (4.40). {*Hint*: Use Exercise 5(c).}

20. Derive (4.42) and (4.43). {*Hint*: For $k < n$, use Gauss quadrature, and for $k = n$ insert the expression for $p_n(f;t)$ from Exercise 19 into the formula for a_n in (4.41). Also use the fact that the elementary Lagrange interpolation polynomials sum up to 1.}
21. Derive (4.44).
22. Prove (4.46). {*Hint*: Use Exercise 4.}
23. (a) Prove that the Lanczos vectors are mutually orthonormal.
    (b) Show that the vectors $\{\boldsymbol{h}_j\}_{j=0}^{n}$, $n < N$, form an orthonormal basis of the *Krylov space*

$$\mathcal{K}_n(\boldsymbol{A}, \boldsymbol{h}_0) = \mathrm{span}(\boldsymbol{h}_0, \boldsymbol{A}\boldsymbol{h}_0, \ldots, \boldsymbol{A}^n\boldsymbol{h}_0).$$

    (c) Prove that

$$\boldsymbol{h}_j = p_j(\boldsymbol{A})\boldsymbol{h}_0, \quad j = 0, 1, \ldots, N,$$

    where p_j is a polynomial of degree j satisfying the three-term recurrence relation

$$\gamma_{j+1}p_{j+1}(\lambda) = (\lambda - \alpha_j)p_j(\lambda) - \gamma_j p_{j-1}(\lambda), \qquad j = 0, 1, \ldots, N-1,$$
$$p_0(\lambda) = 1, \quad p_{-1}(\lambda) = 0.$$

    {*Hint*: Use mathematical induction.}
24. Prove that the polynomial p_k of Exercise 23(c) is equal to the orthonormal polynomial $\tilde{\pi}_k(\,\cdot\,; \mathrm{d}\rho_N)$. {*Hint*: Use the spectral resolution of $\boldsymbol{A}$ and Exercises 23(a) and (c).}
25. Derive the bounds for $(\boldsymbol{A}^{-1})_{ii}$ hinted at in the last sentence of Example 11.

5 Approximation

5.1 Polynomial Least Squares Approximation

Classical Least Squares Problem

We are given N data points (t_k, f_k), $k = 1, 2, \ldots, N$, and wish to find a polynomial $\hat{p}_n$ of degree $\leq n$, $n < N$, such that a weighted average of the squared errors $[p(t_k) - f_k]^2$ is as small as possible among all polynomials p of degree n,

$$\sum_{k=1}^{N} w_k[\hat{p}_n(t_k) - f_k]^2 \leq \sum_{k=1}^{N} w_k[p(t_k) - f_k]^2 \quad \text{for all } p \in \mathbb{P}_n. \tag{5.1}$$

Here, $w_k > 0$ are positive weights, which allow placing more emphasis on data points that are reliable, and less emphasis on others, by choosing them larger resp. smaller. If the quality of the data is uniformly the same, then equal weights, say $w_k = 1$, are appropriate.

The problem as formulated suggests a discrete N-point measure

$$\mathrm{d}\lambda_N(t) = \sum_{k=1}^{N} w_k \delta(t - t_k), \quad \delta = \text{Dirac delta function}, \tag{5.2}$$

in terms of which the problem can be written in the compact form

$$\|\hat{p}_n - f\|^2_{\mathrm{d}\lambda_N} \leq \|p - f\|^2_{\mathrm{d}\lambda_N} \quad \text{for all } p \in \mathbb{P}_n. \tag{5.3}$$

The polynomials $\pi_k(\,\cdot\,) = \pi_k(\,\cdot\,; \mathrm{d}\lambda_N)$ orthogonal (not necessarily monic) with respect to the discrete measure (5.2) provide an easy solution: one writes

$$p(t) = \sum_{i=0}^{n} c_i \pi_i(t), \quad n < N, \tag{5.4}$$

and obtains for the squared error, using the orthogonality of π_k,

$$\begin{aligned} E_n^2 &= \left(\sum_{i=0}^{n} c_i\pi_i - f, \sum_{j=0}^{n} c_j\pi_j - f \right) = \sum_{i,j=0}^{n} c_i c_j (\pi_i, \pi_j) - 2\sum_{i=0}^{n} c_i (f, \pi_i) + \|f\|^2 \\ &= \sum_{i=0}^{n} \left(\|\pi_i\| c_i - \frac{(f, \pi_i)}{\|\pi_i\|} \right)^2 + \|f\|^2 - \sum_{i=0}^{n} \frac{(f, \pi_i)^2}{\|\pi_i\|^2}. \end{aligned} \tag{5.5}$$

(All norms and inner products are understood to be relative to the measure $\mathrm{d}\lambda_N$.) Evidently, the minimum is attained for $c_i = \hat{c}_i(f)$, where

$$\hat{c}_i(f) = \frac{(f, \pi_i)}{\|\pi_i\|^2}, \quad i = 0, 1, \ldots, n, \tag{5.6}$$

are the "Fourier coefficients" of f relative to the orthogonal system $\pi_0, \pi_1, \ldots, \pi_{N-1}$. Thus,

$$\hat{p}_n(t) = \sum_{i=0}^{n} \hat{c}_i(f) \pi_i(t; \mathrm{d}\lambda_N). \tag{5.7}$$

In Matlab, the procedure is implemented in the `OPQ` routine

```
[phat,c]=least_squares(n,f,xw,ab,d)
```

The given function values f_k are input through the `N`×1 array `f`, the abscissae t_k and weights w_k through the `N`×2 array `xw`, and the measure $\mathrm{d}\lambda_N$ through the (`N+1`)×2 array `ab` of recurrence coefficients (the routine determines `N` automatically from the size of `xw`). The 1×(`n+1`) array `d` is the vector of leading coefficients of the orthogonal polynomials. The procedure returns as output the `N`×(`n+1`) array `phat` of the values $\hat{p}_\nu(t_k)$, $0 \leq \nu \leq n$, $1 \leq k \leq N$, and the (`n+1`)×1 array `c` of the Fourier coefficients.

Example 12. Equally weighted least squares approximation on $N = 10$ equally spaced points on $[-1, 1]$.

Matlab program:

```
N=10; k=(1:N)'; d=ones(1,N);
xw(k,1)=-1+2*(k-1)/(N-1); xw(:,2)=2/N;
ab=r_hahn(N-1); ab(:,1)=-1+2*ab(:,1)/(N-1);
ab(:,2)=(2/(N-1))^2*ab(:,2); ab(1,2)=2;
[phat,c]=least_squares(N-1,f,xw,ab,d);
```

Demo#5 The program is applied to the function $f(t) = \ln(2+t)$ on $[-1, 1]$, and selected least squares errors $\hat{E}_n$ are compared in the table below with maximum errors E_n^∞ (taken over 100 equally spaced points on $[-1, 1]$).

n	$\hat{E}_n$	E_n^∞
0	4.88(−01)	6.37(−01)
3	2.96(−03)	3.49(−03)
6	2.07(−05)	7.06(−05)
9	1.74(−16)	3.44(−06)

If $n = N-1$, the least squares error $\hat{E}_{N-1}$ is zero, since the N data points can be interpolated exactly by a polynomial of degree $\leq N-1$. This is confirmed in the first tabular entry for $n = 9$. The infinity errors are only slightly larger than the least squares errors, except for $n = 9$.

Constrained Least Squares Approximation

It is sometimes desirable to impose constraints on the least squares approximation, for example to insist that at certain points s_j the error should be exactly zero. Thus, the polynomial $p \in \mathbb{P}_n$ is subject to the constraints

$$p(s_j) = f_j, \quad j = 1, 2, \ldots, m; \quad m \leq n, \tag{5.8}$$

but otherwise is freely variable. For simplicity we assume that none of the s_j equals one of the support points t_k. (Otherwise, the procedure to be described requires some simple modifications.)

In order to solve the constrained least squares problem, let

$$p_m(f;t) = p_m(f; s_1, \ldots, s_m; t), \quad \sigma_m(t) = \prod_{j=1}^{m} (t - s_j), \tag{5.9}$$

be respectively the polynomial of degree $m-1$ interpolating f at the points s_j and the constraint polynomial of degree m. We then write

$$p(t) = p_m(f;t) + \sigma_m(t) q(t). \tag{5.10}$$

This clearly satisfies the constraints (5.8), and q is a polynomial of degree $n-m$ that can be freely varied. The problem is to minimize the squared error

$$\|f - p_m(f; \cdot) - \sigma_m q \|^2_{\mathrm{d}\lambda_N} = \int_{\mathbb{R}} \left[\frac{f(t) - p_m(f;t)}{\sigma_m(t)} - q(t) \right]^2 \sigma^2_m(t) \mathrm{d}\lambda_N(t)$$

over all polynomials q of degree $n-m$. This is an *unconstrained* least squares problem, but for a new function f^* and a new measure $\mathrm{d}\lambda^*_N$,

$$\text{minimize}: \ \|f^* - q\|_{\mathrm{d}\lambda^*_N}, \quad q \in \mathbb{P}_{n-m}, \tag{5.11}$$

where

$$f^*(t) = \frac{f(t) - p_m(f;t)}{\sigma_m(t)}, \quad \mathrm{d}\lambda^*_N(t) = \sigma^2_m(t)\mathrm{d}\lambda_N(t). \tag{5.12}$$

If $\hat{q}_{n-m}$ is the solution of (5.11), then

$$\hat{p}_n(t) = p_m(f;t) + \sigma_m(t)\hat{q}_{n-m}(t) \tag{5.13}$$

is the solution of the constrained least squares problem. The function f^*, incidentally, can be given the form of a divided difference,

$$f^*(t) = [s_1, s_2, \ldots, s_m, t]f, \quad t \in \operatorname{supp} \mathrm{d}\lambda^*_N,$$

as follows from the theory of interpolation. Note also that the discrete orthogonal polynomials $\pi_k(\,\cdot\,; \mathrm{d}\lambda^*_N)$ needed to solve (5.11) can be obtained from the polynomials $\pi_k(\,\cdot\,; \mathrm{d}\lambda_N)$ by m modifications of the measure $\mathrm{d}\lambda_N(t)$ by quadratic factors $(t - s_j)^2$.

Example 13. Bessel function $J_0(t)$ for $0 \le t \le j_{0,3}$.

Here, $j_{0,3}$ is the third positive zero of J_0. A natural constraint is to reproduce the first three zeros of J_0 exactly, that is, $m = 3$ and

$$s_1 = j_{0,1}, \quad s_2 = j_{0,2}, \quad s_3 = j_{0.3}.$$

Demo#6 The constrained least squares approximations of degrees $n = 3, 4, 5$ (that is, $n - m = 0, 1, 2$) using $N = 51$ equally spaced points on $[0, j_{0,3}]$ (end points included) are shown in Fig. 2. The solid curve represents the exact function, the dashdotted, dashed, an dotted curves the approximants for $n = 3, 4$, and 5, respectively. The approximations are not particularly satisfactory and show spurious behavior near $t = 0$.

Example 14. Same as Example 13, but with two additional constraints

$$p(0) = 1, \quad p'(0) = 0.$$

Demo#7 Derivative constraints, as the one in Example 14, can be incorporated similarly as before. In this example, the added constraints are designed to remove the spurious behavior near $t = 0$; they also improve considerably the overall accuracy, as is shown in Fig. 3. For further details on Matlab implementation, see [10, Examples 3.51 and 3.52].

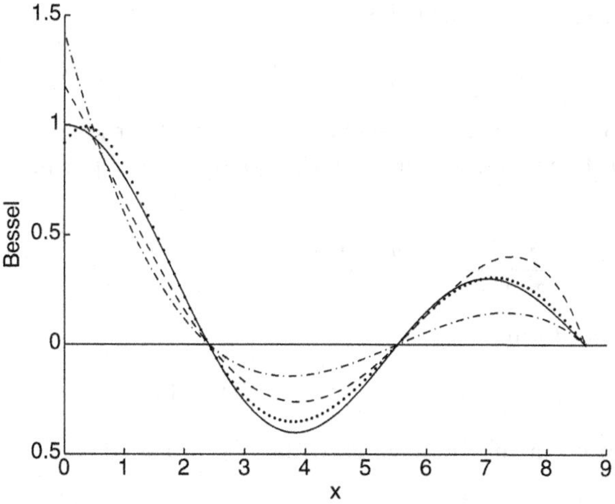

Fig. 2. Constrained least square approximation of the Bessel function

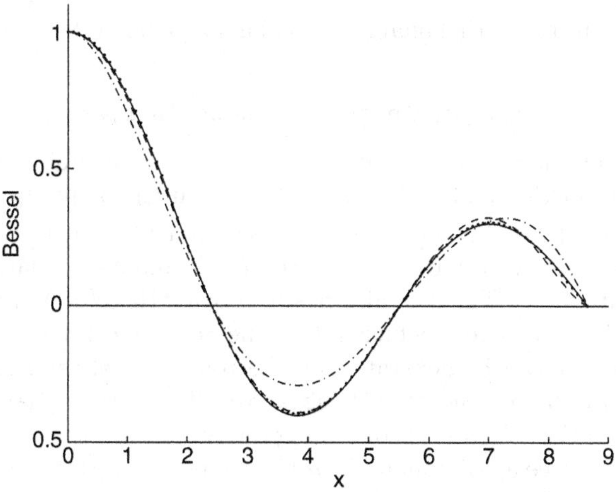

Fig. 3. Derivative-constrained least squares approximation of the Bessel function

Least Squares Approximation in Sobolev Spaces

The task now is to approximate simultaneously functions and some of their first derivatives. More precisely, we want to minimize

$$\sum_{\sigma=0}^{s}\sum_{k=1}^{N} w_k^{(\sigma)}[p^{(\sigma)}(t_k) - f_k^{(\sigma)}]^2$$

over all polynomials $p \in \mathbb{P}_n$, where $f_k^{(\sigma)}$, $\sigma = 0, 1, \ldots, s$, are given function and derivative values, and $w_k^{(\sigma)} > 0$ appropriate weights for each derivative.

These are often chosen such that

$$w_k^{(\sigma)} = \gamma_\sigma w_k, \quad \gamma_\sigma > 0, \quad k = 1, 2, \ldots, N,$$

in terms of one set of positive weights w_k. Evidently,, the problem, analogously to (5.3), can be written in terms of the Sobolev inner product and norm

$$(u, v)_S = \sum_{\sigma=0}^{s} \sum_{k=1}^{N} w_k^{(\sigma)} u^{(\sigma)}(t_k) v^{(\sigma)}(t_k), \quad \|u\|_S = \sqrt{(u, u)_S} \tag{5.14}$$

in the compact form

$$\text{minimize}: \quad \|p - f\|_S^2 \text{ for all } p \in \mathbb{P}_n. \tag{5.15}$$

The solution is entirely analogous to the one provided earlier,

$$\hat{p}_n(t) = \sum_{i=0}^{n} \hat{c}_i(f)\pi_i(t), \quad \hat{c}_i(f) = \frac{(f, \pi_i)_S}{\|\pi_i\|_S^2}, \tag{5.16}$$

where $\{\pi_i\}$ are the orthogonal polynomials of Sobolev type. In Matlab, the procedure is

```
[phat,c]=least_squares_sob(n,f,xw,B)
```

The input parameter `f` is now an $N \times (s+1)$ array containing the N values of the given function and its first s derivatives at the points t_k. The abscissae t_k and the $s+1$ weights $w_k^{(\sigma)}$ of the Sobolev inner product are input via the $N \times (s+2)$ array `xw` (the routine determines N and s automatically from the size of the array `xw`). The user also has to provide the $N \times N$ upper triangular array `B` of the recurrence coefficients for the Sobolev orthogonal polynomials, which for $s = 1$ can be generated by the routine `chebyshev_sob.m` and for arbitrary s by the routine `stieltjes_sob.m`. The output `phat` is an array of dimension (`n+1`)$\times$($N*$(s+1)) containing the N values of the derivative of order σ of the νth-degree approximant $\hat{p}_\nu, \nu \le n$, in positions $(\nu+1, \sigma+1 : s+1 : N*(s+1))$ of the array `phat`. The Fourier coefficients $\hat{c}_i$ are output in the (`n+1`)$\times$1 vector `c`.

Example 15. The complementary error function on $[0, 2]$.

This is the function

$$f(t) = e^{t^2} \operatorname{erfc} t = \frac{2}{\sqrt{\pi}} e^{t^2} \int_t^\infty e^{-u^2} du, \quad 0 \le t \le 2,$$

whose derivatives are easily calculated.

Demo#8 The routine `least_squares_sob.m` is applied to the function f of Example 15 with $s = 2$ and N=5 equally spaced points t_k on $[0, 2]$. All weights are chosen to be equal, $w_k^{(\sigma)} = 1/N$ for $\sigma = 0, 1, 2$. The table below, in the top half, shows selected results for the Sobolev least squares error $\hat{E}_n$

s	n	$\hat{E}_n$	$E^{\infty}_{n,0}$	$E^{\infty}_{n,1}$	$E^{\infty}_{n,2}$
2	0	1.153(+00)	4.759(–01)	1.128(+00)	2.000(+00)
	2	7.356(–01)	8.812(–02)	2.860(–01)	1.411(+00)
	4	1.196(–01)	1.810(–02)	5.434(–02)	1.960(–01)
	9	2.178(–05)	4.710(–06)	3.011(–05)	3.159(–04)
	14	3.653(–16)	1.130(–09)	1.111(–08)	1.966(–07)
0	0	2.674(–01)	4.759(–01)	1.128(+00)	2.000(+00)
	2	2.245(–02)	3.865(–02)	3.612(–01)	1.590(+00)
	4	1.053(–16)	3.516(–03)	5.160(–02)	4.956(–01)

and the maximum errors $E^{\infty}_{n,0}$, $E^{\infty}_{n,1}$, $E^{\infty}_{n,2}$ (over 100 equally spaced points on $[0,2]$) for the function and its first two derivatives. In the bottom half are shown the analogous results for ordinary least squares approximation ($s = 0$) when $n \leq N - 1$. (It makes no sense to consider $n > N - 1$.) Note that the Sobolev least squares error $\hat{E}_{3N-1}$ is essentially zero, reflecting the fact that the Hermite interpolation polynomial of degree $3N - 1$ interpolates the data exactly. In contrast, $\hat{E}_n = 0$ for $n \geq N-1$ in the case of ordinary least squares.

As expected, the table shows rather convincingly that Sobolev least squares approximation approximates the derivatives decidedly better than ordinary least squares approximation, when applicable, and even the function itself, when n is sufficiently large.

5.2 Moment-Preserving Spline Approximation

There are various types of approximation: those that control the maximum pointwise error; those that control some average error (like least squares error); and those, often motivated by physical considerations, that try to preserve the moments of the given function, or at least as many of the first moments as possible. It is this last type of approximation that we now wish to study. We begin with piecewise constant approximation on the whole positive real line $\mathbb{R}_+$, then proceed to spline approximation on $\mathbb{R}_+$, and end with spline approximation on a compact interval.

Piecewise Constant Approximation on $\mathbb{R}_+$

The piecewise constant approximants to be considered are

$$s_n(t) = \sum_{\nu=1}^{n} a_\nu H(t_\nu - t), \quad t \in \mathbb{R}_+, \tag{5.17}$$

where $a_\nu \in \mathbb{R}$, $0 < t_1 < t_2 < \cdots < t_n$, and H is the Heaviside function

$$H(u) = \begin{cases} 1 & \text{if } u \geq 0, \\ 0 & \text{otherwise.} \end{cases}$$

The problem is, for given $f \in C^1(\mathbb{R}_+)$, to find, if possible, the a_ν and t_ν such that

$$\int_0^\infty s_n(t)t^j \mathrm{d}t = \mu_j, \quad j = 0, 1, \ldots, 2n-1, \tag{5.18}$$

where

$$\mu_j = \int_0^\infty f(t)t^j \mathrm{d}t, \quad j = 0, 1, \ldots, 2n-1, \tag{5.19}$$

are the *moments* of f, assumed to exist.

The solution can be formulated in terms of Gauss quadrature relative to the measure

$$\mathrm{d}\lambda(t) = -tf'(t)\mathrm{d}t \quad \text{on } \mathbb{R}_+. \tag{5.20}$$

Indeed, if $f(t) = o(t^{-2n})$ as $t \to \infty$, then the problem has a unique solution if and only if $\mathrm{d}\lambda$ in (5.20) admits an n-point Gauss quadrature formula

$$\int_0^\infty g(t)\mathrm{d}\lambda(t) = \sum_{\nu=1}^{n} \lambda_\nu^G g(\tau_\nu^G), \quad g \in \mathbb{P}_{2n-1}, \tag{5.21}$$

satisfying $0 < \tau_1^G < \tau_2^G < \cdots < \tau_n^G$. If that is the case, then the desired knots t_ν and coefficients a_ν are given by

$$t_\nu = \tau_\nu^G, \quad a_\nu = \frac{\lambda_\nu^G}{\tau_\nu^G}, \quad \nu = 1, 2, \ldots, n. \tag{5.22}$$

A Gauss formula (5.21) always exists if $f' < 0$ on $\mathbb{R}_+$, that is, $\mathrm{d}\lambda(t) \geq 0$.

For the proof, we use integration by parts,

$$\int_0^T f(t)t^j \mathrm{d}t = \frac{1}{j+1} t^{j+1} f(t) \Big|_0^T - \frac{1}{j+1} \int_0^T f'(t)t^{j+1}\mathrm{d}t, \quad j \leq 2n-1,$$

and let $T \to \infty$. The integrated part on the right goes to zero by assumption on f, and the left-hand side converges to the jth moment of f, again by assumption. Therefore, the last term on the right also converges, and since $-tf'(t) = \mathrm{d}\lambda(t)$, one finds

$$\mu_j = \frac{1}{j+1} \int_0^\infty t^j \mathrm{d}\lambda(t), \quad j = 0, 1, \ldots, 2n-1.$$

This shows in particular that the first $2n$ moments of $\mathrm{d}\lambda$ exist, and therefore, if $\mathrm{d}\lambda \geq 0$, also the Gauss formula (5.21).

On the other hand, the approximant s_n has moments

$$\int_0^\infty s_n(t)t^j \mathrm{d}t = \sum_{\nu=1}^{n} a_\nu \int_0^{t_\nu} t^j \mathrm{d}t = \frac{1}{j+1} \sum_{\nu=1}^{n} a_\nu t_\nu^{j+1},$$

so that the first $2n$ moments μ_j of f are preserved if and only if

$$\sum_{\nu=1}^{n}(a_\nu t_\nu)t_\nu^j = \int_0^\infty t^j \mathrm{d}\lambda(t), \quad j = 0, 1, \ldots, 2n-1.$$

This is equivalent to saying that the knots t_ν are the Gauss nodes in (5.21), and $a_\nu t_\nu$ the corresponding weights.

Example 16. Maxwell distribution $f(t) = \mathrm{e}^{-t^2}$ on $\mathbb{R}_+$.

Here,

$$\mathrm{d}\lambda(t) = 2t^2\mathrm{e}^{-t^2}\mathrm{d}t \quad \text{on } \mathbb{R}_+,$$

which is a positive measure obtained (up to the factor 2) by twice modifying the half-range Hermite measure by a linear factor t. The first $n+2$ recurrence coefficients of the half-range Hermite measure can be computed by a discretization method. Applying to these recurrence coefficients twice the routine `chri1.m`, with zero shift, then yields the recurrence coefficients $\alpha_k(\mathrm{d}\lambda)$, $\beta_k(\mathrm{d}\lambda)$, $k \le n-1$, and hence the required n-point Gauss quadrature rule (5.21) for $\mathrm{d}\lambda$. The result for $n = 5$ is depicted in Fig. 4.

Spline Approximation on $\mathbb{R}_+$

The approximant s_n in (5.17) can be interpreted as a spline function of degree 0. We now consider spline functions $s_{n,m}$ of degree $m > 0$,

$$s_{n,m}(t) = \sum_{\nu=1}^{n} a_\nu (t_\nu - t)_+^m, \quad t \in \mathbb{R}_+, \tag{5.23}$$

where u_+^m is the truncated power $u_+^m = u^m$ if $u \ge 0$, and $u_+^m = 0$ if $u < 0$. Given the first $2n$ moments (5.19) of f, the problem again is to determine $a_\nu \in \mathbb{R}$ and $0 < t_1 < t_2 < \cdots < t_n$ such that

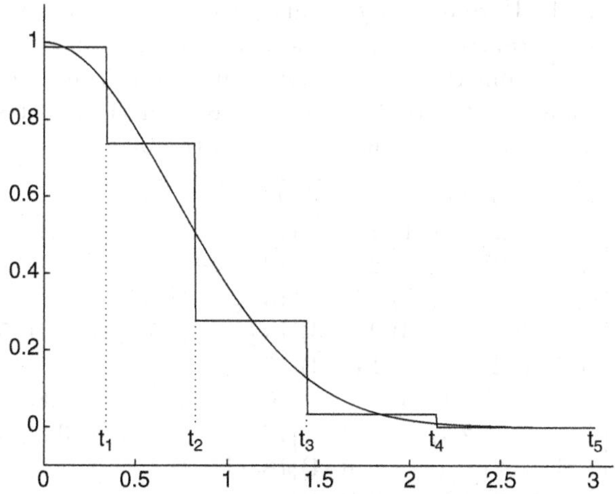

Fig. 4. Piecewise constant approximation of the Maxwell distribution

$$\int_0^{\infty} s_{n,m}(t) t^j \mathrm{d}t = \mu_j, \quad j = 0, 1, \ldots, 2n-1. \tag{5.24}$$

By a reasoning similar to the one in the previous subsection, but more complicated, involving m integrations by part, one proves that for $f \in C^{m+1}(\mathbb{R}_+)$ and satisfying $f^{(\mu)}(t) = o(t^{-2n-\mu})$ as $t \to \infty$, $\mu = 0, 1, \ldots, m$, the problem has a unique solution if and only if the measure

$$\mathrm{d}\lambda^{[m]}(t) = \frac{(-1)^{m+1}}{m!} t^{m+1} f^{(m+1)}(t) \mathrm{d}t \quad \text{on } \mathbb{R}_+ \tag{5.25}$$

admits an n-point Gauss quadrature formula

$$\int_0^{\infty} g(t)\, \mathrm{d}\lambda^{[m]}(t) = \sum_{\nu=1}^{n} \lambda_\nu^G g(\tau_\nu^G) \quad \text{for all } g \in \mathbb{P}_{2n-1} \tag{5.26}$$

satisfying $0 < \tau_1^G < \tau_2^G < \cdots < \tau_n^G$. If that is the case, the knots t_ν and coefficients a_ν are given by

$$t_\nu = \tau_\nu^G, \quad a_\nu = \frac{\lambda_\nu^G}{[\tau_\nu^G]^{m+1}}, \quad \nu = 1, 2, \ldots, n. \tag{5.27}$$

Note that $\mathrm{d}\lambda^{[m]}$ in (5.25) is a positive measure, for each $m \geq 0$, and hence (5.26) exists, if f is completely monotonic on $\mathbb{R}_+$, that is, $(-1)^\mu f^{(\mu)}(t) > 0$, $t \in \mathbb{R}_+$, for $\mu = 0, 1, 2, \ldots$.

Example 17. Maxwell distribution $f(t) = \mathrm{e}^{-t^2}$ on $\mathbb{R}_+$, revisited.

We now have

$$\mathrm{d}\lambda^{[m]}(t) = \frac{1}{m!} t^{m+1} H_{m+1}(t) \mathrm{e}^{-t^2}\, \mathrm{d}t \quad \text{on } \mathbb{R}_+,$$

where H_{m+1} is the Hermite polynomial of degree $m+1$. Here, $\mathrm{d}\lambda^{[m]}$ if $m > 0$ is no longer of constant sign on $\mathbb{R}_+$, and hence the existence of the Gauss rule (5.26) is in doubt. Numerical exploration, using discretization methods, yields the situation shown in the table below, where a dash indicates the presence of a negative Gauss node τ_ν^G, and an asterisk the presence of a pair

n	$m=1$	$m=2$	$m=3$	n	$m=1$	$m=2$	$m=3$
1	6.9(−2)	1.8(−1)	2.6(−1)	11	—	1.1(−3)	1.1(−4)
2	8.2(−2)	—	2.3(−1)	12	—	—	*
3	—	1.1(−2)	2.5(−3)	13	7.8(−3)	6.7(−4)	*
4	3.5(−2)	6.7(−3)	2.2(−3)	14	8.3(−3)	5.6(−4)	8.1(−5)
5	2.6(−2)	—	1.6(−3)	15	7.7(−3)	—	7.1(−5)
6	2.2(−2)	3.1(−3)	*	16	—	4.9(−4)	7.8(−5)
7	—	2.4(−3)	*	17	—	3.8(−4)	3.8(−5)
8	1.4(−2)	—	3.4(−4)	18	5.5(−3)	3.8(−4)	*
9	1.1(−2)	1.7(−3)	2.5(−4)	19	5.3(−3)	—	*
10	9.0(−3)	1.1(−3)	—	20	5.4(−3)	3.1(−4)	*

of conjugate complex Gauss nodes. In all cases computed, there were never more than one negative Gauss node, or more than one pair of complex nodes. The numbers in the table represent the maximum errors $\|s_{n,m} - f\|_\infty$, the maximum being taken over 100 equally spaced points on $[0, \tau_n^G]$.

Spline Approximation on a Compact Interval

The problem on a compact interval, say $[0, 1]$, is a bit more involved than the problem on $\mathbb{R}_+$. For one, the spline function $s_{n,m}$ may now include a polynomial p of degree m, which was absent before since no moment of p exists on $\mathbb{R}_+$ unless $p \equiv 0$. Thus, the spline approximant has now the form

$$s_{n,m}(t) = p(t) + \sum_{\nu=1}^{n} a_\nu (t_\nu - t)_+^m, \quad p \in \mathbb{P}_m, \quad 0 \le t \le 1, \tag{5.28}$$

where $a_\nu \in \mathbb{R}$ and $0 < t_1 < t_2 < \cdots < t_n < 1$. There are two problems of interest:

Problem I. Find $s_{n,m}$ such that

$$\int_0^1 s_{n,m}(t) t^j \mathrm{d}t = \mu_j, \quad j = 0, 1, \ldots, 2n + m. \tag{5.29}$$

Since we have $m + 1$ additional parameters at our disposal (the coefficients of p), we can impose $m + 1$ additional moment conditions.

Problem II. Rather than matching more moments, we use the added degree of freedom to impose $m + 1$ "boundary conditions" at the end point $t = 1$. More precisely, we want to find $s_{n,m}$ such that

$$\int_0^1 s_{n,m}(t) t^j \mathrm{d}t = \mu_j, \quad j = 0, 1, \ldots, 2n - 1 \tag{5.30}$$

and

$$s_{n,m}^{(\mu)}(1) = f^{(\mu)}(1), \quad \mu = 0, 1, \ldots, m. \tag{5.31}$$

It is still true that a solution can be given in terms of quadrature formulae, but they are now respectively generalized Gauss–Lobatto and generalized Gauss–Radau formulae relative to the measure (see [5, 6])

$$\mathrm{d}\lambda^{[m]}(t) = \frac{(-1)^{m+1}}{m!} f^{(m+1)}(t) \mathrm{d}t \quad \text{on } [0, 1]. \tag{5.32}$$

Problem I, in fact, has a unique solution if and only if the *generalized Gauss–Lobatto formula*

$$\begin{aligned} \int_0^1 g(t) \mathrm{d}\lambda^{[m]}(t) &= \sum_{\mu=0}^{m} [\lambda_0^{(\mu)} g^{(\mu)}(0) + (-1)^\mu \lambda_{n+1}^{(\mu)} g^{(\mu)}(1)] \\ &+ \sum_{\nu=1}^{n} \lambda_\nu^L g(\tau_\nu^L), \quad g \in \mathbb{P}_{2n+2m+1}, \end{aligned} \tag{5.33}$$

exists with $0 < \tau_1^L < \cdots < \tau_n^L < 1$. In this case,

$$t_\nu = \tau_\nu^L, \quad a_\nu = \lambda_\nu^L, \quad \nu = 1, 2, \ldots .n, \tag{5.34}$$

and p is uniquely determined by

$$p^{(\mu)}(1) = f^{(\mu)}(1) + (-1)^m m! \lambda_{n+1}^{(m-\mu)}, \quad \mu = 0, 1, \ldots, m. \tag{5.35}$$

Similarly, Problem II has a unique solution if and only if the *generalized Gauss–Radau formula*

$$\int_0^1 g(t) \mathrm{d}\lambda^{[m]}(t) = \sum_{\mu=0}^{m} \lambda_0^{(\mu)} g^{(\mu)}(0) + \sum_{\nu=1}^{n} \lambda_\nu^R g(\tau_\nu^R), \quad g \in \mathbb{P}_{2n+m}, \tag{5.36}$$

exists with $0 < \tau_1^R < \cdots < \tau_n^R < 1$. Then

$$t_\nu = \tau_\nu^R, \quad a_\nu = \lambda_\nu^R, \quad \nu = 1, 2, \ldots .n, \tag{5.37}$$

and (trivially)

$$p(t) = \sum_{\mu=0}^{m} \frac{f^{(\mu)}(1)}{\mu!} (t-1)^\mu. \tag{5.38}$$

In both cases, complete monotonicity of f implies $\mathrm{d}\lambda \geq 0$ and the existence of the respective quadrature formulae. For their construction, see Exercises 12 and 13 of §4.

5.3 Slowly Convergent Series

Standard techniques of accelerating the convergence of slowly convergent series are based on linear or nonlinear sequence transformations: the sequence of partial sums is transformed somehow into a new sequence that converges to the same limit, but a lot faster. Here we follow another approach, more in the spirit of these lectures: the sum of the series is represented as a definite integral; a sequence of quadrature rules is then applied to this integral which, when properly chosen, will produce a sequence of approximations that converges quickly to the desired sum.

An easy way (and certainly not the only one) to obtain an integral representation presents itself when the general term of the series, or part thereof, is expressible in terms of the Laplace transform (or some other integral transform) of a known function. Several instances of this will now be described.

Series Generated by a Laplace Transform

The series

$$S = \sum_{k=1}^{\infty} a_k \tag{5.39}$$

to be considered first has terms a_k that are the Laplace transform

$$(\mathcal{L}f)(s) = \int_0^\infty \mathrm{e}^{-st} f(t)\mathrm{d}t$$

of some known function f evaluated at $s = k$,

$$a_k = (\mathcal{L}f)(k), \quad k = 1, 2, 3, \ldots . \tag{5.40}$$

In this case,

$$\begin{aligned} S &= \sum_{k=1}^{\infty} \int_0^\infty \mathrm{e}^{-kt} f(t)\mathrm{d}t \\ &= \int_0^\infty \sum_{k=1}^{\infty} \mathrm{e}^{-(k-1)t} \cdot \mathrm{e}^{-t} f(t)\mathrm{d}t \\ &= \int_0^\infty \frac{1}{1-\mathrm{e}^{-t}} \mathrm{e}^{-t} f(t)\mathrm{d}t \end{aligned}$$

that is,

$$S = \int_0^\infty \frac{t}{1-\mathrm{e}^{-t}} \frac{f(t)}{t} \mathrm{e}^{-t}\mathrm{d}t. \tag{5.41}$$

There are at least three different approaches to evaluate this integral numerically: one is Gauss–Laguerre quadrature of $(t/(1 - \mathrm{e}^{-t}))f(t)/t$ with $\mathrm{d}\lambda(t) = \mathrm{e}^{-t}\mathrm{d}t$ on $\mathbb{R}_+$; another is rational/polynomial Gauss–Laguerre quadrature of the same function; and a third Gauss–Einstein quadrature of the function $f(t)/t$ with $\mathrm{d}\lambda(t) = t\mathrm{d}t/(\mathrm{e}^t - 1)$ on $\mathbb{R}_+$. In the last method, the weight function $t/(\mathrm{e}^t - 1)$ is widely used in solid state physics, where it is named after Einstein (coming from the Einstein-Bose distribution). It is also, incidentally, the generating function of the Bernoulli polynomials.

Example 18. The Theodorus constant

$$S = \sum_{k=1}^{\infty} \frac{1}{k^{3/2} + k^{1/2}} = 1.860025\ldots .$$

This is a universal constant introduced by P.J. Davis (1993) in connection with a spiral attributed to the ancient mathematician Theodorus of Cyrene.

Here we note that

$$\frac{1}{s^{3/2} + s^{1/2}} = s^{-1/2} \frac{1}{s+1} = \left(\mathcal{L} \frac{1}{\sqrt{\pi t}} * \mathrm{e}^{-t}\right)(s),$$

where the star stands for convolution. A simple computation yields (5.40) with

$$f(t) = \frac{2}{\sqrt{\pi}} F(\sqrt{t}),$$

where

$$F(x) = \mathrm{e}^{-x^2} \int_0^x \mathrm{e}^{t^2}\,\mathrm{d}t$$

is Dawson's integral.

Demo#9 To make $f(t)$ regular at $t = 0$, we divide by $\sqrt{t}$ and write

$$\begin{aligned} S &= \frac{2}{\sqrt{\pi}} \int_0^\infty \frac{t}{1-\mathrm{e}^{-t}} \frac{F(\sqrt{t})}{\sqrt{t}}\, t^{-1/2}\mathrm{e}^{-t}\mathrm{d}t \\ &= \frac{2}{\sqrt{\pi}} \int_0^\infty \frac{F(\sqrt{t})}{\sqrt{t}}\, t^{-1/2} \frac{t}{\mathrm{e}^t - 1}\,\mathrm{d}t. \end{aligned}$$

To the first integral we apply Gauss–Laguerre quadrature with $\mathrm{d}\lambda(t) = t^{-1/2}\mathrm{e}^{-t}\mathrm{d}t$ on $\mathbb{R}_+$, or rational Gauss–Laguerre with the same $\mathrm{d}\lambda$, and to the second integral Gauss–Einstein quadrature (modified by the factor $t^{-1/2}$). The errors committed in these quadrature methods are shown in the table below.

n	Gauss-Laguerre	rational Gauss-Laguerre	Gauss-Einstein
1	9.6799(–03)	1.5635(–02)	1.3610(–01)
4	5.5952(–06)	1.1893(–08)	2.1735(–04)
7	4.0004(–08)	5.9689(–16)	3.3459(–07)
10	5.9256(–10)		5.0254(–10)
15	8.2683(–12)		9.4308(–15)
20	8.9175(–14)		4.7751(–16)
	timing: 10.8	timing: 8.78	timing: 10.4

The clear winner is rational Gauss–Laguerre, both in terms of accuracy and run time.

Example 19. The Hardy–Littlewood function

$$H(x) = \sum_{k=1}^{\infty} \frac{1}{k} \sin\frac{x}{k}, \qquad x > 0.$$

It can be shown that

$$a_k := \frac{1}{k} \sin\frac{x}{k} = (\mathcal{L}f(t;x))(k),$$

where

$$f(t;x) = \frac{1}{2\mathrm{i}} \left[I_0(2\sqrt{\mathrm{i}xt}) - I_0(2\sqrt{-\mathrm{i}xt}) \right]$$

and I_0 is the modified Bessel function. This gives rise to the two integral representations

$$H(x) = \int_0^\infty \frac{t}{1-\mathrm{e}^{-t}} \frac{f(t;x)}{t} \mathrm{e}^{-t}\mathrm{d}t = \int_0^\infty \frac{f(t;x)}{t} \frac{t}{\mathrm{e}^t - 1}\,\mathrm{d}t.$$

Among the three quadrature methods, Gauss–Einstein performs best, but all suffer from internal cancellation of terms in the quadrature sum. The problem becomes more prominent as the number n of terms increases. In this case, other methods can be applied, using the Euler-Maclaurin formula [11].

Fig. 5 shows the behavior of $H(x)$ in the range $0 \le x \le 100$.

"Alternating" Series Generated by a Laplace Transform

These are series in which the general terms are Laplace transforms with alternating signs of some function f, that is, series (5.39) with

$$a_k = (-1)^{k-1}(\mathcal{L} f)(k), \quad k = 1, 2, 3, \dots . \tag{5.42}$$

An elementary computation similar to the one carried out in the previous subsection will show that

$$S = \int_0^\infty \frac{1}{1+\mathrm{e}^{-t}} f(t)\mathrm{e}^{-t}\mathrm{d}t = \int_0^\infty f(t)\frac{1}{\mathrm{e}^t+1}\,\mathrm{d}t. \tag{5.43}$$

We can again choose between three quadrature methods: Gauss–Laguerre quadrature of the function $f(t)/(1+\mathrm{e}^{-t})$ with $\mathrm{d}\lambda(t) = \mathrm{e}^{-t}\mathrm{d}t$, rational/polynomial Gauss–Laguerre of the same function, and Gauss-Fermi quadrature of $f(t)$ with $\mathrm{d}\lambda(t) = \mathrm{d}t/(\mathrm{e}^t+1)$ involving the Fermi function $1/(\mathrm{e}^t+1)$ (also used in solid state physics).

Example 20. The series

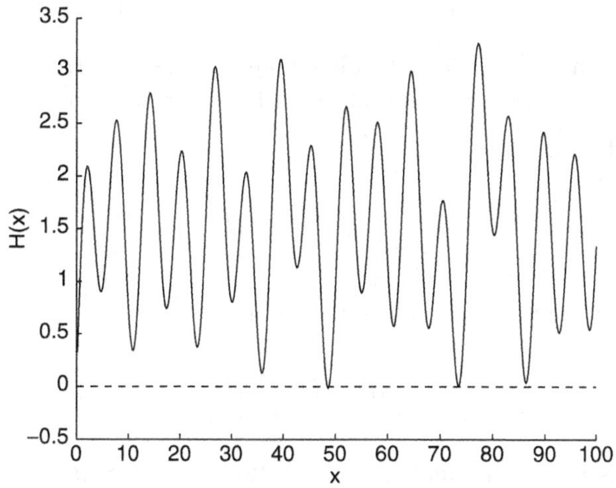

Fig. 5. The Hardy-Littlewood function

$$S=\sum_{k=1}^{\infty}\frac{(-1)^{k-1}}{k}\,\mathrm{e}^{-1/k}.$$

One can show that the function f in question here is $f(t)=J_0(2\sqrt{t})$, with J_0 the Bessel function of order zero. Errors obtained by the three quadrature methods are displayed in the table below, showing the clear superiority of Gauss–Fermi quadrature.

n	Gauss-Laguerre	rational Gauss-Laguerre	Gauss-Fermi
1	1.6961(–01)	1.0310(–01)	5.6994(–01)
4	4.4754(–03)	4.6605(–05)	9.6454(–07)
7	1.7468(–04)	1.8274(–09)	9.1529(–15)
10	3.7891(–06)	1.5729(–13)	2.8163(–16)
15	2.6569(–07)	1.5490(–15)	
20	8.6155(–09)		
40	1.8066(–13)		
	timing: 12.7	timing: 19.5	timing: 4.95

Series Generated by the Derivative of a Laplace Transform

These are series (5.39) in which

$$a_k=-\frac{\mathrm{d}}{\mathrm{d}s}(\mathcal{L}f)(s)\bigg|_{s=k},\quad k=1,2,3,\ldots. \tag{5.44}$$

In this case one finds

$$S=\int_0^{\infty}\frac{t}{1-\mathrm{e}^{-t}}\,f(t)\mathrm{e}^{-t}\mathrm{d}t=\int_0^{\infty}f(t)\frac{t}{\mathrm{e}^{t}-1}\,\mathrm{d}t, \tag{5.45}$$

and Gauss–Laguerre, rational/polynomial Gauss–Laguerre, and Gauss–Einstein quadrature are again options as in Examples 18 and 19.

Example 21. The series

$$S=\sum_{k=1}^{\infty}(\tfrac{3}{2}k+1)k^{-2}(k+1)^{-3/2}.$$

The relevant function f is calculated to be

$$f(t)=\frac{\operatorname{erf}\sqrt{t}}{\sqrt{t}}\cdot t^{1/2},$$

where erf is the error function $\operatorname{erf}x=(2/\sqrt{\pi})\int_0^x \mathrm{e}^{-t^2}\mathrm{d}t$. Numerical results analogous to those in the two previous tables are shown below.

n	Gauss-Laguerre	rational Gauss-Laguerre	Gauss-Einstein
1	4.0125(–03)	5.1071(–02)	8.1715(–02)
4	1.5108(–05)	4.5309(–08)	1.6872(–04)
7	4.6576(–08)	1.3226(–13)	3.1571(–07)
10	3.0433(–09)	1.2087(–15)	5.4661(–10)
15	4.3126(–11)		1.2605(–14)
20	7.6664(–14)		
30	3.4533(–16)		
	timing: 6.50	timing: 10.8	timing: 1.58

The run time is best for Gauss–Einstein quadrature, though the error is worse than for the closest competitor, rational Gauss–Laguerre.

Series Occurring in Plate Contact Problems

The series of interest here is

$$R_p(z) = \sum_{k=0}^{\infty} \frac{z^{2k+1}}{(2k+1)^p}, \quad z \in \mathbb{C},\ |z| \le 1,\ p = 2 \text{ or } 3. \tag{5.46}$$

Rather than expressing the whole general term of the series as a Laplace transform, we do this only for the coefficient,

$$\frac{1}{(k+\frac{1}{2})^p} = (\mathcal{L}f)(k), \quad f(t) = \frac{1}{(p-1)!}\, t^{p-1}\mathrm{e}^{-t/2}. \tag{5.47}$$

Then

$$\begin{aligned} R_p(z) &= \frac{z}{2^p} \sum_{k=0}^{\infty} \frac{z^{2k}}{(k+\frac{1}{2})^p} \\ &= \frac{z}{2^p} \sum_{k=0}^{\infty} z^{2k} \int_0^{\infty} \mathrm{e}^{-kt} \cdot \frac{t^{p-1}\mathrm{e}^{-t/2}}{(p-1)!}\, \mathrm{d}t \\ &= \frac{z}{2^p(p-1)!} \int_0^{\infty} \sum_{k=0}^{\infty} (z^2\mathrm{e}^{-t})^k \cdot t^{p-1}\mathrm{e}^{-t/2} \mathrm{d}t \\ &= \frac{z}{2^p(p-1)!} \int_0^{\infty} \frac{1}{1-z^2\mathrm{e}^{-t}}\, t^{p-1}\mathrm{e}^{-t/2} \mathrm{d}t, \end{aligned}$$

that is,

$$R_p(z) = \frac{z}{2^p(p-1)!} \int_0^{\infty} \frac{t^{p-1}\mathrm{e}^{t/2}}{\mathrm{e}^t - z^2}\, \mathrm{d}t, \quad z^{-2} \in \mathbb{C}\backslash[0,1]. \tag{5.48}$$

The case $z = 1$ can be treated directly by using the connection with the zeta function, $R_p(1) = (1-2^{-p})\zeta(p)$. Assume therefore $z \ne 1$. When $|z|$ is close to 1, the integrand in (5.48) is rather ill-behaved near $t = 0$, exhibiting a steep boundary layer. We try to circumvent this by making the change of variables $e^{-t} \mapsto t$ to obtain

$$R_p(z) = \frac{1}{2^p(p-1)!z} \int_0^1 \frac{t^{-1/2}[\ln(1/t)]^{p-1}}{z^{-2}-t} \, dt.$$

This expresses $R_p(z)$ as a Cauchy integral of the measure

$$d\lambda^{[p]}(t) = t^{-1/2}[\ln(1/t)]^{p-1} dt.$$

Since by assumption, z^{-2} lies outside the interval $[0,1]$, the integral can be evaluated by the continued fraction algorithm, once sufficiently many recurrence coefficients for $d\lambda^{[p]}$ have been precomputed. For the latter, the modified Chebyshev algorithm is quite effective. The first 100 coefficients are available for $p=2$ and $p=3$ in the `OPQ` files `absqm1log1` and `absqm1log2` to 25 resp. 20 decimal digits.

Example 22.

$$R_p(x), \ p=2 \text{ and } 3, \ x=.8, .9, .95, .99, .999 \text{ and } 1.0.$$

Numerical results are shown in he table below and are accurate to all digits

x	$R_2(x)$	$R_3(x)$
.8	0.87728809392147	0.82248858052014
.9	1.02593895111111	0.93414857586540
.95	1.11409957792905	0.99191543992243
.99	1.20207566477686	1.03957223187364
.999	1.22939819733	1.05056774973
1.000	1.233625	1.051795

shown. Full acuracy cannot be achieved for $x \geq .999$ using only 100 recurrence coefficients of $d\lambda^{[p]}$.

Example 23.

$$R_p(e^{i\alpha}), \ p=2 \text{ and } 3, \ \alpha = \omega\pi/2, \ \omega = .2, .1, .05, .01, .001 \text{ and } 0.0.$$

Numerical results are shown in the table below.

p	ω	$\mathrm{Re}(R_p(z))$	$\mathrm{Im}(R_p(z))$
2	.2	0.98696044010894	0.44740227008596
3		0.96915102126252	0.34882061265337
2	0.1	1.11033049512255	0.27830297928558
3		1.02685555765937	0.18409976778928
2	0.05	1.17201552262936	0.16639152396897
3		1.04449441539672	0.09447224926029
2	0.01	1.22136354463481	0.04592009281744
3		1.05140829197388	0.01928202831056
2	0.001	1.232466849	0.006400460
3		1.051794454	0.001936923
2	0.000	1.2336	0.0000
3		1.0518	0.0000

Here, too, full accuracy is not attainable for $\omega \leq 0.001$ with only 100 recurrence coefficients. Curiously, the continued fraction algorithm seems to converge also when $z = 1$, albeit slowly.

Series Involving Ratios of Hyperbolic Functions

More of a challenge are series of the type

$$T_p(x;b) = \sum_{k=0}^{\infty} \frac{1}{(2k+1)^p} \frac{\cosh(2k+1)x}{\cosh(2k+1)b}, \quad 0 \leq x \leq b,\ b > 0,\ p = 2,3, \tag{5.49}$$

which also occur in plate contact problems. Here, we first expand the ratio of hyperbolic cosines into an infinite series,

$$\begin{aligned} &\frac{\cosh(2k+1)x}{\cosh(2k+1)b} \\ &\quad = \sum_{n=0}^{\infty} (-1)^n \left\{ \mathrm{e}^{-(2k+1)[(2n+1)b-x]} + \mathrm{e}^{-(2k+1)[(2n+1)b+x]} \right\}, \end{aligned} \tag{5.50}$$

insert this in (5.49) and apply the Laplace transform technique of the previous subsection. This yields, after an elementary computation (using an interchange of the summations over k and n),

$$T_p(x,b) = \frac{1}{2^p(p-1)!} \sum_{n=0}^{\infty} (-1)^n \mathrm{e}^{(2n+1)b} [\varphi_n(-x) + \varphi_n(x)], \tag{5.51}$$

where

$$\varphi_n(s) = \mathrm{e}^s \int_0^1 \frac{\mathrm{d}\lambda^{[p]}(t)}{\mathrm{e}^{2[(2n+1)b+s]} - t}, \quad -b \leq s \leq b. \tag{5.52}$$

The integral on the right is again amenable to the continued fraction algorithm for $\mathrm{d}\lambda^{[p]}$, which for large n converges almost instantaneously. Convergence of the series (5.51) is geometric with ratio e^{-b}.

Exercises to §5 (Stars indicate more advanced exercises.)

1. With $\pi_0, \pi_1, \ldots, \pi_{N-1}$ denoting the discrete orthogonal polynomials relative to the measure $\mathrm{d}\lambda_N$, and $\hat{c}_i(f)$ the Fourier coefficients of f with respect to these orthogonal polynomials, show that

$$\sum_{i=0}^{n} |\hat{c}_i(f)|^2 \|\pi_i\|^2 \leq \|f\|^2, \quad n < N,$$

with equality holding for $n = N - 1$.

2. Prove the following alternative form for the Fourier coefficients,

$$\hat{c}_i(f) = \frac{1}{\|\pi_i\|^2}\left(f - \sum_{j=0}^{i-1} \hat{c}_j(f)\pi_j, \pi_i\right), \quad i = 0, 1, \ldots, n,$$

and discuss its possible advantages over the original form.
3. Discuss the modifications required in the constrained least squares approximation when ν ($0 \le \nu \le m$) of the points s_j are equal to one of the support points t_k.
4. What are $p_m(f; \cdot)$, f^*, and σ_m in Example 13?
5. Calculate the first and second derivative of the complementary error function of Example 15.
6*. Prove the unique solvability of the problem (5.24) under the conditions stated in (5.25)–(5.26), and, in the affirmative case, derive (5.27).
7. Derive the measure $d\lambda^{[m]}$ for the Maxwell distribution of Example 17.
8. Derive the formula for f in Example 18.
9. Derive the formula for f in Example 19.
10. Derive (5.43).
11. Derive the formula for f in Example 20.
12. Derive (5.45).
13. Derive the formula for f in Example 21.
14. Supply the details for deriving (5.51).

References

1. M. Abramowitz and I.A. Stegun (eds), *Handbook of Mathematical Functions*, Dover Publications, New York, 1992.
2. S. Chandrasekhar, *Radiative Transfer*, Oxford University Press, 1950.
3. B. Danloy, *Numerical construction of Gaussian quadrature formulas for* $\int_0^1 -\mathrm{Log}\, x\, x^\alpha f(x)\,dx$ *and* $\int_0^\infty E_m(x) f(x)\,dx$, Math. Comp. **27** (1973), 861–869.
4. P.J. Davis and P. Rabinowitz, *Some geometrical theorems for abscissas and weights of Gauss type*, J. Math. Anal. Appl. **2** (1961), 428–437.
5. M. Frontini, W. Gautschi, G.V. Milovanović, *Moment-preserving spline approximation on finite intervals*, Numer. Math. **50** (1987), 503–518.
6. M. Frontini, W. Gautschi, G.V. Milovanović, *Moment-preserving spline approximation on finite intervals and Turán quadratures*, Facta Univ. Ser. Math. Inform. **4** (1989), 45–56.
7. W. Gautschi, *Anomalous convergence of a continued fraction for ratios of Kummer functions*, Math. Comp. **31** (1977), 994–999.
8. W. Gautschi, *On the computation of special Sobolev-type orthogonal polynomials*, Ann. Numer. Math. **4** (1997), 329–342.
9. W. Gautschi, *The interplay between classical analysis and (numerical) linear algebra—a tribute to Gene H. Golub*, Electr. Trans. Numer. Anal. **13** (2002), 119–147.

10. W. Gautschi, *Orthogonal Polynomials: Computation and Approximation*, Numerical Mathematics and Scientific Computation, Oxford University Press, Oxford, 2004.
11. W. Gautschi, *The Hardy-Littlewood function: an exercise in slowly convergent series*, J. Comput. Appl. Math. **179** (2005), 249–254.
12. W. Gautschi and M. Zhang, *Computing orthogonal polynomials in Sobolev spaces*, Numer. Math. **71** (1995), 159–183.
13. I.S. Gradshteyn and I.M. Ryzhik, *Tables of Integrals, Series, and Products* (6th edn), Academic Press, San Diego, CA, 2000.
14. W.B. Gragg and W.J. Harrod, *The numerically stable reconstruction of Jacobi matrices from spectral data*, Numer. Math. **44** (1984), 317–335.
15. F. Peherstorfer, K. Petras, *Stieltjes polynomials and Gauss-Kronrod quadrature for Jacobi weight functions*, Numer. Math. **95** (2003), 689–706.
16. G. Szegő, *Orthogonal Polynomials* (4th edn), AMS Colloq. Publ. 23, Amer. Math. Soc., Providence, RI, 1975.
17. P. Verlinden, *Stable rational modification of a weight*, Numer. Algorithms **22** (1999), 183–192.

Equilibrium Problems of Potential Theory in the Complex Plane

Andrei Martínez Finkelshtein *

Departamento de Estadística y Matemática Aplicada, Universidad de Almería, 04120, Almería, Spain
and
Instituto Carlos I de Fisica Teórica y Computacional, Universidad de Granada, Spain
e-mail: andrei@ual.es

Summary. This is a short introduction to the theory of the logarithmic potential in the complex plane. The central ideas are the concepts of energy and equilibrium. We prove some classical results characterizing the equilibrium distribution and discuss the extension of these notions to more general settings when an external field or constraints on the distribution are present. The tools provided by potential theory have a profound impact on different branches of analysis. We illustrate these applications with two examples from approximation theory and complex dynamics.

* Supported, in part, by a research grant from the Ministry of Sciences and Technology (MCYT) of Spain (MTM2005-08648-C02-01), by Junta de Andalucía, Grupo de Investigación FQM229, by INTAS Research Network on Constructive Complex Approximation (INTAS 03-51-6637), and by NATO Collaborative Linkage Grant PST.CLG.979738.

1 Background

1.1 Introduction

This is a very brief introduction to the logarithmic potential in the complex plane, and to the very core of the theory: equilibrium problems. The ambitious goal is to cover some relevant aspects in a 10 hours mini-course.

There are different possible starting points for the development of this theory, such as:

- Partial differential equations;
- Complex analysis: harmonic and subharmonic functions, exceptional sets;
- Variational methods;
- Approximation theory, analytic properties of polynomials and rational functions,

and more. These points of view are obviously related and in a certain sense, equivalent, but bias the aspects you consider more or less relevant. At any rate, this will not affect the exposition here, since this introduction will touch only the core of the theory. Only some of the results will be proved. For a more detailed account we recommend the following literature (the list is far from being exhaustive):

- The books [6], [13] are classical and contain all the facts about the standard theory of logarithmic potential.
- The book [10] is a modern exposition of the same theory; highly recommended!
- The book [11] has an appendix containing the main facts from the theory, beside many applications in approximation.
- There is a chapter in [8] with a concise survey of the potential theory in $\mathbb{C}$. It is also a nice introduction to the applications in modern approximation theory.
- The book [12] is the first full account of the potential theory with an external field; also highly recommended!

1.2 Background or What You Should Bring to Class

- **Borel Measures.** A positive measure on $\mathbb{C}$ is called a *Borel measure* if it is defined and finite on every compact subset of $\mathbb{C}$. The *support* of a measure μ, denoted by $\mathrm{supp}(\mu)$, is the complement of the largest open set of measure zero. If $K \subset \mathbb{C}$ is a compact subset of the complex plane, then by $\mathcal{M}(K)$ we denote the set of all finite (positive) Borel measures μ with $\mathrm{supp}(\mu) \subset K$. The *Dirac delta* δ_a supported at $a \in \mathbb{C}$ is the unit measure defined by

$$\delta_a(K) = \begin{cases} 1, & \text{if } a \in K, \\ 0, & \text{otherwise} \end{cases} \qquad \text{for any Borel set } K \subset \mathbb{C}.$$

If f is a continuous function in $\mathbb{C}$,

$$\int f(z)\, d\delta_a(z) = f(a)\,.$$

- **Weak-* Topology.** In what follows $\mathcal{C}^m(K)$ denotes the set of all continuous functions with at least m continuous derivatives on a compact set $K \subset \mathbb{C}$ and $\mathcal{C}(K) = \mathcal{C}^0(K)$. A sequence $\{\mu_n\} \subset \mathcal{M}(K)$ converges to a measure $\mu \in \mathcal{M}(K)$ in a *weak-* sense* (denoted by $\mu_n \xrightarrow{*} \mu$) if

$$\lim_n \int f(z)\, d\mu_n(z) = \int f(z)\, d\mu(z)\,, \quad \forall f \in \mathcal{C}(K)\,.$$

According to the Riesz representation theorem, $\mathcal{M}(K)$ with the weak-* topology is the dual space of $\mathcal{C}(K)$. We will use the fact that every weakly bounded set in $\mathcal{M}(K)$ is compact, known also as **Helly's selection theorem** (cf. [12, p. 3]): *every sequence $\{\mu_n\} \subset K$ with a uniformly bounded mass ($\sup_n \mu_n(K) < +\infty$) has a weak-* convergent subsequence.*
- **Harmonic Functions.** A real valued function $u(z)$, $z = x + iy$, defined in a domain $D \subset \mathbb{C}$ is called harmonic in D if $u \in \mathcal{C}^2(D)$ and

$$\Delta u(z) \stackrel{\text{def}}{=} u_{xx}(z) + u_{yy}(z) = 0\,, \quad z \in D\,. \tag{1.1}$$

If D is simply connected, this is equivalent to the existence of a holomorphic function f in D such that $u = \mathrm{Re}\, f$.
One of the features of harmonic functions is the **mean-value property** (cf. [12, p. 7]): *if u is harmonic in an open disk $|z-a| < r$ and continuous on $|z-a| \le r$, then*

$$u(a) = \frac{1}{2\pi} \int_0^{2\pi} u(a + re^{i\theta})\, d\theta\,. \tag{1.2}$$

One of the consequences is the **maximum** (or minimum) **principle** (cf. [12, p. 8]): *if u is harmonic in a domain D and attains there either its minimum or its maximum value, then u is constant in D.*

- **Green's Formula.** Given a domain D, we denote by ∂D its boundary. If ∂D is a $\mathcal{C}^1$ curve and $u, v \in \mathcal{C}^2(\overline{D})$, then

$$\iint_D (v\Delta u - u\Delta v)\, dxdy = -\oint_{\partial D} \left(v\,\frac{\partial u}{\partial n} - u\,\frac{\partial v}{\partial n} \right) ds\,,$$

where $\partial/\partial n$ is differentiation in the direction of the *inner* normal of D. Let us highlight some straightforward consequences:

– If u is harmonic in D, then

$$\iint_D u\Delta v\, dxdy = \oint_{\partial D} \left(v\,\frac{\partial u}{\partial n} - u\,\frac{\partial v}{\partial n} \right) ds\,. \tag{1.3}$$

– If both u and v are harmonic in D, this formula takes the form

$$\oint_{\partial D} v\,\frac{\partial u}{\partial n}\, ds = \oint_{\partial D} u\,\frac{\partial v}{\partial n}\, ds\,. \tag{1.4}$$

– If the support of v is a compact set in D, then

$$\iint_D v\Delta u\, dxdy = \iint_D u\Delta v\, dxdy\,. \tag{1.5}$$

- **Sokhotsky-Plemelj's Theorem.** Suppose that γ is an oriented analytic curve or arc in $\mathbb{C}$. The orientation induces a $+$side (on the left) and a $-$side on γ. We denote also by

$$\gamma^o = \gamma \setminus \{\text{points of self intersection and end points}\}\,.$$

If Ω is a domain, $\gamma \subset \Omega$, then for f analytic in $\Omega \setminus \gamma$ we define

$$f_+(t) = \lim_{\substack{z\to t\\ z \text{ on the } +\text{side}}} f(z)\,, \qquad f_-(t) = \lim_{\substack{z\to t\\ z \text{ on the } -\text{side}}} f(z)\,.$$

Let v be a function defined and Hölder continuous on γ. Then the Cauchy-type integral

$$f(z) \stackrel{\text{def}}{=} \frac{1}{2\pi i}\oint_\gamma \frac{v(t)}{t-z}\, dt\,, \quad z \in \mathbb{C}\setminus\gamma\,,$$

defines an analytic function in $\mathbb{C}\setminus\gamma$, and the following relation holds:

$$f_+(t) - f_-(t) = v(t)\,, \quad t \in \gamma^o. \tag{1.6}$$

2 Logarithmic Potentials: Definition and Properties

2.1 Superharmonic Functions

A function $u : K \to \mathbb{R} \cup \{+\infty\}$ is called *lower semi-continuous* (l.s.c.) at a point $z \in \mathbb{C}$ if

$$\varliminf_{t \to z,\, t \in K} u(t) \geq u(z)\,,$$

(in other words, for any $\{t_n\} \subset K$ such that $t_n \to z$ and $\lim_n u(t_n)$ exists, we have $\lim_n u(t_n) \geq u(z)$). Furthermore, u is l.s.c. in a set $K \subset \mathbb{C}$ if it is l.s.c. at every $z \in K$. It is possible to prove that equivalently u is l.s.c. in $\mathbb{C}$ if and only if the set

$$\{z \in \mathbb{C} : u(z) > \alpha\}$$

is open for every $\alpha \in \mathbb{R}$.

Exercise 1. Prove the equivalence above.

Example 2.1. The function

$$f(z) = \begin{cases} 0, & \text{if } |z| \leq 1, \\ 1, & \text{if } |z| > 1, \end{cases}$$

is lower semi-continuous.

Exercise 2. A function u is upper semi-continuous if u is lower semi-continuous. Prove that

upper semi-continuous + lower semi-continuous = continuous

Finally, the third equivalent definition is the following (cf. [12, p. 1]): *u is lower semi-continuous on a compact set K if and only if it is the pointwise limit of an increasing sequence of continuous functions.*

Definition 2.1. *A function $u : D \to \mathbb{R} \cup \{+\infty\}$ on a domain D is called* superharmonic *on D if $u \not\equiv +\infty$, is lower semi-continuous on D, and for every open disk $|z-a| < r$ contained in D,*

$$u(a) \geq \frac{1}{2\pi} \int_0^{2\pi} u(a + re^{i\theta})\, d\theta \tag{2.1}$$

(compare with (1.2)).

A function u is called subharmonic *if $-u$ is superharmonic.*

Exercise 3. Prove that if f is a holomorphic function in a domain D, then $-\log|f|$ is superharmonic (and harmonic where $f \neq 0$).

There are other equivalent definitions of a superharmonic function. For instance, we can replace (2.1) by the existence of a "harmonic minorant" (harmonic function v, bounded from above by u on the boundary and preserving the same property in the whole domain). Another important characterization is valid for smooth functions $u \in \mathcal{C}^2(D)$: u is superharmonic in D if and only if

$$\Delta u(z) \leq 0\,, \quad z \in D \tag{2.2}$$

(compare with (1.1)). We will discuss this property again soon.

Superharmonic functions inherit part of the properties of the harmonic functions. For instance:

Theorem 2.1 (Minimum Principle). *Let D be a bounded domain and v a superharmonic function in D such that*

$$\varliminf_{t\to z,\, t\in D} v(t) \geq M\,,$$

for every $z \in \partial D$. Then $v(z) > M$ for $z \in D$ unless v is constant.

In other words, a non-constant superharmonic function cannot achieve its minimum inside a domain.

2.2 Definition of the Logarithmic Potential

Superharmonic functions are closely related to logarithmic potentials of positive measures. Let μ be a finite positive Borel measure of compact support. Its *logarithmic potential* is defined by

$$V^{\mu}(z) = \int \log \frac{1}{|z-t|}\, d\mu(t)\,. \tag{2.3}$$

Remark 2.1. This definition makes sense also for signed measures or charges on $\mathbb{C}$, represented in general as a difference of two positive measures.

Theorem 2.2. *The potential $V^{\mu}(z)$ of a positive measure is superharmonic in $\mathbb{C}$, harmonic in $\mathbb{C} \setminus \operatorname{supp}(\mu)$, and*

$$V^{\mu}(z) = \mu(\mathbb{C}) \log \frac{1}{|z|} + \mathcal{O}\left(\frac{1}{z}\right) \quad \textit{as } z \to \infty\,.$$

The proof, which consists of a verification of the properties defining a superharmonic and a harmonic function, can be found in [12, pp. 21–22]. For the last part, observe that for $z \neq 0$,

$$V^{\mu}(z) = \mu(\mathbb{C}) \log \frac{1}{|z|} + \int \log \frac{1}{|1-t/z|}\, d\mu(t)\,.$$

Example 2.2. If $a \in \mathbb{C}$ then

$$V^{\delta_a}(z) = \log \frac{1}{|z-a|}\,.$$

Analogously, if $a_1, \ldots, a_n \in \mathbb{C}$ and

$$\mu = \sum_{k=1}^{n} \delta_{a_k},$$

then

$$V^\mu(z) = -\log|p(z)|, \qquad p(z) = (z-a_1)(z-a_2)\dots(z-a_n). \tag{2.4}$$

This example is a key to the tight connection of logarithmic potentials with polynomials.

Exercise 4. Prove that for each $r > 0$,

$$\frac{1}{2\pi}\int \log\frac{1}{|z-re^{i\theta}|}\,d\theta = \begin{cases} \log 1/r, & \text{if } |z| \le r, \\ \log 1/|z|, & \text{if } |z| > r. \end{cases}$$

Observe that the logarithmic potential of the unit Lebesgue measure on a circle $|z| = r$ is constant in the disc $|z| \le r$.

By Theorem 2.2, we see that every potential of a positive measure is a superharmonic function. Furthermore, obviously the sum of a potential with a harmonic function is still superharmonic. Surprisingly enough, these are the only superharmonic functions that exist. This is the content of the **Riesz decomposition theorem** (see [12, p. 100]): *if u is superharmonic in a domain D, then there exists a positive measure $\mu \in \mathcal{M}(D)$ such that in every subdomain D' ($\overline{D'} \subset D$),*

$$u(z) = g(z) + V^\mu(z),$$

where g is a harmonic function (depending on D').

2.3 Some Principles for Potentials

We will mention also two useful properties satisfied by logarithmic potentials of positive measures.

Theorem 2.3 (Continuity Principle). *Let μ be a finite Borel measure on $\mathbb{C}$ with compact support $K \stackrel{\text{def}}{=} \operatorname{supp}(\mu)$. If $z_0 \in K$, then*

$$\varlimsup_{z\to z_0} V^\mu(z) = \varlimsup_{z\to z_0,\, z\in K} V^\mu(z).$$

Furthermore,

$$\lim_{z\to z_0,\, z\in K} V^\mu(z) = V^\mu(z_0) \quad \Rightarrow \quad \lim_{z\to z_0} V^\mu(z) = V^\mu(z_0).$$

In other words, $V^\mu\big|_K$ is continuous at $z = z_0$ if and only if V^μ is continuous at this point.

Proof. See Theorem 3.1.3, p. 54 in [10] or Theorem II.3.5 in [12]. □

Theorem 2.4 (Maximum Principle). *Let μ be a finite Borel measure on $\mathbb{C}$ with compact support $K \stackrel{\text{def}}{=} \operatorname{supp}(\mu)$. Then*

$$V^\mu(z) \le M \text{ for } z \in K \quad \Rightarrow \quad V^\mu(z) \le M \text{ for } z \in \mathbb{C}.$$

Proof. See Theorem 3.1.4, p. 55 in [10] or Corollary II.3.3 in [12]. □

2.4 Recovering a Measure from its Potential

Let D be a domain in $\mathbb{C}$. In (2.2) we have seen that if u is superharmonic (in particular, if it is a potential of a positive measure), then $\Delta u \le 0$. Using Green's theorem we can find a more precise statement, for which we need to generalize the concept of the laplacian.

Let $\mathcal{C}_0^\infty$ be the set of functions $v : D \to \mathbb{R}$ from $\mathcal{C}^\infty$ whose support is a compact set in D. We have seen in (1.5) that

$$\iint_D v \Delta u \, dxdy = \iint_D u \Delta v \, dxdy .$$

This formula is valid under the assumption that $u \in \mathcal{C}^2$. But if u is a general superharmonic function, it is still locally integrable ([10, Theorem 2.5.1]), and we may use the identity above as the definition of the *generalized laplacian*[2] Δu of u; namely, as long as f verifies

$$\iint_D v f \, dxdy = \iint_D u \Delta v \, dxdy, \qquad \forall v \in \mathcal{C}_0^\infty ,$$

we say that $f = \Delta u$. This definition is consistent. Moreover, we have the following remarkable theorem:

Theorem 2.5. *Let μ be a finite positive Borel measure on $\mathbb{C}$ with compact support. Then*

$$\Delta V^\mu = -2\pi\mu.$$

Proof. We just need to prove that for any $v \in \mathcal{C}_0^\infty$,

$$\int_{\mathbb{C}} V^\mu \Delta v \, dxdy = -\int_{\mathbb{C}} 2\pi v \, d\mu .$$

But

$$\begin{aligned} \int_{\mathbb{C}} V^\mu \Delta v \, dxdy &= \int_{\mathbb{C}} \left(\int_{\mathbb{C}} \log \frac{1}{|z-t|} \, d\mu(t) \right) \Delta v \, dxdy \\ &= -\int_{\mathbb{C}} \left(\int_{\mathbb{C}} \log |z-t| \, \Delta v \, dxdy \right) d\mu(t) \end{aligned}$$

[2] More formally, the generalized laplacian is the Radon measure Δu on D.

(for a justification of the use of Fubini's theorem here, see e.g., [10, p. 74]).

Now, using formula (1.3), we have with $u(z) = \log|z-t|$,

$$\int_{\mathbb{C}} \log|z-t|\, \Delta v\, dxdy = \lim_{\varepsilon\to 0} \int_{|z-t|>\varepsilon} \log|z-t|\, \Delta v\, dxdy$$
$$= \lim_{\varepsilon\to 0} \oint_{|z-t|=\varepsilon} \left(v\, \frac{\partial u}{\partial n} - u\, \frac{\partial v}{\partial n} \right) ds\,.$$

Changing the variables to polar, $z = t + re^{i\theta}$, so that on the circle $|z-t| = r$ we have $ds = |dz| = r$ and $\partial/\partial n = -d/dr$, we get

$$\int_{\mathbb{C}} \log|z-t|\, \Delta v\, dxdy = \lim_{\varepsilon\to 0} \int_0^{2\pi} \left(v(t+re^{i\theta}) - r\log r\, \frac{\partial v}{\partial r}(t+re^{i\theta}) \right)\Big|_{r=\varepsilon} d\theta$$
$$= 2\pi v(t)\,,$$

and the statement follows. □

An immediate corollary of this theorem is that V^μ is harmonic in $\mathbb{C} \setminus \operatorname{supp}(\mu)$. Then, there exists an analytic (possibly, multivalued) function $\mathcal{V}$ in $\mathbb{C}\setminus\operatorname{supp}(\mu)$ such that $\operatorname{Re}\mathcal{V} = V^\mu$ in this domain. However, the derivative $\mathcal{V}'$ of $\mathcal{V}$ is *single-valued*, and given by

$$\mathcal{V}'(z) = \int \frac{1}{t-z}\, d\mu(t)\,.$$

By analogy with the standard integrals in complex analysis, this is called the *Cauchy transform* of the measure μ.

Assume now that the support $\operatorname{supp}(\mu)$ is a compact analytic curve or arc in $\mathbb{C}$. If μ is absolutely continuous, we may recover μ' using the Sokhotski-Plemelj formula (1.6):

$$\mu'(z) = \frac{1}{2\pi i}\left(\mathcal{V}'_+(z) - \mathcal{V}'_-(z)\right). \tag{2.5}$$

Example 2.3. By Exercise 4 the logarithmic potential of $d\nu = \frac{1}{2\pi}\, d\theta$ on $\{|z| = r\}$ satisfies

$$V^\nu(z) = \begin{cases} \log 1/r, & \text{if } |z| < r, \\ \log 1/|z|, & \text{if } |z| > r. \end{cases}$$

(observe that this information is sufficient to obtain that $\operatorname{supp}(\nu) \subset \{|z| = r\}$ using Theorem 2.5). We may define

$$\mathcal{V}(z) \stackrel{\text{def}}{=} \begin{cases} \log 1/r, & \text{if } |z| < r, \\ \log 1/z, & \text{if } |z| > r\,, \end{cases} \qquad \Rightarrow \qquad \mathcal{V}'(z) = \begin{cases} 0, & \text{if } |z| < r, \\ -1/z, & \text{if } |z| > r\,. \end{cases}$$

In consequence, with the counter clockwise orientation of the circle,

$$\nu'(z) = \frac{1}{2\pi i}\left(\mathcal{V}'_+(z) - \mathcal{V}'_-(z)\right) = \frac{1}{2\pi i z} \qquad \Rightarrow \qquad d\nu(z) = \frac{dz}{2\pi i z}\,.$$

With the parametrization $z = re^{i\theta}$, $dz = zid\theta$, so that $d\nu = \frac{1}{2\pi}\, d\theta$, recovering the normalized Lebesgue measure on $\{|z| = r\}$.

3 Energy and Equilibrium

3.1 Logarithmic Energy

Let us assume that $K \subset \mathbb{C}$ is a compact subset of the complex plane. The *logarithmic energy* $I(\mu)$ of a Borel measure $\mu \in \mathcal{M}(K)$ is defined as

$$I(\mu) \stackrel{\text{def}}{=} \iint \log \frac{1}{|z-t|} \, d\mu(z) d\mu(t) . \tag{3.1}$$

By Fubini's theorem, we can write also

$$I(\mu) = \int V^{\mu}(z) \, d\mu(z) . \tag{3.2}$$

This definition (that again can be extended to signed measures on $\mathbb{C}$) has the following physical sense: think of μ as being a charge distribution on $\mathbb{C}$. Then $V^{\mu}(z)$ represents the potential energy at z due to μ, and so the right hand side in (3.2) is just the (doubled) total energy of μ.

Obviously, $I(\cdot)$ is a functional on the space $\mathcal{M}(K)$ equipped with the weak-* topology. Unfortunately it is not continuous; however, it has several sufficiently nice properties.

Proposition 3.1. *If* $\mu \in \mathcal{M}(K)$ *then*

$$-\infty < I(\mu) \leq +\infty .$$

Proof. It follows from the representation (3.2). □

Proposition 3.2. *The functional* $I : \mathcal{M}(K) \to (-\infty, +\infty]$ *is lower semicontinuous. In other words, if a sequence* $\{\mu_n\} \subset \mathcal{M}(K)$ *converges to a measure* $\mu \in \mathcal{M}(K)$ *in a weak-* sense, then*

$$I(\mu) \leq \varliminf_{n} I(\mu_n) .$$

For the proof of this proposition we will use the technique of the truncated logarithmic kernel.

We define the *truncated logarithmic kernel* as

$$k^{(\eta)}(z) \stackrel{\text{def}}{=} \begin{cases} \log 1/|z|, & \text{if } |z| \geq \eta, \\ \log 1/\eta, & \text{if } |z| < \eta . \end{cases} \tag{3.3}$$

It has the following properties: for $\eta > 0$,

(P.1) $k^{(\eta)} \in C(\mathbb{C})$;
(P.2) For $z \in \mathbb{C}$,

$$k^{(\eta)}(z) \leq \log \frac{1}{|z|} ; \tag{3.4}$$

(P.3) For $z \in \mathbb{C}$,

$$k^{(\eta)}(z) \nearrow \log \frac{1}{|z|} \quad \text{as } \eta \downarrow 0 \,. \tag{3.5}$$

Proof (of Proposition 3.2). Let us use the truncated kernel (3.3) with $\eta > 0$. If $\mu_n \xrightarrow{*} \mu$, by (P.1),

$$\lim_n \iint k^{(\eta)}(z-t)\, d\mu_n(z) d\mu_n(t) = \iint k^{(\eta)}(z-t)\, d\mu(z) d\mu(t) \,.$$

Using property (P.2) we have that

$$\underline{\lim}_n \iint \log \frac{1}{|z-t|}\, d\mu_n(z) d\mu_n(t) \geq \iint k^{(\eta)}(z-t)\, d\mu(z) d\mu(t) \,.$$

Since the left hand side does not depend on η, we may take limits as $\eta \downarrow 0$. The statement of the proposition follows from property (P.3) and the monotone convergence theorem. □

Let us denote by $\mathcal{M}^*(K) \subset \mathcal{M}(K)$ the subset of all positive measures in $\mathcal{M}(K)$ with finite energy. Obviously, any measure concentrated at a single point (Dirac delta) has an infinite energy, so that $\mathcal{M}^*(K) \subsetneq \mathcal{M}(K)$. An immediate consequence of Proposition 3.2 is that $\mathcal{M}^*(K)$ is closed in the weak-* topology.

As it was mentioned, formula (3.1) makes sense for signed measures or charges on $\mathbb{C}$. The following lemma, whose proof is technical, can be found for instance in [12, Lemma 1.8, p. 29]:

Proposition 3.3. *Let $\mu, \nu \in \mathcal{M}^*(K)$ be such that $\mu(K) = \nu(K)$. Then*

$$I(\mu - \nu) \geq 0$$

and it is zero if and only if $\mu = \nu$.

A positive Borel measure $\mu \in \mathcal{M}(K)$ is called *unit* (or *probability*) *measure* on K if $\mu(K) = 1$. In what follows we denote by $\mathcal{M}_1(K) \subset \mathcal{M}(K)$ the set of unit measures on K, and let $\mathcal{M}_1^*(K) = \mathcal{M}_1(K) \cap \mathcal{M}^*(K)$. Obviously, $\mathcal{M}_1(K)$ and $\mathcal{M}_1^*(K)$ are convex subsets of $\mathcal{M}(K)$. The next result shows that I is a *convex functional* on $\mathcal{M}_1^*(K)$.

Proposition 3.4. *If $\mu, \nu \in \mathcal{M}_1^*(K)$, then*

$$I(\mu - \nu) = 2I(\mu) + 2I(\nu) - 4I\left(\frac{\mu+\nu}{2}\right), \tag{3.6}$$

and

$$I\left(\frac{\mu+\nu}{2}\right) \leq \frac{I(\mu) + I(\nu)}{2}, \tag{3.7}$$

Proof. Identity (3.6) is proved by direct computation. Furthermore, we may rewrite it as

$$I\left(\frac{\mu+\nu}{2}\right) = \frac{1}{2}\, I(\mu) + \frac{1}{2}\, I(\nu) - \frac{1}{4}\, I\left(\mu-\nu\right),$$

and (3.7) follows from Proposition 3.3. □

Exercise 5. Is $d(\mu,\nu) \stackrel{\text{def}}{=} \sqrt{I(\mu-\nu)}$ a distance in $\mathcal{M}_1^*(K)$?

3.2 Extremal Problem, Equilibrium Measure and Capacity

Let us consider the following value:

$$\rho = \inf_{\mu \in \mathcal{M}_1(K)} I(\mu)\,. \tag{3.8}$$

Obviously $\rho \in (-\infty, +\infty]$; if $\rho = +\infty$ we say that K is a *polar set.* For instance, $K = \{0\}$ is a polar set in $\mathbb{C}$. The constant ρ is called the *Robin constant* of K. Furthermore, the (logarithmic) *capacity* of the set K is defined as

$$\operatorname{cap}(K) = \begin{cases} e^{-\rho}, & \text{if } \rho < +\infty, \\ 0, & \text{if } \rho = +\infty. \end{cases}$$

For a non-closed set $U \subset \mathbb{C}$ we may define its capacity as

$$\operatorname{cap}(U) \stackrel{\text{def}}{=} \sup\{\operatorname{cap}(K) :\, K \subset U,\, K \text{ compact}\}\,.$$

This definition is consistent (i.e., does not depend on the compact subsets K exhausting U).

The capacity will play a role of a fine measure of the sets on the plane. We will see soon that it is not a measure in the sense of Lebesgue theory, but it is more natural from the point of view of analytic functions. If a certain property holds at every point of the set K except in its polar subset (or subset of capacity zero), we say that this property holds *"quasi everywhere"* (or *q.e.*).

How "small" is a set of zero capacity (or a polar set)?

Proposition 3.5. *Let μ be a Borel measure on $\mathbb{C}$ with compact support and finite energy. Then*

$$E \text{ is a Borel polar set} \quad \Rightarrow \quad \mu(E) = 0\,.$$

In particular, every Borel polar set has (plane) Lebesgue measure zero. In other words,

$$\textit{quasi everywhere} \quad \Rightarrow \quad \textit{almost everywhere}$$

with respect to plane Lebesgue measure.

This theorem shows that in a certain sense the capacity is a "finer" way of measuring a size of a set than the Borel measures. We will see soon that in general

$$\text{quasi everywhere} \quad \not\Leftarrow \quad \text{almost everywhere}$$

Proof. The key fact about Borel measures is that they are *regular* in the following sense: for every Borel set E, and for every $\varepsilon > 0$, there exist an open set U and a compact set F such that $F \subset E \subset F$, and $\mu(F \setminus U) < \varepsilon$. In other words, $\mu(E)$ can be approximated arbitrarily well by measures of compact subsets from inside, and open sets from outside. Any (finite) Borel measure is regular (see e.g. [10, Theorem A.2.2]).

In our setting assume that E is a Borel set such that $\mu(E) > 0$. To establish the first assertion it is sufficient to show that E is not polar. Regularity of μ implies immediately that we can find a compact subset $K \subset E$ such that $\mu(K) > 0$. To finish our proof it is sufficient to build a measure living on K with a finite energy. We achieve it by taking the restriction of μ to K, $\tilde{\mu} = \mu\big|_K$. Indeed, $\tilde{\mu}$ is not identically zero, and we can estimate its energy. Let us denote by d the diameter of $\operatorname{supp}(\mu)$; then we have

$$\begin{aligned}
I(\tilde{\mu}) &= \iint_{K\times K} \log \frac{1}{|z-t|}\, d\mu(z) d\mu(t) \\
&= \iint_{K\times K} \log \frac{d}{|z-t|}\, d\mu(z) d\mu(t) - \log(d) \iint_{K\times K} d\mu(z) d\mu(t) \\
&= \iint_{K\times K} \log \frac{d}{|z-t|}\, d\mu(z) d\mu(t) - \log(d) \mu(K)^2 .
\end{aligned}$$

Since for $z, t \in \operatorname{supp}(\mu)$ we have $|z-t| \le d$, the integrand above is non-negative on $\operatorname{supp}(\mu) \times \operatorname{supp}(\mu)$, and we can continue the identities with an inequality:

$$\begin{aligned}
I(\tilde{\mu}) &\le \iint_{\mathbb{C}\times\mathbb{C}} \log \frac{d}{|z-t|}\, d\mu(z) d\mu(t) - \log(d) \mu(K)^2 \\
&= \iint_{\mathbb{C}\times\mathbb{C}} \log \frac{1}{|z-t|}\, d\mu(z) d\mu(t) + \log(d) \mu(\mathbb{C})^2 - \log(d) \mu(K)^2 \\
&= I(\mu) + \log(d) \mu(\mathbb{C})^2 - \log(d) \mu(K)^2 < +\infty .
\end{aligned}$$

This proves the first part. In order to establish the assertion for the plane Lebesgue measure, dA, it is sufficient to prove that for any disk $D_r = \{z \in \mathbb{C} : |z| \le r\}$ and for $d\mu = dA\big|_{D_r}$ we have $I(\mu) < +\infty$. This is a consequence of the following bound:

Exercise 6. Let $D_r = \{z \in \mathbb{C} : |z| \le r\}$, $r > 0$, and $d\mu = dA\big|_{D_r}$. Then for $z \in D_r$,

$$V^{\mu}(z) \le 2\pi r^2 - \pi r^2 \log(r^2) .$$

Suggestion: use polar coordinates.

This concludes the proof. □

Corollary 3.1. *A countable union of Borel polar sets is polar. In particular, every countable subset of $\mathbb{C}$ is polar.*

Proof. Suppose that $E = \cup_n E_n$, and $\mathrm{cap}(E_n) = 0$. Let $\mu \in \mathcal{M}(E)$ be compactly supported. By Proposition 3.5, if $I(\mu) < +\infty$, then $\mu(E_n) = 0$ for each n. Using the σ-additivity of μ, we get that $\mu(E) = 0$. This shows that E is polar.

The second part is a direct consequence of the fact that $\{z\}$ is a polar set for every $z \in \mathbb{C}$. □

But not every polar set is countable, as the following example shows.

Example 3.1. Let $\mathbf{s} = (s_1, s_2, \dots)$ be a sequence of numbers such that $0 < s_n < 1$ for all $n \in \mathbb{N}$. We construct a (generalized) Cantor set as follows. Define $C(s_1) = [0,1] \setminus (a,b)$, where (a,b) is symmetric with respect to $1/2$ and $b - a = s_1$. Recursively, we obtain $C(s_1, s_2, \dots, s_n)$ from $C(s_1, s_2, \dots, s_{n-1})$ by removing from the middle of each interval of $C(s_1, s_2, \dots, s_{n-1})$ equal symmetric intervals, in such a way that

$$\frac{\text{total length of } C(s_1, s_2, \dots, s_n)}{\text{total length of } C(s_1, s_2, \dots, s_{n-1})} = 1 - s_n \,. \tag{3.9}$$

Then $\{C(s_1, s_2, \dots, s_n)\}$ is a decreasing sequence of compact subsets of $[0,1]$, and we may define

$$C(\mathbf{s}) \stackrel{\text{def}}{=} \bigcap_{n \geq 1} C(s_1, s_2, \dots, s_n) \,.$$

It is known that, independently of the sequence $\mathbf{s}$, this set is compact, perfect (coincides with the set of its accumulation points), and uncountable. For the capacity of this set we have the following estimate (see [10, Theorem 5.3.7]):

$$\mathrm{cap}\,(C(\mathbf{s})) \leq \frac{1}{2} \prod_{j=1}^{\infty} (1 - s_n)^{1/2^n} \,.$$

Thus, taking for instance $s_n = 1 - 2^{-2^n}$, we get $\mathrm{cap}\,(C(\mathbf{s})) = 0$, as promised.

If K is a non-polar set ($\mathrm{cap}(K) = 0$), it is natural to ask whether "inf" in (3.8) may be replaced by "min", and whether the minimizing measure is unique (for a polar set every measure from $\mathcal{M}_1(K)$ is minimizing). We will see that the answer is yes. Remember that in physics any equilibrium state is related to a minimum of the energy of the system. Following the physical analogy, we will call the minimizing measure the *equilibrium measure* of K, and the corresponding potential, the *equilibrium potential.* Furthermore, if we recall that the force in a potential field is given by the difference of potentials, it is natural to expect that the equilibrium potential will be constant in K. Again this is almost true, or better to say, true q.e. Let us summarize the main properties of the equilibrium measure.

Let K be a non-polar compact set. There exists a unique measure $\lambda_K \in \mathcal{M}_1(K)$, called the *equilibrium measure* (*Robin measure*) of K, such that

$$\rho = \inf_{\mu \in \mathcal{M}_1(K)} I(\mu) \left(\stackrel{\text{def}}{=} \log \frac{1}{\operatorname{cap}(K)} \right) = I(\lambda_K).$$

Furthermore, λ_K is the unique measure in $\mathcal{M}_1^*(K)$ such that

$$\int V^{\lambda_K}(z)\, d(\nu - \lambda_K)(z) \geq 0, \quad \text{for all } \nu \in \mathcal{M}_1^*(K). \tag{3.10}$$

Finally,

$$V^{\lambda_K}(z) \geq \rho \text{ quasi everywhere on } K, \tag{3.11}$$

and

$$V^{\lambda_K}(z) \leq \rho \text{ on } \operatorname{supp}(\lambda_K). \tag{3.12}$$

Consequently, (3.12) holds everywhere in $\mathbb{C}$, and

$$V^{\lambda_K}(z) = \rho \text{ quasi everywhere on } K. \tag{3.13}$$

The statement, contained in (3.11)–(3.13), is known as *Frostman's theorem.* In what follows, we will reserve the notation of λ_K for the equilibrium measure of the set K.

Let us prove the facts listed above in several steps.

Proof (of the existence). By definition of "inf", there exists a sequence $\{\mu_n\} \subset \mathcal{M}_1(K)$ such that

$$\lim_n I(\mu_n) = \rho = \log \frac{1}{\operatorname{cap}(K)}.$$

By weak compactness of the unit ball in $\mathcal{M}(K)$, there exist a measure $\mu \in \mathcal{M}_1(K)$ and a subsequence μ_{n_k} such that $\mu_{n_k} \xrightarrow{*} \mu$. By Proposition 3.2,

$$I(\mu) \leq \varliminf_n I(\mu_{n_k}) = \rho,$$

and by definition of the Robin constant, $I(\mu) = \rho$. □

Proof (of uniqueness). Assume that μ and ν are two (unit) equilibrium measures of K, so that

$$I(\mu) = I(\nu) = \rho.$$

Then by (3.7), $(\mu + \nu)/2$ is also an equilibrium measure of K. But by (3.6), $I(\mu - \nu) = 0$. Now it remains to use Proposition 3.3 to get that $\mu = \nu$. □

Proof (of (3.10)*).* Since K is non-polar, $I(\lambda_K) < +\infty$, and $\lambda_K \in \mathcal{M}_1^*(K)$. For $\varepsilon \in [0,1]$, $\mu, \nu \in \mathcal{M}_1^*(K)$,

$$I(\varepsilon\nu + (1-\varepsilon)\mu) - I(\mu) = 2\varepsilon \int V^{\mu}(z)\, d(\nu - \mu)(z) + \varepsilon^2 I(\nu - \mu), \tag{3.14}$$

which can be checked by a direct computation. Taking $\mu = \lambda_K$ we have the inequality $I(\varepsilon\nu + (1-\varepsilon)\mu) - I(\mu) \geq 0$, valid for $\varepsilon \downarrow 0$, and (3.10) follows. On the other hand, taking $\varepsilon = 1$ in (3.14) we get

$$I(\nu) - I(\mu) = 2\int V^{\mu}(z)\,d(\nu-\mu)(z) + I(\nu-\mu) \geq 2\int V^{\mu}(z)\,d(\nu-\mu)(z)\,,$$

where we have used Proposition 3.3. Hence, (3.10) implies that $I(\nu) - I(\mu) \geq 0$, so that by uniqueness, $\mu = \lambda_K$. □

Proof (of Frostman's theorem). Assume that (3.11) is not true, so that there exists a subset $e \subset K$ such that $\mathrm{cap}(e) > 0$ and

$$V^{\lambda_K}(z) < \rho\,, \quad z \in e\,.$$

Since $\mathrm{cap}(e) > 0$, by the definition of capacity, there exists a positive Borel measure $\nu \in \mathcal{M}_1^*(e) \subset \mathcal{M}_1^*(K)$. Integrating the inequality above with respect to ν, we get

$$\int V^{\lambda_K}(z)\,d\nu(z) < \rho = \int V^{\lambda_K}(z)\,d\lambda_K(z) \quad \Leftrightarrow \quad \int V^{\lambda_K}(z)\,d(\nu-\lambda_K)(z) < 0\,,$$

which contradicts (3.10).

In order to prove (3.12) we need to use the fact that a potential is lower semi-continuous, according to which the set $\{V^{\lambda_K}(z) > \rho\}$ is relatively open. Hence, if we assume that $e \stackrel{\text{def}}{=} \{V^{\lambda_K}(z) > \rho\} \cap \mathrm{supp}(\lambda_K) \neq \emptyset$, then $\lambda_K(e) > 0$. But then, taking into account (3.10), we have

$$\begin{aligned} I(\lambda_K) &= \int V^{\lambda_K}(z)\,d\lambda_K(z) \\ &= \int_e V^{\lambda_K}(z)\,d\lambda_K(z) + \int_{\mathrm{supp}(\lambda_K)\setminus e} V^{\lambda_K}(z)\,d\lambda_K(z) > \rho\,, \end{aligned}$$

which contradicts the definition of λ_K. Hence, $V^{\lambda_K}(z) \leq \rho$ on $\mathrm{supp}(\lambda_K)$. Finally, the validity of this inequality in $\mathbb{C}$ follows from the maximum principle for potentials, Theorem 2.4. □

Remark 3.1. In a certain sense, (3.11)–(3.13) are just Euler-Lagrange equations corresponding to the minimization of the functional $I(\cdot)$.

Proposition 3.6. *Let $\nu \in \mathcal{M}_1(\mathbb{C})$ have a compact support, and K be a compact subset of $\mathbb{C}$. Then*

$$\inf_{z\in K} V^{\nu}(z) \leq I(\lambda_K)\,.$$

Furthermore, if $\nu \in \mathcal{M}_1(K)$, then also

$$\sup_{z\in K} V^{\nu}(z) \geq I(\lambda_K)\,.$$

Proof. In what follows, we denote by $\overline{\mathbb{C}} \stackrel{\text{def}}{=} \mathbb{C} \cup \{\infty\}$ the extended complex plane. The first inequality is a consequence of the minimum principle for superharmonic functions. Indeed, $V^\nu - V^{\lambda_K}$ is superharmonic in $\overline{\mathbb{C}} \setminus K$, and

$$(V^\nu - V^{\lambda_K})(z) \geq V^\nu(z) - I(\lambda_K) \text{ on } K\,.$$

Hence, if $V^\nu(z) > I(\mu)$ on K, then according to the minimum principle (Theorem 2.1), $(V^\nu - V^{\lambda_K})(z) > 0$ should hold in $\overline{\mathbb{C}} \setminus K$. But this is a contradiction with the fact that $(V^\nu - V^{\lambda_K})(\infty) = 0$.

The second inequality is a direct consequence of (3.10). □

Remark 3.2. We see that for $\nu \in \mathcal{M}_1(K)$,

$$\inf_{z \in K} V^\nu(z) \leq I(\lambda_K) \leq \sup_{z \in K} V^\nu(z)\,,$$

i.e., in general, the potential "oscillates" around the Robin constant.

Let us give two more properties of the equilibrium measure, whose proof is beyond the framework of this mini-course.

Proposition 3.7 (Characterization of λ_K). *Properties* (3.11) *and* (3.13) *uniquely characterize the equilibrium measure λ_K in the following sense: if $\sigma \in \mathcal{M}_1^*(K)$ and V^σ coincides with a constant c quasi-everywhere on* $\operatorname{supp}(\sigma)$ *and is $\geq c$ quasi-everywhere on K, then $\sigma = \lambda_K$ and $c = \rho$.*

The proof uses another important property of the logarithmic potential, called principle of domination, and can be found in [12, Section I.3].

Example 3.2. In Exercise 4 we have found that for $r > 0$, the potential of the unit Lebesgue measure supported on the circle $\{z \in \mathbb{C} : |z| = r\}$ is constant on the whole closed disk $\{z \in \mathbb{C} : |z| \leq r\}$. By the previous proposition, this means that we are dealing with the equilibrium measure of the closed disk $\{z \in \mathbb{C} : |z| \leq r\}$. By the way, it is also the equilibrium measure of the circle $\{z \in \mathbb{C} : |z| = r\}$.

Observe that in the previous example the extremal measure is concentrated on the boundary of the disk, something that is in accordance with our intuition of the electrostatic equilibrium. This is not fortuitous. In order to describe the support of λ_K in a general situation we need the following definition.

Definition 3.1. *Let K be a connected and non-polar compact set, and let Ω_∞ be the* outer domain, *that is, the connected component of its complement containing ∞. Then $\operatorname{Pc}(K) \stackrel{\text{def}}{=} \mathbb{C} \setminus \Omega_\infty$ is the* polynomial convex hull[3] *of K. The boundary $\partial \operatorname{Pc}(K) = \partial \Omega_\infty$ of the polynomial convex hull of K is a subset of ∂K, and is called the* outer boundary *of K.*

[3] In other words, $\operatorname{Pc}(K)$ is the union of K and its "holes".

Proposition 3.8 (Support of λ_K). *The equilibrium measure λ_K is supported on the outer boundary $\partial\,\mathrm{Pc}(K)$ of K. In particular,*

$$\lambda_K = \lambda_{\partial\Omega_\infty} \qquad \text{and} \qquad \operatorname{cap}(K) = \operatorname{cap}(\partial\Omega_\infty)\,.$$

Proof. A proof can be found in [12, p. 52]. □

3.3 Link with Conformal Mapping and Green Function

Given a compact set K with $\operatorname{cap}(K) > 0$, the *Green function* of K with pole at infinity is the unique function $g_K(z;\infty)$ defined in the outer domain Ω_∞ with the following properties[4]:

a) g_K is nonnegative and harmonic in Ω_∞,
b) $g_K(z,\infty) - \log|z|$ is harmonic in a neighborhood of $z = \infty$, and
c) $\lim_{z\to z_0,\, z\in\Omega} g_K(z,\infty) = 0$ for q.e. $z_0 \in \partial\Omega_\infty$.

One can extend g_Ω to the whole $\mathbb{C}$ by stipulation

$$g_K(z,\infty) \stackrel{\text{def}}{=} \begin{cases} 0, & \text{if } z \in \text{interior of } \mathrm{Pc}(K), \\ \overline{\lim\limits_{t\to z,\, t\in\Omega}}\, g_K(t,\infty), & \text{if } z \in \partial\,\mathrm{Pc}(K). \end{cases}$$

Then $g_K(\cdot,\infty)$ becomes a non-negative subharmonic function (see for instance [12, Section II.4] for more details).

There is a tight connection between the Green function and the equilibrium potential. Indeed, let us consider

$$V^{\lambda_K}(z) + g_K(z,\infty)\,,$$

defined in the whole plane $\mathbb{C}$. This function is harmonic in $\mathbb{C}\setminus K$, and taking into account that λ_K is a unit measure, by Theorem 2.2, it is harmonic also in a neighborhood of $z=\infty$. Moreover, $V^{\lambda_K}(z)+g_K(z,\infty) = \log(1/\operatorname{cap}(K))$ q.e. on K. Hence, we may apply the (generalized) minimum principle for harmonic functions to obtain the following important relations:

$$V^{\lambda_K}(z) + g_K(z,\infty) = \log\frac{1}{\operatorname{cap}(K)}\,, \qquad z \in \mathbb{C}\,. \tag{3.15}$$

In particular,

$$\lim_{z\to\infty}\left(g(z,\infty) - \log|z|\right) = \log\frac{1}{\operatorname{cap}(K)}\,. \tag{3.16}$$

Exercise 7. Prove that formula (3.16) follows from (3.15).

[4] In the literature, this function is sometimes called the Green function of Ω_∞ and denoted by $g_{\Omega_\infty}(z,\infty)$. The Green's function is a fundamental solution of the Laplace operator, but this is another story.

Corollary 3.2 (Monotonicity of the Capacity). *If $K_1 \subset K_2$ are of positive capacity, then* $\mathrm{cap}(K_1) \le \mathrm{cap}(K_2)$.

Proof. Consider

$$v(z) \stackrel{\text{def}}{=} g_{K_1}(z,\infty) - g_{K_2}(z,\infty)\,.$$

It is a harmonic and non-negative function in $\overline{\mathbb{C}} \setminus K_2$ (why?), and by the maximum principle, $v(\infty) \ge 0$. Using (3.16), the statement follows. □

Hence, if we know how to compute the Green function, we have a way to find the capacity. In turn, the Green function can be obtained from the Riemann conformal mapping. Remember that since $\mathrm{Pc}(K)$ is simply connected, according to the Riemann theorem, there exists a conformal mapping Φ of Ω_∞ onto the exterior of the unit disc $|w| > 1$ such that in a neighborhood of infinity,

$$w = \Phi(z) = c_1 z + c_0 + \frac{c_{-1}}{z} + \dots\,, \qquad c_1 > 0\,. \tag{3.17}$$

Then it is immediate to check that

$$g_K(z,\infty) = \begin{cases} \log|\Phi(z)|\,, & z \in \Omega_\infty, \\ 0, & \text{otherwise.} \end{cases}$$

In particular, in the representation (3.17) above,

$$c_1 = \frac{1}{\mathrm{cap}(K)}\,.$$

Exercise 8. Prove that if $K = \{z \in \mathbb{C} : |z| = r\}$, then $\mathrm{cap}(K) = r$.

Remark 3.3. Combining Proposition 3.8 and Corollary 3.2 we can show that the capacity lacks the main property of a measure: additivity. Indeed, if we take $D = \{|z| < 1\}$, then

$$\mathrm{cap}(D) = \sup\{\mathrm{cap}(\{|z| \le r\}) : r < 1\} = 1,$$

so that if $C = \{|z| = 1\}$, then $C \cap D = \emptyset$, but $\mathrm{cap}(D) = \mathrm{cap}(C) = \mathrm{cap}(C \cup D) = 1$.

Example 3.3. Let $K = [-1,1]$. It is well known that

$$\Phi(z) = z + \sqrt{z^2 - 1}$$

(the inverse of the Zhoukowski function), where the branch of the square root in $\mathbb{C} \setminus [-1,1]$ is fixed by the condition $\sqrt{x^2-1} > 0$ for $x > 1$. In particular,

$$\Phi(z) = 2z + \mathcal{O}(1), \quad z \to \infty,$$

and by (3.16),

$$\mathrm{cap}([-1,1]) = \frac{1}{2} = \frac{\text{length of } [-1,1]}{4}\,.$$

Exercise 9. Prove, using the idea above, that

$$\operatorname{cap}([a,b]) = \frac{b-a}{4}.$$

From (3.15) it follows that for the interval $K=[-1,1]$,

$$\begin{aligned} V^{\lambda_K}(z) = -g_K(z,\infty) + \log 2 &= -\log\left|\frac{z+\sqrt{z^2-1}}{2}\right| \\ &= -\operatorname{Re}\left(\log\frac{z+\sqrt{z^2-1}}{2}\right), \qquad z\notin K. \end{aligned}$$

Let us try to compute the equilibrium measure of the interval $K=[-1,1]$ using the connection with the Cauchy type integrals. Using the notation of (2.5), we have

$$\mathcal{V}(z) = -\log\frac{z+\sqrt{z^2-1}}{2} \quad\Rightarrow\quad \mathcal{V}'(z) = -\frac{1}{\sqrt{z^2-1}}.$$

As expected, $\mathcal{V}'$ is holomorphic in $\mathbb{C}\setminus K$; observe also that since $V^{\lambda_K}\sim -\log|z|$ as $z\to\infty$, we have that

$$\mathcal{V}'(z)\sim -\frac{1}{z}, \quad z\to\infty, \qquad\Leftrightarrow\qquad \sqrt{z^2-1}>0 \text{ for } z>1, \tag{3.18}$$

which selects the branch of $\mathcal{V}'$ in $\mathbb{C}\setminus K$.

Assume that λ_K is absolutely continuous with respect to the Lebesgue measure, and denote by $v(x)=\lambda_K'(x)$. Then, with the standard orientation of the interval, by (2.5),

$$\begin{aligned} v(x) &= \frac{1}{2\pi i}\left(\mathcal{V}'_+(x)-\mathcal{V}'_-(x)\right) \\ &= -\frac{1}{2\pi i}\left(\frac{1}{{}_+\sqrt{x^2-1}} - \frac{1}{{}_-\sqrt{x^2-1}}\right) \\ &= -\frac{1}{\pi i}\frac{1}{{}_+\sqrt{x^2-1}}. \end{aligned}$$

In order to find out what are the boundary values on the upper side of the interval we use (3.18) and obtain on $(-1,1)$,

$$\frac{1}{{}_+\sqrt{x^2-1}} = \frac{1}{{}_+\sqrt{x-1}}\,\frac{1}{{}_+\sqrt{x+1}} \in -i\,\mathbb{R}_+,$$

and finally,

$$v(x) = \frac{1}{\pi}\frac{1}{\sqrt{1-x^2}} > 0, \quad x\in(-1,1).$$

There are explicit formulas for the capacity of many "standard" sets, as well as some bounds in terms of a measure (Lebesgue or other) of the set. An interested reader is referred to [10, Chapter 5].

Let us finally make an observation that will be used later. Obviously, capacity is invariant by translation of K (why?). Furthermore, if $K \to cK$, $c > 0$, is a homotopy, then an analogous transformation, $\mu(z) \to c\,\mu(z/c)$, builds the equilibrium measure of cK, which allows to find $\text{cap}(cK)$. In fact, let us prove that the capacity of a set (but, obviously, not its measure) can be computed relatively simply after a polynomial transformation of the plane. This result will be used later, in Section 4.2.

Theorem 3.1. *Let K be a compact set, and let*

$$q(z) = \sum_{j=0}^{d} a_j z^j, \quad \textit{with } a_d \neq 0. \tag{3.19}$$

Then

$$\text{cap}(q^{-1}(K)) = \left(\frac{\text{cap}(K)}{|a_d|} \right)^{1/d}. \tag{3.20}$$

Here we use the standard notation of $q^{-1}(K)$ for the pre-image of K by q, that is, $q^{-1}(K) \stackrel{\text{def}}{=} \{z \in \mathbb{C} : q(z) \in K\}$.

Proof. Denote $\widetilde{K} \stackrel{\text{def}}{=} q^{-1}(K)$. Let Ω_∞ be the outer domain for K, and $\widetilde{\Omega}_\infty$ the outer domain for $\widetilde{K}$. It is not difficult to check that $q(\widetilde{\Omega}_\infty) = \Omega_\infty$ and $q(\partial\widetilde{\Omega}_\infty) = \partial\Omega_\infty$. The key fact of the proof is the following formula relating the Green's functions of K and $\widetilde{K}$:

$$g_K(q(z), \infty) = d\, g_{\widetilde{K}}(z, \infty). \tag{3.21}$$

Exercise 10. Prove this formula, using the uniqueness of the Green function (or equivalently, the maximum principle for harmonic functions).

We can use this in order to relate the capacities; indeed, by (3.21),

$$g_K(q(z), \infty) - d \log|z| = d(g_{\widetilde{K}}(z, \infty) - \log|z|)\,.$$

We may rewrite the left hand side as

$$(g_K(q(z), \infty) - \log|q(z)|) + \log|q(z)/z^d|\,.$$

Taking limits on both sides as $z \to \infty$, and using that $q(z) \to \infty$ also, by (3.16) we get the identity

$$\log \frac{1}{\text{cap}(K)} + \log|a_d| = d \log \frac{1}{\text{cap}(\widetilde{K})}$$

and (3.20) follows. □

Exercise 11. If $0 \le a < b$ and $K = [-b,-a] \cup [a,b]$, prove that $\mathrm{cap}(K) = \sqrt{b^2 - a^2}/2$. (Suggestion: use $q(z) = z^2$).

Exercise 12. Show that for $q(z) = z^2 - 2$, $q([-2,2]) = [-2,2]$. Give an alternative proof of the fact that $\mathrm{cap}([-2,2]) = 1$.

Exercise 13. Find the logarithmic capacity of a lemniscate $\{z \in \mathbb{C} : |q(z)| = r\}$, $r > 0$, for q given in (3.19).

3.4 Equilibrium in an External Field

The classical theory of equilibrium exposed so far has several generalizations, motivated by a growing number of applications of the potential theory techniques in mathematics. We will mention only one of them, as an illustration.

We have seen that the extremal measure λ_K, defined in Section 3.2, models to a certain extent the equilibrium state of a unit positive charge on a conductor K. A natural question is how all this changes if in addition there is an external electrostatic field present. Beside of "pure" interest, these problems found (starting from the mid 80's of the 20th century) many applications in approximation theory.

Let us describe briefly the mathematical model corresponding to the existence of an external field. We restrict our attention to the case of a bounded set K. The best reference for this part is [12].

Definition 3.2. *A weight function w on K is* admissible *if it is upper semi-continuous on K and the set $K_0 \stackrel{\text{def}}{=} \{z \in K : w(z) > 0\}$ has positive capacity*[5]. *The corresponding* external field Q *is defined by*

$$w(z) \stackrel{\text{def}}{=} \exp(-Q(z)) .$$

For a Borel measure $\mu \in \mathcal{M}(K)$ we define the weighted energy $I_w(\mu)$ *by*

$$I_w(\mu) \stackrel{\text{def}}{=} I(\mu) + 2 \int Q \, d\mu ,$$

where $I(\mu)$ is the (standard) energy of μ.

The equilibrium measure in the external field is obtained again by minimizing the energy functional $I_w(\cdot)$ in the class of probability measures $\mathcal{M}_1(K)$ supported on K. Let us summarize the main properties of the resulting measure (cf. Section 3.2):

Theorem 3.2. (see [12, Section I.1]). *Let w be an admissible weight on a compact set K, and let*

$$\tau_w \stackrel{\text{def}}{=} \inf_{\mu \in \mathcal{M}_1(K)} I_w(\mu) .$$

Then the following properties hold:

[5] In particular, $\mathrm{cap}(K) > 0$.

a) τ_w *is finite.*
b) There exists a unique measure $\lambda_{K,w} \in \mathcal{M}_1^*(K)$ *such that*

$$I_w(\lambda_{K,w}) = \tau_w \,.$$

c) The support $\mathcal{S}_{K,w} \stackrel{\text{def}}{=} \operatorname{supp}(\lambda_{K,w})$ *is compact, contained in* K_0 *(defined above) and* $\operatorname{cap}(\mathcal{S}_{K,w}) > 0$.
d) Setting

$$\rho_w \stackrel{\text{def}}{=} \tau_w - \int Q \, d\lambda_{K,w} \,,$$

the inequality

$$V^{\lambda_{K,w}}(z) + Q(z) \geq \rho_w$$

holds quasi everywhere on K*, and the inequality*

$$V^{\lambda_{K,w}}(z) + Q(z) \leq \rho_w$$

holds for all $z \in \mathcal{S}_{K,w}$.

The constant ρ_w is called the *modified Robin constant* of K in the external field w.

The result (and the proof) is very similar to what we have seen for $Q \equiv 0$. The main feature here is the fact that the support $\mathcal{S}_{K,w}$ is *not known in advance*, and due to the external field Q, is not necessarily in the outer boundary of K anymore. This is, in a certain sense, a problem with free boundaries, widely studied for PDE's, and it is known to be difficult.

In consequence, the support $\mathcal{S}_{K,w}$ is usually the key for the determination of the equilibrium measure $\lambda_{K,w}$ and other constants. There are several techniques that allow us to locate the support; perhaps the better known is the *F-functional* introduced by Mhaskar and Saff in [7]. It is one of the most powerful tools to find the polynomial convex hull $\operatorname{Pc}(\mathcal{S}_{K,w})$ of the support (cf., Definition 3.1).

Definition 3.3. *Let* K *be a compact set,* $\operatorname{cap}(K) > 0$*. The* F-functional *is defined on* K *as follows: for every compact subset* $U \subset K$ *of positive capacity,*

$$F(U) \stackrel{\text{def}}{=} \log \operatorname{cap}(U) - \int Q \, d\lambda_U \,,$$

where λ_U *is the (Robin) equilibrium measure of* U.

Since $\operatorname{cap}(U) = \operatorname{cap}(\operatorname{Pc}(U))$ and $\operatorname{supp}(\lambda_U) \subset \operatorname{Pc}(U)$, we have that $F(U) = F(\operatorname{Pc}(U))$.

Theorem 3.3. *(see Theorem IV.1.5 in [12]). Let* w *be an admissible weight on a compact set* K*. Then the following hold.*

a) For every compact subset $U \subset K$ *of positive capacity,* $F(U) \leq F(\mathcal{S}_{K,w})$.
b) $F(\mathcal{S}_{K,w}) = F(\operatorname{Pc}(\mathcal{S}_{K,w})) = -\rho_w$*, where* ρ_w *is the modified Robin constant.*

c) $\mathrm{Pc}(\mathcal{S}_{K,w})$ *is the smallest polynomially convex set maximizing the F-functional.*

In particular, if K has an empty interior and connected complement (e.g. $K \subset \mathbb{R}$), then $\mathcal{S}_{K,w}$ is the smallest compact set of positive capacity maximizing the F-functional. In many applications a simple convexity argument shows that $\mathcal{S}_{K,w}$ is convex; hence, if $K \subset \mathbb{R}$, we have that $\mathcal{S}_{K,w} = [a, b]$ and the end points a, b are the main parameters of the problems.

Proof (Scheme of the proof of Theorem 3.3). Let $U \subset K$ be of positive capacity. Then by d) in Theorem 3.2,

$$V^{\lambda_{K,w}}(z) \geq -Q(z) + \rho_w$$

holds q.e. on U. Since $\lambda_U \in \mathcal{M}^*(K)$, by Proposition 3.5, page 90, this inequality holds also λ_U-almost everywhere. Integrating it with respect to λ_U we get

$$\iint \log \frac{1}{|z-t|} \, d\lambda_{K,w}(t) d\lambda_U(z) \geq -\int Q \, d\lambda_U(z) + \rho_w \, .$$

Changing the order of integration and recalling that $V^{\lambda_U} \leq \log(1/\operatorname{cap}(U))$ (cf. (3.12)), we get

$$\log \frac{1}{\operatorname{cap}(U)} = \int \log \frac{1}{\operatorname{cap}(U)} \, d\lambda_{K,w}(t) \geq -\int Q \, d\lambda_U(z) + \rho_w$$

so that

$$F(U) \leq -\rho_w \, .$$

Furthermore, if $U = \mathcal{S}_{K,w}$, we can replace the inequalities above by "=", and b) follows. Finally, if for some U, $F(U) = -\rho_w$, then by the arguments above we must have

$$V^{\lambda_U}(z) = \log(1/\operatorname{cap}(U)) \quad \lambda_{K,w}\text{-almost everywhere,}$$

which will imply that $\mathrm{Pc}(\mathcal{S}_{K,w}) = \mathrm{Pc}(U)$. □

Let us show with one example how Theorem 3.3 can be used to determine the extremal support in some cases.

Assume that $K = [-1, 1]$ and w is an admissible weight on K such that Q is convex on $(-1, 1)$.

Exercise 14. Prove that for every positive measure $\mu \in \mathcal{M}(K)$, with $K = [-1, 1]$, the potential V^μ is strictly convex in $K \setminus \operatorname{supp}(\mu)$.

With this result in hand, we can reason as follows: by assumption, $V^{\lambda_{K,w}} + Q$ is a strictly convex function in $K \setminus \mathcal{S}_{K,w}$. Assume that there are two points $p_1, p_2 \in \mathcal{S}_{K,w}$, $-1 < p_1 < p_2 < 1$, such that between them there is no further point of $\mathcal{S}_{K,w}$. Since $V^{\lambda_{K,w}} + Q$ is strictly convex in (p_1, p_2), and at p_1, p_2 the inequality

$$V^{\lambda_{K,w}}(z) + Q(z) \le \rho_w$$

holds, we get that on (p_1, p_2) we have

$$V^{\lambda_{K,w}}(z) + Q(z) < \rho_w \,,$$

which contradicts the inequality on K. Thus, $\mathcal{S}_{K,w}$ *must be an interval,* say $\mathcal{S}_{K,w} = [a,b]$. Now the determination of the extremal support becomes a simple maximization problem of a function on $K \times K$:

$$F(\mathcal{S}_{K,w}) = F([a,b]) = \max_{\alpha,\beta \in K} F([\alpha,\beta])\,.$$

Since

$$\operatorname{cap}([\alpha,\beta]) = \frac{\beta-\alpha}{4}\,, \quad d\lambda_{[\alpha,\beta]}(x) = \frac{1}{\pi}\,\frac{1}{\sqrt{(\beta-x)(x-\alpha)}}\,,$$

we obtain that

$$F([\alpha,\beta]) = \log\left(\frac{\beta-\alpha}{4}\right) - \frac{1}{\pi}\int_\alpha^\beta \frac{Q(x)}{\sqrt{(\beta-x)(x-\alpha)}}\,dx\,.$$

Exercise 15. Prove that if in the situation above, $-1 < a < b < 1$, then the end points of the support $\mathcal{S}_{K,w}$ satisfy the following integral equations:

$$\frac{1}{\pi}\int_\alpha^\beta Q'(x)\sqrt{\frac{x-a}{b-x}}\,dx = 1\,,$$
$$\frac{1}{\pi}\int_\alpha^\beta Q'(x)\sqrt{\frac{b-x}{x-a}}\,dx = -1\,.$$

Remember finally that formulas (3.15) and (3.16) turned out to be very useful for the computation of the equilibrium measure and equilibrium potential without external field. In order to find an analogue of these relations we must recall the definition of the *Dirichlet problem.* In a simplified form (and suited for our needs) it is like follows: assume we have a compact set K on a complex plane, $\operatorname{cap}(K) > 0$, and let f be l.s.c. function defined on $\partial\,\mathrm{Pc}(K) = \partial\Omega_\infty$. We need to find a function u, harmonic in Ω_∞, and such that

$$\lim_{z\to t,\, z\in\Omega_\infty} u(z) = f(t)\,, \quad \text{for } t \text{ q.e. in } \partial\Omega_\infty\,. \tag{3.22}$$

A solution to this problem is given by the so-called *Perron function*

$$H_K(f;z) = \inf\left\{ v \text{ superharmonic in } \Omega_\infty : \ \varliminf_{z\to t,\, z\in\Omega_\infty} v(z) \ge f(t), t\in\partial\Omega_\infty \right\}.$$

In many cases it is important to know when we can drop "quasi-everywhere" in the formulation (3.22). An important tool for that is the extended Green function, introduced in Section 3.3:

$$g_K(z,\infty) \stackrel{\text{def}}{=} \begin{cases} 0, & \text{if } z \in \text{interior of } \operatorname{Pc}(K), \\ \overline{\lim}_{t\to z,\, t\in\Omega} g_K(t,\infty), & \text{if } z \in \partial \operatorname{Pc}(K) = \partial\Omega_\infty. \end{cases}$$

Definition 3.4. *A point $t \in \partial\Omega_\infty$ is called* regular *(with respect to the Dirichlet problem) if the extended Green function $g_K(z,\infty)$, defined above, is continuous at $z = t$. A compact K is called* regular *if every $t \in \partial\Omega_\infty$ is regular.*

Alternatively, we can say that $t \in \partial\Omega_\infty$ is regular if $g_K(t,\infty) = 0$.

Characterizations of the regular points of a compact is a very important topic in potential theory (see e.g. [10, Section 4.2]). For our purpose it will be sufficient to keep in mind that if $\partial\Omega_\infty$ is a Jordan curve, K is regular.

The following result can be found, for instance, in [10, Section 4.1]:

Proposition 3.9. *If K is a regular compact, then for every function f continuous on $\partial\Omega_\infty$ there exists a unique solution u of the Dirichlet problem; it is given by $u = H_K(f;\cdot)$ and satisfies*

$$\lim_{z\to t,\, z\in\Omega_\infty} u(z) = f(t)\,, \quad \forall t \in \partial\Omega_\infty\,.$$

It is well known that if K is the unit circle, the solution to the Dirichlet problem can be given in terms of the *Poisson integral*,

$$u(z) = \frac{1}{2\pi}\int_0^{2\pi} \operatorname{Re}\left(\frac{ze^{i\theta}+1}{ze^{i\theta}-1}\right) f\left(e^{i\theta}\right)\, d\theta\,.$$

For a general K we can use the Riemann mapping $\Phi(z)$ of Ω_∞ onto $|w| > 1$, taking into account that the Dirichlet problem is invariant under the conformal mapping:

$$H_K(f;z) = H_{\Phi(K)}(\Phi \circ f; \Phi(z))\,.$$

Example 3.4. Let $K = [-1,b] \subset \mathbb{R}$, $c > b$. Then

$$\Phi(z) = \frac{2z+1-b+2\sqrt{(z+1)(z-b)}}{b+1}\,, \tag{3.23}$$

with $\sqrt{(x+1)(x-b)} > 0$ for $x > b$. Let $f(x) = \log(c-x) = \log|c-x|$, $x \in K$. Then

$$H_K(f;z) = \log|z-c| - \log\left|\frac{\Phi(z)-\Phi(c)}{\Phi(z)\Phi(c)-1}\right| - g_K(z,\infty)\,. \tag{3.24}$$

Exercise 16. Prove that function $H_K(f;z)$ defined in (3.24) solves the Dirichlet problem stated above.

Now we are ready to obtain an analogue of formulas (3.15) and (3.16) for the equilibrium potential with an external field:

Theorem 3.4. *Let* $\operatorname{cap}(K) > 0$ *and* w *be an admissible weight. Then*

$$V^{\lambda_{K,w}}(z) + H_{\mathcal{S}_{K,w}}(Q; z) + g_{\mathcal{S}_{K,w}}(z, \infty) = \rho_w\,, \quad z \in \Omega_\infty\,. \tag{3.25}$$

The proof is analogous to that of formula (3.15).

Finally, let us consider the following example:

Example 3.5. Assume that there is an external field acting on $K = [-1, 1]$, created by a unit positive charge at $c > 1$. We want to find the corresponding equilibrium measure.

In our notation,

$$Q(x) = V^{\delta_c}(x) = -\log|c - x|\,, \quad w(x) = |c - x|\,.$$

Observe that Q is convex, so that $\mathcal{S}_{K,w}$ is an interval. Physical intuition tells us that $-1 \in \mathcal{S}_{K,w}$, so that $\mathcal{S}_{K,w} = [-1, b]$, $-1 < b \leq 1$. The F-functional now has the form

$$F([-1, b]) = \log\frac{b+1}{4} - \frac{1}{\pi}\int_{-1}^{b} \log\frac{1}{c-x}\frac{1}{\sqrt{(x+1)(b-x)}}\,dx\,.$$

But equation

$$\begin{aligned}\frac{d}{db}F([-1, b]) &= \frac{1}{b+1}\left(1 - \frac{1}{\pi}\int_{-1}^{b}\frac{1}{c-x}\sqrt{\frac{x+1}{b-x}}\,dx\right)\\ &= \frac{1}{b+1}\left(2 - \sqrt{\frac{c+1}{c-b}}\right) = 0\end{aligned}$$

has a unique solution in $(-1, 1)$,

$$b_0 = \frac{3c-1}{4}\,,$$

as long as $1 < c < 5/3$; hence, $S_{K,w} = [-1, b]$, with $b = \min\{b_0, 1\}$. In other words,

$$S_{K,w} = \begin{cases} [-1, 1], & \text{if } c \geq 5/3,\\ [-1, b_0], & \text{if } 1 < c < 5/3\,. \end{cases} \tag{3.26}$$

Once we have found the support of $\lambda_{K,w}$, we can find its potential using formulas (3.24)–(3.25): for $z \notin [-1, 1]$,

$$\begin{aligned}V^{\lambda_{K,w}}(z) &= -H_K(Q; z) - g_K(z, \infty) + \rho_w\\ &= \log|z - c| - \log\left|\frac{\Phi(z) - \Phi(c)}{\Phi(z)\Phi(c) - 1}\right| - 2g_K(z, \infty) + \rho_w\,,\end{aligned}$$

with Φ given in (3.23). Taking into account that $g_K(z, \infty) = \log|\Phi(z)|$, we see that

$$V^{\lambda_{K,w}}(z) = \operatorname{Re}\left(\log(z-c) - \log\left(\frac{\Phi(z)-\Phi(c)}{\Phi(z)\Phi(c)-1}\right) - 2\log(\Phi(z))\right) + \text{const}\,.$$

Hence, we may take

$$\mathcal{V}(z) = \log(z-c) - \log\left(\frac{\Phi(z)-\Phi(c)}{\Phi(z)\Phi(c)-1}\right) - 2\log(\Phi(z))$$

and use formula (2.5):

$$\lambda'_{K,w}(z) = \frac{1}{2\pi i}\left(\mathcal{V}'_+(z) - \mathcal{V}'_-(z)\right)\,.$$

Exercise 17. Complete the computation and prove that the equilibrium measure $\lambda_{K,w}$ for $K=[-1,1]$ and $w(t)=|c-t|$, $c>1$, is

$$d\lambda_{K,w}(x) = \begin{cases} \dfrac{1}{\pi}\dfrac{1}{\sqrt{1-x^2}}\left[2+\dfrac{\sqrt{c^2-1}}{x-c}\right], & x\in(-1,1),\quad 5/3<c\,, \\ \dfrac{2}{\pi(c-x)}\sqrt{\dfrac{b-x}{x+1}}, & x\in(-1,b),\quad 1<c\le 5/3\,, \end{cases}$$

where $b=(3c-1)/4$.

Exercise 18. Assume now that $w(t)=|c-t|$ with $c\in(-1,1)$; analyze for which values of c the support of $\lambda_{K,w}$ is still connected (an interval).

3.5 Other Equilibrium Problems. Equilibrium with Constraints

Let us try to summarize what we have seen so far, from a point of view of the electrostatics.

- A positive charge on a plane conductor K (assuming that the interaction is inversely proportional to the distance) reaches an equilibrium, modelled by the equilibrium measure λ_K, characterized by the minimum energy ρ in this class. Furthermore, on the conductor the charge is free to move, so necessarily the potential (whose gradient is the force) is constant at least where the charge "lives" (support).
- If an external field Q acts on the conductor K, the situation changes, the equilibrium is modelled by the equilibrium measure $\lambda_{K,w}$. K being a conductor, the charge is still free to move, hence the potential *plus the external field* (whose gradient is again the force) is constant where the charge "lives" (support).

It is not too difficult to imagine further generalizations of these models:

- We have a set of conductors K_j ("plates"), each equipped with a corresponding positive (or even signed) charge. The potential on a certain K_j now receives the contribution from the charge on the plate (which is free to move within the conductor) plus the action of the remaining charges (bounded to their corresponding plates). The equilibrium now (which by physical considerations must exist) is modelled by the *vector equilibrium*, introduced by Gonchar and Rakhmanov in [5]. It is applied in the analysis of so-called Hermite-Padé approximants, which are a valuable tool in analytic number theory (see e.g. [8]).
- In each of the situations described above the interaction between charges may obey a different law. From the analytic point of view it would mean that we must use a different (not necessarily logarithmic) kernel in the definition of the potential (2.3). For instance, we may use the Green function as a kernel, which leads to the *Green* potential, connected with the vector equilibrium and applied in rational approximation. If the interaction decays with the square of the distance, we get the *Coulomb* or gravitational potential, etcetera.

Finally, we may consider the case when K is not totally conducting. Indeed, if the conductivity of K is limited from above by a given charge distribution σ then in order to find the equilibrium we must minimize the total energy in the class of unit measures μ satisfying an additional constraint: $\mu \leq \sigma$, in any reasonable sense (for instance, that $\sigma - \mu$ is a positive measure on K). This is the *equilibrium with constraints*, studied for instance in [4]. We may guess now that there are qualitatively new features in this situation. For instance, where the constraint is not saturated ($\sigma - \mu > 0$), everything should resemble the unconstrained case; in particular, the potential (plus eventually the external field) are constant on the support of the equilibrium measure. However, when the constraint *is* saturated ($\sigma = \mu$), we might have less measure than necessary to level the potential to its equilibrium constant.

In order to be more specific, we say that, given an admissible weight w on K (in a sense of Definition 3.2), a measure $\sigma \in \mathcal{M}^*(K)$ is an *admissible constraint* if $\operatorname{supp}(\sigma) = K$ and $\sigma(K_0) > 1$, where $K_0 \stackrel{\text{def}}{=} \{z \in K : w(z) > 0\}$.

Next we define the class of measures

$$\mathcal{M}_1^\sigma(K) \stackrel{\text{def}}{=} \{\mu \in \mathcal{M}_1(K) : 0 \leq \mu \leq \sigma\},$$

where by $\mu \leq \sigma$ we mean that $\sigma - \mu$ is a positive Borel measure. The analogue of the Robin's constant for this problem is given by

$$\tau_w^\sigma \stackrel{\text{def}}{=} \inf_{\mu \in \mathcal{M}_1^\sigma(K)} I_w(\mu).$$

As before, we can prove that there *exists a unique measure* $\lambda_{K,w}^\sigma$ such that $\tau_w^\sigma = I_w(\lambda_{K,w}^\sigma)$; it is called a σ- constrained extremal measure for the weight w. Furthermore, there exists a constant ρ_w^σ such that

$$V^{\lambda^{\sigma}_{K,w}}(z) + Q(z) \geq \rho^{\sigma}_{w} \quad \text{holds } (\sigma - \lambda^{\sigma}_{K,w})\text{- a.e}\,,$$
$$V^{\lambda^{\sigma}_{K,w}}(z) + Q(z) \leq \rho^{\sigma}_{w} \quad \text{holds for all } \operatorname{supp}(\lambda^{\sigma}_{K,w})\,.$$

In other words, K splits into two regions:

- *Non-saturated regions*, where $\lambda^{\sigma}_{K,w} < \sigma$. Here the measure behaves like an unconstrained equilibrium measure, and in particular,

$$V^{\lambda^{\sigma}_{K,w}}(z) + Q(z) \begin{cases} = \rho^{\sigma}_{w} & \text{q.e. on } \operatorname{supp}(\lambda^{\sigma}_{K,w})\,, \\ \geq \rho^{\sigma}_{w} & \text{in } K \setminus \operatorname{supp}(\lambda^{\sigma}_{K,w})\,. \end{cases}$$

- *Saturated regions*, subsets of $\operatorname{supp}(\lambda^{\sigma}_{K,w})$ where $\lambda^{\sigma}_{K,w} = \sigma$. Here we have not enough measure to level up the potential and we expect

$$V^{\lambda^{\sigma}_{K,w}}(z) + Q(z) < \rho^{\sigma}_{w}\,.$$

Notice that if $\sigma \geq \lambda^{\sigma}_{K,w}$, then the extremal measure will not depend on the constraint and will coincide with $\lambda_{K,w}$. In this sense, the constrained energy problem generalizes the weighted energy problem.

4 Two Applications

4.1 Analytic Properties of Polynomials

Let us start with a very classical lemma about the growth of a polynomial of a given degree. It is a mere illustration of the power of the properties of superharmonic functions.

Lemma 4.1 (Bernstein-Walsh). *Let K be a compact in $\mathbb{C}$, and p_n a polynomial of degree $n \in \mathbb{N}$. If*

$$\|p_n\|_K \stackrel{\text{def}}{=} \max_{z\in K} |p_n(z)| \leq M\,,$$

then

$$|p_n(z)| \leq M\, e^{n g_K(z,\infty)}, \qquad z \in \mathbb{C}\,.$$

Taking into account (3.15), we can rewrite the last inequality as

$$|p_n(z)| \leq \frac{M}{(\operatorname{cap}(K))^n}\, e^{-nV^{\lambda_K}(z)}\,.$$

Proof. It is a direct consequence of the minimum principle for superharmonic functions, Theorem 2.1. Indeed, by Exercise 3 we know that $-\log|p_n|$ is superharmonic in $\mathbb{C}$. Let us consider

$$v(z) \stackrel{\text{def}}{=} -\log|p_n(z)| + n\, g_K(z,\infty) \tag{4.1}$$

in the outer domain Ω_∞. It is superharmonic in Ω_∞ and harmonic in a neighborhood of $z = \infty$. Furthermore, it satisfies

$$\underline{\lim}_{t \to z,\, t \in \Omega_\infty} v(t) \geq -\log M\,.$$

By Theorem 2.1, $v(z) > -\log M$ in $z \in \Omega_\infty$. The rest follows from the maximum principle for holomorphic functions. □

Corollary 4.1. *If $P_n(z) = z^n + \ldots$ is a monic polynomial of degree n and K is a non-polar compact set, then*

$$\|P_n\|_K \geq (\operatorname{cap}(K))^n\,.$$

Proof. Let us again use the function v defined in (4.1). With the assumptions of the corollary, v is harmonic in a neighborhood of $z = \infty$, and by (3.16), $v(\infty) = -n \log \operatorname{cap}(K)$. By the minimum principle,

$$\min_{z \in \partial\Omega_\infty} v(z) \leq n \log \frac{1}{\operatorname{cap}(K)}\,,$$

which concludes the proof. □

In relation with this corollary, it is natural to study the monic polynomials that achieve the least possible norm. This is a natural generalization of the Chebyshev polynomials. Assume that $\operatorname{cap}(K) > 0$. The nth *Chebyshev polynomial* T_n is the monic polynomial $T_n(z) = z^n + \ldots$ such that

$$\|T_n\|_K = \inf_{P_n(z) = z^n + \ldots} \|P_n\|_K\,. \tag{4.2}$$

What are the features of the zero distribution of the Chebyshev polynomials? We can guess that the answer will be given in terms of the equilibrium measure. Indeed, using the normalized zero counting measures of T_n and P_n, μ and ν respectively, we can rewrite the defining property (4.2) as

$$\min_{z \in K} V^\mu(z) \geq \min_{z \in K} V^\nu(z)\,.$$

In other words, the minimum of the potential is maximized in the class of measures. This reminds us of the property of the equilibrium measure λ_K given in Proposition 3.6. All these arguments can be made rigorous:

Theorem 4.1. *Let K be a non-polar compact set with empty interior. Denote by ν_n the normalized zero counting measure of the Chebyshev polynomial T_n. Then*

$$\nu_n \xrightarrow{*} \lambda_K \quad \textit{as} \quad n \to \infty\,.$$

See the proof in [12, Section III.4].

In many applications it is important to consider sequences of *weighted* polynomials, of the form

$$p_n(z)w^n(z),$$

where w is an admissible weight on a compact K in the sense of the previous chapter. The supremum norm behavior of these weighted polynomials is roughly as follows: the supremum norm "lives" on a subset of K that is independent of n and p_n, and the behavior outside this subset is typically exponentially small. More precisely, we have the following generalization of the Bernstein-Walsh lemma:

Theorem 4.2. *Let $w : K \to [0, +\infty)$ be an admissible weight. If p_n is a polynomial of degree n and*

$$|w^n(z)p_n(z)| \leq M \quad \text{for } z \in \mathcal{S}_{K,w}. \tag{4.3}$$

Then

$$|p_n(z)| \leq M \exp\left(n(-V^{\lambda_{K,w}}(z) + \rho_w)\right) \quad \text{for } z \in \mathbb{C}. \tag{4.4}$$

Furthermore, the inequality in (4.3) *holds on K.*

Exercise 19. Prove the theorem above. Use the proof of the Bernstein-Walsh lemma as an inspiration.

Observe that from (4.4) it follows that

$$|w^n(z)p_n(z)| \leq M \exp\left(n(-V^{\lambda_{K,w}}(z) - Q(z) + \rho_w)\right) \quad \text{for } z \in K.$$

Remember that by part d) of Theorem 3.2 we have that the inequality

$$V^{\lambda_{K,w}}(z) + Q(z) \geq \rho_w$$

holds quasi everywhere on K, and hence

$$|w^n(z)p_n(z)| \leq M \quad \text{q.e. on } K.$$

In other words, if we neglect the (possible) polar set where this is not true, the (essential) norm is reached on the set $\mathcal{S}_{K,w}$. We can make it formal defining the *essential norm* as

$$\|f\|_K^* \stackrel{\text{def}}{=} \inf\{L > 0 : |f(z)| \leq L \text{ q.e. on } K\}.$$

Then a direct consequence of the previous theorem is the following fact:

$$\deg(p_n) \leq n \quad \Rightarrow \quad \|w^n p_n\|_K^* = \|w^n p_n\|_{\mathcal{S}_{K,w}}^*$$

which can be phrased as *"the essential norm of the weighted polynomial lives on the support $\mathcal{S}_{K,w}$"*.

Exercise 20. What is the lower bound for $\inf_{P_n(z)=z^n+\dots} \|w^n P_n\|_K$? Can you guess what is the limit zero distribution of the monic polynomials T_n of the minimum weighted norm on K (*weighted Chebyshev polynomials*)?

Example 4.1. Assume that $K = [-1, 1]$ and we want to study the asymptotic behavior of the sequence of monic polynomials P_n of degree $2n$, with half (n) of the zeros fixed at $z = 1$, and providing a minimal uniform norm on K within its class. This is a generalization of Chebyshev polynomials, known also as *incomplete polynomials.* It can be written as a weighted extremal problem: writing $P_n(z) = (z-1)^n p_n(z)$, $p_n(z) = z^n + \dots$, we need to find

$$\inf_{p_n(z)=z^n+\dots} \|(z-1)^n p_n(z)\|_{[-1,1]} .$$

If we recall now the results obtained in the Example 3.5, we can say several things about the analytic properties of these polynomials. For instance, from (3.26) with $c = 1$ it follows that always

$$\|(z-1)^n p_n(z)\|_{[-1,1]} = \|(z-1)^n p_n(z)\|_{[-1,1/2]} .$$

Let me finally mention very briefly the role of the equilibrium in the asymptotics of orthogonal polynomials. Remember that the link between both theories is given by formula (2.4): for $p(z) = (z-a_1)(z-a_2)\dots(z-a_n)$,

$$V^{\nu(p)}(z) = -\log|p(z)|, \qquad \nu(p) = \sum_{k=1}^{n} \delta_{a_k} = \text{zero counting measure of } p .$$

Now we can establish the following "Ansatz":

- Orthogonality of a monic polynomial P_n with respect to a weight w on K $\Rightarrow$ extremality of the L^2 norm of P_n $\Rightarrow$ asymptotic extremality of the L^∞ norm of P_n (under certain conditions on w) $\Rightarrow$ max-min property of $|P_n|$ $\Rightarrow$ max-min property of $V^{\nu(P_n)}$ $\Rightarrow$ $\nu(P_n) \sim \lambda_K$.
- Orthogonality of a monic polynomial P_n with respect to a *varying* weight w^n on K $\Rightarrow$ extremality of the *weighted* L^2 norm of P_n $\Rightarrow$ asymptotic extremality of the *weighted* L^∞ norm of P_n (under certain conditions on the measure) $\Rightarrow$ max-min property of $|w^n P_n|$ $\Rightarrow$ max-min property of $V^{\nu(P_n)} + Q \Rightarrow \nu(P_n) \sim \lambda_{K,w}$.
- Orthogonality of a monic polynomial P_n with respect to a *discrete* varying "weight" $w^n(x)d\mu(x)$ on K $\Rightarrow$ extremality of the *weighted* L^2 norm of P_n $\Rightarrow$ asymptotic extremality of the *weighted* L^∞ norm of P_n (under certain conditions on the measure).
  But attention! We have an additional restriction: between two consecutive mass points of the orthogonality measure we can find at most one zero of P_n.
  So, our problem $\Rightarrow$ max-min property of $|w^n P_n|$ *assuming* $\nu(P_n) \lesssim w^n(x)d\mu$ $\Rightarrow$ max-min property of $V^{\nu(P_n)} + Q$ assuming $\nu(P_n) \lesssim w^n(x)d\mu$ $\Rightarrow \nu(P_n) \sim \lambda^\sigma_{K,w}$, if $w^n(x)d\mu \sim \sigma$.

Almost all these "guesses" can be converted into rigorous theorems. Many of these results have been obtained in the eighties. However, very recently the equilibrium potential started to play a fundamental role also for the pointwise or "strong" asymptotics of orthogonal polynomials, obtained in the framework of the matrix Riemann-Hilbert approach to their analysis; for instance, it is used in the construction of the so-called g-functions.

It was Rakhmanov in his pioneering work [9] who realized that the equilibrium with constraints is decisive in the study of the polynomials of discrete orthogonality, used for instance in numerical analysis (see e.g., [2] or [1]). A more detailed discussion of this topic will be carried out in the lecture notes of Beckermann [3] in this volume.

4.2 Complex Dynamics

Let $q(z) = \sum_{j=0}^{d} a_j z^j$ be a polynomial of degree d. We are interested in the *dynamics* of q, namely the behavior of the iterates $q^{\circ n} \stackrel{\text{def}}{=} q \circ q \circ \cdots \circ q$ as $n \to \infty$.

Definition 4.1. *The* attracting basin *of* ∞ *of* q *is the set*

$$\Omega_\infty \stackrel{\text{def}}{=} \{z \in \overline{\mathbb{C}} : q^{\circ n}(z) \to \infty \text{ as } n \to \infty\}$$

(where $\overline{\mathbb{C}}$ *is the extended complex plane).*

Example 4.2. For $q(z) = z$, we have $\Omega_\infty = \{z \in \overline{\mathbb{C}} : |z| > 1\}$.

To consider the non-trivial cases, we assume $d \geq 2$. Then it is easy to see that there exists an $R > 0$ such that $|q(z)| \geq 2|z|$ for $|z| > R$. Put $U_0 = \{z \in \mathbb{C} : |z| > R\}$. If $z \in U_0$, then $|q(z)| \geq 2|z| > R$, so that $q(z) \in U_0$, which means that $z \in q^{-1}(U_0)$. In other words, $U_0 \subset q^{-1}(U_0)$. For $n \in \mathbb{N}$, define

$$U_n = q^{-1}(U_{n-1}) .$$

Lemma 4.2. *With the definition above,*

$$U_0 \subset U_1 \subset \cdots \subset U_n , \tag{4.5}$$

and

$$\Omega_\infty = \bigcup_n U_n . \tag{4.6}$$

Proof. (4.5) is verified by direct computation. For (4.6) the main observation is that $z \in \Omega_\infty$ if and only if $q^{\circ n}(z) \in U_0$ for a certain $n \in \mathbb{N}$. □

Corollary 4.2. Ω_∞ *is a connected open set that is* completely invariant*:* $q^{-1}(\Omega_\infty) = \Omega_\infty$.

Definition 4.2. *The* Julia set *of q is*

$$J \stackrel{\text{def}}{=} \partial\Omega_\infty .$$

In other words, Ω_∞ is the outer domain of the Julia set, explaining the reason for keeping the notation. The set $\mathbb{C} \setminus \Omega_\infty$ is sometimes called the *"filled-in" Julia set.*

The Julia set J can also be defined (informally) as the set of points for which nearby points do not exhibit similar behavior under repeated iterations of q. It is a compact subset of the plane, being also completely invariant: $q^{-1}(J) = J$. The Julia set plays a fundamental role in the dynamics of q.

Exercise 21. Prove that J is completely invariant.

Let us consider some examples for the classical situation, studied by G. Julia himself, $q(z) = z^2 - c$:

- If $c = 0$, then $q^{\circ n}(z) = z^{2^n}$, and so

$$\Omega_\infty = \{z : |z| > 1\} \quad \text{and} \quad J = \{z : |z| = 1\} .$$

- If $c = 2$, then it is possible to show that $q^{\circ n}(w + 1/w) = w^{2^n} + 1/w^{2^n}$, from which it follows that

$$\Omega_\infty = \overline{\mathbb{C}} \setminus [-2, 2] \quad \text{and} \quad J = [-2, 2]$$

(cf. Exercise 12).
- If $c = 0.12 - 0.74i$ or $c = 0.76 + 0.24i$, the Julia sets are the boundary of the black regions in Figure 1. The set J on the right is totally disconnected; Julia sets with this property are referred to as *Fatou dust.*

As we can appreciate from the last example, in general Julia sets have a very complicated structure, a fact that explains their popularity (and beauty). Although there are several approaches in studying these sets, let us explore briefly the connection with potential theory.

The first result is that a Julia set has always positive capacity. In fact, its capacity is surprisingly simple to compute; however, little is known about its area (plane Lebesgue) measure.

Theorem 4.3. *If $q(z) = \sum_{j=0}^{d} a_j z^j$, with $d \geq 2$ and $a_d \neq 0$, then for its Julia set,*

$$\operatorname{cap}(J) = \frac{1}{|a_d|^{1/(d-1)}} .$$

Proof. Let us start with the following technical trick. If $m(z) = \alpha z + \beta$ is a polynomial of degree 1, let us call the *conjugate* polynomial of q the following polynomial,

$$\widetilde{q} = m \circ q \circ m^{-1} ,$$

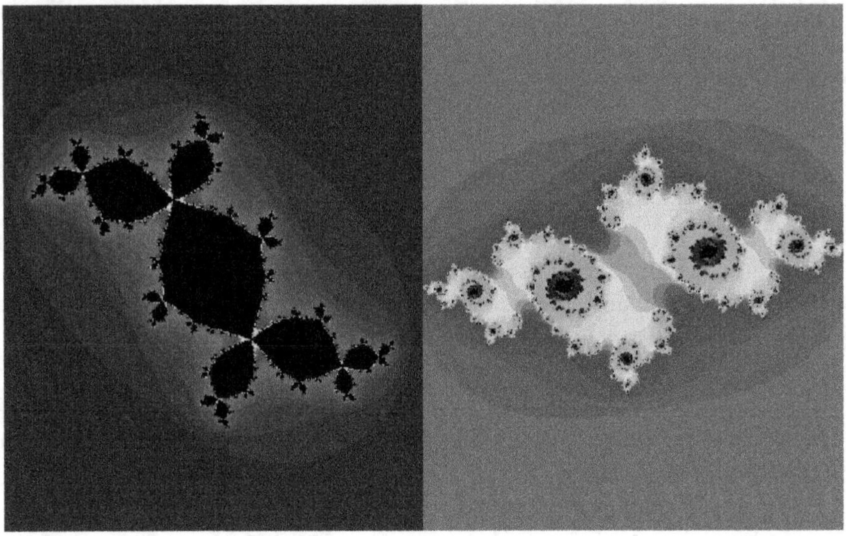

Fig. 1. Julia sets for $c = 0.12 - 0.74i$ (left) and $c = 0.76 + 0.24i$ (right)

which is again a polynomial of degree d. It is easy to see that $\widetilde{q}^{\circ n} = m \circ q^{\circ n} \circ m^{-1}$, so that both q and $\widetilde{q}$ have essentially the same dynamics. Furthermore, if we denote by $\widetilde{\Omega}_\infty$ and $\widetilde{J}$ the attracting basin of ∞ and the Julia set for $\widetilde{q}$, respectively, then

$$\widetilde{\Omega}_\infty = m(\Omega_\infty) \quad \text{and} \quad \widetilde{J} = m(J)\,.$$

By taking $\alpha = a_d^{1/(d-1)}$ and $\beta = a_{d-1}/d$ we obtain $\widetilde{q}(z) = z^d + \mathcal{O}(z^{d-2})$. So, in what follows and without loss of generality we assume that

$$q(z) = z^d + \sum_{j=0}^{d-2} a_j z^j\,. \tag{4.7}$$

For this q we must prove that $\operatorname{cap}(J) = 1$. The main tool will be formula (3.20).

Let us use the construction of the sets U_n of Lemma 4.2, and define $K_n = \mathbb{C} \setminus U_n$. On one hand, we have that $U_n \uparrow \Omega_\infty$, so that[6] $\operatorname{cap}(\mathbb{C} \setminus U_n) \to \operatorname{cap}(\mathbb{C} \setminus \Omega_\infty)$. On the other hand, $K_n = q^{-1}(K_{n-1})$, and by (3.20),

$$\operatorname{cap}(K_n) = \operatorname{cap}(K_{n-1})^{1/d} = \cdots = \operatorname{cap}(K_0)^{1/d^n}\,.$$

[6] Here we use implicitly the following property of the capacity: if K_n are compact sets, $K_1 \supset K_2 \supset \ldots$, and $K = \cap_n K_n$, then $\operatorname{cap}(K_n) \to \operatorname{cap}(K)$ (see [10, Theorem 5.1.3]).

Hence, $\operatorname{cap}(K_n) \to 1$ as $n \to \infty$, proving that $\operatorname{cap}(\Omega_\infty) = 1$. □

Since $\operatorname{cap}(J) > 0$, the Green function $g_J(z, \infty)$ exists. Formula (3.21) and the invariance of the Julia set allow to establish also the following relation for this Green function:

$$g_J(q(z), \infty) = d\, g_J(z, \infty) . \tag{4.8}$$

A nice consequence of this is that the Green function has a purely dynamical interpretation:

Proposition 4.1. *With the notation above,*

$$g_J(z, \infty) = \lim_{n\to\infty} d^{-n} \log |q^{\circ n}(z)| , \quad z \in \Omega_\infty ,$$

the convergence being locally uniform in Ω_∞.

Proof. By the definition of the Green function,

$$g_J(z, \infty) = \log |z| + \mathcal{O}(1), \quad \text{as } z \to \infty .$$

In particular, for $z \in \Omega_\infty$,

$$g_J(q^{\circ n}(z), \infty) = \log |q^{\circ n}(z)| + \mathcal{O}(1), \quad \text{as } n \to \infty .$$

But by formula (4.8),

$$g_J(z, \infty) = d^{-n} g_J(q^{\circ n}(z), \infty) = d^{-n} \log |q^{\circ n}(z)| + \mathcal{O}(d^{-n}), \quad \text{as } n \to \infty .$$

This ends the proof. □

Finally, since $\operatorname{cap}(J) > 0$, we may consider the (Robin) equilibrium measure λ_J of J. By uniqueness, it is also q-invariant, and has the following dynamical interpretation:

Theorem 4.4. *Let $a \in J$, and $\{\nu_n\}$ be the normalized counting measures of the a-values of the polynomials $\{q^{\circ n}\}$:*

$$\nu_n \overset{\text{def}}{=} \frac{1}{d^n} \sum_{q^{\circ n}(\zeta)=a} \delta_\zeta ,$$

where the sum takes into account the multiplicity of the zeros of $q^{\circ n} - a$. Then

$$\nu_n \overset{*}{\longrightarrow} \lambda_J \quad \text{as } n \to \infty ,$$

where λ_J is the equilibrium measure of the Julia set J.

A proof can be found in [10, p. 196].

5 Conclusions, or What You Should Take Home

We have seen many formulas and have proved some theorems. This is a tip of the iceberg, of a beautiful and powerful theory of the logarithmic potential in the complex plane. What is essential here? Any summary risks to omit some important facts and to overestimate others. But here I dare to highlight the following ones:

- Superharmonic functions and logarithmic potentials of a positive measure are two sides of the same class of functions (Riesz decomposition theorem, page 85). This is a connection of the theory with complex analysis.
- There exists a unique equilibrium measure of a compact set, minimizing the energy. It "lives" on its outer boundary and makes the corresponding potential constant inside.
- The minimal energy of a unit measure gives the definition of capacity of a set, which is a natural way of measuring plane sets in many problems.
- Green functions, conformal mapping, equilibrium potential and capacity are very related concepts.
- An introduction of an external field in the equilibrium problem produces a new important feature: now the support of the extremal measure is a priori unknown and is the main parameter of the problem.
- The natural connection of the potential theory on $\mathbb{C}$ with polynomials (and approximation theory) is via the formula connecting the logarithm of the absolute value of the polynomial with the potential of its zero counting measure.
- Many extremal problems in approximation theory, complex dynamics, numerical analysis, mathematical physics, etc., can be solved in terms of some related extremal problems for energy-type functionals acting on measures on the plane, which at the end of the day lead us to consider equilibrium problems (in a classical or generalized sense) for logarithmic (or other) potentials.

References

1. B. Beckermann and A. B. J. Kuijlaars, *Superlinear convergence of conjugate gradients*, SIAM J. Num. Anal. **39** (2001), 300–329.
2. B. Beckermann and E. B. Saff, *The sensitivity of least squares polynomial approximation*, Int. Ser. of Num. Math. **131** (1999), 1–19.
3. B. Beckermann, *Discrete orthogonal polynomials and superlinear convergence of Krylov subspace methods in numerical linear algebra*, Lecture Notes in Mathematics (this volume), Springer-Verlag, 2005.
4. P. Dragnev and E. B. Saff, *Constrained energy problems with applications to orthogonal polynomials of a discrete variable*, J. d'Analyse Mathématique **72** (1997), 229–265.
5. A. A. Gonchar and E. A. Rakhmanov, *The equilibrium problem for vector potentials*, Uspekhi Mat. Nauk **40**, no. 4 (244) (1985), 155–156.

6. N. S. Landkof, *Foundations of Modern Potential Theory*, Grundlehren der Mathematischen Wissenschaften, Springer-Verlag, Berlin, 1972.
7. H. N. Mhaskar and E. B. Saff, *Where does the sup norm of a weighted polynomial live? (A generalization of incomplete polynomials)*, Constr. Approx. **1** (1985), 71–91.
8. E. M. Nikishin and V. N. Sorokin, *Rational Approximations and Orthogonality*, Trans. Math. Monographs **92**, Amer. Math. Soc., Providence, RI, 1991.
9. E. A. Rakhmanov, *Equilibrium measure and the distribution of zeros of the extremal polynomials of a discrete variable*, Mat. Sb. **187** (1996), no. 8, 109–124 (in Russian); Sb. Math. **187** (1996), 1213–1228.
10. Th. Ransford, *Potential Theory in the Complex Plane*, Mathematical Society Student Texts **28**, Cambridge University Press, 1995.
11. H. Stahl and V. Totik, *General Orthogonal Polynomials*, Encyclopedia of Mathematics and its Applications **43**, Cambridge University Press, 1992.
12. E. B. Saff and V. Totik, *Logarithmic Potentials with External Fields*, Grundlehren der Mathematischen Wissenschaften **316**, Springer-Verlag, Berlin, 1997.
13. M. Tsuji, *Potential Theory in Modern Function Theory*, Chelsea, New York, 2nd edition, 1975.

Discrete Orthogonal Polynomials and Superlinear Convergence of Krylov Subspace Methods in Numerical Linear Algebra

Bernhard Beckermann *

Laboratoire Paul Painlevé UMR 8524 (ANO-EDP), UFR Mathématiques – M3,
Université des Sciences et Technologies de Lille,
F-59655 Villeneuve d'Ascq Cedex, France
e-mail: bbecker@math.univ-lille1.fr

Summary. We give a theoretical explanation for the superlinear convergence behavior observed while solving large symmetric systems of equations using the Conjugate Gradient (CG) method or other Krylov subspace methods. We present a new bound on the relative error after n iterations. This bound is valid in an asymptotic sense, when the size N of the system grows together with the number n of iterations. This bound depends on the asymptotic eigenvalue distribution and on the ratio n/N. Similar bounds are given for the task of approaching eigenvalues of large symmetric matrices via Ritz values.

Our findings are related to some recent results concerning asymptotics of discrete orthogonal polynomials due to Rakhmanov and Dragnev & Saff, followed by many other authors. An important tool in these investigations is a constrained energy problem in logarithmic potential theory.

The present notes are intended to be self contained (even if the proofs are sometimes incomplete and we refer to the original literature for details): the first part about Krylov subspace methods should be accessible for people from the orthogonal polynomial community and also for those who do not know much about numerical linear algebra. In the second part we gather the necessary tools from logarithmic potential theory and recall the basic results on the nth root asymptotics of discrete orthogonal polynomials. Finally, in the third part we discuss the fruitful relationship between these two fields and give several illustrating examples.

* Supported in part by INTAS Research Network on Constructive Complex Approximation (INTAS 03-51-6637).

1 Background in Numerical Linear Algebra

1.1 Introduction

The Conjugate Gradient (CG) method is widely used for solving systems of linear equations $Ax = b$ with a positive definite symmetric matrix A. The CG method is popular as an iterative method for large systems, stemming, e.g., from the discretisation of boundary value problems for elliptic PDEs. The rate of convergence of CG depends on the distribution of the eigenvalues of A. A well-known upper bound for the error e_n in the A-norm after n steps is

$$\frac{||e_n||_A}{||e_0||_A} \le 2\left(\frac{\sqrt{\kappa}-1}{\sqrt{\kappa}+1}\right)^n \tag{1.1}$$

where e_0 is the initial error and the condition number $\kappa = \lambda_{\max}/\lambda_{\min}$ is the ratio of the two extreme eigenvalues of A. In practical situations, this bound is too pessimistic, and one observes an increase in the convergence rate as n increases. This phenomenon is known as superlinear convergence of the CG method. It is the purpose of this work to give an explanation for this behavior in an asymptotic sense, following [39, 10, 11, 12].

As we will see in Section 1.4 below, the CG convergence behavior is determined by asymptotics of discrete orthogonal polynomials, and can be bounded above in terms of asymptotics of discrete L_∞ extremal polynomials. More generally, consider the extremal polynomials $T_{n,p}(z) = z^n + \textit{lower powers}$ with regard to some discrete L_p–norm

$$||w_n \cdot T_{n,p}||_{L_p(E_n)} = \min\{||w_n \cdot P||_{L_p(E_n)} : P(z) = z^n + \textit{lower powers}\}, \tag{1.2}$$

where

$$||f||_{L_\infty(E_n)} := \sup_{z\in E_n} |f(z)|, \quad ||f||_{L_p(E_n)} := \left[\sum_{z\in E_n} |f(z)|^p\right]^{1/p}, \tag{1.3}$$

$0 < p < \infty$, with E_n being suitable finite or countable subsets of the complex plane, $\#E_n \geq n+1$, and $w_n(z)$, $z \in E_n$, being (sufficiently fast decreasing) positive numbers.

For the case $p = 2$ of monic discrete orthogonal polynomials, examples include the discrete Chebyshev polynomials [58] (choose $w_n = 1$, $E_n = \{0, 1, ..., n\}$) or other classical families like Krawtchouk or Meixner polynomials [22, 24, 45], see for instance the review in [43]. A study of asymptotics of such polynomials has some important applications, e.g., in coding theory, in random matrix theory [37], or in the study of the continuum limit of the Toda lattice [23].

It was Rakhmanov [58] who first observed that a particular constrained (weighted) energy problem in complex potential theory (see Section 2.2) may furnish a method for calculating the nth root asymptotics of extremal polynomials with respect to so–called *ray sequences* obtained by a suitable renormalization of the sets E_n. Further progress has been made by Dragnev and Saff for real sets E_n being uniformly bounded [24]; they also obtained asymptotics for discrete L_p–norms with $0 < p \leq \infty$. Generalizations for unbounded real sets E_n and exponentially decreasing weights have been discussed by Kuijlaars and Van Assche [45] ($0 < p \leq \infty$) and Kuijlaars and Rakhmanov [43] ($p = 2$). Damelin and Saff [22] studied the case $p = \infty$ for more general classes of weights. Complex possibly unbounded sets E_n and even more general weights have been discussed in [5], where it is also shown that two conjectures of Rakhmanov [43] are true concerning some separation assumption for the sets E_n.

We will explain in Section 2.3 below how some energy problem with constraint and external field will enable us to describe the nth root asymptotics of the polynomials $T_{n,p}$ and the norms $||w_N \cdot T_{n,p}||_{L_p(E_N)}$. What makes the asymptotic analysis difficult is the fact that a polynomial can be small on a discrete set without being uniformly small in the convex hull of this discrete set. To illustrate this observation, we have chosen $E = \{j/20 : j = 1, ..., 20\}$ and the trivial weight $w = 1$, and have drawn the normalized extremal polynomials $T_{n,\infty}/||T_{n,\infty}||_{L_\infty(E)}$ for $n = 5, 10, 18$ in Figure 1. We see that, for $n = 5$, the polynomial is uniformly small on $[1/20, 20/20]$, but this is no longer true for $n = 10$ or $n = 18$.

For the same reason, the classical CG error bound (1.1) gives satisfactory results for small iterations, but can be a crude overestimation in a later stage. Indeed, for small n, a polynomial $p \in P_n$ with $p(0) = 1$ that is small on the spectrum of A has to be uniformly small on the full interval $[\lambda_{\min}, \lambda_{\max}]$ as well. When n gets larger, however, a better strategy for p is to have some of its zeros very close to some of the eigenvalues of A, thereby annihilating the value

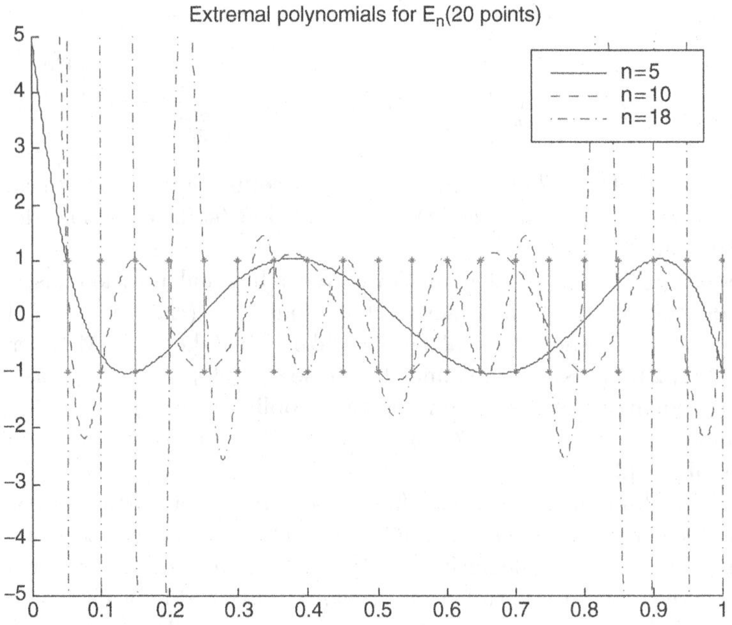

Fig. 1. The polynomials $T_{n,\infty}$ (after normalization) for $n = 5, 10, 18$ for E consisting of 20 equidistant points and trivial weight $w = 1$

of p at those eigenvalues, while being uniformly small on a subcontinuum of $[\lambda_{\min}, \lambda_{\max}]$ only.

As an illustration we look at the case of a matrix A with 100 equally spaced eigenvalues $1, 2, \ldots, 100$. The error curve computed for this example is the solid line in Figure 2. See also [25, page 560]. The classical error bound given by (1.1) with $\kappa = 100$ is the straight line in Figure 2. For smaller values of n, the classical error bound gives an excellent approximation to the actual error. The other curve (the one with the dots) is the asymptotic bound for the error proposed in [10, Corollary 3.2]. This curve follows the actual error especially well for $n \geq 40$, the region of superlinear convergence.

The observations made above have been well known in the numerical linear algebra community, see for instance the monographs [27, 32, 54, 59, 70] or the original articles [3, 4, 31, 63, 64]. Eigenvalues far away from the rest of the spectrum (so-called *outliers*) have been treated in a separate manner improving (1.1) [3, 4]. The strategy described above to get a polynomial being small on the (discrete) spectrum was known as convergence of some Ritz values [31, 63, 64]. In addition, the researchers have been aware of the fact that logarithmic potential theory helps in describing or bounding the rate of convergence [25]. There was also a vague idea about what is a "favorable eigenvalue distribution" in order to get a pronounced superlinear convergence

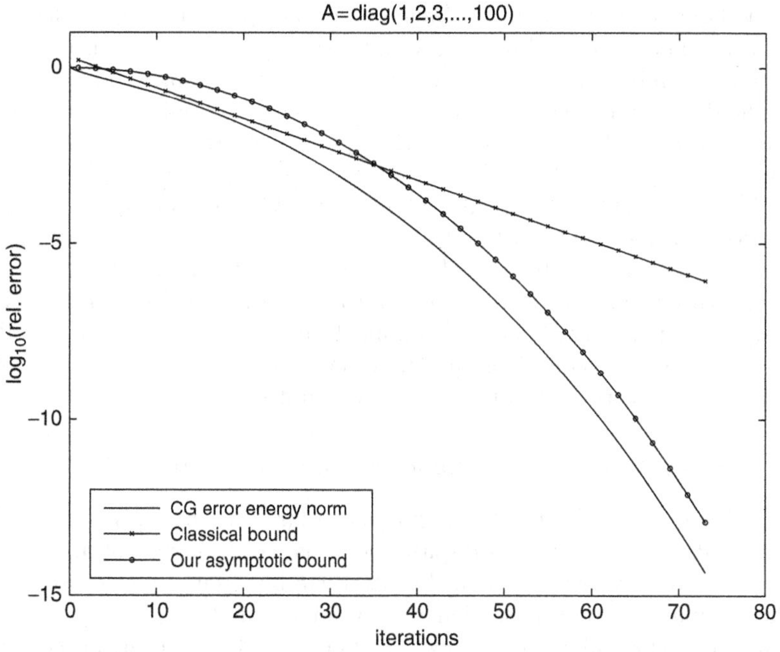

Fig. 2. The CG error curve versus the two upper bounds for the system $Ax = b$ with $A = \text{diag}\,(1, 2, \ldots, 100)$, random solution x, and initial residual $r_0 = (1, ..., 1)^T$

[25, 70]. However, precise analytic formulas seemed to occur for the first time only in [39, 10, 11, 12].

Properly speaking, the concept of superlinear convergence for the CG method applied to a single linear system does not make sense. Indeed, in the absence of roundoff errors, the iteration will terminate at the latest after N steps if N is the size of the system. Also the notion that the eigenvalues are distributed according to some continuous distribution is problematic when considering a single matrix. Therefore we are not going to consider a single matrix A, but instead a sequence $(A_N)_N$ of positive definite symmetric matrices. The matrix A_N has size $N \times N$, and we are interested in asymptotics for large N. These matrices need to have an *asymptotic eigenvalue distribution.*

The rest of this manuscript is organized as follows: In §1.2 we present several Krylov subspace methods and fix the notation. Subsequently, we introduce polynomial language for explaining the link between convergence theory for Krylov subspace methods, and classical extremal problems in the theory of orthogonal polynomials. We shortly describe the general case in §1.3, and then analyze in more detail the case of hermitian matrices in §1.4.

Following [39, 10, 11, 12, 6], we describe in §2 and §3 how logarithmic potential theory may help to analyze the convergence of Krylov subspace methods. Some facts about the weighted energy problem are recalled in §2.1, but here some additional reading would be helpful, see for instance the contribution by Andrei Martínez Finkelshtein [47] in this volume. The constrained weighted energy problem is discussed in some more detail in §2.2, and used in §2.3 in order to describe nth root asymptotics of discrete L_p extremal polynomials.

The link to the convergence of Krylov subspace methods for hermitian matrices is presented in §3.1 and §3.2, where also several illustrating numerical examples are given. The aim of §3.3 and §3.4 is to show that many classes of structured matrices have an asymptotic eigenvalue distribution. We will consider in particular matrices coming from the discretization of (elliptic) partial differential equations in $\mathbb{R}^2$. A generalization to higher dimension is possible, but for the sake of simplicity we omit details.

1.2 Conjugate Gradients, Lanczos, and Ritz Values

For solving $Ax = b$ with A being a sparse large matrix of size $N \times N$, one often makes use of Krylov subspace methods which only require to compute matrix vector products with A, the latter can be often implemented in a very efficient manner. In this section we do not attempt to give a complete account of Krylov subspace methods, the interested reader should consult the monographs [27, 32, 54, 59, 70]. We just recall the basic definitions and some elementary properties on the rate of convergence. Here, very much in the spirit of the Lille research group and in particular of Claude Brezinski (see also [27]), we will use polynomial language, which should make the theory also more accessible for people coming from orthogonal polynomials.

In what follows we will always suppose exact arithmetic and ignore errors due to floating point operations. In particular, we will find that several Krylov subspace methods are mathematically equivalent for symmetric A. However, their implementation differs quite a lot, and thus the results may change in a floating point environment. The link between convergence of Krylov space methods and loss of precision is subject of actual research, see for instance the recent work of Strakos or Meurant, e.g., [51, 67, 66].

A Krylov subspace method consists of computing a sequence $x_0, x_1, x_2, ...$ of approximate solutions of $Ax = b$, with residual

$$r_n = r(x_n) = b - Ax_n.$$

The philosophy behind these methods is that (x_n) "converges quickly" to the solution $A^{-1}b$, i.e., x_n is a "good" approximation already for $n \ll N$. The iterates satisfy

$$x_n \in x_0 + \mathcal{K}_n(A, r_0)$$

with the Krylov space

$$\begin{aligned} \mathcal{K}_n(A, c) &= \text{span}\{c, Ac, A^2c, ..., A^{n-1}c\} \\ &= \{p(A)c : p \text{ a polynomial of degree } \leq n-1\}. \end{aligned}$$

Notice that $r_n \in r_0 + A\mathcal{K}_n(A, r_0)$, and thus

$$r_n = \frac{q_n(A)r_0}{q_n(0)} \tag{1.4}$$

for a certain polynomial q_n of degree n. The Krylov subspace method in question is now defined by imposing on the residual either some minimization property (Minimal Residual MinRES, Generalized Minimal Residual GMRES, Conjugate Gradient CG) or some orthogonality property (projection methods, Full Orthogonalization Method FOM, Lanczos, Conjugate Residual CR).

Definition 1.1. *The nth iterate x_n^{GMRES} of* GMRES *is the unique argument realizing*

$$\min\{||r(x)|| : x \in x_0 + \mathcal{K}_n(A, r_0)\}.$$

The nth iterate x_n^{FOM} of FOM *is defined by*

$$r_n^{\mathrm{FOM}} \perp \mathcal{K}_n(A, r_0).$$

For some vector y, the nth iterate x_n^L of the Lanczos method is defined by

$$r_n^L \perp \mathcal{K}_n(A^*, y)$$

(in case of $y = r_0$ we speak of the symmetric Lanczos method).

For real symmetric positive definite A, the nth iterate x_n^{CG} of the method of conjugate gradients CG *is the unique argument realizing*

$$\min\{||r(x)||_{A^{-1}} : x \in x_0 + \mathcal{K}_n(A, r_0)\}, \qquad ||c||_{A^{-1}} = \sqrt{c^* A^{-1} c}.$$

If A is invertible, and symmetric positive definite, respectively, the functions $x \mapsto ||r(x)||^2$, and $x \mapsto ||r(x)||^2_{A^{-1}}$, respectively, are strictly convex, and thus the iterates of CG and GMRES exist and are unique. In contrast, it may happen that the nth iterate of FOM does not exist, see Corollary 1.2.

Exercise 1. Show that there exists $N' = N'(A, r_0)$ such that, for all $n \geq 0$,

$$\dim \mathcal{K}_n(A, r_0) = \min\{n, N'\}.$$

Hint: first try diagonal A, and use the fact the matrix $(r_0, Ar_0, ..., A^{n-1}r_0)$ is some diagonal matrix times some Vandermonde matrix. Then try diagonalisable A.

Exercise 2. If $x_0 = 0$, and A is invertible, show that $A^{-1}b \in \mathcal{K}_{N'}(A, r_0)$.

It follows from the preceding two exercises that, for $x_0 = 0$ and $n = N'$, the iterates x_n^{FOM}, x_n^{GMRES} and x_n^{CG} give the exact solution of $Ax = b$, but of course we hope that we have a good approximation already much earlier.

There is a link between the size of the residuals of FOM and GMRES given by the following result. A proof is immediate once we have the representation (1.4) in terms of orthogonal polynomials, see Section 1.3.

Exercise 3. Show that

$$\frac{1}{||r_n^{\text{GMRES}}||^2} = \sum_{j=0}^{n} \frac{1}{||r_j^{\text{FOM}}||^2}.$$

Definition 1.2. *The Arnoldi basis $v_1, v_2, ..., v_{N'}$ is such that, for all $n = 1, ..., N'$, the vectors $v_1, ..., v_n$ form an orthonormal basis of $\mathcal{K}_n(A, r_0)$ (obtained by the Arnoldi method: orthogonalize Av_n against $v_1, ..., v_n$, and divide the resulting vector by its norm). We also define the matrices*

$$V_n := (v_1, v_2, ..., v_n) \in \mathbb{C}^{N\times n},$$
$$J_n := V_n^* A V_n \in \mathbb{C}^{n\times n}, \qquad \widehat{J}_n = V_{n+1}^* A V_n \in \mathbb{C}^{(n+1)\times n}.$$

Finally, the eigenvalues of the projected matrix J_n are called nth Ritz values of A.

Exercise 4. Show that J_n and $\hat{J}_n$ are upper Hessenberg (all elements at position (j, k) with $k < j - 1$ are equal to zero). Furthermore, show that

$$n < N' : \qquad V_{n+1}\widehat{J}_n = AV_n. \tag{1.5}$$

Finally, in the case of hermitian A, show that J_n is symmetric (and tridiagonal).

Remark 1.1. By construction we have for $n = N'$ that $V_{N'}J_{N'} = AV_{N'}$. As a consequence, denoting by $\Lambda(B)$ the spectrum of some matrix B, we have that the columns of $V_{N'}$ span an A-invariant subspace, and $\Lambda(J_{N'}) \subset \Lambda(A)$.

Remark 1.2. In case of real symmetric A, it follows from Definition 1.1 that the symmetric Lanczos method and FOM are mathematically equivalent, i.e., $x_n^{FOM} = x_n^L = x_n^{CR}$, the latter denoting the iterates of the conjugate residual method. Also, in this case the GMRES method reduces to the so-called method MinRES (minimal residuals). Finally, we will show in Corollary 1.5 below that, in case of symmetric positive definite A, the CG iterates coincide with the symmetric Lanczos iterates.

1.3 Krylov Subspace Methods and Discrete Orthogonal Polynomials: Non Symmetric Data

One may show that the nth residual polynomial q_n of (1.4) of the Lanczos method is given by the denominator of the nth Padé approximant at infinity of the rational function

$$\pi_N(z) = y^*(zI - A)^{-1} r_0 = \sum_{j=0}^{\infty} z^{-j-1} y^* A^j r_0, \tag{1.6}$$

see for instance [15, §3.6] or for the symmetric case [29]. Hence there is a link between Lanczos method and formal orthogonal polynomials (polynomials being orthogonal with respect to some linear form).

In this Section we will concentrate on FOM/GMRES for non symmetric A. Denote by $\mathcal{P}$ the set of polynomials with complex coefficients, and by $\mathcal{P}_n$ the set of polynomials of degree at most n with complex coefficients. For two polynomials P, Q, we consider the sesquilinear form

$$\prec P, Q \succ = (P(A) r_0)^* Q(A) r_0.$$

The following exercise shows that we have a scalar product

Exercise 5. Let $N' = N'(A, r_0)$ as in Exercise 1. Show that for all $P \in \mathcal{P}_{N'-1} \setminus \{0\}$ we have $\prec P, P \succ > 0$, and that there exists a unique monic polynomial Q of degree N' with $\prec Q, Q \succ = 0$.

As a consequence, we can define uniquely orthonormal polynomials p_n, $n = 0, 1, ..., N'$, satisfying

$$\begin{aligned} j = 0, ..., N' - 1: \ & p_j(z) = k_j z^j + \text{ lower powers}, \qquad k_j > 0, \\ j, k = 0, ..., N' - 1: \ & \prec p_j, p_k \succ = \delta_{j,k}, \\ \text{for all } P \in \mathcal{P}: \ & \prec P, p_{N'} \succ = 0, \qquad p_{N'}(z) = z^{N'} + \text{ lower powers} \end{aligned}$$

(we put $k_{N'} = 1$). These orthonormal polynomials are known to satisfy a (full) recurrence: there exists an upper Hessenberg matrix $J_{N'}$ such that

$$\begin{aligned} z(p_0, p_1, ..., p_{N'-1})(z) = & \ (p_0, p_1, ..., p_{N'-1})(z) J_{N'} \\ & + \frac{k_{N'-1}}{k_{N'}} p_{N'}(z)(0, ..., 0, 1). \end{aligned} \tag{1.7}$$

The Hessenberg matrix J_n occurred already earlier in Definition 1.2. Indeed, this is not an inconsistency in notation, as it becomes clear from the following

Exercise 6. The Arnoldi basis is given by the vectors

$$v_j := p_{j-1}(A) r_0, \quad j = 1, ..., N',$$

and $p_{N'}(A) r_0 = 0$. In particular, the matrices $J_{N'}$ in Definition 1.2 and in (1.7) coincide, and the matrix J_n in Definition 1.2 is just the nth principal submatrix of $J_{N'}$. Finally, for $1 \le n \le N'$ we have

$$z(p_0, p_1, ..., p_{n-1})(z) = (p_0, p_1, ..., p_{n-1})(z) J_n + \frac{k_{n-1}}{k_n} p_n(z)(0, ..., 0, 1). \tag{1.8}$$

As a (more or less immediate) consequence of Exercise 6, we have the following two interpretations in terms of orthogonal polynomials. A proof is left to the reader.

Corollary 1.1. *The nth Ritz values of A are given by the zeros of the orthonormal polynomial p_n.*

As we will see below, $\prec \cdot, \cdot \succ$ can be a discrete Sobolev inner product, its support being a subset of the spectrum of A. Approaching the spectrum of A by Ritz values means that we approach the support of some scalar product by the zeros of the underlying orthogonal polynomials, something familiar for people from the OP community (at least if the support is real, see Section 1.4).

Corollary 1.2. *The nth iterate of* FOM *exists if and only if $p_n(0) \neq 0$. In this case*

$$r_n^{\text{FOM}} = \frac{p_n(A)r_0}{p_n(0)}, \qquad \frac{1}{||r_n^{\text{FOM}}||} = |p_n(0)|.$$

We may also describe the residuals of GMRES in terms of orthonormal polynomials. For this we need some preliminary remarks.

Definition 1.3. *The nth Szegő kernel of the scalar product $\prec \cdot, \cdot \succ$ is defined by*

$$K_{n,2}(x, y) = \sum_{j=0}^{n} \overline{p_j(x)} p_j(y).$$

It is a well-known fact (see for instance the "bible" of Szegő [68]) that

$$\min_{P \in \mathcal{P}_n} \frac{\prec P, P \succ}{|P(0)|^2} = \frac{1}{K_{n,2}(0, 0)}, \tag{1.9}$$

is attained for the polynomial $P(z) = K_{n,2}(0, z)$. By construction of the scalar product $\prec \cdot, \cdot \succ$ and by (1.4) we have for $n < N'$

$$\min_{P \in \mathcal{P}_n} \frac{\prec P, P \succ}{|P(0)|^2} = \min\{||r(x)||^2 : x \in x_0 + \mathcal{K}_n(A, r_0)\},$$

leading to the following characterization

Corollary 1.3. *For the nth iterate of* GMRES, *$n = 0, 1, ..., N' - 1$ we have*

$$r_n^{\text{GMRES}} = \frac{K_{n,2}(0, A)r_0}{K_{n,2}(0, 0)}, \qquad \frac{1}{||r_n^{\text{GMRES}}||^2} = K_{n,2}(0, 0).$$

Exercise 7. Let $\mu_{j,k} := (A^j r_0)^* A^k r_0$. Show that

$$\frac{||r_1^{\text{GMRES}}||^2}{||r_0^{\text{GMRES}}||^2} = \frac{\mu_{0,0}\mu_{1,1} - |\mu_{0,1}|^2}{\mu_{0,0}\mu_{1,1}}.$$

Conclude that the following algorithm (called GMRES(1))

choose any y_0

for $k = 0, 1, \ldots$ until "convergence" do

compute y_{k+1} by one iteration of GMRES with starting vector $x_0 = y_k$

converges ($y_k \to A^{-1}b$ for $k \to \infty$) for all b and y_0 if and only if for all $y \neq 0$ we have $y^* A y \neq 0$.
Hint: if you do not find a direct proof, look in [32].

We end this section by showing that the scalar product can be (but does not need to be) a Sobolev inner product with finite support, possibly in the complex plane. Finally, we show that in case of a normal matrix A we have an important simplification, leading to discrete orthogonal polynomials with possibly complex support.

Example 1.1. For some parameter $\rho > 0$, consider the matrix

$$A = XBX^{-1}, \qquad B = \begin{bmatrix} 1 & 0 & 0 & 0 \\ 1 & 1 & 0 & 0 \\ 0 & 0 & -1 & 0 \\ 0 & 0 & 1 & -1 \end{bmatrix}, \qquad X = \begin{bmatrix} 1 & 0 & 0 & 0 \\ \rho & 1 & 0 & 0 \\ 0 & 0 & 1 & 0 \\ \rho & 0 & 0 & 1 \end{bmatrix},$$

and $r_0 = Xc$, $c = (1, 0, 1, 0)^t$. Notice that B is in Jordan form, and hence $\Lambda(A) = \{-1, 1\}$. It is not difficult to show that, for any polynomial Q,

$$Q(A)r_0 = XQ(B)c = X \begin{bmatrix} Q(1) \\ Q'(1) \\ Q(-1) \\ Q'(-1) \end{bmatrix} = \begin{bmatrix} Q(1) \\ \rho Q(1) + Q'(1) \\ Q(-1) \\ \rho Q(1) + Q'(-1) \end{bmatrix}.$$

Thus we obtain the following simplification for the scalar product

$$\begin{aligned} \prec P, Q \succ = \overline{P(1)} Q(1) + \overline{\rho P(1) + P'(1)} (\rho Q(1) + Q'(1)) \\ + \overline{P(-1)} Q(-1) + \overline{\rho P(1) + P'(-1)} (\rho Q(1) + Q'(-1)), \end{aligned}$$

which reduces to a (discrete) Sobolev inner product in the case $\rho = 0$.

One may indeed show that the scalar product always can be represented in terms of a linear combination of the values and the derivatives of P, Q at the points of the spectrum of $J_{N'}$. Let us have a closer look at a case where the derivatives do not occur.

Theorem 1.1. *If A is normal (i.e., $A^*A = AA^*$), with matrix of right eigenvectors given by X_N and its spectrum by $\Lambda(A) = \{\lambda_1, ...\lambda_N\}$ then with $X_N^{-1} r_0 = (\beta_j)_j$,*

$$P, Q \in \mathcal{P} : \qquad \prec P, Q \succ = \sum_{j=1}^{N} |\beta_j|^2 \overline{P(\lambda_j)} Q(\lambda_j). \tag{1.10}$$

Conversely, if the scalar product has such a representation then at least the projected matrix $J_{N'}$ is normal, and $\Lambda(J_{N'}) = \{\lambda_j : \beta_j \neq 0\} \subset \Lambda(A)$.

Proof. If A is normal then its matrix of eigenvectors X_N is unitary, that is,

$$X_N^* X_N = I, \qquad X_N^* A X_N = \Lambda = \text{diag}\,(\lambda_1, ..., \lambda_N).$$

Observing that for some polynomial P we have

$$P(A) = X_N P(\Lambda) X_n^*, \qquad \text{with} \;\; P(\Lambda) = \text{diag}\,(P(\lambda_1), ..., P(\lambda_N)),$$

we obtain by writing $\beta = X_n^* r_0 = X_n^{-1} r_0$

$$P, Q \in \mathcal{P}_n : \qquad \prec P, Q \succ = \beta^* P(\Lambda)^* Q(\Lambda) \beta = \sum_{j=1}^{N} |\beta_j|^2 \overline{P(\lambda_j)} Q(\lambda_j).$$

In order to show the converse result, suppose that there are N^* distinct λ_j with $\beta_j \neq 0$, say, the terms corresponding to $\lambda_1, ..., \lambda_{N^*}$. Then we may rewrite the sum in (1.10) as

$$P, Q \in \mathcal{P} : \qquad \prec P, Q \succ = \sum_{j=1}^{N^*} |\beta_j^*|^2 \overline{P(\lambda_j)} Q(\lambda_j)$$

for suitable coefficients β_j^* and distinct $\lambda_1, ..., \lambda_{N^*}$. From Exercise 5 together with (1.7) we learn that $N^* = N'$, and $p_{N'}(z) = (z - \lambda_1)...(z - \lambda_{N'})$. It follows from Corollary 1.1 that $J_{N'}$ has the distinct eigenvalues $\lambda_1, ..., \lambda_{N'}$, and the relation $\Lambda(J_{N'}) \subset \Lambda(A)$ was established in Remark 1.1. Consider the matrix

$$Y := (p_k(\lambda_j))_{j=1,...,N',k=0,...,N'-1},$$

then from (1.7) we know that $Y J_{N'} = DY$, $D = \text{diag}\,(\lambda_1, ..., \lambda_{N'})$, and from the representation of the scalar product together with (1.7) we learn that

$$Y^* \,\text{diag}\,(|\beta_1^*|^2, ..., |\beta_{N'}^*|^2)\, Y = I.$$

Hence, up to a certain normalization of the rows, the left eigenvector matrix of $J_{N'}$ is unitary, implying that $J_{N'}$ is normal. □

Analyzing in more detail the preceding proof we obtain the following

Corollary 1.4. *If A is normal then also $J_{N'}$ is normal, with $\Lambda(J_{N'}) \subset \Lambda(A)$. Furthermore, $J_{N'}$ has N' distinct eigenvalues, and*

$$P, Q \in \mathcal{P}: \qquad \prec P, Q \succ = \sum_{\lambda \in \Lambda(J_{N'})} w(\lambda)^2 \overline{P(\lambda)} Q(\lambda),$$

$$w(\lambda)^2 = \frac{1}{K_{N',2}(\lambda,\lambda)} > 0, \qquad \sum_{\lambda \in \Lambda(J_{N'})} w(\lambda)^2 = ||r_0||^2. \tag{1.11}$$

1.4 Krylov Subspace Methods and Discrete Orthogonal Polynomials: Symmetric Data

From now on we will always suppose (except for Remark 1.3) that our matrix of coefficients A is hermitian, i.e., $A = A^*$, with spectrum $\lambda_1 \leq \cdots \leq \lambda_N$. From Exercise 4 it follows that $J_{N'}$ with $N' = N'(A, r_0)$ as in Exercise 1 is an hermitian upper Hessenberg matrix with positive entries $k_n/k_{n+1} > 0$ on the first subdiagonal. However, such a matrix is necessarily tridiagonal, of the form

$$J_n = \begin{bmatrix} b_0 & a_0 & 0 & \cdots & \cdots & 0 \\ a_0 & b_1 & a_1 & 0 & \cdots & 0 \\ 0 & a_1 & b_2 & a_2 & \ddots & \vdots \\ \vdots & \ddots & \ddots & \ddots & \ddots & 0 \\ 0 & \cdots & 0 & a_{n-3} & b_{n-2} & a_{n-2} \\ 0 & \cdots & \cdots & 0 & a_{n-2} & b_{n-1} \end{bmatrix}, \qquad b_n \in \mathbb{R},\ a_n = \frac{k_n}{k_{n+1}} > 0. \tag{1.12}$$

Thus (1.7) becomes a three term recurrence: for $n = 0, ..., N' - 1$

$$z p_n = a_n p_{n+1} + b_n p_n + a_{n-1} p_{n-1}, \qquad p_0(z) = \frac{1}{||r_0||}, \qquad p_{-1}(z) = 0. \tag{1.13}$$

In particular it follows that the orthonormal polynomials p_n have real coefficients, and are orthonormal with respect to the linear functional c acting on the space of polynomials via

$$c(P) = \prec 1, P \succ = r_0^* P(A) r_0, \tag{1.14}$$

i.e., $c(p_j p_k) = \delta_{j,k}$. Also, since hermitian matrices are in particular normal, we have the representation (1.11) for the scalar product (and thus for the linear functional), showing that there is classical orthogonality on the real line (Favard's theorem) with respect to a positive measure with finite support $\Lambda(J_{N'})$ (or a positive linear functional).

In the following theorem we put together some elementary properties of such orthogonal polynomials. If you do not remember them, please try to prove them (not necessarily in the indicated order) or look them up in any standard book about OP.

Theorem 1.2. *(a) For $1 \leq n \leq N'$, the zeros of p_n are simple and real, say,*

$$x_{1,n} < x_{2,n} < \dots < x_{n,n}.$$

(b) Interlacing property: The zeros of p_n and p_{n+1} interlace

$$1 \leq j \leq n < N': \qquad x_{j,n+1} < x_{j,n} < x_{j,n+1}.$$

(c) Separation property: If $1 \leq j < n < n' \leq N'$ then there exists j' such that

$$x_{j,n} \leq x_{j',n'} < x_{j+1,n}.$$

(d) Christoffel-Darboux: For $n < N'$ and $x, y \in \mathbb{C}$

$$K_{n,2}(x,y) = \sum_{j=0}^{n} p_j(x)p_j(y) = a_n \frac{p_{n+1}(x)p_n(y) - p_{n+1}(y)p_n(x)}{x - y}$$

and for $x \in \mathbb{R}$

$$K_{n,2}(x,x) = \sum_{j=0}^{n} p_j(x)p_j(x) = a_n(p'_{n+1}(x)p_n(x) - p_{n+1}(x)p'_n(x)) > 0.$$

(e) Associated linear functionals: For $\gamma \in \mathbb{R}\setminus\Lambda(J_{N'})$, consider the linear functional $\widetilde{c}(P) = c(\widetilde{P})$, $\widetilde{P}(z) = (z-\gamma)P(z)$. Then $K_{n,2}(\gamma,\cdot)$ is an nth orthogonal polynomial with respect to $\widetilde{c}$.

(f) Gaussian quadrature: For any $0 < n < N'$ and any polynomial of degree not exceeding $2n-1$

$$c(P) = \sum_{j=1}^{n} \frac{P(x_{j,n})}{K_{n-1,2}(x_{j,n}, x_{j,n})}.$$

(g) Stieltjes functions and Padé approximation: The rational function

$$\pi_n(z) = ||r_0||^2 \sum_{j=1}^{n} \frac{1}{K_{n-1,2}(x_{j,n}, x_{j,n})} \frac{1}{z - x_{j,n}}$$
$$= ||r_0||^2 (e_0, (zI_n - J_n)^{-1} e_0)$$

with e_0 being the first canonical vector of suitable size is of denominator degree n, of numerator degree $n-1$, has real simple poles and positive residuals, and

$$\pi_{n+1}(z) - \pi_n(z) = \frac{1}{a_n p_n(z) p_{n+1}(z)} = \mathcal{O}(z^{-2n-1})_{z\to\infty}.$$

In particular, π_n is the nth Padé approximant at infinity of the Stieltjes function $z \mapsto r_0^*(zI - A)^{-1} r_0$, and $\pi_{N'}$ coincides with this function. Finally, p_n is an nth Padé denominator. □

Theorem 1.2(g) gives a link between Padé and Krylov subspace methods, compare with the remarks around (1.6). More precisely, we have the following link between CG, FOM (symmetric Lanczos) and Padé.

Corollary 1.5. *For symmetric positive definite A, the methods* CG, FOM *and the symmetric Lanczos method are mathematically equivalent, more precisely,*

$$||r_n^{\mathrm{CG}}|| = \frac{1}{|p_n(0)|}, \qquad ||r_n^{\mathrm{CG}}||_{A^{-1}}^2 = \pi_n(0) - \pi_{N'}(0).$$

Proof. Define the scalar product $\prec \cdot, \cdot \succ^*$ by replacing in (1.11) the term $w(\lambda)^2$ by $w^*(\lambda)^2 = w(\lambda)^2/\lambda$. According to Definition 1.1 the residual polynomial q_n of CG in (1.4) is the (up to scaling) unique solution of the extremal problem

$$\min_{P \in \mathcal{P}_n} \frac{\prec P, P \succ^*}{|P(0)|^2} = \frac{1}{K_{n,2}^*(0,0)} \tag{1.15}$$

with the Szegő function $K_{n,2}^*$ of the new scalar product. By (1.9) the minimum is attained by $q_n(z) = K_{n,2}^*(0,z)$, the latter being proportional to p_n by Theorem 1.2(e). Hence, from (1.4) and Corollary 1.2 we may conclude that $x_n^{\mathrm{CG}} = x_n^{\mathrm{FOM}}$, implying in particular our claim for $||r_n^{\mathrm{CG}}||$.

In order to show the link between CG and Padé, recall that $(zI - A)V_{N'} = V_{N'}(zI - J_{N'})$, and hence

$$\begin{aligned} (r_j^{\mathrm{FOM}})^*(zI - A)^{-1} r_k^{\mathrm{FOM}} &= \frac{v_{j+1}^*(zI - A)^{-1} v_{k+1}}{p_j(0)p_k(0)} \\ &= \frac{[(zI - J_{N'})^{-1}]_{j+1,k+1}}{p_j(0)p_k(0)}. \end{aligned}$$

It is a well-known fact that the inverse of a Jacobi matrix can be expressed in terms of Padé approximants

$$[(zI - J_{N'})^{-1}]_{j+1,k+1} = p_j(z)p_k(z)[\pi_{N'}(z) - \pi_{\max\{j,k\}}(z)],$$

see for instance [73, §60] or the survey paper [49]. Combining these two findings for $z = 0$ leads to our claim. □

The representation of the CG error as Padé error has been applied successfully by Golub, Meurant, Strakos and others [28, 29, 50] to estimate/bound from below the nth CG error after having computed the $(n+p)$th iterate for $p > 1$

$$||r_n^{CG}||_{A^{-1}}^2 \geq \pi_n(0) - \pi_{n+p}(0) = \sum_{j=n}^{n+p-1} \left|\frac{1}{a_j p_j(0) p_{j+1}(0)}\right|.$$

In particular, the authors show that these a posteriori bounds are reliable even in finite precision arithmetic. A similar result has already been mentioned implicitly in the original paper of Hestenes and Stiefel [36].

We should mention that Theorem 1.2(a),(b),(c) provides already a quite precise idea about how Ritz values do approach the real spectrum $\Lambda(J_{N'}) \subset \Lambda(A)$, or, equivalently, how poles of Padé approximants approach the poles of a rational function with positive residuals. However, up to now there is no information about the rate of convergence. For this rate of convergence of Ritz values we have the following result which roughly says that, provided that a certain polynomial extremal problem depending on λ_k gives a "small" value, there is at least one Ritz value "close" to λ_k. We do not claim that there is an eigenvalue "close" to each Ritz value. Indeed there exist examples with $\Lambda(A) = \Lambda(-A)$ where 0 is a Ritz value for each odd n which might be far from $\Lambda(A)$.

Theorem 1.3. [6] *If $\lambda_k \leq x_{1,n}$ then*

$$x_{1,n} - \lambda_k = \min\Big\{ \frac{\sum_{\lambda \in \Lambda(J_{N'}) \setminus \{\lambda_k\}} w(\lambda)^2 (\lambda - x_{1,n}) |q(\lambda)|^2}{w(\lambda_k)^2 |q(\lambda_k)|^2} : \deg q < n, q(\lambda_k) \neq 0 \Big\}.$$

Here the minimum is attained by the polynomial $q(x) = p_n(x)/(x - x_{1,n})$.

If $\lambda_k \in [x_{1,n}, x_{n,n}]$, say, $x_{\kappa-1,n} \leq \lambda_k \leq x_{\kappa,n}$, then

$$(\lambda_k - x_{\kappa-1,n})(x_{\kappa,n} - \lambda_k) = \min\Big\{ \frac{\sum_{\lambda \in \Lambda(J_{N'}) \setminus \{\lambda_k\}} w(\lambda)^2 (\lambda - x_{\kappa-1,n})(\lambda - x_{\kappa,n}) |q(\lambda)|^2}{w(\lambda_k)^2 |q(\lambda_k)|^2} : \deg q < n-1, |q(\lambda_k)| \neq 0 \Big\}.$$

Here the minimum is attained by $q(x) = p_n(x)/[(x - x_{\kappa-1,n})(x - x_{\kappa,n})]$.

Proof. We will show here the first part of the assertion; similar arguments may be applied to establish the second part. If q is a polynomial of degree less than n with $q(\lambda_{k,N}) \neq 0$ and $p(x) = (x - x_{1,n}) \cdot q(x) \cdot \overline{q(\overline{x})}$, then $c(p) \geq 0$ by the Gaussian quadrature formula of Theorem 1.2(f). Hence the right hand side of (1.11) is ≥ 0, and thus

$$x_{1,n} - \lambda_k \leq \frac{\sum_{\lambda \in \Lambda(J_{N'}) \setminus \{\lambda_k\}} w(\lambda)^2 (\lambda - x_{1,n}) |q(\lambda)|^2}{w(\lambda_k)^2 |q(\lambda_k)|^2}.$$

Finally, notice that for the choice $q(x) = p_{n,N}(x)/(x - x_{1,n,N})$ we have $c(p) = 0$ again by Theorem 1.2(f), and thus there is equality in the above estimate. □

There are several possibilities to relate the result of Theorem 1.3 to more classical extremal problems. The term $|\lambda - x_{j,n}|$ could be bounded by $2\,||A||$, and taking into account (1.11) and (1.9), we obtain for instance

$$(\lambda_k - x_{\kappa-1,n})(x_{\kappa,n} - \lambda_k) \le 4\,||A||^2 \Big(\frac{K_{N',2}(\lambda_k,\lambda_k)}{K_{n-2,2}(\lambda_k,\lambda_k)} - 1 \Big), \quad \text{if } x_{\kappa-1,n} \le \lambda_k \le x_{\kappa,n}.$$

It follows that the distance to at least one of the Ritz values become small if all $|p_j(\lambda_k)|^2$ for $j \ge n-1$ are small compared to $1/||r_0||^2 = K_{0,2}(\lambda_k,\lambda_k)$.

Another possibility could be to relate (1.9) to some extremal problem with respect to the maximum norm: for some integer $n \ge 0$, some $z \in \mathbb{C}$ and some compact set $S \subset \mathbb{C}$, consider the quantity

$$E_n(z,S) = \min_{p \in P_n} \max_{\lambda \in S} \frac{|p(\lambda)|}{|p(z)|}. \tag{1.16}$$

Clearly, $E_n(z,S)$ is decreasing in n and increasing in S, and $E_0(z,S) = 1$. Also, for any $a, b \ne 0$ there holds $E_n(a, a+bS) = E_n(0,S)$. The motivation for studying this extremal problem comes from the following observation

Exercise 8. For the inner product (1.11), show the following link between E_n and the Szegő function

$$z \in \mathbb{C}: \qquad \frac{1}{K_{n,2}(z,z)} \le ||r_0||^2 \, E_n(z, \Lambda(J_{N'}))^2.$$

If in addition $\Lambda(A) \subset (0, +\infty)$, show that

$$\frac{1}{|p_n(0)|} \le ||r_0||_{A^{-1}} \, E_n(0, \Lambda(J_{N'})).$$

Also, by dropping the negative terms in the sums occurring in Theorem 1.3 we obtain the following upper bound for the rate of convergence of Ritz values.

Corollary 1.6. *We have the following upper bounds: if $\lambda_k \le x_{1,n}$ then*

$$x_{1,n} - \lambda_k \le 2\,||A||\,\frac{||r_0||^2}{w(\lambda_k)^2} E_{n-1}(\lambda_k, \Lambda(J_{N'}) \setminus (-\infty, x_{1,n}])^2,$$

If $x_{\kappa-1,n} \le \lambda_k \le x_{\kappa,n}$ then

$$(\lambda_k - x_{\kappa-1,n})(x_{\kappa,n} - \lambda_k) \le 4\,||A||^2\,\frac{||r_0||^2}{w(\lambda_k)^2} E_{n-2}(\lambda_k, \Lambda(J_{N'}) \setminus [x_{\kappa-1,n}, x_{\kappa,n}])^2$$

The preceding Theorem gives the idea that eigenvalues λ_k sufficiently away from the rest of the spectrum of $\Lambda(A)$ (so-called outliers) and having a sufficiently large eigencomponent $w(\lambda_k)$ should be well approximated by Ritz values. However, as we will see later, for convergence it will be sufficient that there are not "too many" eigenvalues close to λ_k.

Let us further discuss the extremal problem (1.16). In the case $S \subset \mathbb{R}$ and real z not lying in the convex hull of S, it is known that the polynomial $T_{n,\infty}$ of (1.2) (put $w_n = 1$) is extremal for (1.16), see for instance [27]. The latter is uniquely characterized by a so-called alternant, that is, the extremal polynomial attains its maximum on S at least $n+1$ times, with alternating sign. For the sake of completeness, let us discuss in more detail the case of an interval.

Lemma 1.1. *If $0 < a < b$, the value $E_n(0,[a,b])$ is attained (up to a linear transformation) for the Chebyshev polynomial of the first kind $T_n(\cos(\phi)) = \cos(n\phi)$, and*

$$E_n(0,[a,b]) = \frac{2}{y^n + y^{-n}} \le 2y^n, \qquad y = \frac{\sqrt{b/a}-1}{\sqrt{b/a}+1} < 1.$$

Proof. We first notice that $E_n(0,[a,b]) = E_n(z,[-1,1])$ with $z = (b+a)/(b-a) > 1$. If T_n would not be extremal, then there is a polynomial P of degree n with

$$P(z) = T_n(z), \qquad \text{and } M := ||P||_{L_\infty([-1,1]} < ||T_n||_{L_\infty([-1,1]} = 1.$$

If follows that the polynomial $R := T_n - P$ satisfies $R(z) = 0$, and for $j = 0, 1, ..., n$

$$(-1)^j R(\cos(\frac{\pi j}{n})) = (-1)^j T_n(\frac{\pi j}{n})) - (-1)^j P(\cos(\frac{\pi j}{n}))$$
$$= 1 - (-1)^j P(\cos(\frac{\pi j}{n})) \ge 1 - M > 0.$$

Hence R must have $n+1$ roots, but is a polynomial of degree n, a contradiction. Thus T_n is indeed extremal. It remains to compute its value at z, where we use the recurrence

$$T_{n+1}(z) = (y + y^{-1})T_n(z) - T_{n-1}(z), \quad T_0 = 1, \quad T_1(z) = z = \frac{y+1/y}{2}.$$

□

In [27] one also finds a discussion of the case of S being a union of two intervals. Here the solution may be estimated in terms of Weierstass elliptic functions, see also [1]. A simple upper bound is discussed in the following exercise.

Exercise 9. Using the preceding result, show that, for $0 < a < b$,

$$E_{2n}(0,[-b,-a] \cup [a,b]) \le 2\Big(\frac{b/a-1}{b/a+1}\Big)^n.$$

Derive from this relation an explicit bound for $(\lambda_k - x_{\kappa-1,n})(x_{\kappa,n} - \lambda_k)$ in terms of the distance of λ_k to the rest of the spectrum of A.

A combination of Exercise 8 with Remark 1.2, Corollary 1.3 and Corollary 1.5 (and its proof) leads to the following estimates.

Corollary 1.7. *For hermitian A,*

$$\frac{||r_n^{\mathrm{GMRES}}||}{||r_0^{\mathrm{GMRES}}||} = \frac{||r_n^{\mathrm{MinRES}}||}{||r_0^{\mathrm{MinRES}}||} \leq E_n(0, \Lambda(J_{N'})) \leq E_n(0, \Lambda(A)).$$

Moreover, for symmetric positive definite A

$$\frac{||r_n^{\mathrm{CG}}||_{A^{-1}}}{||r_0^{\mathrm{CG}}||_{A^{-1}}} = \frac{||r_n^{\mathrm{FOM}}||_{A^{-1}}}{||r_0^{\mathrm{FOM}}||_{A^{-1}}} \leq E_n(0, \Lambda(J_{N'})) \leq E_n(0, \Lambda(A)).$$

The bounds of Corollary 1.7 should be considered as worst case bounds since they do not take into account the particular choice of the starting residual. However, it is known [31] that they cannot be sharpened in the following sense: one may give $\tilde{A}, \tilde{r}_0$ with $E_n(0, \Lambda(A)) = E_n(0, \Lambda(\tilde{A}))$ such that there is equality in Corollary 1.7 for these new data.

Notice also that we obtain a proof of (1.1) by replacing $\Lambda(A)$ by its convex hull in Corollary 1.7, and by applying Lemma 1.1. However, following this approach we forget completely about the fine structure of the spectrum. It is the aim of the following sections to analyze more pecisely in terms of logarithmic potential theory how the actual distribution of eigenvalues helps us to improve (1.1).

Remark 1.3. Let us finish this section by commenting briefly on different approaches for bounding the residual of Krylov subspace methods in case of not necessarily normal A. We start from the observation that

$$\frac{||r_n^{\mathrm{GMRES}}||}{||r_0^{\mathrm{GMRES}}||} \leq \min\{\frac{||P(A)||}{|P(0)|} : P \text{ is a polynomial of degree } \leq n\}. \qquad (1.17)$$

For diagonalizable A, the right hand side may be bounded from above by $E_n(0, \Lambda(A))$ times the condition number of the matrix of eigenvectors of A, see for instance [59]. However, for matrices far from being normal, this condition number might be quite large. There are mainly two attempts to overcome this difficulty (see for instance [32] and the references therein), the first is based on the so-called ϵ-pseudo-spectrum

$$\Lambda_\epsilon(A) := \{z \in \mathbb{C} : ||(zI - A)^{-1}|| \geq \frac{1}{\epsilon}\}$$

in terms of $E_n(0, \Lambda_\epsilon(A))$, the second one on the field of values

$$W(A) = \{\frac{y^* A y}{y^* y} : y \in \mathbb{C}^N, y \neq 0\},$$

which by Hausdorff's theorem is a compact and convex set. For a convex set S it can be shown that

$$\exp(-n g_S(0)) \le E_n(0,S) \le 3\exp(-n g_S(0)), \tag{1.18}$$

where g_S denotes the Green function of some compact subset of the complex plane, see Section 2.1. Here the left hand inequality is just the Bernstein-Walsh inequality (see for instance [47] in this volume), and the right hand bound can be shown with help of the corresponding Faber polynomials. A recent and quite deep result of M. Crouzeix [21] (found after the publication of [32]) says that there is a universal constant $C < 34$ such that, for any square matrix A and any polynomial P,

$$||P(A)|| \le C \max_{z\in W(A)} |P(z)|. \tag{1.19}$$

Hence combining (1.17), (1.18), and (1.19) we obtain

$$\frac{||r_n^{\mathrm{GMRES}}||}{||r_0^{\mathrm{GMRES}}||} \le 3C\exp(-n g_{W(A)}(0)),$$

compare with [9]. The constant C can be made smaller by replacing $W(A)$ in (1.19) by some larger convex set, see [9, 8]. A different approach, via estimating directly $||P(A)||$ for a suitable Faber polynomial P, allows even to establish the sharper bound [7]

$$\frac{||r_n^{\mathrm{GMRES}}||}{||r_0^{\mathrm{GMRES}}||} \le 3\exp(-n g_{W(A)}(0).$$

Notice however that such bounds are typically only interesting for small n since they do not allow to describe a superlinear rate of convergence.

2 Extremal Problems in Complex Potential Theory and nth Root Asymptotics of OP

2.1 Energy Problems with External Field

The energy problem with external field has been successfully applied in order to describe asymptotics of orthogonal polynomials on unbounded sets such as Hermite or Freud polynomials. Since this subject has already been discussed in detail in this volume [47], we recall here without proof the basic concepts. Also, for the sake of a simplified presentation, we will restrict ourselves to compact regular real sets and continuous external fields.

Given some compact $\Sigma \subset \mathbb{R}$, we denote by $\mathcal{M}_t(\Sigma)$ the set of Borel measures μ with support $\mathrm{supp}(\mu) \subset \Sigma$ and mass $||\mu|| := \mu(\Sigma) = t$. The logarithmic potential and the energy of a measure $\mu \in \mathcal{M}_t(\Sigma)$ are given by

$$U^\mu(z) = \int \log(\frac{1}{|x-z|})\, d\mu(x), \quad I(\mu) = \iint \log(\frac{1}{|x-y|})\, d\mu(x)\, d\mu(y).$$

Notice that, for a monic polynomial P of degree n, the expression $-\log(|P|^{1/n})$ coincides with the logarithmic potential of some discrete probability measure, which in case of distinct roots has mass $1/n$ at each root of P. On the other hand, such discrete measures are dense in the set of Borel measures, explaining why the tool of logarithmic potential theory is suitable for studying nth root asymptotics.

We define more generally for $\mu, \nu \in \mathcal{M}_t(\Sigma)$ the mutual energy by the expression

$$I(\mu,\nu) = \iint \log(\frac{1}{|x-y|})\, d\mu(x)\, d\nu(y) \in (-\infty, +\infty].$$

The mutual energy is lower semi-continuous, that is, given two sequences $(\mu_n), (\nu_n) \subset \mathcal{M}_t(\Sigma)$, which converge in weak star topology to μ, and ν, respectively (written by $\mu_n \xrightarrow{*} \mu$), we have

$$\liminf_{n\to\infty} I(\mu_n,\nu_n) \geq I(\mu,\nu). \tag{2.1}$$

From this one can deduce as an exercise the *principle of descent*

$$\liminf_{n\to\infty} U^{\mu_n}(z_n) \geq U^{\mu}(z), \quad \text{if } \mu_n \xrightarrow{*} \mu,\ z_n \to z \text{ for } n\to\infty. \tag{2.2}$$

As explained for instance in [47], the capacity of a compact set E is defined by the minimization problem

$$\mathrm{cap}(E) := \exp(-\min\{I(\mu) : \mu \in \mathcal{M}_1(E)\}).$$

If $\mathrm{cap}(E) > 0$, or, equivalently, if there is an $\mu \in \mathcal{M}_1(E)$ with finite energy, then one may prove using the strict convexity of $\mu \mapsto I(\mu)$ on measures with finite energy and Helly's theorem (weak compactness of $\mathcal{M}_x(E)$) that there is a unique measure ω_E called Robin measure realizing the minimum in the definition of the capacity. The Green function of a compact set E is defined by

$$g_E(z) = \log(\frac{1}{\mathrm{cap}(E)}) - U^{\omega_E}(z),$$

behaving at infinity like $\log(|z|) - \log(\mathrm{cap}(E)) + o(1)_{z\to\infty}$, being harmonic in $\mathbb{C} \setminus E$, subharmonic and ≥ 0 in $\mathbb{C}$, and equal to zero quasi everywhere on E, i.e., in $E \setminus E_0$, with $\mathrm{cap}(E_0) = 0$. Conversely, the Green function can be also uniquely characterized by these properties. For instance, for an interval we have (see, e.g., [47])

$$\begin{aligned} g_{[a,b]}(z) &= \log\left|\frac{2z-b-a}{b-a} + \sqrt{(\frac{2z-b-a}{b-a})^2 - 1}\right|, \\ \frac{d\omega_{[a,b]}}{dx}(x) &= \frac{1}{\pi\sqrt{(x-a)(b-x)}}. \end{aligned} \tag{2.3}$$

A compact set E is called regular (with respect to the Dirichlet problem) if g_E is identically zero on E (and hence g_E is continuous in $\mathbb{C}$ by the principle of continuity [60, Theorem II.3.5]). The link with the so-called Wiener condition and the regularity with respect to the Dirichlet problem is a nice piece of harmonic analysis, we refer the interested reader for instance to [60, Appendix A], where it is also shown that regularity is a local property

$$E \text{ regular}, \, x \in E, \, r > 0 \quad \Longrightarrow \quad E \cap \{y \in \mathbb{C} : |y - x| \leq r\} \text{ is regular.} \quad (2.4)$$

Here we only mention a sufficient condition in the following (a little bit difficult) exercise, the interested reader should compare with [55, § 5.4.3].

Exercise 10. Let $E \subset \mathbb{C}$ be compact, and suppose that for each $x \in E$ there exists $C(x), K(x) > 0$ such that, for all $r > 0$ sufficiently small,

$$\operatorname{cap}(\{y \in E : |y - x| \leq r\}) \geq C(x) \, r^{K(x)}.$$

Show that, for any $x \in E$, there exists $(\mu_n) \in \mathcal{M}_1(E)$ with $\mu_n \xrightarrow{*} \delta_x$ (the Dirac measure), and logarithmic potential V^{μ_n} being bounded uniformly in n by some integrable function (use the maximum principle for logarithmic potentials). Also, show that $I(\mu_n, \omega_E) \to U^{\omega_E}(x)$, and deduce that E is regular.

Exercise 11. Using the fact that $\operatorname{cap}([a, b]) = (b - a)/4$, show that a finite union of compact non degenerate intervals is regular. Does it remain regular if one adds an additional point?

Logarithmic potential theory has a nice electrostatic interpretation in $\mathbb{R}^2$ (or for cylinder symmetric configurations in $\mathbb{R}^3$, that is, a mass point in $\mathbb{R}^2$ corresponds to an infinite wire in $\mathbb{R}^3$ with uniform charge). Given a positive unit charge at zero, its electric potential is given by U^{δ_0}. Under this point of view, $\mu \in \mathcal{M}_1(E)$ represents a (static) distribution of a positive unit charge on some set E, with electric potential U^μ and electric energy $I(\mu)$. In physics, the equilibrium state is always described as the one having minimal energy. Thus, ω_E may be considered as the equilibrium distribution of a positive unit charge on a conducting material E, which by physical reasons should have a constant electric potential on E. Also, the fact that $\operatorname{supp}(\omega_E)$ is subset of the outer boundary of E (see for instance [47]) is known in physics as the Faraday principle.

One may wonder about what kind of equilibrium distribution is obtained if there is some additional fixed external field, induced for instance by some negative charge on some isolator outside of E. This problem has already been considered a long time ago by Gauss. Mathematically speaking, we have to solve the following problem:

Definition 2.1. *Let $\Sigma \subset \mathbb{R}$ be a regular compact set, $t > 0$, and $Q \in C(\Sigma)$. For $\mu \in \mathcal{M}_t(\Sigma)$, consider the weighted energy*

$$I^Q(\mu) = I(\mu) + 2\int Q\,d\mu.$$

We consider the problem of finding

$$W_{t,Q,\Sigma} := \inf\{I^Q(\mu) : \mu \in \mathcal{M}_t(\Sigma)\},$$

and, if possible, an extremal measure $\mu_{t,Q,\Sigma}$ *with* $I^Q(\mu_{t,Q,\Sigma}) = W_{t,Q,\Sigma}$.

If the external field Q is repelling or attracting positive charges, it may happen that the support of some extremal measure is a proper subset of Σ. By physical arguments, it should happen that there is a unique equilibrium, and that the corresponding potential is constant on the part of Σ which is charged by the equilibrium measure, and larger than this constant elsewhere on Σ. Indeed, one may give a mathematical proof of this statement.

Theorem 2.1. *Let* $\Sigma \subset \mathbb{R}$ *be a regular compact set,* $t > 0$*, and* $Q \in \mathcal{C}(\Sigma)$*. The extremal measure* $\mu_{t,Q,\Sigma}$ *of Definition 2.1 exists and is unique.*

Moreover, with $w = w_{t,Q,\Sigma} := W_{t,Q,\Sigma} - \int Q\,d\mu_{t,Q,\Sigma}$ *and* $\mu = \mu_{t,Q,\Sigma}$ *there holds*

$$U^\mu(z) + Q(z) \geq w \qquad \text{for } z \in \Sigma, \text{ and} \tag{2.5}$$

$$U^\mu(z) + Q(z) \leq w \qquad \text{for } z \in \operatorname{supp}(\mu). \tag{2.6}$$

Conversely, if there is a measure $\mu \in \mathcal{M}_t(\Sigma)$ *and a constant* w *such that (2.5) and (2.6) hold, then* $\mu = \mu_{t,Q,\Sigma}$ *and* $w = w_{t,Q,\Sigma}$.

Remark 2.1. It follows from (2.5) and (2.6) that the potential of $\mu_{t,Q,\Sigma}$ equals $w_{t,W,\Sigma} - Q$ on $\operatorname{supp}(\mu_{t,Q,\Sigma})$, the latter being continuous. Thus, by the principle of continuity [60, Theorem II.3.5], the potential of $\mu_{t,Q,\Sigma}$ is continuous.

The main ideas for the proof of Theorem 2.1 are discussed in the following exercise.

Exercise 12. Let $\mathcal{N}$ be some closed convex subset of $\mathcal{M}_t(\Sigma)$, containing at least one element μ with $I(\mu) < \infty$.

(a) Show that I^Q is strictly convex.
(b) Using Helly's theorem, show that $W_\mathcal{N} = \min\{I^Q(\mu) : \mu \in \mathcal{N}\}$ is attained for some $\mu_\mathcal{N} \in \mathcal{N}$. Why must such a measure be unique?
(c) By discussing $\nu = s\mu + (1-s)\mu_\mathcal{N}$ for $s \to 0+$, show that $\mu_\mathcal{N}$ is uniquely characterized by the property

$$\mu \in \mathcal{N}: \qquad w_\mathcal{N} := I(\mu_\mathcal{N}) + \int Q\,d\mu_\mathcal{N} \leq I(\mu, \mu_\mathcal{N}) + \int Q\,d\mu.$$

The only property of Theorem 2.1 not being an immediate consequence of the preceding exercise is the fact that the equilibrium measure satisfies the equilibrium conditions (2.5) and (2.6). For showing this property, one applies a principle known as *principle of domination* for logarithmic potentials.

Theorem 2.2. *If $\mu, \nu \in \mathcal{M}(\Sigma)$ with finite energy, if $||\nu|| \leq ||\mu||$, and if, for some constant C, the relation $U^\mu \leq U^\nu + C$ holds μ-almost everywhere, then this relation holds for all $z \in \mathbb{C}$.*

Proof. See [60, Theorem II.3.2]. □

Theorem 2.1 together with Theorem 2.2 allow us to derive the following result known as the weighted Bernstein-Walsh inequality.

Corollary 2.1. *With the assumptions of Theorem 2.1, we consider the weight $w(x) = \exp(-Q(x))$. Then for any polynomial of degree at most n and for all $z \in \mathbb{C}$*

$$\frac{|P(z)|}{||w^n P||_{L_\infty(E)}} \leq \exp(n w_{1,Q,\Sigma} - n U^{\mu_{1,Q,\Sigma}}(z)).$$

Exercise 13. Prove this corollary.

Remark 2.2. For a trivial weight $w = 1$, Corollary 2.1 reduces to the classical Bernstein-Walsh inequality, which in terms of the function E_n of (1.16) can be rewritten as $E_n(z, S) \geq \exp(-n g_S(z))$. However, for applications in numerical linear algebra we need not lower but upper bounds for E_n. Depending on the "smoothness" of the set S, the Bernstein-Walsh inequality is more or less sharp, for instance, from Lemma 1.1 and the explicit formula of (2.3) we learn that $E_n(z, [a, b]) \leq 2\exp(-n g_{[a,b]}(z))$. For general sets S of positive capacity and $z \notin S$ ony may show (see, e.g., [55, Section V.5.3]) that $E_n(z, S)^{1/n}$ tends to $\exp(-g_S(z))$ for $n \to \infty$.

We learn from Corollary 2.1 that the weighted maximum norm of a polynomial P lives on a possibly proper compact subset of Σ, that is, $||w^n P||_{L_\infty(E)} = ||w^n P||_{L_\infty(S^*(1,Q,\Sigma))}$ with

$$S^*(t, Q, \Sigma) := \{z \in E : U^{\mu_{t,Q,\Sigma}}(z) + Q(z) \leq w_{t,Q,\Sigma}\}. \tag{2.7}$$

The link between $S^*(t, Q, \Sigma)$ and the support of the equilibrium measure is studied in [60, Theorem IV.4.1], and in more detail by Buyarov and Rakhmanov [16]. We state parts of their findings without proof.

Theorem 2.3. [16] *Let Σ, Q be as in Theorem 2.1, and define the sets*

$$S(t) := \operatorname{supp}(\mu_{t,Q,\Sigma}), \qquad t \geq 0.$$

The sets $S(t)$ are increasing in t, with $\cap_{\tau > t} S(\tau) = S^*(t, Q, \Sigma)$ for all t, and $S(t) = S^*(t, Q, \Sigma)$ for almost all t. Furthermore, for all $t > 0$ and $z \in \mathbb{C}$,

$$\mu_{t,Q,\Sigma} = \int_0^t \omega_{S(\tau)} \, d\tau, \qquad w_{t,Q,\Sigma} - U^{\mu_{1,Q,\Sigma}}(z) = \int_0^t g_{S(\tau)}(z) \, d\tau.$$

Theorem 2.3 tells us that all extremal quantities are completely determined once one knows $S(t)$ for all $t > 0$. Notice that $S(t)$ may consist of several intervals, or even have a Cantor-like structure. The determination of $S(t)$ for particular classes of external fields Q is facilitated by tools like the F-functional of Rakhmanov-Maskhar-Saff, see [47] or [60, Section IV.1.11], but remains in general a quite difficult task.

A quite interesting example for the preceding findings is the case of an exponential weight $w(x) = \exp(-x^2)$, described below, or more generally Freud weights. This example can be considered as the starting point for the research on weighted energy problems in the last twenty years.

Example 2.1. Given $\alpha > 0$, let $w(x) = \exp(-\gamma_\alpha |x|^\alpha)$ (i.e., $Q(x) = \gamma_\alpha |x|^\alpha$), with

$$\gamma(\alpha) = \int_0^1 \frac{u^{\alpha-1}}{\sqrt{1-u^2}}\, du = \frac{\Gamma(\alpha/2)\Gamma(1/2)}{2\Gamma((\alpha+1)/2)}$$

($\gamma_2 = 1$), and define the probability measure μ with $\mathrm{supp}(\mu) = [-1, 1]$ by the weight function

$$\frac{d\mu}{d\lambda}(\lambda) = s(\alpha, \lambda) = \frac{\alpha}{\pi} \int_{|\lambda|}^1 \frac{u^{\alpha-1}}{\sqrt{u^2 - \lambda^2}}\, du.$$

On shows [60, Theorem IV.5.1] that

$$U^\mu(x) + Q(x) \begin{cases} = w := \log(2) + 1/\alpha \text{ for } x \in [-1, 1], \\ > w = \log(2) + 1/\alpha \text{ for } x \in \mathbb{R} \setminus [-1, 1]. \end{cases}$$

Hence, by Theorem 2.1, for any compact Σ containing $[-1, 1]$ we have $\mu_{1,Q,\Sigma} = \mu$, and $S^*(1, Q, \Sigma) = [-1, 1]$. In particular, for any polynomial P of degree at most n we get from Corollary 2.1 that

$$||w^n P||_{L^\infty(\Sigma)} = ||w^n P||_{L^\infty(\mathbb{R})} = ||w^n P||_{L^\infty([-1,1])}.$$

Using the linear transformation $y = x \cdot (n\gamma_\alpha)^{1/\alpha}$, it follows that

$$||e^{-|y|^\alpha} P(y)||_{L^\infty(\mathbb{R})} = ||e^{-|y|^\alpha} P(y)||_{L^\infty([-(n\gamma_\alpha)^{1/\alpha}, (n\gamma_\alpha)^{1/\alpha}])}.$$

Exercise 14. Relate the findings of Example 2.1 to those of Theorem 2.3.

Remark 2.3. Using the results mentioned in Example 2.1 one may show for instance that the zeros of the Hermite orthogonal polynomials, after division by $\sqrt{n}$, have an asymptotic distribution given by the weight function $s(2, \lambda) = (2/\pi)\sqrt{1 - \lambda^2}$ on $[-1, 1]$. We refer the reader to [60, Section III.6] for more precise results on nth root asymptotics for L_p-extremal polynomials (including the case $p = 2$ of orthogonal polynomials) with respect to varying weights w^n.

We finish this section by discussing two particular examples of external fields given by the negative potential of a measure.

Exercise 15. If $Q(z) = -U^\sigma(z)$ for some $\sigma \in \mathcal{M}(\Sigma)$ with continuous potential, show that $\mu_{t,Q,\Sigma} = \sigma + (t - ||\sigma||)\omega_\Sigma$ for every $t \geq ||\sigma||$.

Things are becoming more exciting if the external field is a negative potential of some measure σ with compact support outside of Σ. Here the extremal measure $\mu_{||\sigma||,-U^\sigma,\Sigma}$ can be considered as a sort of projection of σ onto Σ, more precisely, we obtain the "problem of balayage" [60, Section II.4] studied already by H. Poincaré and Ch. de la Vallée-Poussin: find a measure of the same mass as σ supported on Σ with potential coinciding (up to some constant) with U^σ on Σ. In terms of electrostatics, we look for a positive unit charge on some conductor being in equilibrium with some fixed negative unit charge (on some isolator).

Exercise 16. Let $\Sigma \subset \mathbb{R}$ be compact and regular, $t > 0$, $Q(z) = -U^\sigma(z)$ for some measure σ with compact support $\mathrm{supp}(\sigma) \not\subset \Sigma$, finite energy, and potential continuous on Σ (the latter being true for instance if $\mathrm{supp}(\sigma) \cap \Sigma$ is empty), and write shorter $S(t) = \mathrm{supp}(\mu_{t,Q,\Sigma})$.

(a) Show that $\mathrm{cap}(S(t)) > 0$. Use the maximum principle for subharmonic functions for showing that $g_{S(t)}(z) > 0$ for all $z \notin S(t)$.
(b) Let $\Delta \subset \Sigma$ some Borel set with $\mathrm{cap}(\Delta) = 0$, and $\nu \in \mathcal{M}(\Sigma)$ with finite energy. Show that $\nu(\Delta) = 0$.
(c) By applying Theorem 2.2, show that

$$z \in \mathbb{C}: \qquad U^{\mu_{t,Q,\Sigma}}(z) + Q(z) \leq w_{t,Q,\Sigma} + (||\sigma|| - t)\, g_{S(t)}(z).$$

(d) Using the maximum principle for subharmonic functions and (c), show that $S(t) = \Sigma$ for all $t \geq ||\sigma||$. Deduce that the balayage problem onto Σ has a unique solution given by $\mu_{||\sigma||,-U^\sigma,\Sigma}$.
(e) In the case $t < ||\sigma||$, show that $\mu_{t,Q,\Sigma} + (||\sigma|| - t)\,\omega_{S(t)}$ is the balayage of σ onto $S(t)$.

Exercise 17. Let $\Sigma \subsetneqq \Sigma' \subset \mathbb{R}$ be compact and regular. What is the balayage measure of $\omega_{\Sigma'}$ onto Σ?

For a regular compact $\Sigma \subset \mathbb{R}$, we may define the Green function with pole at $a \in \mathbb{C} \setminus \Sigma$ by the (balayage) formula of a Dirac measure

$$g_\Sigma(z, a) = w_{1,Q,\Sigma} - U^{\mu_{1,Q,\Sigma}}(z) - Q(z), \qquad Q(z) = -U^{\delta_a}(z),$$

compare with [60, Section II.4]. One may show [60, Eqn. (II.4.31)] that it is possible to integrate the preceding formula with respect to a: provided that $Q(z) = -U^\sigma(z)$, we have

$$w_{||\sigma||,Q,\Sigma} - U^{\mu_{||\sigma||,Q,\Sigma}}(z) - Q(z) = \int g_\Sigma(z, a)\, d\sigma(a). \tag{2.8}$$

Provided that explicit formulas for $g_\Sigma(\cdot,\cdot)$ are available, this formula can be exploited to derive explicit formulas for the density of the balayage measure, by recovering the measure from its potential [60, Chapter II.1]. E.g., for $x \in \Sigma = [a,b]$, an interval, and $\sigma(\Sigma) = 0$, corresponding formulas are given in [60, Corollary IV.4.12]

$$\frac{d\mu_{||\sigma||,-U^\sigma,[a,b]}}{dx}(x) = \frac{1}{\pi}\int \frac{\sqrt{|y-a|\,|y-b|}}{|y-x|\,\sqrt{(x-a)(b-x)}}\,d\sigma(y). \tag{2.9}$$

2.2 Energy Problems with Constraint and External Field

Discrete Chebyshev polynomials are orthonormal with respect to the scalar product

$$\prec P, Q \succ = \sum_{z \in E_N} w_N(z)^2\,\overline{P(z)}\,Q(z)$$

with $E_N = \{0, 1, ..., N\}$ and $w_N(z) = 1$. Further systems of "classical" discrete orthogonal polynomials contain

$$\text{Meixner polynomials} \;:\; E_N = \{0,1,2,...\}, \qquad w_N(k)^2 = \frac{c^k (b)_k}{k!}$$

$$\text{Charlier polynomials} \;:\; E_N = \{0,1,2,...\}, \qquad w_N(k)^2 = \frac{c^k e^{-c}}{k!}$$

Krawtchouk polynomials, discrete Freud polynomials, discrete Hahn polynomials, see for instance [20, 24, 22, 45, 43] and the references therein.

It was Rakhmanov [58] who first observed that the nth root of the nth discrete Chebyshev polynomial (and other discrete orthogonal polynomials) for so-called ray sequences, that is, $n, N \to \infty$ in such a manner that $n/N \to t \in (0,1)$, can be described in terms of a constrained weighted equilibrium problem in logarithmic potential theory. Recall from Chapter 1.3 that asymptotics of discrete Chebyshev polynomials are closely related to the convergence behavior of Krylov subspace methods applied to a matrix with equally spaced eigenvalues, and to the convergence of its Ritz values. Other domains of applications for asymptotics of discrete orthogonal polynomials include coding theory and discrete dynamical systems.

Similar to the approach for Hermite polynomials (c.f., Remark 2.3), for obtaining nth root asymptotics it is first required to scale the set E_N by some appropriate power of N. The resulting supports will then have an asymptotic distribution for $N \to \infty$ which can be described by some Borel measure σ. Furthermore, after scaling, the weights w_N will behave like $w(z)^N$ for some appropriate weight which can be written as $w = \exp(-Q)$. The constrained energy problem considered by Rakhmanov [58] consists in minimizing the logarithmic energy $I(\mu)$, where μ is some probability measure satisfying in addition the constraint that $\sigma - \mu$ is some nonnegative measure. The set of such measures will be denoted by

$$\mathcal{M}_t^\sigma := \{\mu \geq 0 : ||\mu|| = t, \sigma - \mu \geq 0\}$$

where $0 < t \leq ||\sigma||$. In our context it will be useful to introduce a weighted analogue of this problem. Its unique solution has been characterized by Dragnev and Saff [24, Theorem 2.1 and Remark 2.3], and further investigated by several other authors. We summarize some of their findings in Theorem 2.4 below, here additional regularity assumptions on σ and Q enable us to obtain a simplified statement.

Theorem 2.4 (see [24]). *Let Q be a continuous real–valued function on some closed set $\Sigma \subset \mathbb{C}$, $w := \exp(-Q)$, and, if Σ is unbounded, suppose that $Q(z) - \log|z| \to +\infty$ for $|z| \to \infty$. Furthermore, let σ be a positive measure with $\mathrm{supp}(\sigma) \subset \Sigma$, such that, for any compact $K \subset \mathrm{supp}(\sigma)$, the restriction $\sigma|_K$ of σ to K has a continuous potential. Finally, let $0 < t < ||\sigma||$.*

Then for the extremal problem

$$W_{t,Q,\sigma} := \inf\{I^Q(\mu) : \mu \in \mathcal{M}_t^\sigma\}$$

there exists a unique measure $\mu_{t,Q,\sigma} \in \mathcal{M}_t^\sigma$ with $W_{t,Q,\sigma} = I^Q(\mu_{t,Q,\sigma})$, and this extremal measure has compact support. Furthermore, there exists a constant $w = w_{t,Q,\sigma}$ such that for $\mu = \mu_{t,Q,\sigma}$ we have the equilibrium conditions

$$U^\mu(z) + Q(z) \geq w \qquad \text{for } z \in \mathrm{supp}(\sigma - \mu), \text{ and} \tag{2.10}$$

$$U^\mu(z) + Q(z) \leq w \qquad \text{for } z \in \mathrm{supp}(\mu). \tag{2.11}$$

Conversely, if $\mu \in \mathcal{M}_t^\sigma$ has compact support and satisfies the equlibrium conditions (2.10), (2.11), for some constant w then $\mu = \mu_{t,Q,\sigma}$.

In terms of electrostatics, we may consider $\mu_{t,Q,\sigma}$ as the equilibrium distribution on $\mathrm{supp}(\sigma)$ of a positive charge of mass t in the presence of an external field Q, but here $\mathrm{supp}(\sigma)$ is no longer conducting: indeed $\mu \leq \sigma$ imposes a constraint on the maximum charge per unit. As a consequence, the corresponding weighted potential is no longer constant on the whole part of $\mathrm{supp}(\sigma)$ charged by our extremal measure: we may have a strictly smaller weighted potential at the part $\mathrm{supp}(\sigma) \setminus \mathrm{supp}(\sigma - \mu_{t,Q,\sigma})$ where the constraint is active. However, in the free part $\mathrm{supp}(\sigma - \mu_{t,Q,\sigma}) \cap \mathrm{supp}(\mu_{t,Q,\sigma})$ the weighted potential is still constant.

We should comment on the proof of Theorem 2.4. For showing existence and uniqueness of the extremal measure, we can follow the reasoning of Exercise 12, at least for compact Σ (the growth condition on Q can be shown to imply that it is sufficient to consider compact Σ). A proof of the equivalent characterization by the equilibrium conditions (2.10) and (2.11) uses Exercise 12(c) and Theorem 2.2, as well as the following observation.

Exercise 18. Let ν be a measure with compact support and continuous potential. Use twice the principle of descent for showing that any measure $\mu \geq 0$ with $\mu \leq \nu$ also has a continuous potential (c.f., [58]). Show also that μ has no mass points.

Remark 2.4. The extremal constant $w_{t,Q,\sigma}$ is not necessarily unique [24, Example 2.4], but will be unique if $\mathrm{supp}(\sigma)$ is connected, or, more generally, if $\mathrm{supp}(\mu_{t,Q,\sigma})$ and $\mathrm{supp}(\sigma - \mu_{t,Q,\sigma})$ have a non-empty intersection.

Example 2.2. For the constraint $\frac{d\sigma}{dx}(x) = \alpha x^{\alpha-1}$ on $\mathrm{supp}(\sigma) = [0, +\infty)$, $\alpha > 1/2$, and the external field $Q(x) = \gamma \cdot x^{\alpha}$, it is shown in [45, Theorem 2.1] that

$$\mu_{t,Q,\sigma} = \int_0^t \omega_{[a(\tau),b(\tau)]}\, d\tau, \tag{2.12}$$

$$w_{t,Q,\sigma} - U^{\mu_{t,Q,\sigma}}(z) = \int_0^t g_{[a(\tau),b(\tau)]}(z)\, d\tau, \tag{2.13}$$

and $\mathrm{supp}(\mu_{t,Q,\sigma}) = [0, b(t)]$, $\mathrm{supp}(\mu_{t,Q,\sigma}) \cap \mathrm{supp}(\sigma - \mu_{t,Q,\sigma}) = [a(t), b(t)]$, where $0 \leq a(t) = t^{\alpha} a_0 < b(t) = t^{\alpha} b_0$ are solutions of the system

$$0 = \frac{1}{\pi} \int_{a(t)}^{b(t)} \frac{Q'(x)\, dx}{\sqrt{(b(t) - x)(x - a(t))}} - \int_{x \leq a(t)} \frac{d\sigma(x)}{\sqrt{(\beta(t) - x)(\alpha(t) - x)}}, \tag{2.14}$$

$$t = \frac{1}{\pi} \int_{a(t)}^{b(t)} \frac{xQ'(x)\, dx}{\sqrt{(b(t) - x)(x - b(t))}} - \int_{x \leq a(t)} \frac{x\, d\sigma(x)}{\sqrt{(b(t) - x)(a(t) - x)}}. \tag{2.15}$$

Exercise 19. In case of compact $\mathrm{supp}(\sigma)$, show the following property of duality

$$\mu_{t,Q,\sigma} + \mu_{||\sigma||-t,\tilde{Q},\sigma} = \sigma, \qquad \text{where } \tilde{Q} := -Q - U^{\sigma}.$$

Let us compare our extremal problem to the unconstrained one of Definition 2.1. In case of a "sufficiently large" constraint we clearly see by comparing Theorem 2.4 with Theorem 2.1 that the following implication holds

$$\sigma \geq \mu_{t,Q,\Sigma} \qquad \Longrightarrow \qquad \mu_{t,Q,\sigma} = \mu_{t,Q,\Sigma}. \tag{2.16}$$

Of course, the same conclusion is true if $\mathrm{supp}(\sigma - \mu_{t,Q,\sigma}) = \Sigma$. In what follows we will consider sometimes the case of a trivial external field $Q = 0$ (and hence compact $\mathrm{supp}(\sigma)$). Here the following result will be helpful relating the constrained energy problem with trivial weight to an unconstrained weighted extremal problem and more precisely to a balayage problem.

Lemma 2.1. *Under the assumptions of Theorem 2.4, if $Q = 0$ then*

$$\sigma - \mu_{t,0,\sigma} = \mu_{||\sigma||-t,\tilde{Q},\tilde{\Sigma}}, \qquad \textit{where } \tilde{Q} := -U^{\sigma} \textit{ and } \tilde{\Sigma} := \mathrm{supp}(\sigma),$$

and moreover

$$\mathrm{supp}(\mu_{t,0,\sigma}) = \mathrm{supp}(\sigma).$$

Proof. Write shorter $S := \mathrm{supp}(\mu_{t,0,\sigma})$. Recall that g_S equals zero on S up to some set of capacity zero. From (2.11) with $Q = 0$ and Exercise 16(b) we conclude that

$$U^{\mu_{t,0,\sigma}}(z) + g_S(z) \leq w := w_{t,0,\sigma}$$

holds $\mu_{t,0,\sigma}$–everywhere on S, and the principle of domination of Theorem 2.2 tells us that the above inequality is true for all $z \in \mathbb{C}$. In particular, for $z \notin S$ we have $U^{\mu_{t,0,\sigma}}(z) < w$ by Exercise 16(a). Comparing with (2.10) shows that $z \notin \mathrm{supp}(\sigma - \mu_{t,0,\sigma}) \supset \mathrm{supp}(\sigma) \setminus S$. Hence $S = \mathrm{supp}(\sigma)$, as claimed in the assertion. Moreover, from (2.10) and (2.11) we see that $\sigma - \mu_{t,0,\sigma}$ satisfies the equilibrium conditions (2.5), (2.6) corresponding to the external field $\tilde{Q}$ on $\tilde{\Sigma}$ with the normalization $||\sigma|| - t$, and hence is equal to $\mu_{||\sigma||-t,\tilde{Q},\tilde{\Sigma}}$ by Theorem 2.1. □

Exercise 20. Suppose that $(0,T) \ni t \mapsto S(t) \subset \mathbb{C}$ with $S(t)$ compact and regular decreasing sets, i.e., $S(t') \subset S(t)$ for $t' > t$, and consider the constraint

$$\sigma(x) = \int_0^T \omega_{S(t)}(x)\, dt, \qquad T = ||\sigma||. \tag{2.17}$$

Show that

$$w_{t,0,\sigma} - U^{\mu_{t,0,\sigma}}(z) = \int_0^t g_{S(\tau)}(z)\, d\tau. \tag{2.18}$$

Hint: verify equilibrium conditions.

Remark 2.5. As shown in [10, Theorem 2.1], it follows from Theorem 2.3 and Lemma 2.1 that the following more general statement is valid:

In case of a trivial external field $Q = 0$, the compact sets $S(t) := \mathrm{supp}(\sigma - \mu_{t,0,\sigma})$ are decreasing in t, any constraint σ has the integral representation (2.17), and formula (2.18) is true.

Remark 2.6. It is an open problem of establishing integral formulas of Buyarov-Rakhmanov type in the case of the constrained weighted extremal problem for general external fields Q. However, besides the preceding remark, there is another case where such formulas may be established (compare with [42, Lemma 3.1, Theorem 3.3, Proof of Lemma 6.2], [23, Chapter 4], [40, Proposition 4.1] and [12]):

Let $\mathrm{supp}(\sigma) = [A,B]$, $Q(A) = 0$, and suppose that the functions Q and $\widetilde{Q}$ defined by

$$\widetilde{Q}(x) = -Q(x) - U^{\sigma}(x)$$

are continuous in $[A,B]$ and have a continuous derivative in (A,B). Suppose in addition that the functions $x \mapsto (x-A)Q'(x)$ and $x \mapsto (B-x)\widetilde{Q}'(x)$ are increasing functions on $[A,B]$. Then (2.12) and (2.13) hold, with $A \leq a(t) < b(t) \leq B$ defined by (2.14), (2.15).

Some other results on the interval case for the constrained unweighted energy problem are given in [41, Theorem 2], [39, Theorem 5.1], [10, Lemma 3.1], in particular one may find (systems of) integral equations for determining the endpoints of $S(t)$, and links with the technique of balayage since, by Exercise 16 and Lemma 2.1, the measure $\sigma - \mu_{t,0,\sigma} + t\omega_{S(t)}$ coincides with the balayage of σ onto $S(t)$.

2.3 Asymptotics for Discrete Orthogonal Polynomials

Let $T_{n,p}$ be the extremal polynomials of (1.2). In this section we describe how the nth root asymptotic of the extremal constants $||w_N \cdot T_{n,p}||_{L_p(E_N)}$ for ray sequences $n, N \to \infty$, $n/N \to t$, as well as the asymptotic distribution of zeros of $T_{n,p}$ may be expressed in terms of the solution of the constrained weighted energy problem of Section 2.2.

For some discrete set E_N we define the corresponding counting measure

$$\nu_N(E_N) = \frac{1}{N} \sum_{z \in E_N} \delta_z,$$

a discrete measure where each element of E_N is charged by the mass $1/N$. Similarly, given a polynomial P with set of zeros Z, we write $\nu_N(P) := \nu_N(Z)$ for the corresponding normalized zero counting measure (where we count zeros according to their multiplicities). As usual, for a sequence of discrete sets $(E_N)_N$ we write $\nu_N(E_N) \xrightarrow{*} \sigma$ if for any continuous function f with compact support there holds

$$\lim_{N\to\infty} \int f(z) d\nu_N(z) = \int f(z)\, d\sigma(z), \quad \text{where} \quad \int f(z) d\nu_N(z) = \frac{1}{N} \sum_{z \in E_N} f(z).$$

Finally, for discrete sets E_N, F_N we define the discrete mutual energy

$$I_N(E_N, F_N) = \frac{1}{N^2} \sum_{x \in E_N} \sum_{y \in F_N, y \neq x} \log(\frac{1}{|x-y|}),$$

the mutual energy between two systems E_N and F_N of positive masspoints.

Exercise 21. Suppose that $\nu_N(E_N) \xrightarrow{*} \mu$, $\nu_N(F_N) \xrightarrow{*} \nu$. Show the lower semi-continuity

$$\liminf_{N\to\infty} I_N(E_N, F_N) \geq I(\mu, \nu).$$

Hint: consider the regularized kernel $(x, y) \mapsto \min\{\eta, \log(\frac{1}{|x-y|})\}$ for $\mathbb{R} \ni \eta \to +\infty$.

Weak asymptotics of discrete L_p-extremal polynomials have been a subject of a number of publications, see [24, Theorem 3.3] (for real compact Σ), [22, Theorem 2.5] (for $p = \infty$ and real Σ), [45, Theorem 7.4 and Lemma 8.3] (for

$0 < p \leq \infty$ and real Σ), [43, Theorem 7.1] (for $p = 2$ and real Σ, see also [44]), and finally [5, Theorem 1.3] (for $0 < p \leq \infty$ and complex supports). We summarize these findings in the following (a bit technical) theorem.

Theorem 2.5. [5] *Let $0 < p \leq \infty$. Furthermore, let Σ, σ, Q, $w = \exp(-Q)$ be as in Theorem 2.4, and $t \in (0, ||\sigma||)$ with $\mathrm{supp}(\mu_{t,Q,\sigma}) \cap \mathrm{supp}(\sigma - \mu_{t,Q,\sigma}) \neq \emptyset$.*

Suppose that the sets $E_N \subset \Sigma$ and the weights $w_N(z) \geq 0$, $z \in E_N$, $N \geq 0$, satisfy the conditions

$$\nu_N(E_N) \xrightarrow{*} \sigma, \tag{2.19}$$

$$\limsup_{N\to\infty} \sup_{z\in E_N} \frac{w_N(z)^{1/N}}{\tilde{w}(z)} \leq 1, \tag{2.20}$$

for some $\tilde{w} \in \mathcal{C}(\Sigma)$ with $|z|\tilde{w}(z) \to 0$ for $|z| \to \infty$ and if for any compact K there holds

$$\limsup_{N\to\infty} \sup_{z\in E_N\cap K} \frac{w_N(z)^{1/N}}{w(z)} \leq 1, \tag{2.21}$$

then

$$\limsup_{n,N\to\infty, n/N\to t} ||w_N \cdot T_{n,p}||_{L_p(E_N)}^{1/N} \leq \exp(-w_{t,Q,\sigma}), \tag{2.22}$$

and for any $z \in \mathbb{C}$ satisfying

$$\lim_{N\to\infty} U^{\nu_N(V(z)\cap E_N)}(z) = U^{\sigma|_{V(z)}}(z) \tag{2.23}$$

for some open neighborhood $V(z)$ of z with $\sigma(\partial V(z)) = 0$ there holds

$$\limsup_{n,N\to\infty, n/N\to t} \left[\min_{\deg P \leq n} \frac{||w_N \cdot P||_{L_p(E_N)}}{|P(z)|} \right]^{1/N} \leq \exp(U^{\mu_{t,Q,\sigma}}(z) - w_{t,Q,\sigma}), \tag{2.24}$$

If moreover there exists some bounded open neigborhood V of $\mathrm{supp}(\mu_{t,Q,\sigma})$ with

$$\lim_{N\to\infty} I_N(V \cap E_N, V \cap E_N) = I(\sigma|_V), \tag{2.25}$$

if for any compact K there holds

$$\lim_{N\to\infty} \sup_{z\in E_N\cap K} |w_N(z)^{1/N} - w(z)| = 0, \tag{2.26}$$

and if in the case $p < \infty$ there exists a $p' \in (0, p)$ with

$$\limsup_{N\to\infty} \left[|| [z\, w]^N ||_{L_{p'}(E_N)} \right]^{1/N} < \infty, \tag{2.27}$$

then (2.22) and (2.24) are sharp, more precisely, we have

$$\lim_{n,N\to\infty, n/N\to t} ||w_N \cdot T_{n,p}||_{L_p(E_N)}^{1/N} = \exp(-w_{t,Q,\sigma}), \tag{2.28}$$

$$\lim_{n,N\to\infty,n/N\to t}\left[\min_{\deg P\le n}\frac{||w_N\cdot P||_{L_p(E_N)}}{|P(z)|}\right]^{1/N}$$

$$= \exp(U^{\mu_{t,Q,\sigma}}(z) - w_{t,Q,\sigma}), \quad (2.29)$$

with z as in (2.23), and finally, if the two-dimensional Lebesgue measure of Σ is zero, there holds

$$\nu_N(T_{n,p}) \xrightarrow{*} \mu_{t,Q,\sigma} \quad \textit{for } n,N\to\infty, n/N\to t. \quad (2.30)$$

The statement of this theorem simplifies considerably for compact Σ (why?). Since E_N then has $\mathcal{O}(N)$ elements, we can use classical inequalities between Hölder norms showing that $L_p(E_N)$–norms for two different p are equivalent up to a factor being some power of N (which of course will vanish once we take Nth roots). Therefore it is evident that the right hand side of (2.22), (2.24), (2.28), (2.29), and (2.30) do not depend on p.

We will comment in Remark 2.7 below on the different assumptions and variations proposed by different authors and subsequently give the main ideas of the proof of Theorem 2.5 for compact Σ (for a proof for general Σ we refer the reader to [5, Theorem 1.3 and Theorem 1.4(c)]). Let us first have a look at the examples mentioned in the introduction of Section 2.2.

Example 2.3. [**Discrete Chebyshev Polynomials**] After scaling (dividing the support by N) we obtain assumption (2.19) with σ being the Lebesgue measure on $[0,1]$, having a continuous potential. Here conditions (2.20), (2.21), (2.26) and (2.27) are trivially true with $Q=0$. The interested reader may check that also condition (2.25) holds, and that, by (2.3),

$$\sigma(x) = \int_0^1 \omega_{S(t)}(x)\,dt, \qquad S(t) = [\frac{1}{2} - \frac{\sqrt{1-t^2}}{2}, \frac{1}{2} + \frac{\sqrt{1-t^2}}{2}].$$

Thus explicit formulas for $\mu_{t,0,\sigma}$ are given in Exercise 20, in particular one obtains from (2.29) for $p=+\infty$ (compare with [10, Corollary 3.2])

$$\lim_{n,N\to\infty,n/N\to t} \log(E_n(0,\{1/N,2/N,...,N/N\})^{1/N})$$

$$= w_{t,0,\sigma} - U^{\mu_{t,0,\sigma}}(0) = -\frac{(1+t)\log(1+t)+(1-t)\log(1-t)}{2}. \quad (2.31)$$

Example 2.4. [**Meixner Polynomials**] Here, again after division of the support by N, we obtain for σ the Lebesgue measure on $[0,+\infty)$. By Stirling's formula, we have for $x=k/N$ and $w(x)=\exp(x\log(c)/2)$

$$\frac{w_N(x)^{1/N}}{w(x)} = c^{-x/2}[\frac{c^{Nx}\Gamma(Nx+b)}{\Gamma(Nx+1)}]^{1/(2N)} = 1+o(1)_{N\to\infty}$$

uniformly for x in some compact. Hence (2.20), (2.21), (2.26) and (2.27) are true with $Q(x) = x\,\log(1/c)/2$. The corresponding equilibrium measure $\mu_{t,Q,\sigma}$ has been given in Example 2.2. Here we may explicitly solve the system (2.14), (2.15), of integral equations, and $a_0 = (1-\sqrt{c})/(1+\sqrt{c}) = 1/b_0$.

Remark 2.7. Let us shortly comment on the different assumptions of Theorem 2.5 and relate them with related conditions proposed by other authors. Conditions (2.19), (2.21) and (2.26) allow to relate our discrete L_p norm to the extremal problem with data σ and Q.

Conditions (2.20) and (2.27) ensure the finiteness of $||w_n P||_{L_p(E_n)}$ for a polynomial of degree at most n, at least for sufficiently large n. Such an additional condition is required for $p < \infty$ for controlling the contribution to the L_p norm of in modulus large elements of E_n. Stronger sufficient conditions for (2.27) in case of unbounded Σ have been discussed in [45] and [5, Lemma 2.7].

Finally, by considering the example $E_N = F_N \cup (e^{-N} + F_N)$, $F_N = \{0, 1/N, 2/N, ..., N/N\}$ it becomes clear that our asymptotic bounds (2.22), (2.24) cannot be sharp since they do not take into account the clustering of points of the support, compare also the discussion in [43, Section 8]. Rakhmanov [58] considered the additional separation condition

$$\liminf_{N\to\infty} \inf_{x,y\in E_N\cap K, x\neq y} N\cdot|x-y| > 0$$

for all compact sets K. The weaker condition

$$\lim_{N\to\infty} \max_{y\in K\cap E_N} \left| \prod_{x\in E_N\cap K, x\neq y} |y-x|^{1/N} - \exp(-U^{\sigma|_K}(y)) \right| = 0$$

for any compact K was proposed in [24] (see also [22, 45, 43] for some generalizations). It may be shown [24, Lemma 3.2] that, e.g., sets of zeros of suitable orthogonal polynomials satisfy this condition. One may show that any of these two conditions imply (2.25). This latter separation condition (2.25) was conjectured to be sufficient by Rakhmanov at the Sevilla OPSF conference [43, Conjectures 2 and 3], and proved to be sufficient later in [5].

We end this section by giving the main ideas of the proof of Theorem 2.5 for compact Σ, compare also with [10, Theorem 2.1 and Theorem 2.2] for $Q = 0$ and [12, Theorem 2.2] for general Q. For general Σ the reader may consult the statements [5, Theorem 1.3 and Theorem 1.4(c)] and their proofs.

Proof. (of (2.22), (2.24)). Given $\epsilon > 0$, it is sufficient to construct a sequence of monic polynomials p_N of degree $n = n(N)$ with $n(N)/N \to t$ for $N \to \infty$, such that

$$\limsup_{N\to\infty} ||w_N p_N||^{1/N}_{L_\infty(E_N)} \leq e^{\epsilon - w_{t,Q,\sigma}}, \quad \lim_{N\to\infty} |p_N(0)|^{1/N} = e^{-U^{\mu_{t,Q,\sigma}}(0)}, \tag{2.32}$$

where we suppose that (2.23) holds for $z = 0$ and some $V(0)$ (notice that (2.23) is true for any $z \notin \Sigma$ by assumption (2.19) since then $x \mapsto \log(|x-z|)$ is continuous in Σ). We will choose the zeros of p_N in E_N.

First notice that, by assumption (2.21), it is sufficient to show (2.32) for $w_N = w^N$. The main idea of the proof is that one is able to discretize $\mu_{t,Q,\sigma}$ with help of points in E_N: there exist sets E_N^* with

$$\text{card}(E_N^*) = n(N), \qquad E_N^* \subset E_N, \qquad \nu_N(E_N^*) \xrightarrow{*} \mu_{t,Q,\sigma}, \tag{2.33}$$

for $N \to \infty$ (see [10, Lemma A.1] for real Σ and [5, Lemma 2.1(d)]). Consider the polynomial p_N with simple zeros given by the elements of E_N^*, and the compact set

$$K_\epsilon = \{\lambda \in \Sigma : U^{\mu_{t,Q,\sigma}}(\lambda) + Q(z) \leq w_{t,Q,\sigma} - \epsilon\}, \tag{2.34}$$

then from Theorem 2.4 and from the uniqueness of the extremal constant $w_{,Q,\sigma}$ we know that $\mu_{t,Q,\sigma}(K_\epsilon) = \sigma(K_\epsilon) < t$. Hence only $o(N)$ elements of $E_N \cap K_\epsilon$ are not in $E_N^* \cap K_\epsilon$, but more than $o(N)$ elements lie in $E_N^* \setminus K_\epsilon$. Hence, by possibly exchanging $o(N)$ elements we may add to (2.33) the additional requirement that

$$E_N \cap K_\epsilon = E_N^* \cap K_\epsilon,$$

implying that

$$||w^N p_N||_{L_\infty(E_N)}^{1/N} = ||w^N p_N||_{L_\infty(E_N \setminus K_\epsilon)}^{1/N} =: \exp(-Q(\zeta_N) - U^{\nu_N(E_N^*)}(\zeta_N))$$

for some $\zeta_N \in E_N \setminus K_\epsilon$. By going to subsequences if necessary, we may suppose that $\zeta_N \to \zeta \in \Sigma$, and hence by the principle of descent (2.2) and by continuity of Q

$$\limsup_{N\to\infty} ||w^N p_N||_{L_\infty(E_N)}^{1/N} \leq \sup_{\zeta \in \Sigma \setminus K_\epsilon} \exp(-Q(\zeta) - U^{\mu_{t,Q,\sigma}}(\zeta)) \leq e^{\epsilon - w_{t,Q,\sigma}}$$

the last inequality following from the definition of K_ϵ. We also have from the principle of descent and from (2.33) that

$$\limsup_{N\to\infty} |p_N(0)|^{1/N} = \limsup_{N\to\infty} \exp(-U^{\nu_N(E_N^*)}(0)) \leq e^{-U^{\mu_{t,Q,\sigma}}(0)}.$$

The assumption $\sigma(\partial V(0)) = 0$ and thus $\mu_{t,Q,\sigma}(\partial V(0)) = 0$ allows us to conclude that

$$\nu_N(E_N^* \setminus V(0)) \xrightarrow{*} \mu_{t,Q,\sigma} - \mu_{t,Q,\sigma}|_{V(0)},$$
$$\nu_N(V(0) \cap (E_N \setminus E_N^*)) \xrightarrow{*} \sigma|_{V(0)} - \mu_{t,Q,\sigma}|_{V(0)},$$

and hence again by the principle of descent and by (2.23)

$$\begin{aligned}\limsup_{N\to\infty} U^{\nu_N(E_N^*)}(0) &= \lim_{N\to\infty} U^{\nu_N(E_N^* \setminus V(0))}(0) + \lim_{N\to\infty} U^{\nu_N(E_N \cap V(0))}(0) \\ &\quad - \liminf_{N\to\infty} U^{\nu_N(V(0)\cap(E_N \setminus E_N^*))}(0) \\ &\geq U^{\mu_{t,Q,\sigma} - \mu_{t,Q,\sigma}|_{V(0)}}(0) + U^{\sigma|_{V(0)}}(0) \\ &\quad - U^{\sigma|_{V(0)} - \mu_{t,Q,\sigma}|_{V(0)}}(0) \\ &= U^{\mu_{t,Q,\sigma}}(0),\end{aligned}$$

showing (2.32). □

Proof. (of (2.28), (2.29)). It is shown implicitly in [5, Lemma 2.1(c) and Lemma 2.2] that we may suppose that the set V in (2.25) satisfies $\sigma(\partial V) = 0$. Let $F_N \subset E_N$ with $\nu_N(F_N) \xrightarrow{*} \mu$. We claim that

$$F_N \subset E_N \cap V, \quad \nu_N(F_N) \xrightarrow{*} \mu \quad \Longrightarrow \quad \lim_{N\to\infty} I_N(F_N, F_N) = I(\mu,\mu). \quad (2.35)$$

Indeed, since $\sigma(\partial V) = 0$, we have that $\nu_N(E_N \cap V) \xrightarrow{*} \sigma|_V$, and hence with $F_N' := (E_N \cap V) \setminus F_N$, $\nu_N(F_N') \xrightarrow{*} \sigma|_V - \mu$ we have

$$\begin{aligned}\limsup_{N\to\infty} I_N(F_N, F_N) &= \limsup_{N\to\infty}\Big(I_N(E_N\cap V, E_N \cap V) \\ &\qquad - I_N(F_N', F_N') - 2I_N(F_N, F_N')\Big) \\ &\le \limsup_{N\to\infty}(E_N \cap V, E_N \cap V) \\ &\qquad - \liminf_{N\to\infty} I_N(F_N', F_N') - 2\liminf_{N\to\infty} I_N(F_N, F_N') \\ &\le I(\sigma|_V) - I(\sigma|_V - \mu) - 2I(\mu, \sigma|_V - \mu) = I(\mu),\end{aligned}$$

where in the last inequality we have applied Exercise 21 and (2.25). From Exercise 21 it also follows that $\liminf_N I_N(F_N, F_N) \ge I(\mu)$, showing that (2.35) holds.

Let $\epsilon > 0$, and K_ϵ as in (2.34). According to the equilibrium conditions in Theorem 2.4 and thanks to continuity we find some open set K with $\mathrm{supp}(\mu_{t,Q,\sigma}) \subset K \subset K_{-\epsilon}$. By possibly replacing K by some smaller set, we may also suppose that $K \subset V$, and that $\sigma(\partial K) = 0$. Finally, notice that $t' := (\sigma - \mu_{t,Q,\sigma})(K) > 0$ since for any ζ in the by assumption non-empty set $\mathrm{supp}(\mu_{t,Q,\sigma}) \cap \mathrm{supp}(\sigma - \mu_{t,Q,\sigma})$ there exists a small neighborhood U of ζ with $U \subset V$, and thus $t' \ge (\sigma - \mu_{t,Q,\sigma})(U) > 0$ by definition of the support.

We now consider the weighted Fekete points, a set Φ_N of $n(N)+1$ elements of $E_N \cap K$ which minimize the expression

$$I_N(\Phi_N, \Phi_N) + 2\int Q\, d\nu_N(\phi_N).$$

By discretizing $\mu_{t,Q,\sigma}$ as in the preceding proof with $n(N)+1$ elements in $E_N \cap K$, we obtain a candidate Φ_N^* with $I_N(\Phi_N^*, \Phi_N^*) + 2\int Q\, d\nu_N(\phi_N^*) \to I^Q(\mu_{t,Q,\sigma})$ according to (2.35). Hence

$$I^Q(\mu_{t,Q,\sigma}) \ge \limsup_{N\to\infty} I_N(\Phi_N, \Phi_N) + 2\int Q\, d\nu_N(\phi_N).$$

On the other hand, by Exercise 21 and Theorem 2.4,

$$\liminf_{N\to\infty} I_N(\Phi_N, \Phi_N) + 2\int Q\, d\nu_N(\phi_N) \ge I^Q(\mu_{t,Q,\sigma}),$$

which by the uniqueness of the extremal measure shows that $\nu_N(\phi_N) \xrightarrow{*} \mu_{t,Q,\sigma}$.

According to (2.22), (2.24) and (2.26), the assertions (2.28) and (2.29) will follow by showing that

$$\liminf_{N\to\infty}\Big[\min_{\deg P\le n(N)} \frac{||w^N P||_{L_\infty(\Phi_N)}}{|P(0)|}\Big]^{1/N} \ge \exp(U^{\mu_{t,Q,\sigma}}(0) - w_{t,Q,\sigma}),$$

$$\liminf_{N\to\infty} \min_{\deg P\le n(N)} ||w^N P||_{L_\infty(\Phi_N)}^{1/N} \ge \exp(-w_{t,Q,\sigma}),$$

However, since Φ_N has $n(N)+1$ elements, both expressions on the left can be written explicitly in terms of Lagrange polynomials, a task which we leave as an exercise. Then it is not difficult to see that the above two formulas follow from the principle of descent, and from a fact which we will show now: for any $z_N \in \Phi_N$ with $E_N^* := \Phi_N \setminus \{z_N\}$ there holds

$$\limsup_{N\to\infty} U^{\nu_N(E_n^*)}(z_N) + Q(z_N) \le w_{t,Q,\sigma} + \epsilon. \tag{2.36}$$

Write $F_N := (E_N \cap K)\setminus E_N^*$, with $n'(N)$ elements, and observe that $\nu_N(F_N \cup E_N) \to \sigma|_K$ and $\nu_N(F_N) \to \sigma|_K - \mu_{t,Q,\sigma}$, where

$$\frac{n'(N)}{N} \to t' = (\sigma - \mu_{t,Q,\sigma})(K)$$

with $t' > 0$, and, by (2.35),

$$\begin{aligned}
&\lim_{N\to\infty} I_N(E_N^*, F_N)\\
&= \lim_{N\to\infty} \frac{I_N(E_N^*\cup F_N, E_N^*\cup F_N) - I_N(E_N^*, E_N^*) - I_N(F_N, F_N)}{2}\\
&= I(\mu_{t,Q,\sigma}, \sigma|_K - \mu_{t,Q,\sigma}).
\end{aligned}$$

It follows from the definition of the Fekete points (replace one element of Φ_N by an element of $(E_N \cap V)\setminus \Phi_N$) that

$$z \in F_N : \qquad U^{\nu_N(E_N^*)}(z_N) + Q(z_N) \le U^{\nu_N(E_N^*)}(z) + Q(z).$$

Therefore, we can bound $U^{\nu_N(E_N^*)}(z_N) + Q(z_N)$ above by

$$\frac{1}{n'(N)} \sum_{z\in F_N} [U^{\nu_N(E_N^*)}(z) + Q(z)] = \frac{N}{n'(N)}[I_N(E_N^*, F_N) + \int Q\, d\nu_N(F_N)],$$

the right-hand term tending for $N \to \infty$ to

$$\begin{aligned}
&\frac{1}{t'}\Big(I(\mu_{t,Q,\sigma}, \sigma|_K - \mu_{t,Q,\sigma}) + \int Q\, d(\sigma|_K - \mu_{t,Q,\sigma})\Big)\\
&\qquad\qquad = \frac{1}{t'}\int \Big(U^{\mu_{t,Q,\sigma}} + Q\Big)\, d(\sigma|_K - \mu_{t,Q,\sigma})
\end{aligned}$$

which according to $\mathrm{supp}(\sigma|_K - \mu_{t,Q,\sigma}) \subset K \subset K_{-\epsilon}$ can be bounded above by $w_{t,Q,\sigma} + \epsilon$, as claimed in (2.36). □

Proof. (of (2.30)). See [5, Theorem 1.3(b)]. □

3 Consequences

3.1 Applications to the Rate of Convergence of CG

As mentioned already in the introduction, we want to provide a better understanding of the superlinear convergence of CG iteration, and in particular to explain the form of the error curve as seen in Figure 2, and in all examples considered below. Recall from Corollary 1.7 the link between the CG error of a positive definite matrix A of size $N \times N$ having the spectrum $\Lambda(A)$, and the quantity $E_n(0, \Lambda(A))$. We will argue that for large N, the error $E_n(0, \Lambda(A))$ in the polynomial minimization problem (1.16) is approximately

$$\frac{1}{N} \log E_n(\Lambda(A)) \approx - \int_0^t g_{S(\tau)}(0) d\tau \qquad (3.1)$$

where $t = n/N \in (0, 1)$ and $S(\tau)$, $\tau > 0$, is a decreasing family of sets, depending on the distribution of the eigenvalues of A. The sets $S(\tau)$ have the following interpretation: $S(\tau)$ is the subcontinuum of $[\lambda_{\min}, \lambda_{\max}]$ where the optimal polynomial of degree $[\tau N]$ is uniformly small.

From Corollary 1.7 and (3.1) we find the improved approximation

$$\frac{\|r_n^{CG}\|_{A^{-1}}}{\|r_0^{CG}\|_{A^{-1}}} \lessapprox \rho_t^n \qquad (3.2)$$

with

$$\rho_t = \exp\left(-\frac{1}{t} \int_0^t g_{S(\tau)}(0) d\tau \right) \qquad (3.3)$$

depending on n, since $t = n/N$. As the sets $S(\tau)$ are decreasing as τ increase, their Green functions $g_{S(\tau)}(0)$, evaluated at 0, increase with τ. Hence the numbers ρ_t decrease with increasing n, and this explains the effect of superlinear convergence (notice that $\log(\rho_{n/N})$ equals the slope at n of the bound on a semi-logarithmic plot).

Indeed, we will only show that (3.2) holds in an asymptotic sense after taking nth roots. However, in order to be able to take limits, we need to consider sequences of matrices A_N having a joint asymptotic eigenvalue distribution. Such sequences of matrices occur naturally in the context of the discretization of elliptic PDEs, by varying the stepsize or some other parameter of discretization, see Section 3.4. We have the following result [10, Theorem 2.1].

Theorem 3.1. *Let $(A_N)_N$ be a sequence of symmetric invertible matrices, A_N of size $N \times N$, satisfying the conditions*

(i) There exists a compact Σ and a positive Borel measure σ such that $\Lambda(A_N) \subset \Sigma$ for all N, and $\nu_N(\Lambda(A_N)) \xrightarrow{*} \sigma$ for $N \to \infty$;
(ii) σ has a continuous potential;
(iii) $U^{\nu_N(\Lambda(A_N))}(0) \to U^{\sigma}(0)$ for $N \to \infty$.

Define $S(t) := \text{supp}(\sigma - \mu_{t,0,\sigma})$, *with the extremal measure* $\mu_{t,0,\sigma}$ *as in Theorem 2.4. Then for* $t \in (0, \|\sigma\|)$, *we have*

$$\limsup_{\substack{n,N\to\infty \\ n/N\to t}} \frac{1}{n} \log E_n(0, \Lambda(A_N)) \leq -\frac{1}{t} \int_0^t g_{S(\tau)}(0)\, d\tau \tag{3.4}$$

Proof. Apply Theorem 2.5 for $p = \infty$, $w_N = 1$, and $Q = 0$: conditions (i), and (iii), are corresponding to (2.19), and (2.23) for $z = 0$, respectively. The conditions (2.20) and (2.21) are trivially true. Then our claim follows from (2.24), where according to Remark 2.5 we may replace $w_{t,0,\sigma} - U^{\mu_{t,0,\sigma}}(0)$ by our integral formula (2.18). □

Remark 3.1. According to Theorem 2.5, if we have the additional separation condition (2.25) on the spacing of eigenvalues, then there is equality in (3.4) in Theorem 3.1. In particular, if follows from the comments after Corollary 1.7 that the asymptotic CG bound on the right-hand side of (3.4) cannot be improved.

Remark 3.2. For determining σ, each λ in $\Lambda(A_N)$ is taken only once, regardless of its multiplicity. Hence it might happen that $\|\sigma\| < 1$. However, Theorem 2.5 remains valid even if one counts multiplicities, since $E_n(\cdot, \cdot)$ becomes larger if one adds points close to multiple eigenvalues to $\Lambda(A)$. The new conditions obtained form (i)–(iii) counting multiplicities will be referred to as (i)', (ii)', and (iii)'.

Remark 3.3. The condition (ii) is not very restrictive. For example, if σ is absolutely continuous with respect to Lebesgue measure with a bounded density then (ii) is satisfied. It is also satisfied if the density has only logarithmic-type or power-type singularities at a finite number of points. On the other hand, condition (ii) is not satisfied if σ has point masses.

In case of simple eigenvalues, condition (iii) may be rewritten as

$$\lim_{N\to\infty} |\det(A_N)|^{1/N} = \exp(-U^\sigma(0)).$$

Comparing (iii) with the principle of descent (2.2), we see that this condition prevents too many eigenvalues close to 0. If (iii) would not hold, then the matrices A_N are ill-conditioned and the estimate (3.2) may very well fail.

Remark 3.4. In [11], the following strategy was considered in order to find a polynomial p_n of degree n being small on $\Lambda(A_N)$ and $p_n(0) = 1$ (and thus to find an upper bound for $E_n(0, \Lambda(A_N))$):

Choose some fixed set S. Each eigenvalue of A_N outside the set S is a zero of p_n. This determines a certain fraction of the zeros of p_n. Clearly the set S has to be sufficiently big so that the number of eigenvalues of A_N outside S is less than n. The other zeros of p_n are free and they are chosen with the aim to minimize $\|p_n\|_{L_\infty(S)}$.

Though this strategy of imitating the CG polynomial seems to be natural, it depends very much on a good choice of the set S, see also the discussion in [31] and [25, Section 6]. Indeed, choosing a large set S means that there only few outliers, and their influence on the supnorm $\|p_n\|_S$ is small. Since we only need a polynomial which is small on the discrete set $\Lambda(A_N)$ but not necessarily in the gaps between the eigenvalues, $\|p_n\|_{L_\infty(S)}$ may be much bigger than $\|p_n\|_{\Lambda(A_N)}$. On the other hand, choosing a small set S means that we fix a lot of zeros of p_n and so we loose a lot of freedom in our choice for minimizing $\|p_n\|_{L_\infty(S)}$.

The main result of [11] is that the above strategy cannot produce a better asymptotic bound on $E_n(0, \Lambda(A_N))$ than Theorem 3.1, and that (under some additional assumptions) $S = S(t)$ leads to the same bound.

Remark 3.5. Provided that the sets $S(t)$ of Theorem 3.1 are intervals, say, $S(t) = [a(t), b(t)]$, we may give an interpretation of the bound (3.2) in terms of *marginal condition numbers*: Since $g_{S(t)}(0)$ is increasing in t, we get from (3.3) that

$$\log(\rho_t^n) = -N \int_0^{n/N} g_{S(\tau)}(0)\, d\tau \leq -\sum_{j=0}^{n-1} g_{S(j/N)}(0),$$

and, by (2.3), estimate (3.2) can be rewritten as

$$\frac{\|e_n^{\mathrm{CG}}\|_A}{\|e_0^{\mathrm{CG}}\|_A} \lessapprox \prod_{j=0}^{n-1} \frac{\sqrt{b(\frac{j}{N})/a(\frac{j}{N})} - 1}{\sqrt{b(\frac{j}{N})/a(\frac{j}{N})} + 1}.$$

Hence the classical bound (1.1) is obtained for constant b/a, and we see that the superlinear convergence behavior is obtained if the marginal condition number $b(\frac{j}{N})/a(\frac{j}{N})$ strictly decreases. Indeed, as we will see in Section 3.2, some extremal eigenvalues will be matched by Ritz values, and can be disregarded for the further convergence behavior.

Remark 3.6. For the moment it is not completely clear how to generalize Theorem 3.1 to the case of matrices with unbounded spectra and asymptotic eigenvalue distribution given by σ with unbounded support. In this case, we certainly have to impose some growth condition on σ around infinity such that the constraint is active around infinity.

Let us give some examples illustrating Theorem 3.1.

Example 3.1. The case of equidistant eigenvalues

$$\Lambda(A_N) = \{\frac{1}{N}, \frac{2}{N}, \ldots, \frac{N}{N}\},$$

leading to σ being the Lebesgue measure on $[0, 1]$, has already been discussed in Example 2.3, see also [10, Section 3]. For CG we obtain the error curve as well as the bounds (1.1), (3.3) as displayed in Figure 2, see Section 1.1.

Example 3.2. Consider the "worst case" eigenvalues

$$\Lambda(A_N) = \{2 + 2\cos(\pi \frac{j}{N+1}) : j = 1, ..., N\},$$

here conditions (i)–(iii) of Theorem 3.1 hold with $\sigma = \omega_{[0,4]}$. Comparing with Exercise 20 we see that $S(t) = [0, 4]$ for $0 < t < 1$, and thus the bound (3.3) is trivial. Indeed, for this example it is known that there are starting residuals such that CG does not lead to a small residual before reaching $n \approx N$.

Example 3.3. For the Toeplitz matrix $A_N := (\gamma^{|j-k|})_{j,k=1,2...,N}$, $0 < \gamma < 1$, of Kac, Murdock and Szegő [38, p. 783] it is shown in [10, Section 4] (see also Section 3.3 below) that conditions (i)',(ii)',(iii)' hold with

$$\sigma(x) = \frac{1}{x}\,\omega_{[a,1/a]}(x) = \int_0^1 \omega_{[a,b(t)]}(x)\,dt,$$

$$a = \frac{1-\gamma}{1+\gamma}, \quad b(t) = \begin{cases} 1/a & \text{for } t \le a \\ a/t^2 & \text{for } t \ge a. \end{cases}$$

Numerical experiments for the symmetric positive definite Toeplitz matrix T_{200} of order 200 of Kac, Murdock and Szegő are given in Figure 3. The four different plots correspond to the choices $\gamma \in \{1/2, 2/3, 5/6, 19/20\}$ of the parameter. Notice that the CG error curve (solid line) of the last two plots is clearly affected by rounding errors leading to loss of orthogonality, whereas the GMRES relative residual curves (dotted line) behave essentially like predicted by our theory. In particular, the classical bound (1.1) (crosses) does no longer describe correctly the size of the relative residual of GMRES for $n \ge 20$ and $\gamma \in \{5/6, 19/20\}$. Experimentally we observe that the range of superlinear convergence starts in the different examples approximately at the iteration indices ≥ 50, 30, 20, and 10, respectively. This has to be compared with the predicted quantity $N \cdot a$ which for the different choices of γ approximately takes the values 66, 40, 29, and 5, respectively. Though these numbers differ slightly, we observe that the new bound (3.3) reflects quite precisely the shape of the relative residual curve, and in particular allows to detect the ranges of linear and of superlinear convergence.

Example 3.4. Consider the two dimensional Poisson equation

$$-\frac{\partial^2 u(x,y)}{\partial x^2} - \frac{\partial^2 u(x,y)}{\partial y^2} = f(x,y)$$

for (x, y) in the unit square $0 < x, y < 1$, with Dirichlet boundary conditions on the boundary of the square. The usual five-point finite difference approximation on the uniform grid

$$(j/(m_x + 1), k/(m_y + 1)), \qquad j = 0, 1, \ldots, m_x + 1, k = 0, 1, \ldots, m_y + 1,$$

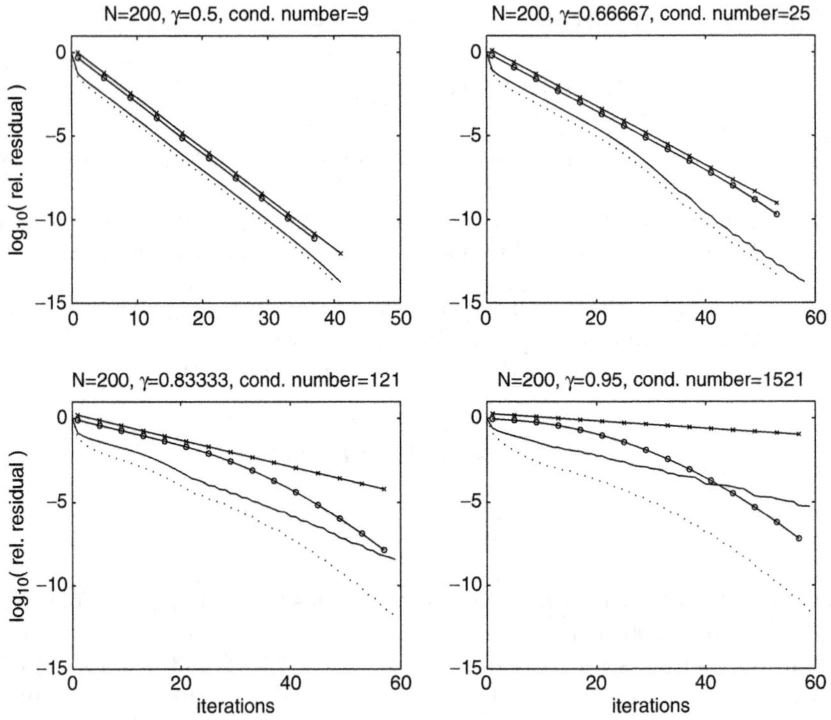

Fig. 3. The error curve of CG (*solid line*) and GMRES (*dotted line*) versus the classical upper bound (*crosses*) and our asymptotic upper bound (*circles*) for the system $T_{200}x = b$, with random solution x, and initial residual $r_0 = (1, ..., 1)^T$. Here T_N is the Kac, Murdock and Szegő matrix, with parameter $\gamma \in \{1/2, 2/3, 5/6, 19/20\}$

leads to a linear system of size $N \times N$ where $N = m_x m_y$. After rescaling, the coefficient matrix of the system may be written as a sum of Kronecker products

$$A_N = \frac{(m_x+1)}{(m_y+1)} B_{m_x} \otimes I_{m_y} + \frac{(m_y+1)}{(m_x+1)} I_{m_x} \otimes B_{m_y} \tag{3.5}$$

where

$$B_m = \begin{bmatrix} 2 & -1 & 0 & \cdots & 0 \\ -1 & 2 & -1 & \ddots & \vdots \\ 0 & \ddots & \ddots & \ddots & 0 \\ \vdots & \ddots & \ddots & \ddots & -1 \\ 0 & \cdots & 0 & -1 & 2 \end{bmatrix}_{m \times m} \tag{3.6}$$

and I_m is the identity matrix of order m. It is well known and easy to verify that the eigenvalues of B_m are

$$\mu_{k,m} = 2 - 2\cos\frac{\pi k}{m+1}, \qquad k = 1, \ldots, m,$$

and that the eigenvalues $\lambda_{j,k}$ of A_N are connected with the eigenvalues of B_m via

$$\lambda_{j,k} = \frac{m_y+1}{m_x+1}\mu_{j,m_x} + \frac{m_x+1}{m_y+1}\mu_{k,m_x}, \quad j = 1, \ldots, m_x, k = 1, \ldots, m_y. \tag{3.7}$$

We consider the limit relation

$$m_x, m_y \to \infty, \qquad \frac{m_x}{m_y} \to \delta \leq 1,$$

then it is not difficult to see using (3.7) that condition (i)' holds with

$$\begin{aligned}
\int f\, d\sigma &= \int_0^1 d\phi \int_0^1 d\psi\, f(2\delta(1-\cos(\pi\phi)) + 2\delta^{-1}(1-\cos(\pi\psi))) \\
&= \frac{1}{\pi^2}\int_0^{4\delta} dx \int_x^{x+4\delta^{-1}} d\lambda \frac{f(\lambda)}{\sqrt{x(4\delta - x)(\lambda - x)(4\delta^{-1} - \lambda + x)}} \\
&= \int_0^{4\delta + 4\delta^{-1}} \sigma'(\lambda) d\lambda
\end{aligned}$$

with

$$\sigma'(\lambda) := \frac{1}{\pi^2}\int_{\max\{0,\lambda-4\delta^{-1}\}}^{\min\{4\delta,\lambda\}} \frac{dx}{\sqrt{x(4\delta - x)(\lambda - x)(4\delta^{-1} - \lambda + x)}}.$$

The substitution $x' = 4\delta - x$ shows that $\sigma'(4\delta + 4\delta^{-1} - \lambda) = \sigma'(\lambda)$, and thus we only need to consider the case where $\lambda \geq 2\delta + 2\delta^{-1}$, and hence $\lambda \geq 4\delta$. We now construct a linear fractional transformation $y = T(x)$ with $T(0) = 0$, $T(4\delta) = 1$, $T(\lambda) = \infty$, and hence

$$T(x) = \frac{x}{\lambda - x}\frac{\lambda - 4\delta}{4\delta}, \qquad \gamma := T(\lambda - 4\delta^{-1}) = \frac{1}{16}(\lambda - 4\delta^{-1})(\lambda - 4\delta) \leq 1,$$

and the substitution $y = T(x)$ leads to

$$\begin{aligned}
\sigma'(\lambda) &= \frac{1}{4\pi^2}\int_{\max\{\gamma,0\}}^1 \frac{dy}{\sqrt{y(1-y)(y-\gamma)}} \\
&= \frac{1}{\pi}\int_{0,\sin(\pi t/2)\geq\gamma}^1 \frac{dy}{\sqrt{16\sin^2(\pi t/2) - 16\gamma}}.
\end{aligned}$$

By substituting γ, we find with $\Delta := 2\delta + 2\delta^{-1}$

$$\sigma = \int_0^1 \omega_{S(t)}\, dt, \quad S(t) = [\Delta - \sqrt{\Delta^2 - 16\sin^2(\frac{\pi t}{2})}, \Delta + \sqrt{\Delta^2 - 16\sin^2(\frac{\pi t}{2})}],$$

and thus the extremal measures by Exercise 20. One may compare our findings for $\delta = 25/40$ with numerical experiments presented in Figure 13 of Example 3.9 below, where both histograms for $m_x = 25$ and $m_y = 40$ and the density function σ' of the limiting distribution are drawn. Notice that σ has the support $S(0) = [0, 4\delta + 4\delta^{-1}]$, and that σ' has logarithmic singularities at 4δ and at $4\delta^{-1}$.

The sets $S(t)$ have been known before only in the case $\delta = 1$, and thus $\Delta = 4$ [10, Section 5]. More precisely, we observe that $\lambda_{j,k} = \lambda_{k,j}$, that is, most of the eigenvalues have multiplicity at least 2. Also, $\lambda_{j,m+1-j} = 4$ for all $j = 1, \ldots, m$, and the eigenvalue 4 has multiplicity m. We suspect that $N/2 + o(N)$ eigenvalues have multiplicity 2. In this case, not only condition (i)' but also condition (i) holds, with the new constraint being half of the old constraint. Since

$$\mu_{t,0,\sigma/2} = \frac{1}{2}\mu_{2t,0,\sigma},$$

one should therefore replace in the above formula for $S(t)$ the term t by $2t$, in order to obtain a sharper error bound. This is confirmed by our numerical experiments presented in Figure 4.

For exponentially decreasing eigencomponents, we may give an improvement of Theorem 3.1 based on Theorem 2.5 for $p = \infty$, compact real Σ and nontrivial external field, see [12, Theorem 2.2].

Theorem 3.2. *Besides the assumptions of Theorem 2.5, suppose that there is a sequence of starting residuals $(r_{0,N})_N$ and a nonnegative $Q \in C(\Sigma)$, $w(\lambda) = \exp(-Q(\lambda))$, such that*

(iv) if the eigenelements of A_N are given by $(\lambda_{j,N}, v_{j,N})$ with $||v_{j,N}|| = 1$, then[2]

$$\limsup_{N\to\infty} \max_j \exp(Q(\lambda_{j,N})) \left[\frac{|(r_{0,N}, v_{j,N})|}{||r_{0,N}||}\right]^{1/N} \leq 1.$$

Then, for every $t \in (0, ||\sigma||)$,

$$\limsup_{\substack{n,N\to\infty \\ n/N\to t}} \frac{1}{N} \log\left(\frac{||e^{\mathrm{CG}}_{n,N}||_{A_N}}{||e^{\mathrm{CG}}_{0,N}||_{A_N}}\right) \leq U^{\mu_{t,Q,\sigma}}(0) - w_{t,Q,\sigma}, \tag{3.8}$$

where $\mu_{t,Q,\sigma}$ and $w_{t,Q,\sigma}$ are as in Theorem 2.4. A similar bound is valid for the nth relative residual of MINRES.

Proof. See (1.15) and Corollary 1.3, and use the fact that from condition (iii) it follows that $[\min_j |\lambda_{j,N}|]^{1/N} \to 1$. □

[2] In case of distinct eigenvalues, the quantity $|(r_{0,N}, v_{j,N})|$ was called before $\beta_{j,N}$.

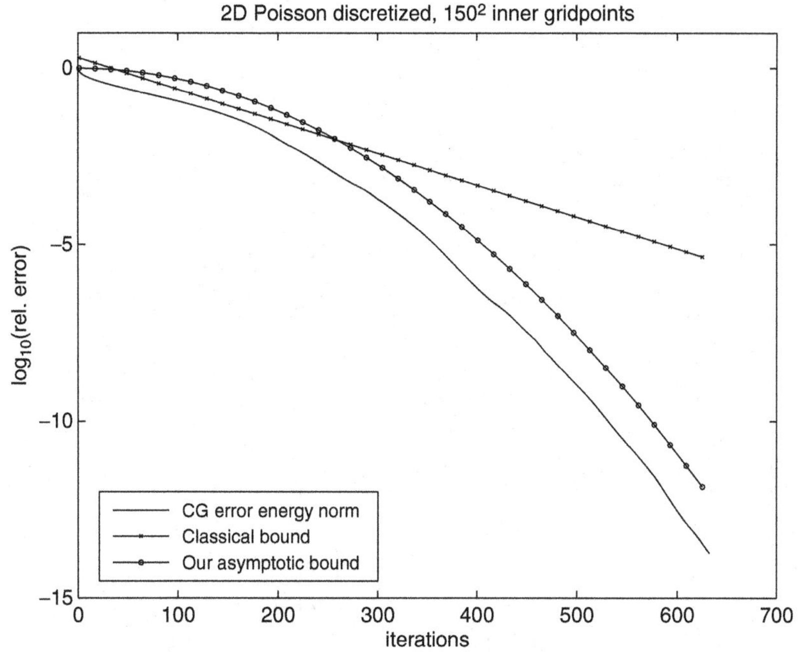

Fig. 4. The CG error curve versus the two upper bounds for the system $A_N x = b$ resulting from discretizing the $2D$ Poisson equation on a uniform grid with $m_x = m_y = 150$. We have chosen a random solution x, and initial residual $r_0 = (1, ..., 1)^T$, and obtain superlinear convergence from the beginning. Notice that the classical upper bound for CG is far too pessimistic for larger iteration indices. For the new bound we have added a factor 1/2 in front of σ since $\lambda_{j,k} = \lambda_{k,j}$, and we suspect that most of the eigenvalues are of multiplicity 2

Example 3.5. As a motivating model problem for Theorem 3.2, we consider the one dimensional Poisson equation $-u''(x) = f(x)$, $x \in [0, 1]$, with homogeneous Dirichlet boundary conditions $u(0) = u(1) = 0$. The usual central finite difference approximation on the uniform grid $j/(N+1)$, $j = 0, 1, \ldots, N+1$, leads to a linear system $A_N x = b_N$ with N equations and unknowns, where $A_N = B_N$ of (3.6), and

$$b_N = (N+1)^2 \cdot \left[f(1/(N+1)) \; f(2/(N+1)) \cdots f(N/(N+1)) \right]^T.$$

Both the one dimensional Poisson problem and the system $A_N x = b_N$ are easy to solve; however, this toy problem can serve to explain convergence behavior observed also in less trivial situations. From Example 3.2 we know that conditions (i)–(iii) are satisfied with $\sigma = \omega_{[0,4]}$, and that, for general starting residual, one obtains poor CG convergence, as confirmed by Figure 5.

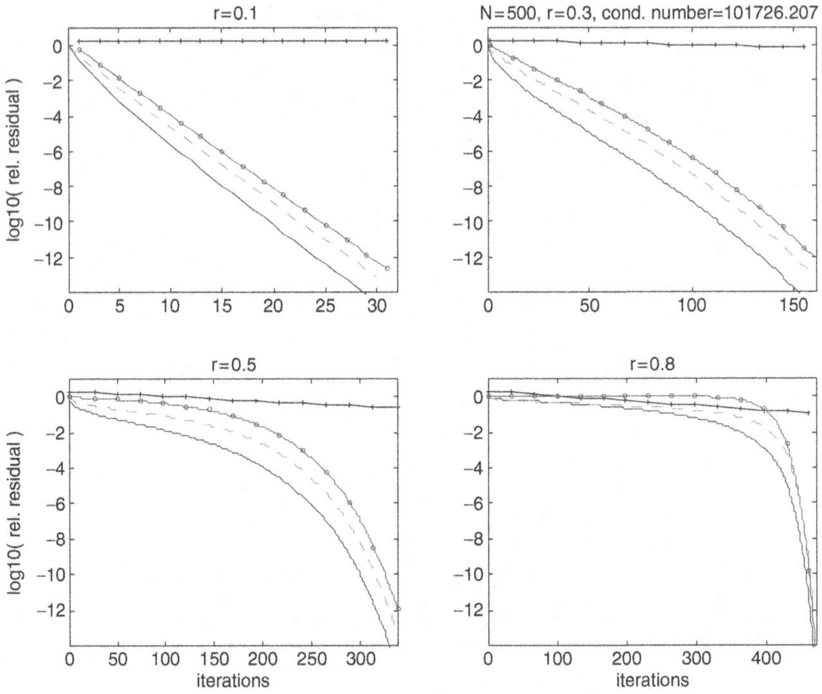

Fig. 5. The one dimensional Poisson problem discretized on a uniform grid ($N = 500$) for $f(x) = \sum_{j=1}^{N} r^j \sin(j\pi x)$, $r = 0.1, 0.3, 0.5, 0.8$. We find the error curve of CG (solid blue line) and the classical bound (1.1) (black line with crosses), and our new asymptotic bound (red line with circles). For comparison we give the MINRES relative residual curve in (dashed green line). Notice that, for $r = 0.8$, there is hardly any superlinear convergence and one has to reach approximately the dimension of the system in order to achieve full precision

Here we will be interested in what happens for the CG starting vector 0 (i.e., $r_{0,N} = b_N$) and particularly smooth functions f, namely

$$f(x) = \sum_{j=1}^{\infty} f_j \sin(\pi j x), \quad x \in [0,1], \quad \text{where} \quad r := \limsup_{j\to\infty} |f_j|^{1/j} \in (0,1).$$

It is shown in [12, Lemma 3.1] that here condition (iv) holds with

$$Q(x) = \frac{\log(1/r)}{\pi} \arccos(\frac{2-\lambda}{2}).$$

Also, the reader may verify that the assumptions of Remark 2.6 hold. As shown in [12, Section 3], here the integral equations (2.14),(2.15) can be solved in terms of the complete elliptic integral $K(\cdot)$ and the Jacobi elliptic functions: if $k = k(r)$ is defined by

$$\frac{\log(1/r)}{\pi} K(k) = K(\sqrt{1-k^2}),$$

then

$$a(t) = 4\text{cn}^2((1-t)K(k);k), \qquad b(t) = a(t)/\text{dn}^2((1-t)K(k);k),$$

and we obtain the asymptotic CG error bound of (2.12).

3.2 Applications to the Rate of Convergence of Ritz Values

In order to approximate eigenvalues of large real symmetric matrices A of order N via the Lanczos method with starting vector $r_0 \in \mathbb{R}^N$, one computes the so-called Ritz values, namely, the eigenvalues $x_{1,n} < ... < x_{n,n}$ of the (tridiagonal) matrix Jacobi matrix J_n, see Definition 1.2 and Corollary 1.1. Depending on the eigenvector components $\beta_1, .., \beta_N$ of the starting vector r_0, some of the eigenvalues $\lambda_1 \leq ... \leq \lambda_N$ of A are well approximated by Ritz values even if n is much smaller than the dimension N. Classical results on convergence and on technical details of the Lanczos method may be found in many textbooks. Let us cite here the well-known Kaniel-Page-Saad estimate for extremal eigenvalues [30, 56, 59, 70] which is a consequence of Corollary 1.6 and Lemma 1.1

$$\left| \frac{x_{1,n} - \lambda_1}{\lambda_N - \lambda_1} \right| \leq \frac{1}{T_{n-1}(1 + 2\frac{\lambda_2 - \lambda_1}{\lambda_N - \lambda_2})^2} \frac{1}{\beta_1^2} \sum_{j=2}^{n} \beta_j^2, \tag{3.9}$$

with T_n the nth Chebyshev polynomial of the first kind. Thus one may expect geometric convergence of the smallest (largest) Ritz value to the smallest (largest) eigenvalue for a fixed matrix A, but the rate of convergence will depend on the size of the eigenvector component β_1, and on the (relative) distance of λ_1 to the other eigenvalues. For an "inner" eigenvalue λ_k lying in the convex hull of the Ritz values, say, $x_{\kappa-1,n} < \lambda_k \leq x_{\kappa,n}$ for some $\kappa = \kappa(k)$, we get by combining Corollary 1.6 and Exercise 9

$$\begin{aligned} \min_{\ell} |\lambda_k - x_{\ell,n}|^2 &\leq |(\lambda_k - x_{\kappa-1,n})(\lambda_k - x_{\kappa,n})| \\ &\leq 2\,b^2 \left[\frac{b/a - 1}{b/a + 1}\right]^{[n/2]-1} \frac{1}{\beta_k^2} \sum_{j=1, j\neq k}^{n} \beta_j^2, \end{aligned} \tag{3.10}$$

where $a = \min\{\lambda_{k+1} - \lambda_k, \lambda_k - \lambda_{k-1}\}$, and $b = \max\{\lambda_N - \lambda_k, \lambda_k - \lambda_1\}$. Notice again that we may only expect an interesting rate of convergence if λ_k is well separated from the rest of the spectrum, and if $|\beta_k|/||r_0||$ is sufficiently large.

There exist (worst case) examples A, r_0 with eigenvalue and eigenvector component distribution such that the bounds (3.9) or (3.10) are (approximately) sharp. However, for matrices occurring in applications one observes quite often that the above bounds greatly overestimate the actual error, even for a judicious choice of the set in Corollary 1.6 (for instance a finite union of intervals representing the parts of the real axis where the spectrum of A is relatively dense).

Trefethen and Bau [70, p. 279] observed a relationship with electric charge distributions, and claimed that the Lanczos iteration tends to converge to eigenvalues in regions of "too little charge" for an equilibrium distribution. This has been made more precise by Kuijlaars [39] who considered, as in the preceding section, a sequence of symmetric matrices A_N which are supposed to have an asymptotic eigenvalue distribution

$$\nu_N(\Lambda(A_N)) \xrightarrow{*} \sigma \quad \text{for } N \to \infty. \tag{3.11}$$

Then, following Trefethen and Bau, Kuijlaars compared $\omega_{\text{supp}(\sigma)}$ and σ, and considered more precisely the constrained energy problem with external field $Q = 0$ of Section 2.2.

In the remainder of this section we will suppose that A_N has N distinct eigenvalues $\lambda_{1,N} < ... < \lambda_{N,N}$ contained all in some compact set Σ. Also, we suppose that the Lanczos method is applied to matrix A_N with starting vector $r_{0,N}$ having eigenvector components $\beta_{1,N}, ..., \beta_{N,N}$, and we are interested in measuring the distance of an eigenvalue $\lambda_{j,N}$ to the set of Ritz values $x_{1,n,N} < ... < x_{n,n,N}$ obtained in the nth iteration of the Lanczos process. We then have the following result

Theorem 3.3. *Suppose that the asymptotic distribution of the spectra of $(A_N)_N$ is given by σ, which has a continuous potential. Let k_N be a sequence of indices such that*

$$\lim_{N\to\infty} \lambda_{k_N,N} = \lambda \tag{3.12}$$

and suppose that

$$\lim_{N\to\infty} \frac{1}{N} \sum_{j\neq k_N} \log|\lambda_{k_N,N} - \lambda_{j,N}| = \int \log|\lambda - \lambda'|\, d\sigma(\lambda'), \tag{3.13}$$

$$\liminf_{N\to\infty} \left[\frac{|\beta_{k_N,N}|}{||r_{0,N}||}\right]^{1/N} =: \rho \in (0,1]. \tag{3.14}$$

Then

$$\limsup_{\substack{n,N\to\infty \\ n/N\to t}} \min_j |\lambda_{k_N,N} - x_{j,n,N}|^{1/N} \leq \frac{1}{\rho} \exp\Big(U^{\mu_{t,0,\sigma}}(\lambda) - w_{t,0,\sigma}\Big). \tag{3.15}$$

If moreover there exists a nonnegative function $Q \in \mathcal{C}(\Sigma)$ with

$$\limsup_{N\to\infty} \sup_j \exp(Q(\lambda_{j,N})) \left[\frac{|\beta_{k_N,N}|}{||r_{0,N}||}\right]^{1/N} \leq 1,$$

then

$$\limsup_{\substack{n,N\to\infty\\ n/N\to t}} \min_j |\lambda_{k_N,N} - x_{j,n,N}|^{1/N} \le \frac{1}{\rho}\exp\Big(U^{\mu_{t,Q,\sigma}}(\lambda) - w_{t,Q,\sigma}\Big). \tag{3.16}$$

Proof. We may suppose without loss of generality that $\lambda = 0$. Using the estimate of Corollary 1.6, we apply Theorem 2.5 for $p = \infty$ to the sets $E_N := \{\lambda_{j,N} - \lambda_{k_N,N} : j \neq k_N\}$. □

Remark 3.7. Recall from Remark 2.5 that we may replace $w_{t,0,\sigma} - U^{\mu_{t,0,\sigma}}(\lambda)$ in (3.15) by some mean of the Green functions $g_{S(t)}(\lambda)$, see (2.18). Hence, for $\rho = 1$, the right-hand side of (3.15) is strictly negative for $\lambda \notin \bigcap_{\tau<t} \mathrm{supp}(\sigma - \mu_{\tau,0,\sigma})$, in correspondence with the heuristic observation of Trefethen and Bau.

Remark 3.8. In case of (3.16) one may observe a further phenomenon which for the data of Example 3.5 is shown in Figure 6 and which is not fully covered

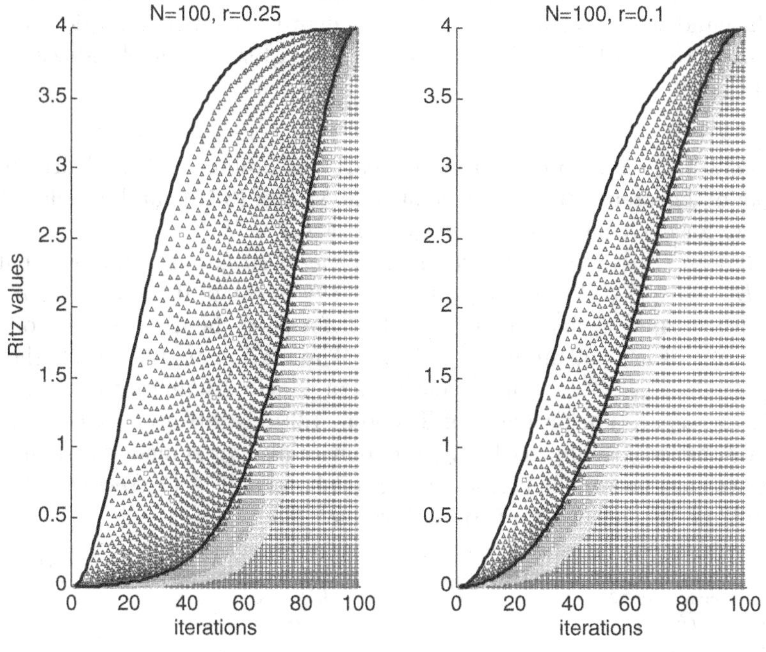

Fig. 6. Convergence of Ritz values for our $1D$ Poisson model problem with particular smooth right hand side. Here $N = 100$, $x_{0,N} = 0$, and $f(x) = \sum_{j=1}^{N} r^j \sin(j\pi x)$, $r \in \{1/4, 1/10\}$. The two black curves indicate the graphs of a, b. We draw in the nth column, $1 \le n \le N$, the position of the nth Ritz values within the interval $[0, 4]$. Here the color/symbol indicates the distance of the Ritz value to the set of eigenvalues of A_N, compare with (3.17). Notice that nearly all Ritz values are red in the range $[0, a(t)]$, $t = n/N$, that there are no Ritz values in $[b(t), 4]$, and that in the range $[a(t), b(t)]$ hardly any Ritz value converged (up to some exceptions by "accident")

by Theorem 3.3: all eigenvalues in $\mathrm{supp}(\mu_{t,Q,\sigma}) \setminus \mathrm{supp}(\sigma - \mu_{t,Q,\sigma})$ are fit by Ritz values (the constraint is active), and there is hardly any Ritz value in $\mathrm{supp}(\sigma - \mu_{t,Q,\sigma}) \setminus \mathrm{supp}(\mu_{t,Q,\sigma})$.

Remark 3.9. The first condition (3.13) will be true if the eigenvalues $\lambda_{k_N+j,N}$ for $j \neq 0$ do not approach "too fast" $\lambda_{k_N,N}$ when $N \to \infty$. This separation condition has been suggested by Dragnev and Saff [24, Definition 3.1] as a sufficient condition for ensuring nth root asymptotics for discrete orthogonal polynomials. It holds if the distance $|\lambda_{k_N+j,N} - \lambda_{k_N,N}|$ is bounded below by a constant times $|j|/N$ [58] or by some positive power of this quantity; however, it is excluded that two neighboring eigenvalues approach exponentially. Condition (3.14) means that the starting vector $r_{0,N}$ has a sufficiently large eigencomponent for the eigenvalue $\lambda_{k_N,N}$.

Remark 3.10. The statement of (3.15) can be found in [6, Theorem 2.1(a)]. Before, Kuijlaars [39] had established a related inequality with $\rho = 1$, and the right-hand side of (3.15) replaced by its square root. As assumption, Kuijlaars imposed that (3.13) and (3.14) for $\rho = 1$ hold for any set of indices verifying (3.12).

Remark 3.11. According to the first part of Corollary 1.6, we learn from the proof of Theorem 3.3 that the right-hand side of (3.15) can be replaced by its square for extremal eigenvalues $k_N = 1$ or $k_N = N$.

Indeed, this is also true for more general situations: suppose that $\Sigma = [A, B]$, and $B' \in \Sigma$ such that $w_{t,0,\sigma} - U^{\mu_{t,0,\sigma}}(\lambda) > 0$ for $\lambda \in [A, B']$. Furthermore suppose that (3.13) and (3.14) for $\rho = 1$ hold for any set of indices verifying (3.12) and limit $\lambda \in [A, B']$. Then it is not difficult to show that $|\lambda_{k,N} - \lambda_{k+1,N}|^{1/N} \to 1$ for eigenvalues in $[A, B']$. Taking into account (3.15) and the separation property of Theorem 1.2(c), a little bit of combinatorics shows that $x_{j,n,N}$ approaches $\lambda_{j,N}$ exponentially for sufficiently small j. This implies that one of the factors on the left-hand side of the second estimate of Corollary 1.6 can be dropped, and we have that

$$\limsup_{\substack{n,N\to\infty \\ n/N\to t}} \min_j |\lambda_{k_N,N} - x_{j,n,N}|^{1/N} = \limsup_{\substack{n,N\to\infty \\ n/N\to t}} |\lambda_{k_N,N} - x_{k_N,n,N}|^{1/N}$$
$$\leq \exp\Big(2U^{\mu_{t,0,\sigma}}(\lambda) - 2w_{t,0,\sigma}\Big)$$

provided that $\lambda \in [A, B']$.

In the same spirit, one can show that for all but at most one exceptional eigenvalue in any closed sub-interval of Σ with strictly positive $w_{t,0,\sigma} - U^{\mu_{t,0,\sigma}}$ we have this improved rate of convergence. Finally, there are examples where the rate of convergence for the exceptional eigenvalue is given by (3.15). A detailed discussion of these exceptional indices is given in [6].

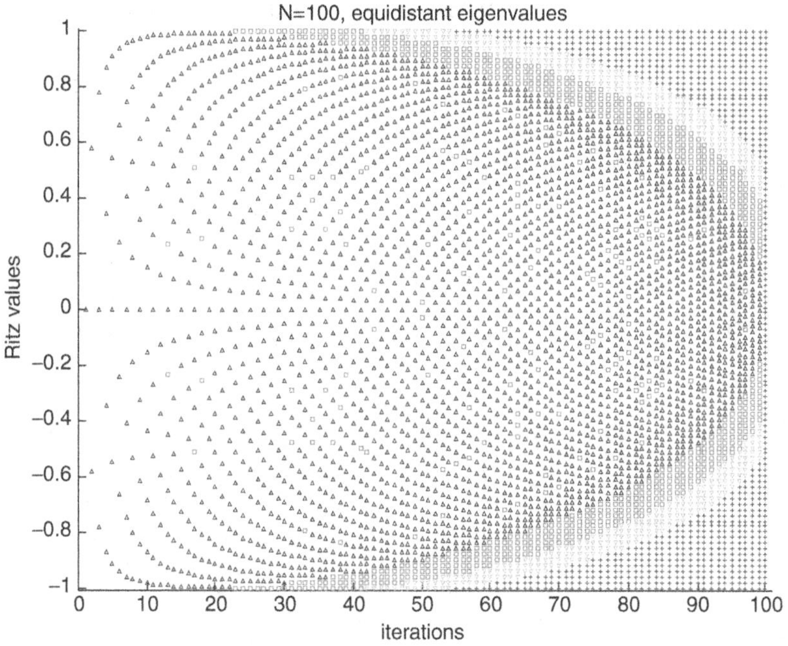

Fig. 7. Convergence of Ritz values for 400 equidistant eigenvalues in $[-1,1]$. We draw in the nth column, $1 \leq n \leq N$, the position of the nth Ritz values within the interval $[0,4]$. Here the color/symbol indicates the distance of the Ritz value to the set of eigenvalues of A_N, compare with (3.17)

Example 3.6. If A_N has equidistant eigenvalues in $[-1,1]$, we found in Example 3.1 that $S(t) = [-\sqrt{1-t^2}, \sqrt{1-t^2}]$. Indeed, as shown in Figure 7, the eigenvalues outside the disk are found by the Lanczos method.

The numerical results displayed in Figure 7 as well as in subsequent experiments have been obtained by the Lanczos method with full reorthogonalization, in order to prevent loss of orthogonality, due to finite precision arithmetic. The following symbols/colors are used to indicate the distance of a given Ritz value to the set of eigenvalues

color	symbol	distance between Ritz value and set of eigenvalues
red	$+$	less than $0.5\,10^{-14}$
yellow	∇	between $0.5\,10^{-14}$ and $0.5\,10^{-8}$
green	$\square$	between $0.5\,10^{-8}$ and $0.5\,10^{-3}$
blue	$\triangle$	larger than $0.5\,10^{-3}$

(3.17)

Example 3.7. For the Poisson problem of Example 3.4 and $m_x = 9$, $m_y = 13$, the Ritz values are displayed in Figure 8. As in the preceding example, the color/symbol is chosen depending on the distance of Ritz values to the set of eigenvalues, as a function of the iteration index $n = 1, 2, \ldots, N = 9*13 = 117$.

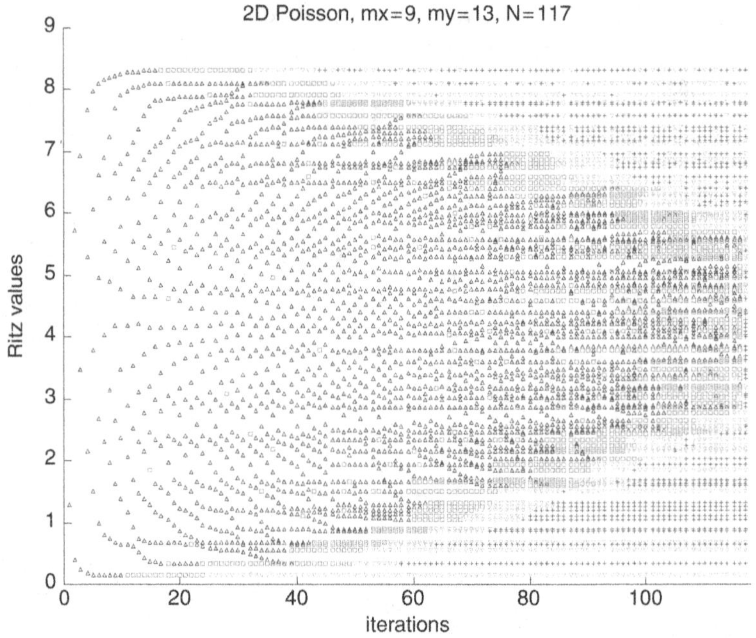

Fig. 8. Convergence of Ritz values for the Poisson problem with $m_x = 9$ and $m_y = 13$. Notice that, even for large $n \approx N$, hardly any eigenvalue in $S(1) = [4\delta, 4\delta^{-1}] = [2.77, 5.78]$ is found by Ritz values

Observe that eigenvalues outside of the set $S(n/N)$ described in Example 3.4 are well approximated by Ritz values, but not those in $S(n/N)$. Notice also that, even for large $n \approx N$, hardly any eigenvalue in $S(1) = [4\delta, 4\delta^{-1}] = [2.77, 5.78]$ is well approximated by Ritz values.

Example 3.8. Consider the eigenvalues

$$\lambda_{j,N} = \cos(\pi \frac{2j-1}{2N}) \cdot |\cos(\pi \frac{2j-1}{2N})|^{\alpha - 1}, \quad \alpha > 0,$$

and eigencomponents $\beta_{j,N} = 1$, having clearly an asymptotic eigenvalue distribution σ with continuous potential. As mentioned already in Example 3.2, no convergence of Ritz values can be expected if $\alpha = 1$. Things become more interesting for $\alpha = 2$ (more eigenvalues close to zero) or for $\alpha = 1/2$ (more eigenvalues close to the endpoints ± 1). This behavior is displayed in Figure 9.

In the case $\alpha = 2$ one may show that for $0 < t \leq 1/\sqrt{2}$ we obtain $S(t) = [-1, 1]$, and there is no geometric convergence of Ritz values. For $t \in (1/\sqrt{2}, 1)$, the sets are strictly decreasing and of the form $S(t) = [-b(t), b(t)]$, but the resulting formulas for $b(t)$ are complicated, we omit details. The convergence

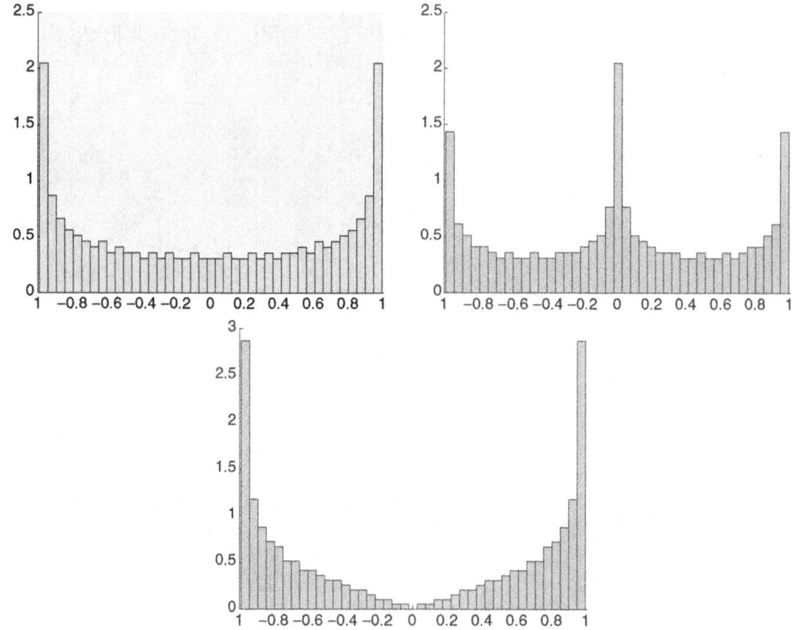

Fig. 9. Bar chart for the eigenvalue distribution of 400 eigenvalues in the case $\alpha = 1$ (equilibrium distribution), $\alpha = 2$, and $\alpha = 1/2$ (from the left to the right)

behavior of the corresponding Ritz values can be found in Figure 10, indeed, for $n \leq N/\sqrt{2}$ hardly any eigenvalue is well approximated by a Ritz value.

For the case $\alpha = 1/2$ it is shown in [6] that

$$S(t) = [-1, -r(t)] \cup [r(t), 1], \qquad r(t) = \frac{1 - \cos(\pi t/2)}{1 + \cos(\pi t/2)}.$$

Notice that the eigenvalues (and the eigenvector components) are symmetric with respect to the origin. Thus $p_{2n-1,N}$ is odd, and $p_{2n,N}$ is even. Moreover, $\lambda_{N+1,2N+1} = x_{n+1,2n+1,2N+1} = 0$, and thus here is a perfect rate of convergence. However, the eigenvalue $\lambda_{N+1,2N+1}$ is approached by the Ritz values $x_{n+1,2n,2N+1} = -x_{n,2n,2N+1}$. Comparing with Remark 3.11 we get an "exceptional" eigenvalue with a smaller rate of convergence. In contrast, for even N no exceptional eigenvalue occurs, even if the Ritz value $x_{n+1,2n+1,2N} = 0$ is not close to the spectrum. This last example contradicts the widely believed fact that first extremal eigenvalues are found by Ritz values.

3.3 Circulants, Toeplitz Matrices and their Cousins

A circulant matrix of order N generated by some exponential polynomial $\phi(\theta) := \phi_0 + \phi_1 e^{i\theta} + ... + \phi_{N-1} e^{(N-1)i\theta}$ is defined by

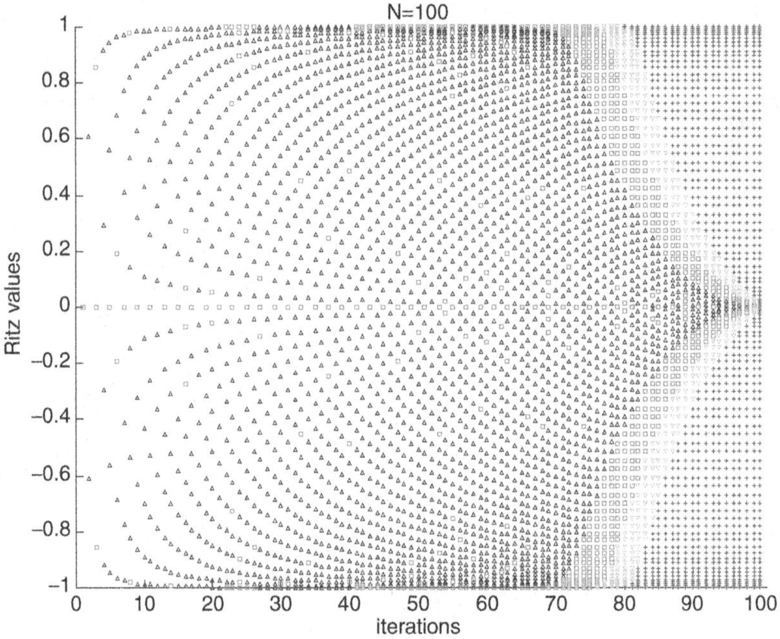

Fig. 10. Convergence of Ritz values for "squares" of 100 Chebyshev eigenvalues in $[-1,1]$ $(\alpha = 2)$

$$C_N(\phi) = \begin{bmatrix} \phi_0 & \phi_1 & \cdots & \phi_{N-1} \\ \phi_{N-1} & \phi_0 & \cdots & \phi_{N-2} \\ \vdots & \vdots & & \vdots \\ \phi_1 & \phi_2 & \cdots & \phi_0 \end{bmatrix}, \tag{3.18}$$

i.e., $C_N(\phi)$ is constant along diagonals. It is easily seen that $C_N(\phi)$ is diagonalized by the unitary FFT matrix of eigenvectors

$$\Omega_N = \frac{1}{\sqrt{N}} \Big[\exp(\frac{2\pi i j k}{N}) \Big]_{j,k=0,1,\ldots,N-1},$$

with corresponding eigenvalues given by $\phi(\frac{2\pi i(k-1)}{N})$, $k = 1, \ldots, N$. Notice also that $C_N(\phi)$ is normal, and in addition hermitian if and only if all eigenvalues are real. One easily checks using the explicit knowledge of the eigenvalues that, if $\phi^{(N)}$ is the partial sum of a exponential power series ϕ being absolutely convergent (such symbols ϕ are called of Wiener class), then for $N \to \infty$

$$\nu_N(C_N(\phi^{(N)})) \xrightarrow{*} \sigma_\phi, \quad \text{where} \quad \int f \, d\sigma_\phi = \frac{1}{2\pi} \int_0^{2\pi} f(\phi(e^{is}))\, ds, \tag{3.19}$$

where here and in what follows we count eigenvalues according to their multiplicities. We speak of a circulant matrix of level two (and by iteration of level t)

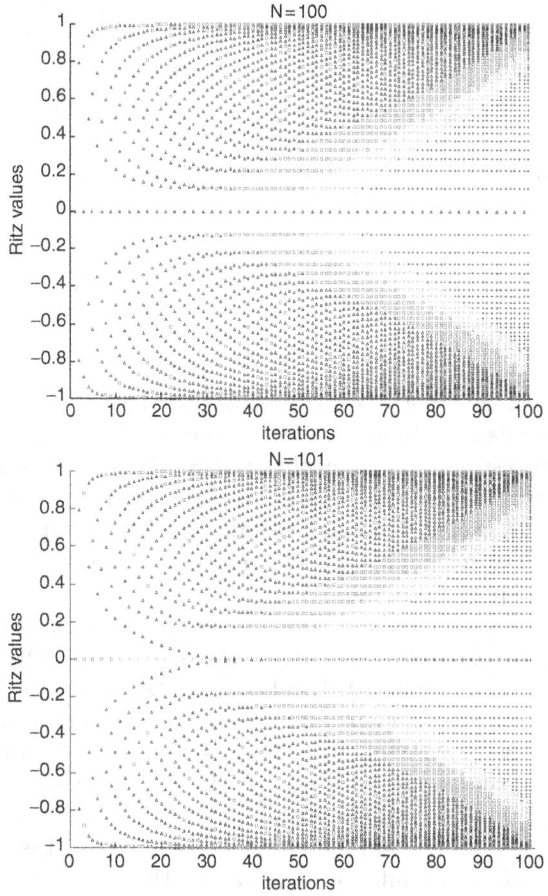

Fig. 11. Convergence of Ritz values for "square roots" of $N \in \{100, 101\}$ Chebyshev eigenvalues in $[-1, 1]$ ($\alpha = 1/2$). In the top plot ($N = 100$) the Ritz values $x_{n+1,2n+1,N} = 0$ for odd n are not close to the spectrum. In the bottom plot ($N = 101$) one observes the phenomena of exceptional eigenvalues

or a circulant-circulant matrix of order $m_x\, m_y$ if there is a block structure with m_x^2 blocks as in (3.18), with each block being itself a circulant matrix. Thus such matrices are induced by a bivariate exponential polynomial ϕ with degree in x being equal to $m_x - 1$ and degree in y being equal to $m_y - 1$, and we write $C_{m_x,m_y}(\phi)$. With the Kronecker product

$$A \otimes B = (A_{j,k} B)_{j,k},$$

we see that the matrix of eigenvectors is given by the unitary matrix $\Omega_{m_x} \otimes \Omega_{m_y}$ and the eigenvalues by the expressions $\phi(\exp(\frac{2\pi i(j-1)}{m_x}), \exp(\frac{2\pi i(k-1)}{m_y}))$, $j = 1, ..., m_x$, $k = 1, ..., m_y$. Thus, as in (3.19), if $\phi^{(m_x,m_y)}$ is the truncation of a bivariate exponential power series ϕ being absolutely convergent, then for $m_x \to \infty$, $m_y \to \infty$,

$$\nu_{m_x m_y}(C_{m_x,m_y}(\phi^{(m_x,m_y)})) \xrightarrow{*} \sigma_\phi, \tag{3.20}$$

with

$$\int f \, d\sigma_\phi = \frac{1}{(2\pi)^2} \iint_{[0,2\pi]^2} f(\phi(\theta)) \, d\theta.$$

Toeplitz matrices are generated by Fourier series

$$T_N(\phi) = \begin{bmatrix} \phi_0 & \phi_1 & \cdots & \phi_{N-1} \\ \phi_{-1} & \phi_0 & \cdots & \phi_{N-2} \\ \vdots & \vdots & & \vdots \\ \phi_{1-N} & \phi_{2N} & \cdots & \phi_0 \end{bmatrix}, \qquad \phi(\theta) = \sum_{j=-\infty}^{\infty} \phi_j e^{ij\theta}, \tag{3.21}$$

which are again constant along diagonals, and hermitian if ϕ is real-valued. Hence any circulant is Toeplitz, but not conversely. We also define Toeplitz-Toeplitz matrices (or level 2 Toeplitz matrices) $T_{m_x,m_y}(\phi)$ induced by some bivariate Fourier series ϕ as a matrix with Toeplitz block structure, each individual block being also of Toeplitz structure. Similarly, we speak of Toeplitz-circulant matrices (Toeplitz block structure with circulant blocks) or circulant-Toeplitz matrices.

It is well-known that these matrices occur in the discretization by finite differences using the five point stencil of the Poisson PDE on $[0,1]^2$, more precisely we have a (banded) Toeplitz-Toeplitz matrix in case of Dirichlet boundary conditions (compare with Example 3.4), and a (banded) circulant-circulant matrix in case of homogeneous Neumann boundary conditions. Toeplitz systems arise also in a variety of other applications, such as signal processing and time series analysis, see [19] and the references cited therein.

For Toeplitz matrices and their level 2 counterparts, it is in general impossible to give explicit formulas for eigenvalues. However, we may find formulas for the asymptotic eigenvalue distribution, compare for instance with [34, pp. 63-65], [14, Theorem 5.10 and Corollary 5.11].

Theorem 3.4. *Let ϕ be a univariate absolutely convergent Fourier series (we say that ϕ is of Wiener class) and real-valued. Then $\nu_N(T_N(\phi)) \to \sigma_\phi$ for $N \to \infty$, with σ_ϕ as in (3.19).*

If ϕ be a bivariate absolutely convergent and real-valued Fourier series then for $m_x, m_y \to \infty$ we have that $\nu_{m_x\, m_y}(T_{m_x,m_y}(\phi)) \to \sigma_\phi$, with σ_ϕ as in (3.20).

In the proof of Theorem 3.4 we will require the following perturbation result of Tyrtyshnikov [71] in a form given by Serra Capizzano [61, Proposition 2.3]. See also Tilli [69].

Theorem 3.5. [61, Proposition 2.3] *Let (A_N) be a sequence of Hermitian matrices where A_N has size $N \times N$. Suppose for every $\epsilon > 0$ there exists N_ϵ such that for every $N \geq N_\epsilon$ a splitting*

$$A_N = B_N(\epsilon) + R_N(\epsilon) + \Delta_N(\epsilon)$$

where $B_N(\epsilon)$, $R_N(\epsilon)$ *and* $\Delta_N(\epsilon)$ *are Hermitian matrices so that for* $N \geq N_\epsilon$,

$$\text{rank } R_N(\epsilon) \leq C_1(\epsilon)N, \quad \text{and} \quad ||\Delta_N(\epsilon)|| \leq C_2(\epsilon)$$

where $C_1(\epsilon)$ *and* $C_2(\epsilon)$ *are positive constants independent of* N *with*

$$\lim_{\epsilon\to 0} C_1(\epsilon) = \lim_{\epsilon\to 0} C_2(\epsilon) = 0.$$

Suppose that, for every $\epsilon > 0$, *the limit* $\nu_N(\Lambda(B_N(\epsilon))) \xrightarrow{*} \sigma_\epsilon$ *for* $N \to \infty$ *exists, and that the limit* $\sigma_\epsilon \xrightarrow{*} \sigma$ *for* $\epsilon \to 0$ *exists, then* $\nu_N(\Lambda(A_N)) \xrightarrow{*} \sigma$ *for* $N \to \infty$.

Proof. Apply the Courant minimax principle and the theorem of Bauer and Fike [30] telling us that if A, B are two hermitian matrices with "small" $||A - B||$, then for each eigenvalue of A there exists an eigenvalue of B which is "close". □

Proof. (of Theorem 3.4) We will give the main idea of proof for the case of a Toeplitz matrix, the arguments for a level 2 Toeplitz matrix are similar. Denote by $\phi^{(N)}$ the Fourier sum obtained from ϕ by taking the $[N^{1/3}]$th partial sum. Since for the p-matrix norm $|| \cdot ||_p$ of a matrix we have

$$(||A||_2)^2 \leq ||A||_\infty \, ||A||_1,$$

and since ϕ is of Wiener class, we find that, for $\epsilon > 0$ and sufficiently large N, we have

$$||\phi - \phi^{(N)}||_{L_\infty([-\pi,\pi])} < \epsilon, \qquad ||T_N(\phi) - T_N(\phi^{(N)})|| < \epsilon.$$

Then $T_N(\phi^{(N)})$ is banded, of bandwidth $\leq N^{1/3}$, and we need to modify at most $N^{2/3}$ entries in order to transform it to a hermitian circular matrix B_N. Since the eigenvalues of this hermitian circular matrix are explicitly known, we obtain the assumptions Theorem 3.5 with $\sigma_\epsilon = \sigma_\phi$, and our claim follows from Theorem 3.5. □

As seen from the above proof, results similar to Theorem 3.4 are true for Toeplitz-circulant or circulant-Toeplitz matrices.

By Theorem 3.4 we see that condition (i)' of Section 3.1 holds for sequences of hermitian (level 2) Toeplitz matrices. Also, condition (ii)' will be true for instance for continuous symbols. Let us shortly comment of condition (iii)' for the case of hermitian positive definite Toeplitz matrices (and hence $\phi \geq 0$). A result of Szegő (see [34, p. 44 and p. 66]) is that

$$\lim_{N\to\infty} \frac{\det(T_N(\phi))}{\det(T_{N-1}(\phi))} = \exp\left(\frac{1}{2\pi}\int_{-\pi}^{\pi} \log\phi(\theta)\, d\theta\right) \tag{3.22}$$

provided that ϕ satisfies the Szegő condition

$$\int_{-\pi}^{\pi} \log\phi(\theta)\, d\theta > -\infty.$$

Notice that this condition can be rewritten as $U^{\sigma_\phi}(0) < +\infty$. Also recall the link to strong asymptotics of orthogonal polynomials on the unit circle (the ratio of determinants in (3.22) is linked to the leading coefficient of such orthonormal polynomials).

It follows from (3.22) that

$$\lim_{N\to\infty} \log(|\det|T_N(\phi)|^{1/N}) = \frac{1}{2\pi}\int_{-\pi}^{\pi} \log\phi(\theta)\,d\theta = \int \log\lambda\,d\sigma(\lambda) \in \mathbb{R},$$

and the condition (iii') is satisfied.

3.4 Discretization of Elliptic PDE's

The asymptotic eigenvalue distribution of matrices obtained by a finite difference discretization of elliptic partial differential equations has been discussed in detail by Serra-Cappizano [62]. Here we will not look for the greatest generality, but just have a look at the particular example of a 2D diffusion equation on some polyhedral domain in $\mathbb{R}^2$, discretized by the classical five point stencil.

Let $\Omega \subset [0,1]^2$ be some open polyhedron, and $b : \Omega \to [0,+\infty)$ piecewise continuous. We solve the diffusion problem

$$\mathrm{div}(b\nabla u) = f \quad \text{on } \Omega$$

plus Dirichlet (Neumann) boundary conditions via central finite differences, with stepsizes

$$\Delta x = \frac{1}{m_x+1}, \qquad \Delta y = \frac{1}{m_y+1},$$

This gives a system of linear equations for the unknowns $u_{j,k} \approx u(j\Delta x, k\Delta y)$ with

$$(j,k) \in \{(j,k) \in \mathbb{Z}^2 \mid (j\Delta x, k\Delta y) \in \Omega\},$$

given by

$$\begin{aligned}
&\frac{m_x+1}{m_y+1}\Big[-b_{j-1/2,k}u_{j-1,k} - b_{j+1/2,k}u_{j+1,k} + (b_{j-1/2,k} + b_{j+1/2,k})u_{j,k}\Big] \\
&+ \frac{m_x+1}{m_y+1}\Big[-b_{j,k-1/2}u_{j,k-1} - b_{j,k+1/2}u_{j,k+1} + (b_{j,k-1/2} + b_{j,k+1/2})u_{j,k}\Big] \\
&= \frac{f_{j,k}}{(m_x+1)(m_y+1)}
\end{aligned}$$

where $b_{j-1/2,k} = c((j-1/2)h, kh)$, etc. Supposing that there are N gridpoints in Ω, we can write this system as $A_N x = b_N$, where it is known that (at least for b strictly positive) $\mathrm{cond}(A_N)$ grows like $\mathcal{O}(N)$. Notice that for $b = 1$ and $\Omega = (0,1)^2$ and Dirichlet boundary conditions we recover the Toeplitz-Toeplitz matrix of Example 3.4. In what follows we will not specify further the (discretisation of the) boundary conditions, since in all cases this only leads to a small rank pertubation of order $\mathcal{O}(\sqrt{N})$, and hence by Theorem 3.5 does not affect the asymptotic eigenvalue distribution.

As in Example 3.4 we consider the limit relation

$$m_x, m_y \to \infty, \quad \frac{m_x}{m_y} \to \delta < 1,$$

Again, for $b = 1$ and $\Omega = (0,1)^2$, the asymptotic eigenvalue distribution has been determined in Example 3.4, compare also with Theorem 3.4 for the symbol

$$\phi(s_1, s_2) = 2\delta(1 - \cos(s_1)) + 2\delta^{-1}(1 - \cos(s_2)). \tag{3.23}$$

In the general case we find the following result, which is a consequence of a more general result of Serra-Cappizano [62].

Theorem 3.6. *Under the above assumptions on Ω and b, we have that, for any continuous function f with compact support,*

$$\lim_{N\to\infty} \int f \, d\nu_N(A_N) = \frac{1}{m(\Omega)} \int_\Omega dx \frac{1}{(2\pi)^2} \iint_{[0,2\pi]^2} ds \, f(b(x) \cdot \phi(s))$$

with ϕ as in (3.23) and $m(\cdot)$ the two-dimensional Lebesgue measure.

Proof. By covering Ω by "small" squares $S_{j,N}$ of equal size tending to zero for $N \to \infty$, we may replace A_N by some block diagonal matrix C_N where entries with row/column index corresponding to points in squares $S_{j,N}$ (for the row) and $S_{k,N}$ (for the column) will be replaced by zero if $S_{j,N} \neq S_{k,N}$ or if $S_{j,N} \cup S_{k,N}$ is not a subset of Ω or if b is not continuous on $S_{j,N} \cup S_{k,N}$. By choosing a correct size of the square, we see that the rank of $A_N - C_N$ is $o(N)$. Denote by B_N the matrix obtained from C_N by replacing all b–values of a diagonal block corresponding to the square $S_{j,N}$ by some constant $b(\xi_{j,N})$ with $\xi_{j,N} \in S_{j,N}$. Then $||B_N - C_N||$ is small by continuity. Applying Theorem 3.4 for each square and summing up all squares we then find that

$$\lim_{N\to\infty} \int f \, d\nu_N(B_N) = \frac{1}{m(\Omega)} \int_\Omega dx \frac{1}{(2\pi)^2} \iint_{[0,2\pi]^2} ds \, f(b(x) \cdot \phi(s)),$$

and our claim follows from Theorem 3.5. □

Remark 3.12. It is interesting to observe that for $b = 1$ we find the same asymptotic eigenvalue distribution as in Example 3.4 independently on the domain Ω.

Remark 3.13. Let $M := \sup_\Omega b$, then from Theorem 3.6 it becomes clear that the asymptotic eigenvalue distribution is described by some measure σ with $\operatorname{supp}(\sigma) = [0, (4\delta + 4\delta^{-1})M] =: \Sigma$. One may also prove that all eigenvalues of A_N lie in Σ. Finally, defining the measure τ by

$$\tau((-\infty, r]) := \frac{m(\{x \in \Omega : b(x) \le r\})}{m(\Omega)},$$

with support given by the essential range of b, and denoting the extremal measure of Example 3.4 by σ_0, we find that σ is obtained by taking the Mellin convolution of τ and σ_0. More precisely, if τ, σ_0 have densities τ', σ_0' then also σ has a density σ', given by

$$\sigma'(y) = \int_{y/M}^{4\delta+4\delta^{-1}} \sigma_0'(x)\tau'(\frac{y}{x})\frac{dx}{x}.$$

If $m := \inf_\Omega b > 0$, we deduce that

$$\sigma'(y) = \frac{1}{4\pi\, m(\Omega)} \int_\Omega \frac{dx}{b(x)} + y\frac{\delta+\delta^{-1}}{32\pi\, m(\Omega)} \int_\Omega \frac{dx}{b(x)^2} + \mathcal{O}(y^2)_{y\to 0}.$$

It is interesting to compare this formula with the Weyl formula for the asymptotic distribution of eigenvalues of the corresponding differential operator.

Example 3.9. We consider the Poisson problem, i.e., $b = 1$, on four different domains $\Omega \subset [0,1]^2$ displayed in Figure 12. On the bottom of Figure 13 one may find histograms for the eigenvalue distribution in the case $m_x = 15$, $m_y = 40$, and hence $\delta = 15/40 = 0.375$. In blue we have drawn the density of the asymptotic eigenvalue distribution, which according to Remark 3.12 is the same for the four domains. In the upper part of Figure 13 we find the convergence history of CG for random starting vector. In all four cases, the actual CG convergence looks quite similar (notice the different scales for the iteration index, since the number of unknowns differs depending on how many grid points are lying in Ω. In all cases we find that the classical and our new asymptotic bound lie above the actual CG error curve, the latter describing quite well the slope of the convergence curve.

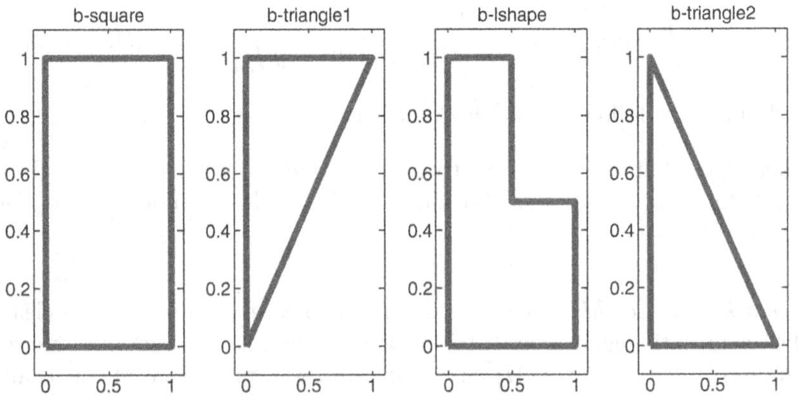

Fig. 12. The four different domains, referred to as b-square, b-triangle-1, b-lshape and b-triangle2 (from the left to the right)

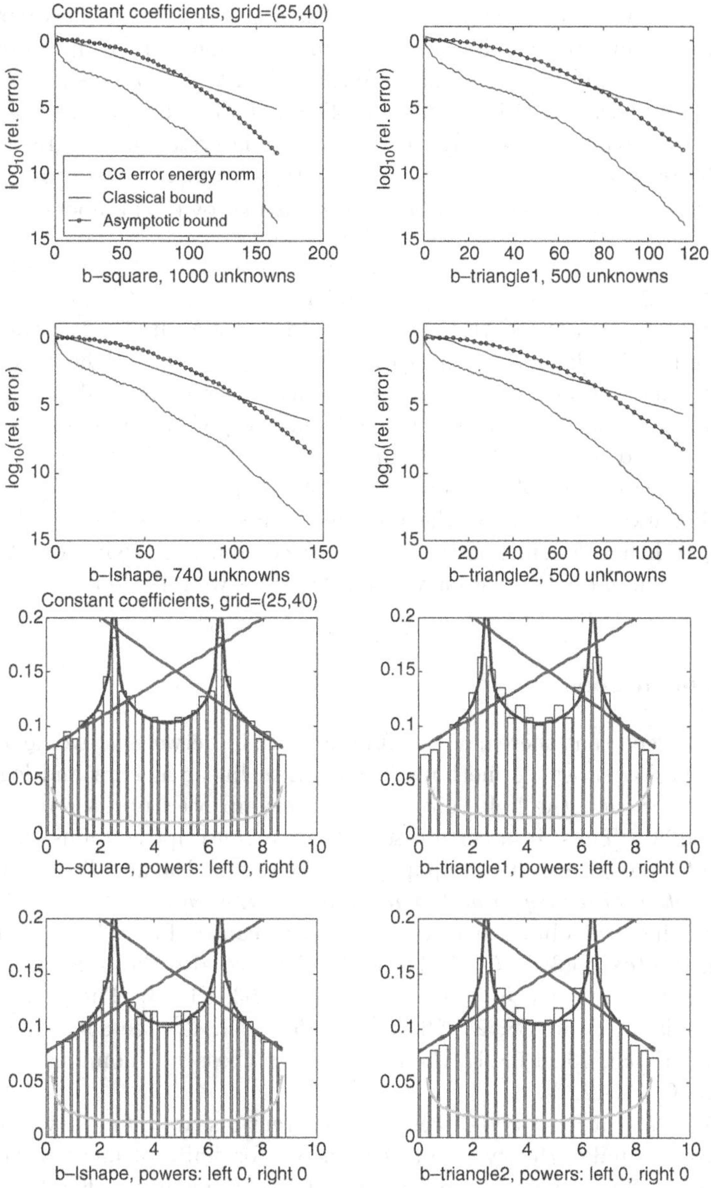

Fig. 13. Convergence rate for the diffusion problem with $b = 1$ on four different domains, the b-square, b-triangle-1, b-lshape and b-triangle2. We have chosen $m_x = 25$ and $m_y = 40$

Here and also in the next example we proceeded as follows to compute numerically the asymptotic convergence bound (3.3): first the quite complicated density of the asymptotic eigenvalue distribution σ of Theorem 3.6 was replaced around both endpoints by the first two nontrivial terms in the Taylor expansion around the endpoints (see the red curves). For this new constraint $\widetilde{\sigma}$, we expect that $S(t)$ is an interval. The endpoints of this interval were obtained by solving numerically the corresponding system of integral equations. One finds as yellow curve the density of the extremal measure $\mu_{t,0,\widetilde{\sigma}}$, with t being the ratio of the last iteration index, divided by the number of unknowns.

Example 3.10. As a second example we consider the diffusion problem with $b(x, y) = 1 + y$ on the same four different domains $\Omega \subset [0, 1]^2$ displayed in Figure 12. On the bottom of Figure 14 one may find histograms for the eigenvalue distribution in the case $m_x = m_y = 40$, and hence $\delta = 1$. In blue we have drawn the density of the asymptotic eigenvalue distribution, which has a shape depending on the domain, especially in a neighborhood of the right end point of $\text{supp}(\sigma)$. In the upper part of Figure 14 we find the convergence history of CG for random starting vector. Notice that the final iteration index divided by the number of unknowns is 0.138, 0.186, 0.158, 0.198, and thus depends on the domain. Again our new asymptotic bound lies above the actual CG error curve, and describes well its slope.

3.5 Conclusions

We have seen that there is a fruitful relationship between convergence behavior of Krylov subspace methods in numerical linear algebra and logarithmic potential theory, the link being given by asymptotics of discrete orthogonal polynomials. Thus, in a certain sense, this manuscript contains the next 2-4 steps of the nice introduction paper [25] of Driscoll, Toh and Trefethen entitled *From potential theory to matrix iteration in six steps.*

The linear algebra theory described here can be found in much more detail in the textbooks [27, 30, 32, 54, 59, 70], see also the original articles on superlinear convergence [3, 4, 31, 52, 53, 56, 63, 64, 74]. The potential theoretic tools are from [58, 24, 16, 22, 41, 45, 5, 40, 42], see also the textbooks [47, 60, 57, 55, 46]. Finally, the link between these two domains is described in [39, 10, 11, 12, 6].

There are at least two directions of current research: first it would be nice to have a similar theory as in Section 3.4 for finite element discretization of elliptic PDEs, including techniques of grid refinements. For P1 elements, some work of Serra-Capizzano and the author is in progress. What is so attractive about the finite element method is that one proceeds by projection, and therefore there are inequalities between the eigenvalues of A_N and of the continuous differential operator. Thus for instance the Weyl formula should tell us much about superlinear convergence for CG.

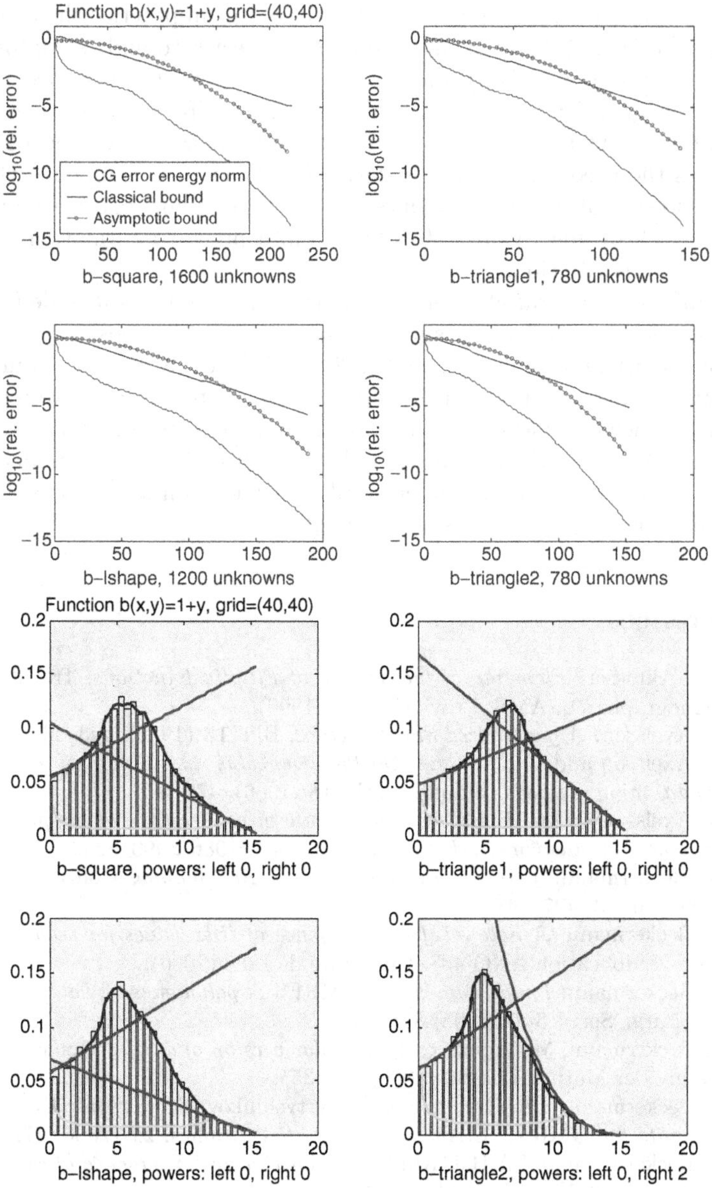

Fig. 14. Convergence rate for the diffusion problem with $b(x, y) = 1 + y$ on four different domains, the b-square, b-triangle-1, b-lshape and b-triangle2. We have chosen $m_x = m_y = 40$

In order to make CG perform better, one uses in practice the technique of preconditioning. A quite involved research project is to find asymptotic eigenvalue distributions for such preconditioned matrices. For instance, there is a whole theory about how one should precondition Toeplitz matrices with the help of circulant matrices, see for instance [19]. However, for level 2 Toeplitz matrices the theory is much less developed. What may happen is that there is a clustering of many eigenvalues around some point, and we should zoom into this clustering in order to obtain more precise information about the eigenvalue distribution.

A different popular class of preconditioning techniques include the incomplete Choleski factorization and its relaxed generalizations, see for instance [59] or the original articles [2, 18, 17, 26, 48, 72]. In [18, 17], the (complicated) triangular matrices in the incomplete Choleski factorization were replaced by circulants, which made it possible to make a more precise analysis for the Poisson problem with periodic boundary conditions. For the model problem of Section 3.4, the asymptotic eigenvalue distribution is determined in some work in progress of Kuijlaars and the author.

References

1. N.I. Akhieser, *Elements of the Theory of Elliptic Functions*, Transl. of Math. Monographs **79**, AMS, Providence RI (1990).
2. O. Axelsson, *A generalized SSOR method*, BIT **13** (1972), 443–467.
3. O. Axelsson and G. Lindskog, *On the eigenvalue distribution of a class of preconditioning methods*, Numer. Math. **48** (1986), 479–498.
4. O. Axelsson and G. Lindskog, *On the rate of convergence of the preconditioned conjugate gradient method*, Numer. Math. **48** (1986), 499–523.
5. B. Beckermann, *On a conjecture of E.A. Rakhmanov*, Constr. Approx. **16** (2000), no. 3, 427–448.
6. B. Beckermann, *A note on the convergence of Ritz values for sequences of matrices*, Publication ANO 408, Université de Lille (2000).
7. B. Beckermann, *Image numérique,* GMRES *et polynômes de Faber*, C. R. Acad. Sci. Paris, Ser. I **340** (2005) 855–860.
8. B. Beckermann, M. Crouzeix, *A lenticular version of a von Neumann inequality*, Archiv der Mathematik **86** (2006) 352–355.
9. B. Beckermann, S.A. Goreinov, E.E. Tyrtyshnikov, *Some remarks on the Elman estimate for* GMRES, SIAM J. Matrix Anal. Applics. **27** (2006) 772–778.
10. B. Beckermann and A.B.J. Kuijlaars, *Superlinear convergence of conjugate gradients*, SIAM J. Num. Anal. **39** (2001), 300–329.
11. B. Beckermann and A.B.J. Kuijlaars, *On the sharpness of an asymptotic error estimate for conjugate gradients*, BIT **41** (2001), 856–867.
12. B. Beckermann and A.B.J. Kuijlaars, *Superlinear CG convergence for special right-hand sides*, Electr. Trans. Num. Anal. **14** (2002), 1–19.
13. B. Beckermann, S. Serra Capizzano, *On the asymptotic spectrum of Finite Element matrix sequences*, manuscript (2005).
14. A. Böttcher and B. Silbermann, *Introduction to Large Truncated Toeplitz Matrices*, Universitext, Springer Verlag, New York, 1999.

15. C. Brezinski, *Biorthogonality and its Applications to Numerical Analysis*, Marcel Dekker, New York, 1992.
16. V.S. Buyarov and E.A. Rakhmanov, *Families of equilibrium measures with external field on the real axis*, Mat. Sb. **190** (1999), no. 6, 11–22 (Russian); Sb. Math. **190** (1999), 791–802.
17. T.F. Chan and H.C. Elman, *Fourier analysis of iterative methods for elliptic problems*, SIAM Review **31** (1989), 20–49.
18. T.F. Chan, *Fourier analysis of relaxed incomplete factorization preconditioners*, SIAM J. Sci. Stat. Comput. **12** (1991), 668–680.
19. R.H. Chan and M.K. Ng, *Conjugate gradient methods for Toeplitz systems*, SIAM Review **38** (1996), 427–482.
20. T. S. Chihara, *An Introduction to Orthogonal Polynomials*, Gordon and Breach, New York, 1978.
21. M. Crouzeix, *Numerical range, holomorphic calculus and applications*, manuscript (2005).
22. S.B. Damelin and E.B. Saff, *Asymptotics of weighted polynomials on varying sets*, manuscript (1998).
23. P. Deift and K.T-R. McLaughlin, *A continuum limit of the Toda lattice*, Mem. Amer. Math. Soc. **131** (1998), no. 624.
24. P.D. Dragnev and E.B. Saff, *Constrained energy problems with applications to orthogonal polynomials of a discrete variable*, J. d'Analyse Math. **72** (1997), 223–259.
25. T.A. Driscoll, K.-C. Toh and L.N. Trefethen, *From potential theory to matrix iteration in six steps*, SIAM Review **40** (1998), 547–578.
26. H.C. Elman, *A stability analysis of incomplete LU factorizations*, Math. Comp. **47** (1986), 191–217.
27. B. Fischer, *Polynomial Based Iteration Methods for Symmetric Linear Systems*, Wiley, Teubner, Stuttgart, 1996.
28. G. Golub, G. Meurant, *Matrices, moments and quadratures II; how to compute the norm of the error in iterative methods*, BIT **37** (1997), 687–705.
29. G.H. Golub, Z. Strakos, *Estimates in quadratic formulas*, Numer. Algorithms **8** (1994), 241–268.
30. G.H. Golub, C.F. Van Loan, *Matrix Computations*, Second Edition, Johns Hopkins University Press, Baltimore, 1993.
31. A. Greenbaum, *Comparisions of splittings used with the conjugate gradient algorithm*, Numer. Math. **33** (1979), 181–194.
32. A. Greenbaum, *Iterative Methods for Solving Linear Systems*, SIAM, Philadelphia PA, 1997.
33. A. Greenbaum and L.N. Trefethen, *GMRES/CR and Arnoldi/Lanczos as matrix approximation problems*, SIAM J. Sci. Comp. **15** (1994), 359–368.
34. U. Grenander and G. Szegő, *Toeplitz Forms and their Applications*, 2nd ed., Chelsea, New York, 1984.
35. I. Gustafsson, *A class of first order factorizations*, BIT **18** (1978), 142–156.
36. M.R. Hestenes, E.L. Stiefel, *Methods of conjugate gradients for solving linear systems*, J. Research Nat. Bur. Standards **49** (1952), 409–436.
37. K. Johansson, *Shape fluctuations and random matrices*, Comm. Math. Phys. **209** (2000), 437–476.
38. M. Kac, W. Murdock, and G. Szegö, *On the eigenvalues of certain Hermitian forms*, J. Ration. Mech. Anal. **2** (1953), 767–800.

39. A.B.J. Kuijlaars, *Which eigenvalues are found by the Lanczos method?* SIAM J. Matrix Anal. Appl. **22** (2000), 306–321.
40. A.B.J. Kuijlaars, *On the finite gap ansatz in the continuum limit of Toda lattice,* Duke Math. J. **104** (2000), 433–462.
41. A.B.J. Kuijlaars and P.D. Dragnev, *Equilibrium problems associated with fast decreasing polynomials,* Proc. Amer. Math. Soc. **127** (1999), 1065–1074.
42. A.B.J. Kuijlaars and K.T-R. McLaughlin, *Long time behavior of the continuum limit of the Toda lattice, and the generation of infinitely many gaps from C^∞ initial data,* Comm. Math. Phys. **221** (2001), 305–333.
43. A.B.J. Kuijlaars and E.A. Rakhmanov, *Zero distributions for discrete orthogonal polynomials,* J. Comp. Appl. Math. **99** (1998), 255–274.
44. A.B.J. Kuijlaars and E.A. Rakhmanov, *Zero distributions for discrete orthogonal polynomials (Corrigendum),* J. Comp. Appl. Math. **104** (1999), 213.
45. A.B.J. Kuijlaars and W. Van Assche, *Extremal polynomials on discrete sets,* Proc. London Math. Soc. **79** (1999), 191–221.
46. N.S. Landkof, *Foundations of Modern Potential Theory.* Grundlehren der Mathematischen Wissenschaften, Vol 190, Springer Verlag, New York, 1972.
47. A. Martínez-Finkelstein, *Equilibrium problems potential theory in the complex plane,* Lecture Notes in Mathematics (this volume), Springer-Verlag, 2005.
48. J.A. Meijerink and H.A. van der Vorst, *An iterative solution method for linear systems of which the coefficient matrix is a symmetric M-matrix,* Math. Comp. **31** (1977), 148–162.
49. G. Meurant, *A review on the inverse of tridiagonal and block tridiagonal symmetric matrices,* SIAM J. Matrix Anal. Appl. **13** (1992), 707–728.
50. G. Meurant, *The computation of bounds for the norm of the error in the conjugate gradient algorithm,* Numer. Algorithms **16** (1998), 77–87.
51. G. Meurant, *Computer Solution of Large Linear Systems,* Studies in Mathematics and its Applications, Vol. 28, North-Holland, 1999.
52. I. Moret, *A note on the superlinear convergence of GMRES,* SIAM J. Numer. Anal. **34** (1997), 513–516.
53. A.E. Naiman, S. Engelberg, *A note on conjugate gradient convergence - Part II,* Numer. Math. **85** (2000), 665–683.
54. O. Nevanlinna, *Convergence of Iterations for Linear Equations,* Birkhäuser, Basel, 1993.
55. E.M. Nikishin and V.N. Sorokin, *Rational Approximations and Orthogonality,* Nauka, Moscow, 1988; Engl. transl. in Translations of Mathematical Monographs **92**, Amer. Math. Soc., Providence, R.I., 1991.
56. C. Paige, B. Parlett, and H. van der Vorst, *Approximate solutions and eigenvalue bounds from Krylov subspaces,* Num. Lin. Alg. Applics. **2** (1995), 115–135.
57. T. Ransford, *Potential Theory in the Complex Plane,* Cambridge University Press, Cambridge, 1995.
58. E.A. Rakhmanov, *Equilibrium measure and the distribution of zeros of the extremal polynomials of a discrete variable,* Mat. Sb. **187** (1996), no. 8, 109–124 (in Russian); Sb. Math. **187** (1996), 1213–1228.
59. Y. Saad, *Iterative Methods for Sparse Linear Systems,* PWS Publishing, Boston, MA, 1996.
60. E.B. Saff and V. Totik, *Logarithmic Potentials With External Fields,* Springer, Berlin, 1997.

61. S. Serra Capizzano, *Distribution results on the algebra generated by Toeplitz sequences: a finite-dimensional approach*, Linear Algebra Appl. **328** (2001), 121–130.
62. S. Serra Capizzano, *Generalized locally Toeplitz sequences: spectral analysis and applications to discretized partial differential equations*, Linear Algebra Appl. **366** (2003), 371–402.
63. G.L.G. Sleijpen and A. van der Sluis, *Further results on the convergence behavior of conjugate-gradients and Ritz values*, Linear Algebra Appl. **246** (1996), 233–278.
64. A. van der Sluis and H.A. van der Vorst, *The rate of convergence of conjugate gradients*, Numer. Math. **48** (1986), 543–560.
65. H. Stahl and V. Totik, *General Orthogonal Polynomials*, Cambridge University Press, Cambridge, 1992.
66. Z. Strakos, *Theory of Convergence and Effects of Finite Precision Arithmetic in Krylov Subspace Methods*, Dr. Sc. Thesis; AS CR, 104+156 pp., Prague, 2001.
67. Z. Strakos, P. Tichy, *On error estimation in the conjugate gradient method and why it works in finite precision computations*, Electron. Trans. Numer. Anal. **13** (2002), 56–80.
68. G. Szegő, *Orthogonal Polynomials*, Amer. Math. Soc. Colloq. Publ. **23**, Providence RI, 1974 (fourth edition).
69. P. Tilli, *Locally Toeplitz sequences: spectral properties and applications*, Linear Algebra Appl. **278** (1998), 91–120.
70. L.N. Trefethen and D. Bau III, *Numerical Linear Algebra*, SIAM, Philadelphia PA, 1997.
71. E.E. Tyrtyshnikov, *A unifying approach to some old and new theorems on distribution and clustering*, Linear Algebra Appl. **232** (1996), 1–43.
72. H.A. van der Vorst, *High performance preconditioning*, SIAM J. Sci. Stat. Comput. **10** (1989), 1174–1185.
73. H.S. Wall, *Analytic Theory of Continued Fractions*, Chelsea, Bronx NY, 1973.
74. R. Winther, *Some superlinear convergence results for the conjugate gradient method*, SIAM J. Numer. Anal. **17** (1980), 14–17.

Orthogonal Rational Functions on the Unit Circle: from the Scalar to the Matrix Case

Adhemar Bultheel[1], **Pablo González-Vera**[2], **Erik Hendriksen**[3], **Olav Njåstad**[4]

[1]Department of Computer Science, Katholieke Universiteit Leuven, Celestijnenlaan 200A, B-3001 Leuven, Belgium
e-mail: Adhemar.Bultheel@cs.kuleuven.be

[2]Department of Mathematical Analysis, University of La Laguna, Tenerife, Spain

[3]Department of Mathematics, University of Amsterdam, The Netherlands

[4]Department of Mathematical Sciences, Norwegian University of Science and Technology,
Trondheim, Norway

Summary. The purpose of these lecture notes is to give a short introduction to the theory of orthogonal rational functions (ORF) on the unit circle. We start with the classical problem of linear prediction of a stochastic process to give a motivation for the study of Szegő's problem and to show that in this context it will turn out that not as much the ORF but rather the reproducing kernels will play a central role. Another example of rational Krylov iteration shows that it might also be interesting to consider ORF on the real line, which we shall not discuss in these lectures.

In a second part we will show that most of the results of the scalar case translate easily to the case of matrix valued orthogonal rational functions (MORF).

There are however many aspects that are intimately related to these ideas that we do not touch upon, like continued fractions, Nevanlinna-Pick interpolation, moment problems, and many other aspects of what is generally known as Schur analysis.

1 Motivation: Why Orthogonal Rational Functions?

We want to give a summary of known results about orthogonal rational functions (ORF) with respect to some measure whose support is contained in the unit circle of the complex plane. But before we start, we want to give some motivation of why it may be interesting to generalize orthogonal polynomials to ORF for which the poles of the successive rational functions of increasing degree are taken from a preselected sequence of complex numbers.

1.1 Linear Prediction

Consider a discrete time stationary stochastic process $\{y_n\}_{n\in\mathbb{Z}}$. The index n denotes a time instance. Stationarity means that $\mu_k = \mathsf{E}\{y_n\overline{y}_{n-k}\}$ is independent of n. We suppose that all the y_n are zero mean, identically distributed. A problem considered already by Wiener is to predict y_n at instant n from observation of the preceding y_{n-k}, $k \geq 1$. Thus we want to find coefficients a_k such that y_n is predicted by $\hat{y}_n = -\sum_{k=1}^{\infty} a_k y_{n-k}$ such that we minimize the energy of the prediction error or innovation process $e_n = \sum_{k=0}^{\infty} a_k y_{n-k}$ where $a_0 = 1$. That is we want to minimize the expected value $\mathsf{E}\{|e_n|^2\}$.

Because in the space spanned by $\{y_n\}_{n\in\mathbb{Z}}$, we can consider $\mathsf{E}\{x\overline{y}\}$ as an inner product of x and y, we can formulate the previous problem by saying that we look for the orthogonal projection of the present onto its past. Modulo some technical details, it can be shown that this problem can be reformulated in an L_2 setting as an infinite dimensional least squares problem, i.e., finding the projection of 1 onto the space H_2^- spanned by $\{z^{-1}, z^{-2}, \ldots\}$ in the Hilbert space $L_2(\mu)$ of the unit circle where the orthogonality measure is the spectral measure of the process (the trigonometric moments of this measure are the μ_k introduced above). The result is that the optimal predictor is given by the spectral factor of the spectral measure. Suppose for simplicity that the spectral measure is absolutely continuous with weight $w(e^{i\omega})$, then the spectral factor is the function $\sigma(z)$ which is the outer spectral factor of w (σ and $1/\sigma$ are analytic outside the unit disk, and satisfy $|\sigma(t)|^2 = w(t)$ for $|t| = 1$).

There are two ways one can think of to solve this problem. Either we project 1 onto subspaces $\mathcal{L}_n^-$ of H_2^- of dimension $n = 1, 2, \ldots$ and let n tend to ∞, or we can solve the trigonometric moment problem to find the spectral measure and subsequently or simultaneously do a spectral factorization.

If we assume that the value of y_n is mostly defined by the near past and much less by a past that is longer ago, then a natural choice is to take $\mathcal{L}_n^- = \text{span}\{z^{-1}, \ldots, z^{-n}\}$. So the optimal predictor of this finite dimensional problem is to find the polynomial $\hat{\varphi}_n(z) = \sum_{k=0}^{n} a_k z^{-k}$ with $a_0 = 1$ that has minimal norm in $L_2(\mu)$. This is known to be the reciprocal of the monic nth orthogonal polynomial. From the classical Szegő theory, it is known that (under appropriate conditions on the measure) these orthogonal polynomials converge (up to a constant) to the outer spectral factor of the orthogonality measure (whose inverse is often called Christoffel function in this context). Thus solving the projection problem leads asymptotically to a spectral factor of the measure. For practical applications though we do not want n to be very large, since that would need long and expensive filters. So we stop this process with a finite n, and hope to have a good approximation. It is not difficult however to find a simple example that requires a very long filter to get a reasonable approximation.

Suppose $y_n = -\sum_{k=1}^{\infty} a^k y_{n-k}$. Thus the predictor is $1 + a/z + a^2/z^2 + \cdots = 1/(1 - a/z)$. If $|a|$ is close to 1, then the sequence a^k will decay very slowly, and we shall need a high degree polynomial $\hat{\varphi}_n$ to obtain a good approximation. If however, we know an estimate $\hat{a}$ of a from observing the FFT of the observations, then a filter expanded in terms of powers of $(1-\hat{a}/z)$ might need only a couple of terms to give a good approximation. For a more complicated behaviour of the spectral density of the process, one might need more poles to get an accurate model and in such a case a rational approximant from $\text{span}\{1, 1/(1-\alpha_1/z), 1/[(1-\alpha_1/z)(1-\alpha_2/z)], \ldots\}$ will be a better option to model the system.

1.2 Krylov Subspace Methods

If A is a linear operator on a Hilbert space H (e.g., a large $N \times N$ matrix operating on $\mathbb{R}^N$) and $v \in H$, then the space $\mathcal{K}_{n+1}(A, v) = \text{span}\{v_0, \ldots, v_n\}$ with $v_k = A^k v$, is called a Krylov subspace. To solve a linear equation $Ax = b$ or an eigenvalue problem for A, the problem is projected onto a Krylov subspace of finite (i.e., low) dimension ($n \ll N$ in the matrix example) and this low dimensional problem is solved to give an approximation to the original problem. To compute the projection, an orthogonal basis is constructed for the Krylov subspace. Clearly, the $(k+1)$st orthogonal vector q_k has to be a combination of the the first $k+1$ vectors in the Krylov subspace. Hence it is of the form $q_k = \varphi_k(A)v$ with $\varphi_k(z)$ a polynomial of degree k.

Exercise 1. Suppose that A is real self adjoint and positive definite, then prove that the orthonormality $q_k^T q_l = \delta_{k,l}$ is equivalent with the orthogonality of the polynomials $\langle \varphi_k, \varphi_k \rangle = \delta_{kl}$ with respect to the inner product defined by $\langle \varphi_k, \varphi_l \rangle = L(\varphi_k \varphi_l)$ where the linear functional L is defined on the space of polynomials by its moments $m_k = L(z^k) = v^T A^k v$. Note that the metric for the standard basis $\{1, z, z^2, \ldots\}$ is a Hankel matrix whose rank can not be larger than N, the size of A.

Thus in the classical Lanczos method for symmetric matrices, the three-term recurrence relation for the orthogonal polynomials leads to a short recurrence between the successive vectors q_k, meaning that q_n can be computed from q_{n-1} and q_{n-2}, which does not need a full Gram-Schmidt orthogonalization.

However, computing the v_k is like an application of the power method and therefore, the v_k will quickly converge to an eigenvector corresponding to a dominant eigenvalue. Thus, if we want an eigenvalue in the neighborhood of α, then we should not iterate with A, but with $B = (A - \alpha I)^{-1}$. The rational Krylov method (RKS) of A. Ruhe [26, 27, 28] allows for a different shift α in every iteration step. Thus $v_k = (A - \alpha_k I)^{-1} v_{k-1}$, or even more generally $v_k = (A - \sigma_k I)(A - \alpha_k I)^{-1} v_{k-1}$, where α_k is used to enforce the influence of the eigenspaces of the eigenvalues in the neighborhood of α_k, while σ_k is used to suppress the influence of the eigenspaces of the eigenvalues in the neighborhood of σ_k. Anyway, this construction of v_k means that we may write v_k as $v_k = r_k(A)v$ with r_k a rational function of the form $p_k(z)/[(z - \alpha_1) \cdots$

$(z-\alpha_k)]$ with p_k a polynomial of degree at most k, so that after orthogonalizing v_k with respect to the previous vectors, we obtain a vector $q_k = \varphi_k(A)v$ where $\varphi_k(z)$ is again a rational function of the same form as r_k. Since the classical moment matrix has a Hankel structure, this theory will be related to orthogonality on the real line.

Skipping all the technical details, it will be obvious that orthogonality of the q_k will lead to some orthogonality of the rational functions φ_k. Again, a simple recurrence of the ORF will lead to an efficient implementation of the RKS.

1.3 Numerical Quadrature

One more example, before we dive into the general theory of ORF. To compute an integral $\int_a^b f(x)w(x)dx$, it is well known that a quadrature formula of the form $\sum_{k=1}^n \lambda_k f(\xi_k)$ will be exact for all polynomials of degree up to $2n-1$ if the nodes of this quadrature formula are the zeros of the polynomial φ_n which is the nth orthogonal polynomial orthogonal with respect to the inner product $\langle f, g\rangle = \int_a^b f(x)g(x)w(x)dx$, and if the weights are given by the corresponding Christoffel numbers $\lambda_k = 1/\sum_{j=0}^{n-1} |\varphi_k(\xi_k)|^2$.

However, if the integrand f does not behave much like a polynomial, some other quadrature formula can be better. For example if we want to integrate $f(x) = \sin(x)/p_n(x)$ over the interval $[0, \pi/2]$ with $w = 1$. If $p_n(x)$ is a real polynomial with complex conjugate pairs of zeros that are close to the interval of integration, then it would be much better to have a quadrature formula that is exact for all rational functions of degree n that have poles close to the zeros of the polynomial p_n. Using ORF for the interval $[a, b]$ with respect to the weight w, with prescribed poles, it is possible to derive formulas that are similar to the Gaussian formulas, i.e., taking for the nodes the zeros of the ORF φ_n and as weights $\lambda_k = 1/\sum_{j=0}^{n-1} |\varphi_k(\xi_k)|^2$. These quadrature formulas will be exact in a certain space of dimension $2n-1$. It is clear that this should lead to much more accurate results in examples like we described above.

2 Orthogonal Rational Functions on the Unit Circle

We give in Sections 2–6 an introduction to the theory of ORF for a measure that is supported on the unit circle of the complex plane. All the results of this part (and more) can be found in the monograph [3].

2.1 Preliminaries

We denote

$$\mathbb{D} = \{z \in \mathbb{C} : |z| < 1\}, \quad \mathbb{T} = \{z \in \mathbb{C} : |z| = 1\}, \quad \mathbb{E} = \{z \in \mathbb{C} : |z| > 1\}.$$

For any function f we introduce the *parahermitian conjugate* f_*, defined by $f_*(z) = \overline{f(1/\overline{z})}$. Note that for $t \in \mathbb{T}$, $f_*(t) = \overline{f(t)}$.

Let μ be a probability measure on $\mathbb{T}$ with infinite support and $L_2(\mu)$ the Hilbert space with inner product $\langle f, g \rangle_\mu = \int f(t)\overline{g(t)}d\mu(t)$. If $d\mu = d\lambda$ is the normalized Lebesgue measure $d\lambda = dt/(2\pi)$ we drop μ from the notation. The trigonometric moments are denoted by $c_k = \int t^{-k}d\mu(t)$, $k \in \mathbb{Z}$.

Introducing the Riesz-Herglotz kernel

$$D(t, z) = \frac{t + z}{t - z}$$

we can associate with μ its Riesz-Herglotz transform

$$\Omega_\mu(z) = ic + \int D(t, z)d\mu(t)$$

with $\Omega_\mu(0) = 1 + ic$. This Ω_μ belongs to the class $\mathcal{C}$ of Carathéodory functions

$$\mathcal{C} = \{f \in H(\mathbb{D}) : \operatorname{Re} f(\mathbb{D}) > 0\}$$

where $H(\mathbb{D})$ denotes the functions analytic in $\mathbb{D}$.

Exercise 2. Prove that

$$\operatorname{Re} \Omega_\mu(z) = \int P(t, z)d\mu(t)$$

with Poisson kernel

$$P(t, z) = \frac{1}{2}[D(t, z) + \overline{D(1/\overline{t}, z)}] = \frac{t(1 - |z|^2)}{(t - z)(1 - \overline{z}t)}, \tag{2.1}$$

which can be simplified for $t \in \mathbb{T}$ as follows

$$P(t, z) = \operatorname{Re} D(t, z) = \frac{1 - |z|^2}{|t - z|^2}, \quad t \in \mathbb{T}, \ z \in \mathbb{D}.$$

Moreover with $c_k = \int t^{-k}d\mu(t)$

$$\Omega_\mu(z) = c_0 + 2\sum_{k=1}^{\infty} c_k z^k, \quad (c_0 = 1)$$

and the nontangential limit of $\operatorname{Re} \Omega_\mu(z)$ when $|z| \to 1$ is a.e. $(d\lambda)$ equal to μ' which is the density function of the absolute continuous part of μ.

Let $H_2(\mathbb{D})$ be the classical Hardy space of functions analytic in $\mathbb{D}$ that have a boundary value on $\mathbb{T}$ that is square integrable. Every function $f \in H_2(\mathbb{D})$ has a canonical inner-outer factorization. This means that it can be written as $f = UF$ with U inner and F outer. An inner function belongs to the class of bounded analytic functions

$$\mathcal{B}(\mathbb{D}) = \{f \in H(\mathbb{D}) : f(\mathbb{D}) \subset \mathbb{D}\}$$

and $|f(t)| = 1$ a.e. on $\mathbb{T}$. A Blaschke product is an example of an inner function. It is defined as $B(z) = \prod_n \zeta_n(z)$ with

$$\zeta_n(z) = z_n \frac{z - \alpha_n}{1 - \overline{\alpha}_n z}, \quad z_n = -\frac{|\alpha_n|}{\alpha_n}, \quad \alpha_n \in \mathbb{D} \setminus \{0\}$$

and $z_n = 1$ if $\alpha_n = 0$. It is well known that a Blaschke product converges (to a finite function not identically zero) if and only if

$$\sum_n (1 - |\alpha_n|) < \infty,$$

which means that the $|\alpha_n|$ should approach 1 fast enough. Any inner function is of the form

$$U(z) = e^{i\gamma} B(z) S(z), \quad \gamma \in \mathbb{R}, \quad S(z) = \exp\left\{-\int D(t,z) d\nu(t)\right\}$$

with ν a bounded positive singular ($\nu' = 0$ a.e.) measure and B a Blaschke product that catches all the zeros of U.

An outer function in $H_2(\mathbb{D})$ is a function of the form

$$F(z) = e^{i\gamma} \exp\left\{\int D(t,z) \log \psi(t) d\lambda(t)\right\}, \quad \gamma \in \mathbb{R}$$

where $\log \psi \in L_1$ and $\psi \in L_2$.

If $\log \mu' \in L_1$ (Szegő's condition), then we may define the spectral factor of μ

$$\sigma(z) = c \exp\left\{\frac{1}{2} \int D(t,z) \log \mu'(t) d\lambda(t)\right\}, \quad c \in \mathbb{T}, \quad z \in \mathbb{D}.$$

It is an outer function, uniquely defined up to a constant $c \in \mathbb{T}$. It has a nontangential limit to the boundary $\mathbb{T}$ that satisfies $|\sigma(t)|^2 = \mu'(t)$, a.e., $t \in \mathbb{T}$. The constant c can be fixed by requiring for example that $\sigma(0) > 0$.

Exercise 3. Prove that

$$|\sigma(z)|^2 = \exp\left\{\int P(t,z) \log \mu'(t) d\lambda\right\}.$$

More about the material in this section can be found in many textbooks a.o. [8, 16, 15, 14, 25, 30].

2.2 The Fundamental Spaces

We select a sequence $\{\alpha_k\}_{k=0}^{\infty} \subset \mathbb{D}$ with $\alpha_0 = 0$ and define the partial Blaschke products

$$B_0 = 1, \quad B_n(z) = B_{n-1}(z)\zeta_n(z) = \prod_{k=1}^{n} z_k \frac{z - \alpha_k}{1 - \overline{\alpha}_k z}, \qquad n \geq 1.$$

The functions $\{B_0, B_1, \ldots, B_n\}$ span the space

$$\mathcal{L}_n = \left\{ \frac{p_n}{\pi_n} : \pi_n(z) = \prod_{k=1}^{n} (1 - \overline{\alpha}_k z), \quad p_n \in \mathcal{P}_n \right\} \tag{2.2}$$

where $\mathcal{P}_n$ is the space of polynomials of degree at most n.

If we set all $\alpha_k = 0$, then $\mathcal{L}_n = \mathcal{P}_n$. In that case the Gram matrix of $\mathcal{L}_n$ for the standard basis is a *Toeplitz matrix* with entries $\left\langle z^i, z^j \right\rangle_\mu = c_{j-i}$ where $i, j = 0, \ldots, n$. If all the α_k are mutually distinct, then the Gram matrix for the basis $\{1, 1/(1 - \overline{\alpha}_1 z), \ldots, 1/(1 - \overline{\alpha}_n z)\}$ has entries

$$\left\langle \frac{1}{1 - \overline{\alpha}_i z}, \frac{1}{1 - \overline{\alpha}_j z} \right\rangle_\mu = \frac{1}{2(1 - \overline{\alpha}_i \alpha_j)} \int \left[\overline{D(t, \alpha_i)} + D(t, \alpha_j) \right] d\mu(t)$$
$$= \frac{1}{2} \frac{\overline{\Omega_\mu(\alpha_i)} + \Omega_\mu(\alpha_j)}{1 - \overline{\alpha}_j \alpha_i}.$$

Such a matrix is called a *Pick matrix*.

Exercise 4. If ∂_w^k denotes the kth derivative with respect to the variable w, prove that

$$\Omega_\mu^{(k)}(w) := \partial_w^k \Omega_\mu(w) = \int \partial_w^k D(t, w) d\mu(t) = 2(k!) \int \frac{t d\mu(t)}{(t - w)^{k+1}}$$

and

$$\overline{\Omega_\mu^{(k)}(w)} = \int \overline{\partial_w^k D(t, w)} d\mu(t).$$

In the more general situation where we want to construct a Gram matrix for the space $\mathcal{L}_n$ for fixed n and where some of the of the α_k are repeated, we can rearrange them so that equal α_k are grouped. We can then take the basis consisting of functions of the form $1/(1 - \overline{\alpha}_k z), \ldots, 1/(1 - \overline{\alpha}_k z)^{\nu_k}$ if α_k is repeated ν_k times, and similarly for the other α_i. Technically, it is rather difficult to write a general expression of an entry in the Gram matrix for such a basis, but with the help op the previous exercise, one can show the following.

Exercise 5. Suppose there are $m + 1$ different $\alpha_0, \ldots, \alpha_m$ (recall $\alpha_0 = 0$) which appear with multiplicity $\nu_0, \ldots, \nu_m$ respectively with $\sum_{i=0}^{m} \nu_i = n+1$, then the Gram matrix for the basis of $\mathcal{L}_n$ that we just mentioned will only depend on $\Omega_\mu^{(k)}(\alpha_i)$, $k = 0, \ldots, \nu_i - 1$, $i = 0, \ldots, m$. We could call this a generalized Pick matrix.

Whatever the basis is that we choose for $\mathcal{L}_n$, it will always be similar to a Toeplitz matrix.

Theorem 2.1. *If $W = [w_0, \ldots, w_n]^T$ is a basis for $\mathcal{L}_n$, with Gram matrix $G_n(W) = \langle W, W \rangle_\mu = [\langle w_k, w_l \rangle_\mu]_{k,l=0}^n$, then there is a Toeplitz matrix T and an invertible matrix V such that $VG_n(W)V^H = T$.*

Proof. Note that we may always choose the basis $\{\ell_k(z) = z^k/\pi_n(z) : k = 0, \ldots, n\}$ with $\pi_n(z)$ as in (2.2). Note that the Gram matrix for this basis is Toeplitz, whose entries are trigonometric moments for the measure $d\mu(t)/|\pi_n(t)|^2$. Since every basis transformation is represented by an invertible matrix, the theorem follows. □

2.3 Reproducing Kernels

Suppose $\{e_0, \ldots, e_n\}$ is a basis for $\mathcal{L}_n$, orthonormal with respect to μ. Then $k_n(z, w) = \sum_{k=0}^n e_k(z)\overline{e_k(w)}$ is a reproducing kernel for $\mathcal{L}_n$, i.e., $\langle f(\cdot), k_n(\cdot, w) \rangle_\mu = f(w)$ for all $f \in \mathcal{L}_n$ and $w \in \mathbb{D}$. If $f \in \mathcal{L}_m$ with $m > n$, then $\langle f(\cdot), k_n(\cdot, w) \rangle_\mu$ gives the orthogonal projection of f onto $\mathcal{L}_n$. Both observations follow immediately by writing f as a linear combination of the e_k and using the orthonormality. Note that the reproducing kernel is unique. The previous expression does not depend on a particular choice of the orthonormal basis. For example, it is immediately seen that if $\{e_k\}$ is an orthonormal basis for $\mathcal{L}_n$, then $\{B_n e_{k*}\}$ is also an orthonormal basis because $B_n e_{k*} \in \mathcal{L}_n$ for $k = 0, \ldots, n$ and

$$\langle B_n e_{k*}, B_n e_{l*} \rangle_\mu = \int |B_n(t)|^2 \overline{e_k(t)} e_l(t) d\mu(t) = \langle e_l, e_k \rangle = \delta_{k,l}.$$

Therefore, by using $B_n(z)e_{k*}(z) = B_n(z)\overline{e_k(1/\overline{z})}$ as orthonormal basis:

$$k_n(z, w) = B_n(z)\overline{B_n(w)} \sum_{k=0}^{n} \overline{e_k(1/\overline{z})} e_k(1/\overline{w}) = B_n(z)\overline{B_n(w)} k_n(1/\overline{w}, 1/\overline{z}).$$

By using a basis transformation, one may express $k_n(z, w)$ in terms of *any* basis for $\mathcal{L}_n$ as follows.

Theorem 2.2. *If $E_n(z) = [e_0(z), \ldots, e_n(z)]^T$ represents any basis for $\mathcal{L}_n$, then the reproducing kernel is given by*

$$k_n(z, w) = \frac{-1}{\det G_n} \det \begin{bmatrix} G_n & E_n(z) \\ E_n(w)^H & 0 \end{bmatrix}$$

where $G_n = \langle E_n, E_n \rangle_\mu$ is the Gram matrix of E_n. The superscript ^{H} refers to the complex conjugate (Hermitian) transpose.

Proof. Let $F_n = VE_n$ with V invertible be a column of orthonormal basis functions. Then $I = \langle F_n, F_n\rangle_\mu = V\langle E_n, E_n\rangle_\mu V^H = VG_nV^H$, so that $G_n^{-1} = V^HV$ and thus

$$k_n(z,w) = F_n(w)^H F_n(z) = E_n(w)^H V^H V E_n(z) = E_n(w)^H G_n^{-1} E_n(z),$$

which proves the theorem. □

From now on, we shall use the notation ϕ_k for the orthonormal basis for $\mathcal{L}_n$ that is ordered such that $\phi_0 = 1$ and $\phi_k \in \mathcal{L}_k \setminus \mathcal{L}_{k-1}$ for $k = 1, 2, \ldots$ Using the partial Blaschke products B_k, which also form a basis, we can write $\phi_n = a_{n,n}B_n + a_{n,n-1}B_{n-1} + \cdots + a_{n,1}B_1 + a_{n,0}$. We shall denote the leading coefficient $a_{n,n}$ (with respect to the basis B_k) by κ_n. Since ϕ_n is uniquely defined up to a constant of modulus 1, we can fix ϕ_n uniquely, for example by assuming that $\kappa_n > 0$, which we shall do throughout this lecture. Note that the trailing coefficient $a_{n,0}$ (with respect to the basis B_k) is given by $\phi_n(\alpha_1)$. To derive a similar notation for the leading coefficient, we need a generalization of the *reciprocal* function. We shall denote this generalization by a superscript *. In general we set for any function $f \in \mathcal{L}_n$: $f^*(z) = B_n(z)f_*(z)$. Note that the superstar notation is ambiguous since it depends on the n that is considered. So a notation like $f^{[n]}$ instead of f^* would be more appropriate. However, in order not to overload the notation, if not explicitly mentioned, it should be clear from the context which n is intended. Note that with this notation we have $\phi_n^*(\alpha_n) = \kappa_n > 0$. We can now immediately write down as a consequence of the previous theorem:

Corollary 2.1. *If the $\phi_n = \kappa_n B_n + \cdots$ with $\kappa_n > 0$ are the orthonormal basis functions for $\mathcal{L}_n$ as introduced above and $k_n(z,w)$ is the reproducing kernel, then $k_n(z,\alpha_n) = \kappa_n\phi_n^*(z)$ and $k_n(\alpha_n,\alpha_n) = \kappa_n^2$.*

Exercise 6. Prove this corollary.

The following Christoffel-Darboux relations hold:

Theorem 2.3. *With the notation just introduced we have*

$$\begin{aligned} k_n(z,w) &= \frac{\phi_{n+1}^*(z)\overline{\phi_{n+1}^*(w)} - \phi_{n+1}(z)\overline{\phi_{n+1}(w)}}{1 - \zeta_{n+1}(z)\overline{\zeta_{n+1}(w)}} \\ &= \frac{\phi_n^*(z)\overline{\phi_n^*(w)} - \zeta_n(z)\overline{\zeta_n(w)}\phi_n(z)\overline{\phi_n(w)}}{1 - \zeta_n(z)\overline{\zeta_n(w)}}. \end{aligned}$$

If $z = w = t \in \mathbb{T}$, then one may pass to the limit which will introduce derivatives:

$$P(t,\alpha_{n+1})k_n(t,t) = t[\phi_{n+1}'(t)\overline{\phi_{n+1}(t)} - \phi_{n+1}^{*\prime}(t)\overline{\phi_{n+1}^*(t)}],$$

with $P(t,z)$ the Poisson kernel (2.1).

Proof. Because the numerators and denominators of the expressions on the right hand side vanish for $z = 1/\overline{w}$, this zero cancels out and the right hand sides are in $\mathcal{L}_n$ as a function of z. Using

$$k_n(z,w) = B_n(z)\overline{B_n(w)} \sum_{k=0}^{n} \phi_{k*}(z)\overline{\phi_{k*}(w)}$$

(which follows from Theorem 2.2) we may write

$$\frac{k_{n+1}(z,w)}{B_{n+1}(z)\overline{B_{n+1}(w)}} - \frac{k_n(z,w)}{B_n(z)\overline{B_n(w)}} = \phi_{(n+1)*}(z)\overline{\phi_{(n+1)*}(w)}.$$

Multiplying by the denominator gives

$$[k_n(z,w) + \phi_{n+1}(z)\overline{\phi_{n+1}(w)}] - \zeta_{n+1}(z)\overline{\zeta_{n+1}(w)}k_n(z,w) = \phi^*_{n+1}(z)\overline{\phi^*_{n+1}(w)},$$

which gives the first formula when z and w do not coincide on $\mathbb{T}$. The other formula is obtained when in the previous formula we replace n by $n-1$ so that

$$k_n(z,w) + \zeta_n(z)\overline{\zeta_n(w)}[\phi_n(z)\overline{\phi_n(w)} - k_n(z,w)] = \phi^*_n(z)\overline{\phi^*_n(w)}$$

from which the proof follows. □

Exercise 7. Prove the confluent case.

It can be seen that this relation does not depend on the fact that all $\alpha_k \in \mathbb{D}$.

Note that if we set $\phi_k = p_k/\pi_k$ with $\pi_k(z) = \prod_{i=0}^{k}(1 - \overline{\alpha}_i z)$, then the p_k do not form an orthogonal polynomial sequence w.r.t. a positive measure, so that the theory of ORF is not quite the same as the theory of orthogonal polynomials w.r.t. a varying measure. With the Christoffel-Darboux formula, setting $z = w$, it is not difficult to derive the following property.

Theorem 2.4. *For all $n \geq 0$, $\phi^*_n(z) \neq 0$ for $z \in \mathbb{D}$ (hence, $\phi_n(z) \neq 0$ for $z \in \mathbb{E}$) and $|\phi_{n+1}(z)/\phi^*_{n+1}(z)| < 1$, $(= 1, > 1)$ for $z \in \mathbb{D}$, $(\mathbb{T}, \mathbb{E})$.*

Exercise 8. Prove this theorem.

2.4 Recurrence Relations

The kernels satisfy a fundamental recurrence relation

Theorem 2.5. *The kernels satisfy (superstar w.r.t. z)*

$$\begin{bmatrix} k^*_n(z,w) \\ k_n(z,w) \end{bmatrix} = t_n(z,w) \begin{bmatrix} k^*_{n-1}(z,w) \\ k_{n-1}(z,w) \end{bmatrix}$$

with

$$t_n(z,w) = c_n \begin{bmatrix} 1 & \overline{\rho_n} \\ \rho_n & 1 \end{bmatrix} \begin{bmatrix} \zeta_n(z) & 0 \\ 0 & 1 \end{bmatrix} \begin{bmatrix} 1 & \overline{\gamma_n} \\ \gamma_n & 1 \end{bmatrix}$$

where

$$\begin{aligned} c_n &= (1 - |\rho_n|^2)^{-1} \\ \rho_n &= \rho_n(w) = \overline{\phi_n(w)}/\phi_n^*(w) \\ \gamma_n &= \gamma_n(w) = -\zeta_n(w)\rho_n(w). \end{aligned}$$

Proof. The Christoffel-Darboux relation implies

$$k_n(z,w) = \phi_n^*(z)\overline{\phi_n^*(w)} + \zeta_n(z)\overline{\zeta_n(w)}k_{n-1}(z,w).$$

Multiply this with $\overline{\rho_n(w)}$ and substitute (superstar w.r.t. z)

$$\phi_n^*(z)\overline{\phi_n^*(w)\rho_n(w)} = k_n^*(z,w) - \zeta_n(z)k_{n-1}^*(z,w)$$

to get

$$k_n(z,w)\overline{\rho_n(w)} = \zeta_n(z)\overline{\zeta_n(w)}k_{n-1}(z,w)\overline{\rho_n(w)} + k_n^*(z,w) - \zeta_n(z)k_{n-1}^*(z,w).$$

Take the superstar conjugate of this relation and solve for $k_n^*(z,w)$ and $k_n(z,w)$ and the result follows. □

The Christoffel-Darboux relation also implies a recurrence relation for the ϕ_n.

Theorem 2.6. *The orthonormal functions satisfy*

$$\begin{bmatrix} \phi_n(z) \\ \phi_n^*(z) \end{bmatrix} = \tilde{t}_n(z) \begin{bmatrix} \phi_{n-1}(z) \\ \phi_{n-1}^*(z) \end{bmatrix}$$

where

$$\begin{aligned} \tilde{t}_n(z) &= \frac{\kappa_n}{\kappa_{n-1}} \begin{bmatrix} 1 & 0 \\ 0 & z_n \end{bmatrix} \begin{bmatrix} \varepsilon_n & \delta_n \\ \overline{\delta}_n & \overline{\varepsilon}_n \end{bmatrix} \begin{bmatrix} \frac{z-\alpha_{n-1}}{1-\overline{\alpha}_n z} & 0 \\ 0 & \frac{1-\overline{\alpha}_{n-1}z}{1-\overline{\alpha}_n z} \end{bmatrix} \\ &= \frac{\kappa_n}{\kappa_{n-1}} \frac{1-\overline{\alpha}_{n-1}z}{1-\overline{\alpha}_n z} \begin{bmatrix} \overline{z}_{n-1}\varepsilon_n & 0 \\ 0 & z_n\overline{\varepsilon}_n \end{bmatrix} \begin{bmatrix} 1 & \overline{\lambda}_n \\ \lambda_n & 1 \end{bmatrix} \begin{bmatrix} \zeta_{n-1}(z) & 0 \\ 0 & 1 \end{bmatrix} \\ &= \sqrt{\frac{1-|\alpha_n|^2}{1-|\alpha_{n-1}|^2}} \frac{1}{\sqrt{1-|\lambda_n|^2}} \frac{1-\overline{\alpha}_{n-1}z}{1-\overline{\alpha}_n z} \begin{bmatrix} \eta_n^1 & 0 \\ 0 & \eta_n^2 \end{bmatrix} \begin{bmatrix} 1 & \overline{\lambda}_n \\ \lambda_n & 1 \end{bmatrix} \begin{bmatrix} \zeta_{n-1}(z) & 0 \\ 0 & 1 \end{bmatrix}, \end{aligned}$$

with

$$\varepsilon_n = z_n \frac{1-\overline{\alpha}_{n-1}\alpha_n}{1-|\alpha_{n-1}|^2} \frac{\overline{\phi_n^*(\alpha_{n-1})}}{\kappa_n}, \qquad \delta_n = \frac{1-\overline{\alpha}_n\alpha_{n-1}}{1-|\alpha_{n-1}|^2} \frac{\phi_n(\alpha_{n-1})}{\kappa_n},$$

$$\lambda_n = \overline{z}_{n-1}\frac{\overline{\delta}_n}{\overline{\varepsilon}_n} = \eta_n^1 \frac{\overline{\phi_n(\alpha_{n-1})}}{\phi_n^*(\alpha_{n-1})} \in \mathbb{D}, \quad \eta_n^1 = z_n\overline{z}_{n-1}\frac{1-\overline{\alpha}_{n-1}\alpha_n}{1-\overline{\alpha}_n\alpha_{n-1}} \in \mathbb{T},$$

$$\eta_n^2 = \overline{z}_{n-1}z_n\overline{\eta}_n^1 = \frac{1-\overline{\alpha}_n\alpha_{n-1}}{|1-\overline{\alpha}_n\alpha_{n-1}|} \frac{\phi_n^*(\alpha_{n-1})}{|\phi_n^*(\alpha_{n-1})|} \in \mathbb{T}.$$

Proof. (sketch) From the superstar conjugate (w.r.t. z) of the Christoffel-Darboux relation we find

$$\frac{\phi_n^*(z)\phi_n(\alpha_{n-1}) - \phi_n(z)\phi_n^*(\alpha_{n-1})}{\zeta_n(\alpha_{n-1}) - \zeta_n(z)} = \phi_{n-1}(z)\kappa_{n-1}.$$

The superstar conjugate of this relation is

$$\frac{\phi_n^*(z)\overline{\phi_n^*(\alpha_{n-1})} - \phi_n(z)\overline{\phi_n(\alpha_{n-1})}}{1 - \zeta_n(z)\overline{\zeta_n(\alpha_{n-1})}} = \phi_{n-1}^*(z)\kappa_{n-1}.$$

Eliminate $\phi_n^*(z)$ between these two relations and the first form of the recurrence for $\phi_n(z)$ is obtained. The second one follows immediately from this because the Christoffel-Darboux relations imply $(|\phi_n^*(z)|^2 - |\phi_n(z)|^2)/(1 - |\zeta_n(z)|^2) > 0$ so that $\phi_n^*(\alpha_{n-1}) \neq 0$ and $\lambda_n \in \mathbb{D}$. For the third one, recall that $\kappa_n = \phi_n^*(\alpha_n)$, and use again the Christoffel-Darboux relation with $z = w = \alpha_n$ to obtain the appropriate factor in front. □

Note that if all the α_k are zero, then we recover the polynomial case. The recurrence relation is just the Szegő recurrence and the λ_n being then equal to $\overline{\phi_n(0)}/\kappa_n$, are the Szegő parameters (and are sometimes called Schur or reflection coefficients).

Since this is derived from the Christoffel-Darboux relation, the first recurrence does not depend on the α_k being in $\mathbb{D}$. However, we can not allow $\overline{\alpha}_n\alpha_{n-1} = 1$, in which case λ_n is not defined. In such a case we call the system *degenerate*. If $\phi_n(\alpha_{n-1}) = 0$, then we call the system *exceptional*. If all the α_k are in $\mathbb{D}$ or all in $\mathbb{E}$, then the system is non-degenerate and $\lambda_n \in \mathbb{D}$.

The vector $[\phi_n, \phi_n^*]^T$ is a solution of the recurrence with initial condition $[\phi_0, \phi_0^*]^T = [1, 1]^T$. This is not the only solution. With the initial condition $[1, -1]^T$ we get another, independent solution. It will be formulated in terms of some functions $\psi_n \in \mathcal{L}_n$ which we shall call the ORF *of the second kind.* We introduce them as follows.

$$\psi_0 = 1, \quad \psi_n(z) = \int D(t, z)[\phi_n(t) - \phi_n(z)]d\mu(t), \quad n \geq 1.$$

Exercise 9. Let f be such that $f_* \in \mathcal{L}_{n-1}$, then prove by orthogonality that

$$\int D(t, z)[f(t)/f(z) - 1]\phi_n(t)d\mu(t) = 0.$$

From this derive that for such an f

$$\psi_n(z) = \int D(t, z)[\phi_n(t)f(t)/f(z) - \phi_n(z)]d\mu(t), \quad n \geq 1. \tag{2.3}$$

Similarly derive that for an $f \neq 0$ such that $f \in \mathcal{L}_n$ and $f(\alpha_n) = 0$ it holds that

$$\psi_n^*(z) = \int D(t, z)\left[\phi_n^*(t)f(t)/f(z) - \phi_n^*(z)\right]d\mu(t), \quad n \geq 1. \tag{2.4}$$

We can now prove that $[\psi_n, -\psi_n^*]^T$ is another solution of the recurrence relation for $[\phi_n, \phi_n^*]^T$, which corresponds to the initial condition $[1, -1]^T$. In other words $[\psi_n, \psi_n^*]^T$ satisfies the same recurrence relation as $[\phi_n, \phi_n^*]^T$, but with λ_n replaced by $-\lambda_n$.

Theorem 2.7. *With the notation of Theorem 2.6 we have*

$$\begin{bmatrix} \phi_n(z) & \psi_n(z) \\ \phi_n^*(z) & -\psi_n^*(z) \end{bmatrix} = \tilde{t}_n(z) \begin{bmatrix} \phi_{n-1}(z) & \psi_{n-1}(z) \\ \phi_{n-1}^*(z) & -\psi_{n-1}^*(z) \end{bmatrix}. \tag{2.5}$$

Proof. We only have to prove the relation for the ψ_n. From the definition and the previous exercise we have for $n > 1$

$$\begin{bmatrix} \psi_{n-1}(z) \\ -\psi_{n-1}^*(z) \end{bmatrix} = -\Omega_\mu(z) \begin{bmatrix} \phi_{n-1}(z) \\ \phi_{n-1}^*(z) \end{bmatrix} + \int D(t,z) \begin{bmatrix} \phi_{n-1}(t) \\ \frac{\zeta_{n-1}(z)}{\zeta_{n-1}(t)} \phi_{n-1}^*(t) \end{bmatrix} d\mu(t).$$

Multiply from the left by $\tilde{t}_n(z)$ and the right hand side becomes

$$-\Omega_\mu(z) \begin{bmatrix} \phi_n(z) \\ \phi_n^*(z) \end{bmatrix} + \int D(t,z) Q(t,z) \begin{bmatrix} \phi_n(t) \\ \phi_n^*(t) \end{bmatrix} d\mu(t)$$

with

$$Q(t,z) = \frac{(z - \alpha_{n-1})(1 - \overline{\alpha}_n t)}{(1 - \overline{\alpha}_n z)(t - \alpha_{n-1})}.$$

Using a technique like in the previous exercise, it can be shown that the integral equals

$$\int D(t,z) \begin{bmatrix} \phi_n(t) \\ \frac{\zeta_n(z)}{\zeta_n(t)} \phi_n^*(t) \end{bmatrix} d\mu(t)$$

so that on the right hand side we get again the same expression as in the starting relation but with $n-1$ replaced by n. This proves the theorem for $n > 1$. □

Exercise 10. Check the theorem for $n = 1$.

Far reaching generalizations of the Christoffel-Darboux relations can be obtained for any couple of solutions (x_n, x_n^+) and (y_n, y_n^+) of this recurrence. We give them without proof.

Theorem 2.8. *Given two solutions (x_n, x_n^+) and (y_n, y_n^+) of this recurrence, we can define*

$$F_n(z,w) = x_n^+(z) y_n(w) - x_n(z) y_n^+(w),$$

and then we have a Liouville-Ostrogradskii formula

$$\frac{F_n(z,w)}{1 - \zeta_n(z)/\zeta_n(w)} - B_n(w) \frac{F_0(z,w)}{1 - \zeta_0(z)/\zeta_0(w)} = -\sum_{k=0}^{n-1} x_k(z) y_k^+(w) \frac{B_n(w)}{B_k(w)}$$

and with the definition

$$G_n(z,w) = x_n^+(z)\overline{y_n^+(w)} - x_n(z)\overline{y_n(w)},$$

we have a Green formula

$$\frac{G_n(z,w)}{1-\zeta_n(z)\overline{\zeta_n(w)}} - \frac{G_0(z,w)}{1-\zeta_0(z)\overline{\zeta_0(w)}} = \sum_{k=0}^{n-1} x_k(z)\overline{y_k(w)}.$$

Choosing (x_n, x_n^+) and/or (y_n, y_n^+) equal to (ϕ_n, ϕ_n^*) or (ψ_n, ψ_n^*) gives several identities which we leave for the reader to discover. We just give one example:

$$\psi_n(z)\phi_n^*(z) + \psi_n^*(z)\phi_n(z) = 2B_n(z)P(z,\alpha_n) \tag{2.6}$$

with $P(z,w)$ the Poisson kernel (2.1). It is obtained by taking $z = w$ and $(x, x^+) = (\phi, \phi^*)$ and $(y, y^+) = (\psi, -\psi^*)$ in the Liouville-Ostrogradskii formula. It can also be obtained by taking determinants in (2.5). Therefore we refer to it as the determinant formula.

Note that the Christoffel-Darboux relation was crucial in our development so far. From this, we derived the recurrence relation for the kernels, and for the ORF which gave rise to the introduction of the second kind functions.

This can be inverted: if we have a recurrence relation for ϕ_n as given above, then they will be ORF with respect to some positive measure on $\mathbb{T}$. This is a Favard type theorem which reads as follows:

Theorem 2.9 (Favard). *Given a sequence $\{\alpha_0 = 0, \alpha_1, \alpha_2, \ldots\} \subset \mathbb{D}$ and suppose that with initial condition $\phi_0 = 1$, the ϕ_n are generated by a recurrence relation of the 3rd form given in Theorem 2.6 with all $\lambda_n \in \mathbb{D}$, and the unimodular constants such that $\phi_n^*(\alpha_n) > 0$, then these ϕ_n will form a sequence of ORF with respect to a probability measure on $\mathbb{T}$.*

We formulate the theorem here as a motivation for the introduction of quadrature formulas and will give its proof later in Section 4.

Because the functions of the second kind satisfy a recurrence relation of this form (the λ_n have to be replaced by $-\lambda_n$), it follows by the Favard theorem that they also are a system of ORF, with respect to a positive measure on $\mathbb{T}$. Thus we can attribute to ψ_n the same properties as we can attribute to the ϕ_n (location of the zeros, Christoffel-Darboux relations, etc.). If $\Omega_\mu \in \mathcal{C}$ is the Riesz-Herglotz transform of μ, then clearly $1/\Omega_\mu$ is also in $\mathcal{C}$. So it can be written as the Riesz-Herglotz transform of some measure ν. It can be shown that if the ϕ_n are orthogonal w.r.t. μ, then the associated functions of the second kind ψ_n, will be orthogonal w.r.t. the associated ν.

Obtaining a constructive proof for this theorem is closely related to the applications we mentioned in the beginning: we construct a measure that will generate an inner product that entails orthogonality in $\mathcal{L}_n$ and then let n tend to infinity. The first part is related to numerical quadrature and interpolation which we shall consider in the next section.

3 Quadrature and Interpolation

3.1 Quadrature

One way to obtain a quadrature formula for the integral $I_\mu\{f\} = \int f(t)d\mu(t)$ is to interpolate f by a function f_n in a number of points $\{\xi_{ni}\}_{i=1}^n \subset \mathbb{T}$, and to approximate $I_\mu\{f\}$ by $I_n\{f\} = I_\mu\{f_n\}$.

Exercise 11. Consider the function $f_n \in \mathcal{L}_{p,q} = \{fg : f \in \mathcal{L}_{p*}, g \in \mathcal{L}_q\}$, $p, q \geq 0$, $p + q = n - 1$, where $\mathcal{L}_{p*} = \{f : f_* \in \mathcal{L}_p\}$. Show that this f_n shall interpolate f in the points $\{\xi_{ni}\}_{i=1}^n \subset \mathbb{T}$ if

$$f_n(t) = \sum_{k=1}^n L_{n,k}(t) f(\xi_{nk}), \quad L_{n,k}(t) = \ell_{n,k}(t)\left(\prod_{i=1}^q \frac{1-\overline{\alpha}_i\xi_{nk}}{1-\overline{\alpha}_i t}\right)\prod_{i=1}^p \frac{\xi_{nk}-\alpha_i}{t-\alpha_i},$$

where $\ell_{n,k}$ are the classical Lagrange polynomials, i.e., $\ell_{n,k}(\xi_{nj}) = \delta_{k,j}$. Note that the $L_{n,k} \in \mathcal{L}_{p,q}$ are rational generalizations for which also $L_{n,k}(\xi_{nj}) = \delta_{k,j}$.

Thus the quadrature formula is $I_n\{f\} = \sum_{k=1}^n \lambda_{n,k} f(\xi_{nk})$, $\lambda_{n,k} = I_\mu\{L_{n,k}\}$. This is called an interpolating quadrature formula, which is obviously exact for all $f \in \mathcal{L}_{p,q}$.

To obtain the largest possible domain in which we get an exact quadrature formula, we have to choose the nodes ξ_{nk} in a particular way. Like in Gaussian quadrature formulas, we could try to choose them as the zeros of ϕ_n, but that is impossible because the ORF ϕ_n have all their zeros inside $\mathbb{D}$, and not on $\mathbb{T}$. However the following result holds.

Theorem 3.1. *The function $Q_n(z) = \phi_n(z) + \tau\phi_n^*(z)$, $\tau \in \mathbb{T}$ has n simple zeros on $\mathbb{T}$ and it is para-orthogonal which means that it is orthogonal to $\mathcal{L}_{n-1} \cap \mathcal{L}_n(\alpha_n)$ where $\mathcal{L}_n(\alpha_n) = \{f \in \mathcal{L}_n : f(\alpha_n) = 0\}$. Moreover $\langle 1, Q_n\rangle_\mu \neq 0 \neq \langle B_n, Q_n\rangle_\mu$.*

Proof. By the Christoffel-Darboux relations we know that $|\phi_n^*|^2 > |\phi_n|^2$ in $\mathbb{D}$ and the opposite inequality holds in $\mathbb{E}$. Thus Q_n can only have zeros in $\mathbb{T}$. These zeros are simple, otherwise, we would have $Q_n(\xi) = Q_n'(\xi) = 0$, which implies that in ξ we have $\phi_n/\phi_n^* = -\tau = -1/\overline{\tau} = \overline{\phi_n^*/\phi_n}$ or $\phi_n'\overline{\phi_n} - (\phi_n^*)'\overline{\phi_n^*} = 0$. But letting z and w approach $\xi \in \mathbb{T}$ in the Christoffel-Darboux formula leads to $\frac{(1-\overline{\alpha}_n\xi)(\xi-\alpha_n)}{1-|\alpha_n|^2}[\phi_n'(\xi)\overline{\phi_n(\xi)} - (\phi_n^*)'(\xi)\overline{\phi_n^*(\xi)}] = k_{n-1}(\xi,\xi) > 0$, which is a contradiction.

Concerning the para-orthogonality we note that $\langle f, \phi_n\rangle_\mu = 0$ if $f \in \mathcal{L}_{n-1}$ while if $f \in \mathcal{L}_n(\alpha_n)$, then it can be written as $B_n g_*$ with $g \in \mathcal{L}_{n-1}$, and thus $\langle f, \phi_n^*\rangle_\mu = \langle B_n g_*, B_n\phi_{n*}\rangle_\mu = \langle \phi_n, g\rangle_\mu = 0$. On the other hand $\langle 1, Q_n\rangle_\mu = \tau\langle 1, \phi_n^*\rangle_\mu = \tau\langle \phi_n, B_n\rangle_\mu \neq 0$, and similarly for the other inequality. □

Now take an arbitrary function $R \in \mathcal{L}_{n-1,n-1}$, suppose we interpolate it with a function $R_n \in \mathcal{L}_{0,n-1} = \mathcal{L}_{n-1}$, using the zeros $\{\xi_{n1}, \ldots \xi_{nn}\}$ of Q_n. Then simply writing it out shows that the interpolation error can be written as

$$E_n(z) = R(z) - R_n(z) = Q_n(z)S(z), \quad S \in \mathcal{L}_{n-1} \cap \mathcal{L}_n(\alpha_n).$$

Thus, because of the para-orthogonality of Q_n, we get an integration error $I_\mu\{E_n\} = 0$, so that the quadrature formula is exact in $\mathcal{L}_{n-1,n-1}$. This is the highest possible degree of exactness that can be obtained with n nodes in the sense that with n nodes on $\mathbb{T}$, it is impossible to integrate exactly all functions in $\mathcal{L}_{n,n-1}$ or $\mathcal{L}_{n-1,n}$. In this sense the quadrature formula is optimal and we shall refer to it as the *rational (or R-) Szegő quadrature formula.* We denote the (discrete) measure having masses λ_{nk} at the points ξ_{nk} by μ_n so that by definition $I_n\{f\} = I_{\mu_n}\{f\}$.

3.2 Interpolation

We have used interpolation to construct a quadrature formula and found that integration with respect to μ and μ_n is the same in $\mathcal{L}_{n-1,n-1}$. This happens if and only if their Riesz-Herglotz transforms take the same values at the points $\{\alpha_0, \ldots, \alpha_n\}$ (taking multiplicity into account) since the respective Gram matrices depend only on their (derivative) values in the poles $\{\alpha_k\}_{k=0}^{n-1}$ (see exercise 5). So let us first find out what the Riesz-Herglotz transform of μ_n is.

Theorem 3.2. *Set* $\Omega_n(z) = \int D(t,z)d\mu_n(t) = -\frac{P_n(z)}{Q_n(z)}$. *Then for* $n \geq 1$

$$P_n(z) = \int D(t,z)[Q_n(t)f(t)/f(z) - Q_n(z)]d\mu(t) = \psi_n(z) - \tau\psi_n^*(z)$$

for any f *such that* $f_* \in \mathcal{L}_{n-1} \cap \mathcal{L}_n(\alpha_n)$.

Proof. If we write the terms explicitly, then it becomes clear that the previous integrand is in $\mathcal{L}_{n-1,n-1}$ for $n \geq 2$, and thus we can replace the integral by the R-Szegő quadrature formula with nodes the zeros of Q_n, giving $P_n(z)f(z) = -f(z)Q_n(z)I_n\{D(\cdot,z)\} = -f(z)Q_n(z)\Omega_n(z)$. The case $n = 1$ is left as an exercise. The expression with ψ_k follows from (2.3) and its superstar. □

Because of this result we could call P_n a para-orthogonal function of the second kind. It is para-orthogonal with respect to the same measure as for which the ψ_n are orthogonal.

The previous theorem has an interesting corollary.

Corollary 3.1. *The weights of the R-Szegő formula are given by*

$$\lambda_{nk} = \frac{1}{2\xi_{nk}} \frac{P_n(\xi_{nk})}{Q_n'(\xi_{nk})} = \frac{1}{\sum_{i=0}^{n-1} |\phi_i(\xi_{nk})|^2} > 0.$$

Proof. Recall $\Omega_n(z) = -P_n(z)/Q_n(z) = I_n\{D(\cdot,z)\} = \sum_{k=1}^n \lambda_{nk} D(\xi_{nk},z)$ and $Q_n(\xi_{nk}) = 0$. Thus after multiplying by $(z-\xi_{nk})$ we have

$$(z-\xi_{nk})\Omega_n(z) = \sum_{j\neq k} \lambda_{nj} D(\xi_{nj},z)(z-\xi_{nk}) + \lambda_{nk} D(\xi_{nk},z)(z-\xi_{nk}).$$

Taking the limit for $z \to \xi_{nk}$ gives the first formula. For the expression with the kernel, use the fact that $Q_n(\xi_{nk}) = 0$, thus $\tau = 1/\overline{\tau} = -\phi_n(\xi_{nk})/\phi_n^*(\xi_{nk})$, and the confluent Christoffel-Darboux formula to get

$$Q_n'(\xi_{nk}) = \frac{P(\xi_{nk},\alpha_n)}{\xi_{nk}\overline{\phi_n(\xi_{nk})}} k_{n-1}(\xi_{nk},\xi_{nk})$$

with $P(t,z)$ the Poisson kernel (2.1), while the determinant formula leads to $P_n(\xi_{nk}) = 2P(\xi_{nk},\alpha_n)/\overline{\phi_n(\xi_{nk})}$, so the corollary is proved. □

We can now derive interpolation properties for $\Omega_n = -P_n/Q_n$:

Theorem 3.3. *Let Ω_n and Ω_μ be the Riesz-Herglotz transforms of μ_n and μ respectively, then $\Omega_\mu(z) - \Omega_n(z) = zB_{n-1}(z)h(z)$, for $n \geq 1$ with h analytic in $\mathbb{D}$.*

Proof. This follows because of exercise 5. □

Since we assumed $\tau \in \mathbb{T}$, we cannot set $\tau = 0$, but we do have the same type of interpolation with an extra interpolation in α_n.

Theorem 3.4. *For the ORF and the ORF of the second kind we have*

$$\Omega_\mu(z)\phi_n(z) + \psi_n(z) = zB_{n-1}(z)h(z), \quad \Omega_\mu(z)\phi_n^*(z) - \psi_n^*(z) = zB_n(z)g(z)$$

with h and g analytic in $\mathbb{D}$.

Proof. Since $[\phi_n\Omega_\mu + \psi_n]/B_{n-1} = \int D(t,z)\phi_n(t)/B_{n-1}(t)d\mu(t)$, we find a function analytic in $\mathbb{D}$ because it is a Cauchy-Stieltjes integral and setting $z = 0$, so that $D(t,0) = 1$, the integral becomes $\langle \phi_n, B_{n-1}\rangle_\mu = 0$. The second relation is similarly proved since now for $z = 0$ we find $\langle \phi_n^*, B_n\rangle_\mu = 0$. □

Note that by the determinant formula, we have for $R_n(z) = \psi_n^*/\phi_n^*$ that for $t \in \mathbb{T}$: $w_n(t) = \operatorname{Re} R_n(t) = \frac{1}{2}[R_n(t) + R_{n*}(t)] = P(t,\alpha_n)/|\phi_n(t)|^2$. Because of the extra interpolation at α_n, one might expect that the absolute continuous measure $w_n(t)d\lambda(t)$ gives the same integrals in $\mathcal{L}_{n,n}$, and it does indeed.

Theorem 3.5. *The inner product in $\mathcal{L}_n$ is the same for the measure μ and for the absolute continuous measure with weight $w_n = P(\cdot,\alpha_n)/|\phi_n|^2$.*

Proof. Since $\int |\phi_n(t)|^2 w_n(t) d\lambda(t) = \int P(t, \alpha_n) d\lambda(t) = 1 = \|\phi_n\|_\mu^2$, the norm is maintained. Moreover

$$\begin{aligned}\langle \phi_n, \phi_k \rangle_{w_n} &= \int P(t, \alpha_n) \frac{\phi_{k*}(t)}{\phi_{n*}(t)} d\lambda(t) \\ &= \int P(t, \alpha_n) \left\{ \frac{\phi_k^*(t) B_n(t)/B_k(t)}{\phi_n^*(t)} \right\} d\lambda(t) = 0,\end{aligned}$$

which follows because the factor in curly brackets is analytic in $\mathbb{D} \cup \mathbb{T}$, so that we may apply Poisson's formula to find zero because $B_n(z)/B_k(z)$ is zero in α_n. Thus ϕ_n has norm 1 and is orthogonal to $\mathcal{L}_{n-1}$ for the weight w_n, and because the recurrence relation then defines all the previous ϕ_k uniquely (provided they have the proper normalization of positive leading coefficient), the theorem follows. □

We note that the para-orthogonality conditions alone do not define the functions Q_n completely. The para-orthogonal functions that we proposed are also $\overline{\tau}$-invariant, which means that $Q_n^* = \overline{\tau} Q_n$. The latter is essential to guarantee that they have n simple zeros on $\mathbb{T}$. It can be shown that this invariance property and the para-orthogonality completely defines the Q_n up to the parameter $\tau \in \mathbb{T}$ and a normalizing constant factor.

3.3 Interpolation and Quadrature Using the Kernels

After we formulated the recurrence relation for the kernels in Theorem 2.4, we gave the recurrence for the ORF in Theorem 2.6, but from there on, the kernels were neglected. However, what has been developed for the ORF, can be repeated and generalized to the kernels. Since the methodology is completely analogous, we leave the proofs in this section as a major exercise.

First of all, we note that the kernels are produced by their recurrence relation when we give the initial conditions $(k_0^*, k_0) = (1, 1)$. As in the case of the ORF, we obtain another independent solution when we start with $(l_0^*, -l_0) = (1, -1)$. The resulting kernels could be called kernels of the second kind. They satisfy

$$\begin{bmatrix} l_n^*(z, w) \\ -l_n(z, w) \end{bmatrix} = t_n(z, w) \begin{bmatrix} l_{n-1}^*(z, w) \\ -l_{n-1}(z, w) \end{bmatrix}$$

with $t_n(z, w)$ as in Theorem 2.4. By our remark after Theorem 2.9, it should be clear that $l_n(z, w) = \sum_{k=0}^n \psi_n(z) \overline{\psi_k(w)}$ is a reproducing kernel for the space $\mathcal{L}_n$ w.r.t. the same measure as for which the ψ_k are the ORF. The interpolation properties are a bit harder to obtain because they require a parameterized (in w) Riesz-Herglotz transform:

$$\Omega_\mu(z,w) = \int \frac{D(t,z)}{P(t,w)} d\mu(t) + c,$$

with

$$c = \frac{\overline{w}\int t d\mu(t) - w \int t^{-1} d\mu(t)}{1-|w|^2} \in i\mathbb{R}, \quad (z,w) \in \mathbb{D}\times\mathbb{D},$$

$D(t,z)$ the Dirichlet and $P(t,w)$ the Poisson kernel. The c is chosen to make $\Omega_\mu(w,w)$ real (and hence it equals 1). Then it is possible to formulate an interpolation property saying that $l_n(z,w) - k_n(z,w)\Omega_\mu(z,w)$ vanishes for all $z \in \{w, \alpha_1, \ldots, \alpha_n\}$. Thus the special role of $\alpha_0 = 0$ has been removed and is played by an arbitrary $w \in \mathbb{D}$. Since $k_n(z,w)$ does not vanish for any $z \in \mathbb{D}\cup\mathbb{T}$ if $w \in \mathbb{D}$, we may consider $\Omega_n(z,w) = l_n(z,w)/k_n(z,w)$, which is the parameterized Riesz-Herglotz transform of the absolutely continuous measure $P(t,w)k_n(w,w)/|k_n(t,w)|^2 d\lambda(t)$. The inner product in $\mathcal{L}_n$ w.r.t. the latter will not depend on w and will give the same results as the inner product w.r.t. $d\mu(t)$.

4 Density and the Proof of Favard's Theorem

4.1 Density

For the density of the Blaschke products in $L_p(\mathbb{T})$, one can easily adapt a result of [1, p.244] to find the following.

Theorem 4.1. *Define the Blachke products as before for $n \geq 0$ and $B_{-n} = 1/B_n$ for $n \geq 1$. Then the system $\{B_n\}_{n\in\mathbb{Z}}$ is complete in $L_p(\mathbb{T})$, for any $p \geq 1$ as well as in the space $C(\mathbb{T})$ of continuous functions on $\mathbb{T}$ if and only if $\sum(1-|\alpha_k|) = \infty$ (the Blaschke product diverges).*
Also the system $\{B_n\}_{n=0}^{\infty}$ is complete in H_p for any $p \geq 1$ if and only if $\sum(1-|\alpha_k|) = \infty$ (the Blaschke product diverges).

Walsh [31, p.305-306] states

Theorem 4.2. *If the Blaschke product diverges and $f \in H_2$, then $f_n \in \mathcal{L}_n$ which interpolates f in $\{w, \alpha_1, \ldots, \alpha_n\} \subset \mathbb{D}$ (w is an arbitrary but fixed number in $\mathbb{D}$) will converge to f uniformly on compact subsets of $\mathbb{D}$, and if f is continuous up to $\mathbb{T}$, we also have uniform convergence on $\mathbb{T}$.*

For a general probability measure, Szegő's condition $\log\mu' \in L_1$ plays a role. For example, from Walsh [31, p.116,186,50,92,144] we may conclude that the polynomials are dense in $L_p(\mu)$ if and only if $\log\mu' \notin L_1$. For the rational case, the situation seems to be less simple.

Theorem 4.3.

1. *If* $\sum_k (1-|\alpha_k|) = \infty$ *then* $\log \mu' \notin L_1 \Leftrightarrow \{B_n\}_{n\geq 0}$ *is complete in* $L_2(\mu)$
2. *If* $\log \mu' \in L_1$ *then* $\sum_k (1-|\alpha_k|) = \infty \Leftrightarrow \{B_n\}_{n\geq 0}$ *is complete in* $H_2(\mu)$
3. *If* $\sum_k (1-|\alpha_k|) = \infty$ *then* $\{B_n\}_{n\in\mathbb{Z}}$ *is complete in* $L_2(\mu)$.

Note that we have no equivalence anymore. The divergence of the Blaschke product implies completeness of $\{B_n\}_{n=0}^{\infty}$ in $H_2(\mu)$ and of $\{B_n\}_{n\in\mathbb{Z}}$ in $L_2(\mu)$, but the converse need not be true. We leave it to the reader to look up the proofs of the above theorems. The precise characterization of completeness of the Blaschke products in not totally cleared out.

4.2 Proof of Favard's Theorem

It is easily proved by induction that the ϕ_n^* generated by the recursion are indeed superstar conjugates of the ϕ_n that it generates and the parameters λ_n are given by their expressions of Theorem 2.6. Moreover, the $\{\phi_k\}_{k=0}^n$ are a set of ORF with respect to the weight $w_n(t) = P(t,\alpha_n)/|\phi_n|^2$. This is shown as follows. Note that for $0 \leq m \leq n$

$$\langle \phi_n, \phi_m \rangle_{w_n} = \int \frac{\phi_{m*}(t)}{\phi_{n*}(t)} P(t,\alpha_n) d\lambda(t) = \left. \frac{B_n(z)}{B_m(z)} \frac{\phi_m^*(z)}{\phi_n^*(z)} \right|_{z=\alpha_n} = \delta_{nm}.$$

Thus $\phi_n \perp_{w_n} \mathcal{L}_{n-1}$, and because an inverse recurrence (obtained by multiplying (2.5) from the left with the inverse of the matrix $\tilde{t}_n(z)$), defines all the previous ORF uniquely, the orthonormality with respect to w_n is proved.

Now for $n \to \infty$, we have a sequence of weights w_n, uniformly bounded ($\int w_n d\lambda = 1$), so that there is a subsequence w_{n_k} which converges weakly. Thus there is some μ such that $\lim_{n\to\infty} \int f(t) w_{n_k}(t) d\lambda(t) = \int f(t) d\mu(t)$ for all functions f continuous on $\mathbb{T}$. If $\sum (1-|\alpha_k|) = \infty$, then a previous completeness result shows that the measure μ is unique because the Riesz representation of a linear functional on $C(\mathbb{T})$ is unique.

5 Convergence

5.1 Orthogonal Polynomials w.r.t. Varying Measures

For the convergence of the ORF we shall rely on convergence results of orthogonal polynomials with respect to varying measures (OPVM). Let $\mathcal{P}_m$ we the space of polynomials of at most degree m and construct a sequence of OPVM for the measure $d\mu(t)/|\pi_n(t)|^2$ with $\pi_n(t) = \prod_{k=1}^n (1-\overline{\alpha}_k z)$. We denote the orthonormal ones by $\phi_{n,k}(z) = v_{n,k} z^k + \cdots$, $v_{n,k} > 0$. By our general theory they satisfy the recurrence $\phi_{n,m}(z) = e_{n,m}[z\phi_{n,m-1}(z) + \overline{\lambda}_{n,m} \phi_{n,m-1}^{\bar{*}}(z)]$ with $\lambda_{n,m} = \overline{\phi_{n,m}(0)}/v_{n,m}$ and $e_{n,m} = (1-|\lambda_{n,m}|^2)^{-1/2} = v_{n,m}/v_{n,m-1}$. Here $\phi_{n,m}^{\bar{*}}(z) = z^n \overline{\phi_{n,m}(1/\overline{z})}$. From these OP we can construct a set of ORF for $\mathcal{L}_n$, but as n increases, the whole set of ORF will change.

Exercise 12. Denote $f_{n,m}(z) = t_{n,m}\phi_{n,m}(z)/\pi_n(z)$ with $t_{n,m} \in \mathbb{T}$ for an appropriate normalization. Prove that the $\{f_{n,m} : m = 0, \ldots, n\}$ is a set of orthonormal rational functions for $\mathcal{L}_n$ with respect to μ. They are obtained by orthonormalizing the basis $\{z^k/\pi_n(z) : k = 0, \ldots, n\}$. Show also that

$$h_n(z) = \frac{z\phi_{n,n}(z) - \frac{\phi_{n,n}(\alpha_n)}{\phi^*_{n,n}(\alpha_n)}\alpha_n\phi^{\bar{*}}_{n,n}(z)}{(z-\alpha_n)\pi_n(z)} \in \mathcal{L}_n \setminus \mathcal{L}_{n-1} \tag{5.1}$$

and that $\langle h_n, g\rangle_\mu = 0$ for all $g \in \mathcal{L}_{n-1}$, so that there is some constant c_n with $|c_n| = |\phi^*_n(0)|$ such that $h_n = c_n\phi_n$.

We need the following conditions for the probability measure μ and the point set $A = \{\alpha_1, \alpha_2, \ldots\}$:

1. $\mu' > 0$ a.e. (λ) (Erdős-Turán condition)
2. $\sum_{n=1}^{\infty}(1 - |\alpha_n|) = \infty$ (BD = Blaschke divergence condition)

The first condition is denoted as $\mu \in$ ET. If it satisfies the stronger Szegő condition $\log\mu' \in L_1$, we denote it by $\mu \in$ SZ. The second condition is denoted as $A \in$ BD, while if it is replaced by the stronger condition that A is compactly contained in $\mathbb{D}$ (i.e., the α_k stay away from the boundary), we denote it as $A \in$ CC. The following results are borrowed from the work of Guillermo López [20, 21, 22].

Theorem 5.1. *If $\mu \in ET$ and $A \in$ BD then with our previous notation*

1. $\lim_{n\to\infty} \lambda_{n,n+k+1} = 0$.
2. $\lim_{n\to\infty} v_{n,n+k+1}/v_{n,n+k} = 1$.
3. $\lim_{n\to\infty} \phi_{n,n+k+1}(z)/\phi_{n,n+k}(z) = z$ *locally uniformly in* $\mathbb{E}$.
4. $\lim_{n\to\infty} \phi^{\bar{*}}_{n,n+k+1}(z)/\phi^{\bar{*}}_{n,n+k}(z) = 1$ *locally uniformly in* $\mathbb{D}$.
5. $\lim_{n\to\infty} \phi^{\bar{*}}_{n,n+k}(z)/\phi_{n,n+k}(z) = 0$ *locally uniformly in* $\mathbb{E}$.

If $\mu \in$ SZ, $A \in$ CC and σ is the outer spectral factor of μ, then

6. $\lim_{n\to\infty} \phi^{\bar{*}}_{n,n+k}(z)/\pi_n(z) = 1/\sigma(z)$ *locally uniformly in* $\mathbb{D}$.

5.2 Szegő's Condition and Convergence

We can now show that the following holds.

Theorem 5.2. *If $\mu \in$ SZ and $A \in$ BD, then we have the following (l.u. means locally uniformly)*

$$\lim_{n\to\infty} k_n(z,w) = \frac{1}{\overline{\sigma(w)}(1-\overline{w}z)\sigma(z)}, \quad \textit{l.u. } (z,w) \in \mathbb{D}\times\mathbb{D},$$

$$\lim_{n\to\infty} (1-\overline{\alpha}_n z)\frac{\phi^*_n(z)}{\phi^*_n(0)} = \frac{\sigma(0)}{\sigma(z)}, \quad \textit{l.u. } z \in \mathbb{D},$$

$$\lim_{n\to\infty} \phi_n(z) = \lim_{n\to\infty} \frac{\phi_n(z)\phi_n^*(0)}{\phi_n^*(z)} = \lim_{n\to\infty} \frac{\phi_n(z)}{\phi_n^*(z)} = 0, \quad l.u.\ z \in \mathbb{D}.$$

If $\mu \in \mathrm{SZ}$, $\sigma(0) > 0$, *and* $A \in \mathrm{CC}$ *then*

$$\lim_{n\to\infty} \eta_n \phi_n^*(z) \frac{1-\overline{\alpha}_n z}{\sqrt{1-|\alpha_n|^2}} = \frac{1}{\sigma(z)}, \quad \eta_n = \frac{|\phi_n^*(0)|}{\phi_n^*(0)} \in \mathbb{T}, \quad l.u.\ z \in \mathbb{D}.$$

Proof. For the first relation, note that since the $f_{n,k}$ form an orthonormal basis for $\mathcal{L}_n$, we have

$$k_n(z,w) = \sum_{k=0}^{n} f_{n,k}(z)\overline{f_{n,k}(w)} = \frac{\sum_{k=0}^{n} \phi_{n,k}(z)\overline{\phi_{n,k}(w)}}{\pi_n(z)\overline{\pi_n(w)}}.$$

Then use Christoffel-Darboux for the ϕ_{nk} and divide by $f_{n,n}^*(z)\overline{f_{n,n}^*(w)}$:

$$\frac{k_n(z,w)}{f_{n,n}^*(z)\overline{f_{n,n}^*(w)}} = \frac{1}{1-\overline{w}z} - \frac{z\overline{w}}{1-\overline{w}z}\left(\frac{\phi_{n,n}(z)}{\phi_{n,n}^{\bar{*}}(z)}\right)\overline{\left(\frac{\phi_{n,n}(w)}{\phi_{n,n}^{\bar{*}}(w)}\right)}. \tag{5.2}$$

The last two factors go to 0, while for some $\eta_n \in \mathbb{T}$, $\lim_{n\to\infty} \eta_n f_{n,n}^*(z) = 1/\sigma(z)$, which proves the first result.

Since $h_n = c_n\phi_n$, we have by our expression for h_n

$$(1-\overline{\alpha}_n z)\phi_n^*(z) = \frac{1}{\overline{c}_n} \frac{\phi_{n,n}^{\bar{*}}(z)}{\pi_n(z)} \left[1 - \overline{\eta}_n \overline{\alpha}_n z \frac{\phi_{n,n}(z)}{\phi_{n,n}^{\bar{*}}(z)}\right], \quad \eta_n = \frac{\phi_{n,n}(\alpha_n)}{\phi_{n,n}^{\bar{*}}(\alpha_n)}. \tag{5.3}$$

Recall that $\phi_{n,n}(z)/\phi_{n,n}^{\bar{*}}(z) \to 0$ l.u. in $\mathbb{D}$, and being a Blaschke product it is bounded by 1 in $\mathbb{D} \cup \mathbb{T}$ so that $|\eta_n| \leq 1$. Hence the term between square brackets goes to 1 as $n \to \infty$. On the other hand, $\phi_{n,n}^{\bar{*}}(z)/\pi_n(z)$ converges to $1/\sigma(z)$ l.u. in $\mathbb{D}$, up to some normalizing constant. But this constant will cancel against the same constant in the denominator which is obtained by setting $z = 0$. This proves the second result.

Taking $z = w$ in the first relation implies $\sum_{k=1}^{n} |\phi_k(z)|^2$ converges for $n \to \infty$. Thus $\lim_n \phi_n(z) = 0$. When multiplying the inverse of the second relation of this theorem with $\phi_n(z)$, and noting that $(1-\overline{\alpha}_n z)^{-1}$ is uniformly bounded if z is in a compact subset of $\mathbb{D}$, we also get the second limit.

For the third one, we note that $|\phi_n^*(0)|$ is uniformly bounded away from 0, because the Christoffel-Darboux relation implies $|\phi_n^*(0)|^2 \geq 1$.

For the last limit, note that η_n is used to normalize ϕ_n^* to make it positive in $z = 0$. By the Christoffel-Darboux relation for $z = w = 0$

$$\frac{|\phi_n^*(0)|^2}{1-|\alpha_n|^2} = k_{n-1}(0,0) + \frac{|\phi_n(0)|^2}{1-|\alpha_n|^2}.$$

Because $\phi_n(0) \to 0$, and the α_k stay away from the circle, the last term goes to 0, and we know that the second one goes to $|\sigma(0)|^{-2}$. Therefore we see that $|\phi_n^*(0)|/\sqrt{1-|\alpha_n|^2} \to 1/\sigma(0)$. In combination with the second limit of this theorem, the result follows. $\square$

When $\mu \notin \mathrm{SZ}$, then σ need not be defined, and the previous relations can not be obtained. A way to avoid the σ is to consider ratio asymptotics, so that the σ cancel out. These are weaker results in the sense that if $\mu \in \mathrm{SZ}$, then the ratio asymptotics are almost immediately obtained from the stronger asymptotics that were previously obtained. We use again the OPVM with measure $\mu/|\pi_n|^2$. Denoting the reproducing kernel for $\mathcal{P}_n$ w.r.t. this measure as $k_{n,n}(z,w)$, it is easily seen that $k_{n,n}(z,0) = v_n \phi_{n,n}^{\bar{*}}(z)$. Also the usual rational kernel for $\mathcal{L}_n$ satisfies $k_n(z,0) = k_{n,n}(z,0)/\pi_n(z)$.

Exercise 13. Applying Theorem 3.5 to the OPVM, prove that the inner product in $\mathcal{L}_n$ w.r.t. $d\mu(t)$ and w.r.t. $v_n^2/|k_n(t,0)|^2 d\lambda(t)$ is the same. Of course, this is also a direct consequence of the results in Section 3.3.

Theorem 5.3. *If $\mu \in \mathrm{ET}$ and $A \in \mathrm{BD}$, then*

$$\lim_{n\to\infty} \frac{k_n(z,0)}{k_{n+1}(z,0)} = 1, \qquad l.u.\ z \in \mathbb{D}.$$

and with the superstar referring to z,

$$\lim_{n\to\infty} k_n^*(z,0)/k_n(z,0) = 0, \qquad l.u.\ z \in \mathbb{D}.$$

Proof. Set $g_n(z) = k_n(z,0)/k_{n+1}(z,0) - v_n/v_{n+1}$, then by the previous exercise we see that

$$\int |g_n(t)|^2 d\lambda(t) = 2\frac{v_n^2}{v_{n+1}^2}\left(1 - \frac{v_n}{v_{n+1}}\right). \tag{5.4}$$

On the other hand, because g_n is analytic in $\mathbb{D} \cup \mathbb{T}$, we have by the Poisson formula $g_n(z) = \int P(t,z) g_n(t) d\lambda(t)$. For $t \in \mathbb{T}$, $|P(t,z)| \le M$ holds uniformly for z in a compact subset of $\mathbb{D}$. Thus we also have $|g_n(z)| \le M \int |g_n(t)| d\lambda(t)$ and also $|g_n(z)|^2 \le M^2 \int |g_n(t)|^2 d\lambda(t)$ so that

$$|g_n(z)|^2 \le 2M^2 \frac{v_n^2}{v_{n+1}^2}\left(1 - \frac{v_n}{v_{n+1}}\right). \tag{5.5}$$

Thus $\frac{k_n(z,0)}{k_{n+1}(z,0)} \to 1$ iff $\frac{v_n}{v_{n+1}} \to 1$. Obviously $k_{n-1}(0,0)/k_n(0,0) \le 1$, but using the Christoffel-Darboux relation we also have

$$\frac{k_{n-1}(0,0)}{k_n(0,0)} = \frac{|\phi_n^*(0)|^2 - |\phi_n(0)|^2}{|\phi_n^*(0)|^2 - |\alpha_n|^2|\phi_n(0)|^2} \ge \frac{|\phi_n^*(0)|^2 - |\phi_n(0)|^2}{|\phi_n^*(0)|^2 - |\phi_n(0)|^2} = 1.$$

So that $\lim_{n\to\infty} v_n^2/v_{n+1}^2 = 1$, which proves the first part.

For the second relation, note that $k_n(z,0) = \phi^{\bar{*}}_{n,n}(z)v_n\eta_n/\pi_n(z)$ and hence $k^*_n(z,0) = \phi_{n,n}(z)v_n\overline{\eta}_n/\pi_n(z)$ with $\eta_n \in \mathbb{T}$. Take the ratio and use $\phi_{n,n}(z)/\phi^{\bar{*}}_{n,n}(z) \to 0$ l.u. in $\mathbb{D}$, to conclude the proof. □

Theorem 5.4. *If $\mu \in \mathrm{ET}$ and $A \in \mathrm{CC}$, then*

$$\lim_{n\to\infty} \frac{\zeta_n(z)\phi_n(z)}{\phi^*_n(z)} = \lim_{n\to\infty} \frac{\phi_n(z)}{\phi^*_n(z)} = \lim_{n\to\infty} \lambda_n = 0, \quad \textit{l.u. in } \mathbb{D}.$$

Denote $\Phi^*_n(z) = \phi^*_n(z)/\phi^*_n(0)$, then under the previous conditions

$$\lim_{n\to\infty} \frac{\Phi^*_{n+1}(z)(1-\overline{\alpha}_{n+1}z)}{\Phi^*_n(z)(1-\overline{\alpha}_n z)} = 1, \quad \textit{l.u. in } \mathbb{D}.$$

Proof. Repeat the proof of the second part in Theorem 5.2, i.e., use exercise 12 giving $\phi_n = c_n h_n$ and its superstar conjugate, to find expressions for $(z-\alpha_n)\phi_n(z)$ and $(1-\overline{\alpha}_n z)\phi^*_n(z)$ and take their ratio. Then using $\phi_{n,n}(z)/\phi^{\bar{*}}_{n,n}(z) \to 0$ from Theorem 5.1 leads to the first conclusion. Note that we need $\phi_{n,n}(\alpha_n)/\phi^{\bar{*}}_{n,n}(\alpha_n)$ to go to zero, which can only be guaranteed when $A \in \mathrm{CC}$. It is also needed for the convergence of $\phi_n(0)/\phi^*_n(0)$ because this will follow form Theorem 5.3 or the previous exercise if A is in a compact subset of $\mathbb{D}$. For the last equality note that $|\lambda_n| = |\phi_n(\alpha_{n-1})/\phi^*_n(\alpha_{n-1})|$.

For the second formula, write the second Christoffel-Darboux formula for $k_n(z,0)$ and its superstar conjugate and eliminate $\phi_n(z)$ to obtain

$$\Phi^*_n(z)(1-\overline{\alpha}_n z) = \frac{\phi^*_n(z)(1-\overline{\alpha}_n z)}{\phi^*_n(0)} = \frac{k_n(z,0)}{v_n^2}\left[1-\overline{\alpha}_n z\frac{\overline{\phi_n(0)}}{\phi^*_n(0)}\frac{k^*_n(z,0)}{k_n(z,0)}\right].$$

Note that the term in square brackets goes to 1 l.u. in $\mathbb{D}$ as $n\to\infty$. Rewrite this for n replaced by $n+1$ and take their ratio. Because $k_{n+1}(z,0)/k_n(z,0)$ and v_{n+1}/v_n go to 1 as $n\to\infty$, the second result follows. □

6 Szegő's Problem

This problem is essentially the problem of linear prediction. Recall that the linear prediction problem can be formulated as finding $f \in H_2(\mu)$ such that $\|f\|^2_\mu$ is minimal with the side condition that $f(0) = 1$. This f represented the prediction error. Note that we have replaced the z of the first section by $1/z$ to be in line with our discussion of ORF, which is obviously a matter of convention. We consider a slightly more general situation where we replace the constraint by $f(w) = 1$ for some $w \in \mathbb{D}$. This is a problem that can be solved in any reproducing kernel Hilbert space.

Theorem 6.1. *Let H be a reproducing kernel Hilbert space with kernel $k(z,w)$. Then the minimum of $\|f\|^2_\mu$ with $f(w) = 1$ is obtained for $f(z) = K(z) \equiv k(z,w)/k(w,w)$ and the minimum is $1/k(w,w)$.*

Proof. Obviously $\|K(z)\|_\mu^2 = 1/k(w,w)$. For any other f satisfying $f(w) = 1$ for which $\|f\|_\mu^2 = m$ we have $0 \leq \|f - K\|_\mu^2 = \|f\|_\mu^2 + 1/k(w,w) - 2\text{Re}\,\langle f, K\rangle_\mu = m - 1/k(w,w)$. Thus $m \geq 1/k(w,w)$. □

From the polynomial case, it is known that the reproducing kernel for $H_2(\mu)$ is given by the Szegő kernel $s(z,w) = [(1-\overline{w}z)\sigma(z)\overline{\sigma(w)}]^{-1}$ where σ is the outer spectral factor of μ, which we suppose to be normalized by $\sigma(0) > 0$. Thus we have a completely predictable sequence if the minimum $|\sigma(0)|^2$ of $\|f\|_\mu^2$, with constraint $f(0) = 1$, is zero, i.e., since $|\sigma(0)|^2 = \exp\{\int \log \mu'(t) d\lambda(t)\}$, if $\int \log \mu' d\lambda = -\infty$, and thus $\log \mu' \notin L_1$. If we do not have complete predictability, it is still a valuable objective to find the best possible predictor by minimizing the prediction error.

Instead of minimizing over the complete past (i.e., all of $H_2(\mu)$), we may be less ambitious and start by minimizing over a finite dimensional subspace, e.g., $\mathcal{L}_n$. There the minimum of the general problem is $k_n(z,w)/k_n(w,w)$.

The advantage of computing these approximants using ORF over the computation with OPVM is that they are easily computed recursively. If the approximation is not good enough for a certain n, then increasing n by 1 requires just one more step of the recurrence relation. For the OPVM, increasing n by 1, would mean that we have to start the computations all over again. Note also that we have at every step an estimate of the prediction error which is $1/k_n(0,0)$ or $1/k_n(w,w)$ for the general problem. Thus if the $|\phi_n(w)|$ do not go to zero fast enough, then $k_n(w,w)$ will go to ∞ and the error will go to zero, but under the conditions of our theorems, the $k_n(w,w)$ will converge, so that the error will not go to zero. It will be bounded from below by $(1-|w|^2)|\sigma(w)|^2$.

We also note the following result:

Theorem 6.2. *Let $\mu \in \text{SZ}$, $w \in \mathbb{D}$ a fixed number, and let $s_w(z) = s(z,w)$ be the Szegő kernel. Consider the problem* $\min \|f - s_w\|_\mu^2$ *over all $f \in \mathcal{L}_n$, then the solution is $k_n(z,w)$ and the minimum is $s_w(w) - k_n(w,w)$. If $A \in \text{BD}$ then* $\lim_{n\to\infty} \|k_n(t,w) - s_w(t)\|_\mu = 0$.

Proof. Suppose $f(w) = a$, then $\|s_w - f\|_\mu^2 = \|s_w\|_\mu^2 + \|f\|_\mu^2 - 2\text{Re} f(w)$. Thus we have to minimize $\|f\|_\mu^2 - 2\text{Re}\, a$ over $f \in \mathcal{L}_n$ with $f(w) = a$ and minimize the result over all possible a. In other words, we have to find the infimum over a of $|a|^2/k_n(w,w) - 2\text{Re}\, a$ and this is given by $a = k_n(w,w)$. This proves the first part. If $A \in \text{BD}$ then as $n \to \infty$, $\mathcal{L}_n$ becomes dense in $H_2(\mu)$, which means that the error in $H_2(\mu)$ tends to 0. □

7 Hilbert Modules and Hardy Spaces

The results on the matrix case (Sections 7–11) can be found in [2, 5, 17, 19, 10, 13, 12, 11, 18]. For the polynomial case and Hilbert modules see also [32, 33, 23, 7, 4, 9, 24].

7.1 Inner Products and Norms

We now consider matrix valued functions: $f : \mathbb{C} \to \mathbb{C}^{N\times N}$. The space of these functions is a left and a right module over $\mathbb{C}^{N\times N}$ (the product is noncommutative). We write $A \geq 0$ to mean that A is nonnegative definite and $A > 0$ means that A is positive definite, while $A \geq B$ is the same as $A - B \geq 0$. By A^H we mean the Hermitian conjugate and $\mathrm{tr}(A)$ is the normalized trace of A, i.e., $\mathrm{tr}(A) = \sum_{k=1}^{N} a_{kk}$. By $S = A^{1/2}$ we denote the Hermitian square root of $A \geq 0$, thus $A = S^2$ while $L = A^{L/2}$, resp. $R = A^{R/2}$ denote a left resp. right square root, meaning $A = LL^H$ resp. $A = R^H R$. Note that L is unique up to a right unitary factor and R up to a left unitary factor.

We introduce a matrix valued and a scalar valued inner product which will define $\mathbb{L}_2$ of square integrable functions that will be both a Hilbert module and a Hilbert space. Let μ be a nonnegative Hermitian measure on $\mathbb{T}$. This means that it is a square matrix whose entries are complex valued measures such that the whole matrix takes values that are nonegative definite. Because $0 \leq M \leq \mathrm{tr}(M)I$ holds for any nonnegative definite matrix, it holds for the scalar trace-measure $\tau = \mathrm{tr}(\mu)$ that $\tau(E) = 0$ implies $\mu(E) = 0$. This means that μ is absolutely continuous with respect to τ and we may define the trace-derivative μ'_τ as the matrix whose entries are $d\mu_{i,j}/d\tau$ and it holds that $\mu(E) = \int_E \mu'_\tau d\tau$. Furthermore it can be shown that $0 \leq \mu'_\tau \leq I$ a.e. τ and also $(\mu'_\tau)^{1/2}$ is measurable. More generally, we define $\int f d\mu g^H$ as $\int f \mu'_\tau g^H d\tau$, thus as a matrix whose entries are $\int [f\mu'_\tau g^H]_{ij} d\tau$. We denote this matrix valued "inner product" as $\langle\langle f, g\rangle\rangle_{\ell,\mu}$ where the ℓ stands for "left" since it is possible to define in a completely analogous way $\langle\langle f, g\rangle\rangle_{r,\mu} = \int f^H d\mu g$. In the sequel we treat only the left version and leave the right version to the reader. For many results however, left and right elements will be interacting. We shall always assume that $\int d\mu = I$.

The class of matrix valued functions for which $\langle\langle f, f\rangle\rangle_{\ell,\mu}$ exists forms a left module over $\mathbb{C}^{N\times N}$ that has some pre-Hilbert space-like properties. Indeed, for any constant square matrix a it holds that

- $\langle\langle f+g, h\rangle\rangle_{\ell,\mu} = \langle\langle f, h\rangle\rangle_{\ell,\mu} + \langle\langle g, h\rangle\rangle_{\ell,\mu}$ and $\langle\langle af, g\rangle\rangle_{\ell,\mu} = a\,\langle\langle f, g\rangle\rangle_{\ell,\mu}$ (linearity)
- $\langle\langle f, g\rangle\rangle_{\ell,\mu} = [\langle\langle g, f\rangle\rangle_{\ell,\mu}]^H$ (symmetry)
- $\langle\langle f, f\rangle\rangle_{\ell,\mu} \geq 0$ with $\mathrm{tr}\,\langle\langle f, f\rangle\rangle_{\ell,\mu} = 0 \Leftrightarrow f = 0$ (positive definite)

To equip this class of functions with some topology for which it has to be complete, we also need a genuine scalar valued norm. This norm will be implied by a scalar valued inner product, which we define as follows: $\langle f, g\rangle_{\ell,\mu} = \mathrm{tr}\,\langle\langle f, g\rangle\rangle_{\ell,\mu} = \int \mathrm{tr}(f\mu'_\tau g^H)d\tau$. The corresponding scalar norm is $|f|_{\ell,\mu} = [\langle f, f\rangle_{\ell,\mu}]^{1/2}$. So we can now define the *Hilbert space* $\mathbb{L}_2(\mu)$ of square integrable matrix valued functions with respect to the scalar valued $\langle\cdot,\cdot\rangle_{\ell,\mu}$, that is complete with respect to the norm $|f|_{\ell,\mu}$. It is also a *Hilbert module* with respect to the matrix valued $\langle\langle\cdot,\cdot\rangle\rangle_{\ell,\mu}$ as we explained above.

Occasionally we also use a Euclidean inner product, namely $(f,g)_E = \mathrm{tr}(fg^H) = \mathrm{tr}(g^H f)$ and a Euclidean norm $|f|_E = (f,f)_E^{1/2} = |f|_F/\sqrt{N}$ where $|f|_F$ is the Frobenius norm and N the size of the matrix.

If $\mu = \lambda I$, we drop μ from the notation. For example $\mathbb{L}_2$ means $\mathbb{L}_2(\lambda I)$ and $\langle\langle f,g\rangle\rangle_\ell$ means $\langle\langle f,g\rangle\rangle_{\ell,\lambda}$ etc. We can define matrix valued Hardy spaces $\mathbb{H}_p$ as the matrix valued functions f, analytic in $\mathbb{D}$ and such that

$$|f|_p = \sup_{0\le r<1} [\int |f(rt)|_E^p d\lambda(t)]^{1/p} < \infty, \qquad 0<p<\infty$$

and for $p=\infty$, $|f|_\infty = \sup_{0\le r<1, t\in\mathbb{T}} |f(rt)|_E < \infty$. As in the scalar case, any $f \in \mathbb{H}_p$ has a nontangential limit to the circle that belongs to $\mathbb{L}_p$ (which has an obvious definition viz., $f \in \mathbb{L}_p \Leftrightarrow |f|_E \in L_p$).

7.2 Carathéodory Function and Spectral Factor

Carathéodory functions are functions analytic in $\mathbb{D}$ with positive real part. Thus with $\mathrm{Re}\,\Omega(z) = \frac{1}{2}[\Omega(z) + \Omega(z)^H] \ge 0$. There is an essentially unique relationship between positive measures and Carathéodory functions, which we represent as $\Omega_\mu(z) = \int D(t,z)d\mu(t) + i\mathrm{Im}\,\Omega_\mu(0)$. We shall assume that $\Omega_\mu(0)$ is real so that according to $\int d\mu = I$ we get $\Omega_\mu(0) = I$. Just like in the scalar case we have $\mathrm{Re}\,\Omega_\mu(z) = \int P(t,z)d\mu(t)$ and this has a nontangential limit to the circle $\mathbb{T}$ which equals $\mu' = d\mu/d\lambda$ a.e. (λ) and its Fourier series is $\Omega_\mu(z) = c_0 + 2\sum_{k=1}^{\infty} c_k z^k$ where the coefficients are the moments $c_k = \int t^{-k} d\mu(t)$.

For the spectral factor, we need to assume that $\log\det\mu' \in L_1$ (Szegő's condition). If that condition holds then there exist left and right outer spectral factors σ^L and σ^R such that $\mu' = \sigma^L\sigma^L_* = \sigma^R_*\sigma^R$ on $\mathbb{T}$. The substar for a matrix function is defined as $f_*(z) = f(1/\overline{z})^H$. These σ^L and σ^R are outer in $\mathbb{H}_2$ in the sense that their determinant is scalar outer in $H_{2/N}$. They are analytic in $\mathbb{D}$ and their determinant does not vanish in $\mathbb{D}$, so that their inverses are also outer in $\mathbb{H}_2$. These σ^L, resp. σ^R are uniquely defined up to a right, resp. left unitary constant factor.

8 MORF and Reproducing Kernels

8.1 Orthogonal Rational Functions

Matrix valued orthogonal rational functions (MORF) can now be obtained in much the same way as in the scalar case. Given a sequence of $\alpha_k \in \mathbb{D}$, we keep the notation of Blaschke factors $\zeta_k(z)$ and Blaschke products $B_n(z)$ from the scalar case and define $\mathcal{L}_n$ as the linear span over $\mathbb{C}^{N\times N}$ of $\{B_k I\}_{k=0}^n$. The block Gram matrix consists of the blocks $\langle\langle B_k, B_l\rangle\rangle_{\ell,\mu} = \langle\langle B_k, B_l\rangle\rangle_{r,\mu}$ and it depends only on $\Omega_\mu(\alpha_k)$ and its derivatives at these points up to an order depending on the multiplicity of α_k in the sequence.

By a block Gram-Schmidt algorithm we can orthormalize the previous basis as follows: assuming $\int d\mu = I$, set $\phi_0^L = I$, and for $n = 1, 2, \ldots$ we set $\tilde{\phi}_n^L = B_n I - \sum_{k=0}^{n-1} \langle\langle B_n I, \phi_k^L \rangle\rangle_{\ell,\mu} \phi_k^L$ and $\phi_n^L = m_n^{-1} \tilde{\phi}_n^L$ where m_n is a left square root of $\langle\langle \tilde{\phi}_n^L, \tilde{\phi}_n^L \rangle\rangle_{\ell,\mu}$. Note that the left square root need not be invertible in general if $\tilde{\phi}_n^L$ is not zero. Then the algorithm will break down, and we say that μ is degenerate. It can be shown that if $\log \det \mu' \in L_1$, then this degeneracy will not occur and all the ORF can be constructed. The basis ϕ_n for the Hilbert module is orthonormal in the sense that $\langle\langle \phi_k^L, \phi_j^L \rangle\rangle_{\ell,\mu} = \delta_{k,j} I$. The ϕ_n are defined up to a constant unitary factor from the left. To fix this, we shall assume that the leading coefficient κ_n^L of ϕ_n^L with respect to the basis $\{B_k I\}$ is Hermitian positive definite (every invertible matrix A can be written as $A = U^L B^L$ with U^L unitary and B^L positive definite [6, p. 22]). The leading coefficient is $\kappa_n^L = [\phi_n^{L*}(\alpha_n)]^H$ where the superstar for $f \in \mathcal{L}_n$ is defined as in the scalar case: $f^* = B_n f_*$. Note that for $f, g \in \mathcal{L}_n$ we have $\langle\langle f, g \rangle\rangle_{\ell,\mu} = \langle\langle f^*, g^* \rangle\rangle_{r,\mu}$.

8.2 Reproducing Kernels

Given a left orthonormal system ϕ_k^L for $\mathcal{L}_n$, we can define a left reproducing kernel $k_n^L(z, w) = \sum_{k=0}^{n} \phi_k^L(w)^H \phi_k^L(z)$ which reproduces because $\langle\langle f, k_n^L(\cdot, w) \rangle\rangle_{\ell,\mu} = f(w)$ for all $f \in \mathcal{L}_n$. The right reproducing kernel is given by $k_n^R(w, z) = \sum_{k=0}^{n} \phi_k^R(z)[\phi_k^R(w)]^H$ and for any $f \in \mathcal{L}_n$ we have $\langle\langle k_n^R(w, \cdot), f \rangle\rangle_{r,\mu} = f(w)$.

Exercise 14. Mimic the scalar proof to derive the following Christoffel-Darboux relations:

$$k_n^L(z, w) = \frac{[\phi_{n+1}^{R*}(w)]^H \phi_{n+1}^{R*}(z) - [\phi_{n+1}^L(w)]^H \phi_{n+1}^L(z)}{1 - \zeta_{n+1}(z)\overline{\zeta_{n+1}(w)}}$$

and

$$k_n^L(z, w) = \frac{[\phi_n^{R*}(w)]^H \phi_n^{R*}(z) - \zeta_n(z)\overline{\zeta_n(w)}[\phi_n^L(w)]^H \phi_n^L(z)}{1 - \zeta_n(z)\overline{\zeta_n(w)}}.$$

Consequently $k_n^L(z, \alpha_n) = \kappa_n^R \phi_n^{R*}(z)$ and $k_n^L(\alpha_n, \alpha_n) = \kappa_n^R [\kappa_n^R]^H$. The right versions are obtained by symmetry.

$$k_n^R(w, z) = \frac{\phi_{n+1}^{L*}(z)[\phi_{n+1}^{L*}(w)]^H - \phi_{n+1}^R(z)[\phi_{n+1}^R(w)]^H}{1 - \zeta_{n+1}(z)\overline{\zeta_{n+1}(w)}}$$

and

$$k_n^R(w, z) = \frac{\phi_n^{L*}(z)[\phi_n^{L*}(w)]^H - \zeta_n(z)\overline{\zeta_n(w)}\phi_n^R(z)[\phi_n^R(w)]^H}{1 - \zeta_n(z)\overline{\zeta_n(w)}}.$$

From this right variant we have $k_n^R(\alpha_n, z) = \phi_n^{L*}(z)\kappa_n^L$ and $k_n^R(\alpha_n, \alpha_n) = [\kappa_n^L]^H \kappa_n^L$.

Theorem 8.1. *The reproducing kernels satisfy the following recurrence relations*

$$s_n(w)\begin{bmatrix} k_n^{R*}(w,z) \\ k_n^L(z,w) \end{bmatrix} = t_n(z,w)\begin{bmatrix} k_{n-1}^{R*}(w,z) \\ k_{n-1}^L(z,w) \end{bmatrix}$$

where

$$s_n(w) = \begin{bmatrix} I - [\rho_n^R(w)]^H \rho_n^L(w) & 0 \\ 0 & I - \rho_n^L(w)[\rho_n^R(w)]^H \end{bmatrix},$$

$$t_n(z,w) = \begin{bmatrix} I & [\rho_n^R(w)]^H \\ \rho_n^L(w) & I \end{bmatrix} \begin{bmatrix} \zeta_n(z) I & 0 \\ 0 & I \end{bmatrix} \begin{bmatrix} I & [\gamma_n^R(w)]^H \\ \gamma_n^L(w) & I \end{bmatrix},$$

with

$$\rho_n^L(w) = [\phi_n^L(w)]^H [\phi_n^{L*}(w)]^{-1}, \quad \rho_n^R(w) = [\phi_n^{R*}(w)]^{-1}[\phi_n^R(w)]^H,$$

and $[\gamma_n^R(w)\ \gamma_n^L(w)] = -\zeta_n(w)[\rho_n^R(w)\ \rho_n^L(w)]$.

Proof. The proof is along the same lines as in the scalar case. One needs the relation $k_n^L(z,w) = B_n(z)\overline{B_n(w)} k_n^R(1/\overline{z}, 1/\overline{w})$. We leave it as an exercise to the reader. □

Note that it follows from the Christoffel-Darboux relations that

$$\begin{aligned} \{I - \rho_n^L(w)[\rho_n^R(w)]^H\}&[\phi_n^{R*}(w)]^H \phi_n^{R*}(w) \\ &= [\phi_n^{R*}(w)]^H \phi_n^{R*}(w) - [\phi_n^L(w)]^H \phi_n^L(w) > 0. \end{aligned}$$

Thus $I - \rho_n^L(w)[\rho_n^R(w)]^H$ can not be singular for $w \in \mathbb{D}$. Note that this implies also that $[\phi_n^{R*}(w)]^H \phi_n^{R*}(w)$ is not singular and hence positive definite, thus $\det \phi_n^{R*}(w) \neq 0$ for all $w \in \mathbb{D}$ and thus it is invertible in $\mathbb{D}$. A similar observation holds for $I - \rho_n^L(w)[\rho_n^R(w)]^H$ and for $\phi_n^{L*}(w)$.

Exercise 15. From this recurrence derive that

$$\left(I - \gamma_n^L(w)[\gamma_n^R(w)]^H\right) k_{n-1}^L(w,w) = \left(I - \rho_n^L(w)[\rho_n^R(w)]^H\right) k_n^L(w,w)$$

and symmetrically

$$\left(I - [\gamma_n^R(w)]^H \gamma_n^L(w)\right) k_{n-1}^R(w,w) = \left(I - [\rho_n^R(w)]^H \rho_n^L(w)\right) k_n^R(w,w).$$

We can bring a bit more symmetry into the recurrence for the kernels by considering normalized versions of the kernels. Suppose $L_n(w) = [k_n^L(w,w)]^{L/2}$ is a left square root and $R_n(w) = [k_n^R(w,w)]^{R/2}$ is a right square root, then we call $K_n^L(z,w) = [L_n(w)]^{-1} k_n(z,w)$ and $K_n^R(w,z) = k_n^R(w,z)[R_n(w)]^{-1}$ the normalized kernels.

Exercise 16. Use the Christoffel-Darboux relations to show that

$$\rho_n^L(w)k_{n-1}^R(w,w) = k_{n-1}^L(w,w)\rho_n^R(w).$$

With the left square root $L_n(w)$ and the right square roots $R_n(w)$ that we just introduced, use this result to prove that

$$[L_{n-1}(w)]^{-1}\rho_n^L(w)[R_{n-1}(w)]^H = [L_{n-1}(w)]^H\rho_n^R(w)[R_{n-1}(w)]^{-1}.$$

Thus we succeeded in symmetrizing the ρ_n^R and ρ_n^L and call the result $\rho_n(w)$ the *symmetrized recursion parameters* from now on. The Christoffel-Darboux relations also imply that $\rho_n(w)$ is strictly contractive in $\mathbb{D}$, i.e., $\rho_n^H\rho_n < I$ and hence also $\rho_n\rho_n^H < I$. Similar observations can be made when ρ is replaced by γ.

We need one more lemma before we can state our symmetrized recurrence relation for the normalized kernels.

Lemma 8.1. *Define $F(\rho) = I - \rho\rho^H$ and $G(\rho) = I - \rho^H\rho$, then it is always possible to choose the square roots $L_n(w)$ and $R_n(w)$ in the definition of $\rho_n(w)$ such that (recall that $A^{1/2}$ denotes the Hermitian square root and $A^{-1/2}$ is its inverse)*

$$[L_{n-1}(w)]^{-1}L_n(w) = F(\rho_n(w))^{-1/2}F(\gamma_n(w))^{1/2}$$

and

$$R_n(w)[R_{n-1}(w)]^{-1} = G(\rho_n(w))^{-1/2}G(\gamma_n(w))^{1/2}.$$

Proof. First note that $F(\gamma_n) = L_{n-1}^{-1}(I - |\zeta_n|^2\rho_n^L[\rho_n^R]^H)L_{n-1}$. Hence

$$L_{n-1}F(\gamma_n)L_{n-1}^H = (I - |\zeta_n|^2\rho_n^L[\rho_n^R]^H)k_{n-1}^L(w,w),$$

which by the previous exercise 15 equals $(I - \rho_n^L[\rho_n^R]^H)k_n^L(w,w)$ while $I - \rho_n^L[\rho_n^R]^H = L_{n-1}F(\rho_n)L_{n-1}^{-1}$. Thus we may conclude that $F(\gamma_n)L_{n-1}^HL_n^{-H} = F(\rho_n)L_{n-1}^{-1}L_n$. Multiply by the inverse of $F(\gamma_n)$ to get $[F(\rho_n)]^{-1}F(\gamma_n) = L_{n-1}^{-1}L_nL_n^HL_{n-1}^{-H}$. This is obviously positive definite, so we can take its Hermitian square root $Q_n = L_{n-1}^{-1}L_nU_n$ with U_n some unitary matrix. This U_n can be included in the left square root L_n. This proves the first relation. The second one follows by symmetry. □

We have now a simplified recursion for the normalized kernels which follows immediately from the previous results.

Theorem 8.2. *Define $F(\rho) = I - \rho\rho^H$ and $G(\rho) = I - \rho^H\rho$ and the Halmos extension*

$$H(\rho) = \begin{bmatrix} G(\rho)^{-1/2} & 0 \\ 0 & F(\rho)^{-1/2} \end{bmatrix}\begin{bmatrix} I & \rho^H \\ \rho & I \end{bmatrix} = \begin{bmatrix} I & \rho^H \\ \rho & I \end{bmatrix}\begin{bmatrix} G(\rho)^{-1/2} & 0 \\ 0 & F(\rho)^{-1/2} \end{bmatrix},$$

the following recurrence holds

$$\begin{bmatrix} K_n^{R*}(w,z) \\ K_n^L(z,w) \end{bmatrix} = \theta_n(z,w) \begin{bmatrix} K_{n-1}^{R*}(w,z) \\ K_{n-1}^L(z,w) \end{bmatrix},$$

with

$$\theta_n(z,w) = H(\rho_n(w))Z_n(z)H(\gamma_n(w)), \quad Z_n(z) = \begin{bmatrix} \zeta_n(z)I & 0 \\ 0 & I \end{bmatrix},$$

where $\rho_n(w)$ *is as defined in the previous lemma and* $\gamma_n(w) = -\zeta_n(w)\rho_n(w)$.

We can, like in the scalar case, define associated kernels $L_n^L(z,w)$ and $L^R(w,z)$, which start from the initial conditions $L_0^{R*} = I = L_0^L$ and then obey the recursion

$$\begin{bmatrix} L_n^{R*}(w,z) \\ -L_n^L(z,w) \end{bmatrix} = \theta_n(z,w) \begin{bmatrix} L_{n-1}^{R*}(w,z) \\ -L_{n-1}^L(z,w) \end{bmatrix},$$

with $\theta_n(z,w)$ like in the previous theorem.

The matrices θ_n have some special properties. For example, it is clear that they are para-J-unitary, i.e., they satisfy

$$\theta_n^* J \theta_n = \theta_n J \theta_n^* = J, \quad J = \begin{bmatrix} I & 0 \\ 0 & -I \end{bmatrix}.$$

This immediately implies:

Corollary 8.1. *For the normalized kernels* K_n^L, K_n^R, L_n^L *and* L_n^R, *the following relations hold*

- $K_n^L(z,w)L_n^{L*}(z,w) + L_n^L(z,w)K_n^{L*}(z,w) = 2B_n(z)I,$
- $K_n^{R*}(w,z)L_n^R(w,z) + L_n^{R*}(w,z)K_n^R(w,z) = 2B_n(z)I,$
- $K_n^{L*}(z,w)K_n^L(z,w) = K_n^R(w,z)K_n^{R*}(w,z),$
- $L_n^{L*}(z,w)L_n^L(z,w) = L_n^R(w,z)L_n^{R*}(w,z).$

Proof. This follows from the fact that

$$\begin{bmatrix} K_n^{R*} & L_n^{R*} \\ K_n^L & -L_n^L \end{bmatrix} = \Theta_n \begin{bmatrix} I & I \\ I & -I \end{bmatrix} \Leftrightarrow \Theta_n = \frac{1}{2} \begin{bmatrix} K_n^{R*} + L_n^{R*} & K_n^{R*} - L_n^{R*} \\ K_n^L - L_n^L & K_n^L + L_n^L \end{bmatrix}$$

where $\Theta_n = \theta_n\theta_{n-1}\cdots\theta_1$. Because all the θ_k are *-J-unitary, the same holds for the product Θ_n. Writing explicitly $\Theta_n J \Theta_n^* = J$ gives the previous results. □

The matrices θ_n, and therefore also the product Θ_n, are also *J-lossless*, which means that as a function of z, its entries are functions in the Nevanlinna class (ratios of H_∞ functions), it is J-contractive in $\mathbb{D}$ and J-unitary on $\mathbb{T}$, thus $\theta_n J \theta_n^H \le J$ for $z \in \mathbb{D}$ while $\theta_n J \theta_n^H = J$ for $z \in \mathbb{T}$.

Among other things, it is shown in [5] that this induces the following properties.

- $K_n^L(z,w)$, $L_n^L(w,z)$, $K_n^R(w,z)$, and $L_n^R(w,z)$ are invertible for $w \in \mathbb{D}$ and $z \in \mathbb{D} \cup \mathbb{T}$. Hence also ϕ_n^{R*} and ϕ_n^{L*} are invertible in $\mathbb{D} \cup \mathbb{T}$.
- $[K_n^L(z,w)]^{-1}L_n^L(z,w)$ and $L_n^R(w,z)[K_n^R(w,z)]^{-1}$ are Carathéodory functions for $w \in \mathbb{D}$.
- $[K_n^L(z,w)]^{-1}$ and $[K_n^R(w,z)]^{-1}$ are in $\mathbb{H}_2$.

9 Recurrence for the MORF

9.1 The Recursion

The proof of the recurrence relation for the MORF is given in different steps which we formulate as lemmas. Note that in these lemmas the ϕ_n^L and ϕ_n^R are temporarily just orthogonal functions. They are not necessarily normalized in the particular way we agreed upon before. In the ultimate Theorem 9.1 we will be back to the usual situation.

Lemma 9.1. *Starting with arbitrary invertible constant matrices ϕ_0^L and ϕ_0^R from $\mathbb{C}^{N\times N}$, define for $k = 1,\ldots,n$*

$$f_k^L(z) = \frac{z-\alpha_{k-1}}{1-\overline{\alpha}_k z}\phi_{k-1}^L(z), \quad f_k^R(z) = \frac{z-\alpha_{k-1}}{1-\overline{\alpha}_k z}\phi_{k-1}^R(z),$$

$$\begin{aligned} D_k^L &= -C_k^L \langle\langle f_k^{R*}, \phi_0^L\rangle\rangle_{\ell,\mu} \langle\langle f_k^L, \phi_0^L\rangle\rangle_{\ell,\mu}^{-1}, \\ D_k^R &= -\langle\langle \phi_0^R, f_k^R\rangle\rangle_{r,\mu}^{-1} \langle\langle \phi_0^R, f_k^{L*}\rangle\rangle_{r,\mu} C_k^R, \end{aligned}$$

$$\phi_k^L(z) = D_k^L f_k^L(z) + C_k^L f_k^{R*}(z), \quad \phi_k^R(z) = f_k^R(z) D_k^R + f_k^{L*}(z) C_k^R,$$

then $\{\phi_k^L\}_{k=0}^n$ and $\{\phi_k^R\}_{k=0}^n$ form a left, resp. right orthogonal basis for $\mathcal{L}_n$. Moreover

$$\begin{aligned} D_k^L &= -C_k^L \langle\langle f_k^{R*}, \phi_p^L\rangle\rangle_{\ell,\mu} \langle\langle f_k^L, \phi_p^L\rangle\rangle_{\ell,\mu}^{-1}, \\ D_k^R &= -\langle\langle \phi_p^R, f_k^R\rangle\rangle_{r,\mu}^{-1} \langle\langle \phi_p^R, f_k^{L*}\rangle\rangle_{r,\mu} \end{aligned}$$

for any $p = 0,\ldots,k-1$ and

$$\begin{aligned} C_k^L &= \frac{1-\overline{\alpha}_k\alpha_{k-1}}{1-|\alpha_{k-1}|^2}\phi_k^L(\alpha_{k-1})[\phi_{k-1}^{R*}(\alpha_{k-1})]^{-1}, \\ C_k^R &= \frac{1-\overline{\alpha}_k\alpha_{k-1}}{1-|\alpha_{k-1}|^2}[\phi_{k-1}^{L*}(\alpha_{k-1})]^{-1}\phi_k^R(\alpha_{k-1}), \end{aligned}$$

$$D_k^L = -z_k\frac{1-\overline{\alpha}_{k-1}\alpha_k}{1-|\alpha_{k-1}|^2}\left[[\phi_{k-1}^{L*}(\alpha_{k-1})]^{-1}\phi_k^{L*}(\alpha_{k-1})\right]^H$$

and

$$D_k^R = -z_k\frac{1-\overline{\alpha}_{k-1}\alpha_k}{1-|\alpha_{k-1}|^2}\left[\phi_k^{R*}(\alpha_{k-1})[\phi_{k-1}^{R*}(\alpha_{k-1})]^{-1}\right]^H.$$

Proof. This is by induction on n. The matrix D_1^L is chosen in such a way that ϕ_1^L is left orthogonal to ϕ_0^L and similarly for ϕ_1^R. Suppose now that the theorem holds up to $n-1$. Then it is clear that $\phi_n^L \in \mathcal{L}_n \setminus \mathcal{L}_{n-1}$. Using $\langle\langle f, g\rangle\rangle_{\ell,\mu} = \langle\langle f^*, g^*\rangle\rangle_{r,\mu}$, it is not difficult to show that it is left orthogonal to

$\frac{z-\alpha_n}{1-\overline{\alpha}_{n-1}z}\mathcal{L}_{n-2} \subset \mathcal{L}_{n-1}$, while D_n^L is chosen such that ϕ_n^L is left orthogonal to ϕ_0^L. Because for any $f \in \mathcal{L}_{n-1}$ clearly $f(z) - f(\alpha_n) \in \frac{z-\alpha_n}{1-\overline{\alpha}_{n-1}z}\mathcal{L}_{n-2}$, the left orthogonality to $\mathcal{L}_{n-1}$ follows.
The relation between C's and D's follows because

$$\langle\langle \phi_k^L, \phi_p^L \rangle\rangle_{\ell,\mu} = \langle\langle \phi_p^R, \phi_k^R \rangle\rangle_{r,\mu} = 0$$

for all $p = 0, \ldots, k-1$. □

Lemma 9.2. *The following recursion also gives left and right MORF (we use the notation of the previous lemma)*

$$\begin{bmatrix} \phi_n^L(z) \\ \phi_n^{R*}(z) \end{bmatrix} = \frac{1-\overline{\alpha}_{n-1}z}{1-\overline{\alpha}_n z} \begin{bmatrix} I & [\lambda_n^L]^H \\ \lambda_n^R & I \end{bmatrix} \begin{bmatrix} \zeta_{n-1}(z)I & 0 \\ 0 & I \end{bmatrix} \begin{bmatrix} \phi_{n-1}^L(z) \\ \phi_{n-1}^{R*}(z) \end{bmatrix},$$

with

$$\lambda_n^L = -\overline{z}_{n-1} \langle\langle \phi_{n-1}^L, \phi_n^{R*} \rangle\rangle_{\ell,\mu}^{-1} \langle\langle \phi_{n-1}^L, \phi_n^L \rangle\rangle_{\ell,\mu},$$
$$\lambda_n^R = -\overline{z}_{n-1} \langle\langle \phi_n^R, \phi_{n-1}^R \rangle\rangle_{r,\mu} \langle\langle \phi_n^{L*}, \phi_{n-1}^R \rangle\rangle_{r,\mu}^{-1}.$$

Proof. Setting

$$\lambda_n^L = -\overline{z}_{n-1}\big[[D_n^L]^{-1}C_n^L\big]^H, \quad \lambda_n^R = -\overline{z}_{n-1}\big[C_n^R[D_n^R]^{-1}\big]^H,$$

it readily follows from the previous relations that the right hand side of this theorem results in

$$\begin{bmatrix} -\overline{z}_{n-1}[D_n^L]^{-1}\phi_n^L(z) \\ -\overline{z}_{n-1}[D_n^R]^{-1}\phi_n^{R*}(z) \end{bmatrix}.$$

and because the ϕ_n^L are left orthogonal, so are the $-\overline{z}_{n-1}[D_n^L]^{-1}\phi_n^L(z)$ and a similar observation holds for the right versions. □

Lemma 9.3. *If we work with orthonormal MORF, then*

$$\lambda_n^L = \lambda_n^R = \overline{z}_{n-1}z_n \frac{1-\overline{\alpha}_{n-1}\alpha_n}{1-\overline{\alpha}_n\alpha_{n-1}} \rho_n(\alpha_{n-1})$$

with $\rho_n(w)$ the symmetrized recursion parameter.

Proof. Recall $\kappa_n^R = \phi_n^{R*}(\alpha_n)$ and $\kappa_n^L = \phi_n^{L*}(\alpha_n)$ and the expressions for the λ's, the C's and the D's to find that

$$\lambda_n^L = \overline{z}_{n-1}z_n \frac{1-\overline{\alpha}_{n-1}\alpha_n}{1-\overline{\alpha}_n\alpha_{n-1}} [\kappa_{n-1}^R]^{-1}[\phi_n^L(\alpha_{n-1})]^H[\phi_n^{L*}(\alpha_{n-1})]^{-1}[\kappa_{n-1}^L]^H$$

and

$$\lambda_n^R = \overline{z}_{n-1}z_n \frac{1-\overline{\alpha}_{n-1}\alpha_n}{1-\overline{\alpha}_n\alpha_{n-1}} [\kappa_{n-1}^R]^H[\phi_n^{R*}(\alpha_{n-1})]^{-1}[\phi_n^R(\alpha_{n-1})]^H[\kappa_{n-1}^L]^{-1}.$$

We may now define matrices

$$N_n^L = \langle\langle \phi_n^L, \phi_n^L \rangle\rangle_{\ell,\mu}^{L/2}, \quad N_n^R = \langle\langle \phi_n^R, \phi_n^R \rangle\rangle_{r,\mu}^{R/2},$$

and multiply ϕ_n^L from the left by $[N_n^L]^{-1}$ and multiply ϕ_n^R from the right by $[N_n^R]^{-1}$, then we obtain orthonormal bases which we shall again denote by ϕ_n^L and ϕ_n^R. Note that the left and right square roots can be chosen such that the leading coefficients are positive definite. Furthermore, from the expression for λ_n^L and for ρ_n^L, we see that

$$\lambda_n^L = \overline{z}_{n-1} z_n \frac{1-\overline{\alpha}_{n-1}\alpha_n}{1-\overline{\alpha}_n \alpha_{n-1}} [\kappa_{n-1}^R]^{-1} \rho_n^L(\alpha_{n-1}) [\kappa_{n-1}^L]^H.$$

Finally use $\rho_n(w) = [K_{n-1}^L(w,w)]^{L/2} \rho_n^L(w) \left[[K_{n-1}^R(w,w)]^{R/2}\right]^H$ and

$$K_n^L(\alpha_n, \alpha_n) = \kappa_n^R [\kappa_n^R]^H, \quad K_n^R(\alpha_n, \alpha_n) = [\kappa_n^L]^H \kappa_n^L$$

to find the expression for λ_n. □

Theorem 9.1. *The MORF have initial conditions $\phi_0^L = \phi_0^R = I$ and satisfy the recurrence*

$$\begin{bmatrix} \phi_n^L(z) \\ \phi_n^{R*}(z) \end{bmatrix} = t_n(z) \begin{bmatrix} \phi_{n-1}^L(z) \\ \phi_{n-1}^{R*}(z) \end{bmatrix},$$

with

$$t_n(z) = \frac{\sqrt{1-|\alpha_n|^2}}{\sqrt{1-|\alpha_{n-1}|^2}} \frac{1-\overline{\alpha}_{n-1} z}{1-\overline{\alpha}_n z} H(\lambda_n) Z_{n-1}(z)$$

where $H(\lambda_n)$ and $Z_n(z)$ are as defined before and

$$\lambda_n = \eta_n \rho_n(\alpha_{n-1}), \quad \eta_n = z_n \overline{z}_{n-1} \frac{1-\overline{\alpha}_{n-1}\alpha_n}{1-\overline{\alpha}_n \alpha_{n-1}} \in \mathbb{T}$$

with $\rho_n(w)$ the symmetrized recursion parameter,

Proof. First note that

$$\begin{bmatrix} \hat{\phi}_n^L(z) \\ \hat{\phi}_n^{R*}(z) \end{bmatrix} = \frac{1-\overline{\alpha}_{n-1} z}{1-\overline{\alpha}_n z} \begin{bmatrix} I & \lambda_n^H \\ \lambda_n & I \end{bmatrix} Z_{n-1}(z) \begin{bmatrix} \phi_{n-1}^L(z) \\ \phi_{n-1}^{R*}(z) \end{bmatrix}$$

gives orthogonal functions, so that it remains to define the normalizing factors

$$N_n = \left[\left\langle\left\langle \hat{\phi}_n^L, \hat{\phi}_n^L \right\rangle\right\rangle_{\ell,\mu}^{L/2} \right]^{-1} \quad \text{and} \quad M_n = \left[\left\langle\left\langle \hat{\phi}_n^{R*}, \hat{\phi}_n^{R*} \right\rangle\right\rangle_{\ell,\mu}^{L/2} \right]^{-1}.$$

From the Christoffel-Darboux relation it follows that

$$\begin{bmatrix} \phi_n^L(\alpha_{n-1}) \\ \phi_n^{R*}(\alpha_{n-1}) \end{bmatrix}^H J \begin{bmatrix} \phi_n^L(\alpha_{n-1}) \\ \phi_n^{R*}(\alpha_{n-1}) \end{bmatrix} = -(1-|\zeta_n(\alpha_{n-1})|^2) \kappa_{n-1}^R [\kappa_{n-1}^R]^H.$$

On the other hand, the recursion gives

$$\begin{bmatrix} \phi_n^L(\alpha_{n-1}) \\ \phi_n^{R*}(\alpha_{n-1}) \end{bmatrix} = \frac{1-\overline{\alpha}_{n-1}z}{1-\overline{\alpha}_n z} \begin{bmatrix} N_n & 0 \\ 0 & M_n \end{bmatrix} \begin{bmatrix} \lambda_n^H[\kappa_{n-1}^R]^H \\ [\kappa_{n-1}^R]^H \end{bmatrix}.$$

Plug this into the previous relation then one gets (we set $\gamma_j = \sqrt{1-|\alpha_j|^2}$)

$$\lambda_n N_n^H N_n \lambda_n^H - M_n^H M_n = -\frac{\gamma_n^2}{\gamma_{n-1}^2} I.$$

For arbitrary unitary matrices U_n and V_n,

$$N_n = \frac{\gamma_n}{\gamma_{n-1}} U_n (I - \lambda_n^H \lambda_n)^{-1/2}, \quad M_n = \frac{\gamma_n}{\gamma_{n-1}} V_n (I - \lambda_n \lambda_n^H)^{-1/2}$$

are solutions. The U_n and V_n can be used to normalize the leading coefficients of ϕ_n^L and ϕ_n^R. □

Again it can be noted that for all $\alpha_k = 0$, the matrix versions of Szegő polynomials appear as a special case. The λ_n which are in this case given by $\rho_n(0)$ and are the block Szegő parameters.

9.2 Functions of the Second Kind

We can define functions of the second kind by setting $\phi_0^L = \psi_0^R = I$ as initial conditions for the recursion

$$\begin{bmatrix} \psi_n^L(z) \\ -\psi_n^{R*}(z) \end{bmatrix} = t_n(z) \begin{bmatrix} \psi_{n-1}^L(z) \\ -\psi_{n-1}^{R*}(z) \end{bmatrix},$$

with t_n as in the previous theorem. Along the same lines as in the scalar case, it can be proved that for $n \geq 1$

$$\psi_n^L(z) = \int D(t,z)[\phi_n^L(t) - \phi_n^L(z)]d\mu(t),$$

$$\psi_n^R(z) = \int D(t,z)d\mu(t)[\phi_n^R(t) - \phi_n^R(z)].$$

Because

$$T_n(z) = t_n(z) \cdots t_1(z) = \frac{1}{2} \begin{bmatrix} \phi_n^L(z) + \psi_n^L(z) & \phi_n^L(z) - \psi_n^L(z) \\ \phi_n^{R*}(z) - \psi_n^{R*}(z) & \phi_n^{R*}(z) + \psi_n^{R*}(z) \end{bmatrix}$$

and $(1-\overline{\alpha}_n z)T_n(z)/\gamma_n$ is again a lossless and *-J-unitary matrix, we may conclude as in the case of the kernels that

- ϕ_n^{R*} and ψ_n^{R*} have no zeros in $\mathbb{D} \cup \mathbb{T}$ (a zero of a matrix means a zero of its determinant).
- $[\phi_n^{R*}]^{-1}\psi_n^{R*}$ and $\psi_n^{R*}[\phi_n^{R*}]^{-1}$ are Carathéodory functions.
- $[\phi_n^{L*}]^{-1}$ and $\psi_n^{R*}]^{-1}$ are in $\mathbb{H}_2$.
- $\phi_{n*}^R(t)\psi_n^R(t) + \psi_{n*}^R(t)\phi_n^R = 2P(t,\alpha_n)I$ for $t \in \mathbb{T}$.
- $\phi_n^L(t)\psi_{n*}^L(t) + \psi_n^L(t)\phi_{n*}^L = 2P(t,\alpha_n)I$ for $t \in \mathbb{T}$.
- $\phi_n^R\phi_n^{R*} = \phi_n^{L*}\phi_n^L$ and $\psi_n^R\psi_n^{R*} = \psi_n^{L*}\psi_n^L$.

10 Interpolation and Quadrature

10.1 The Kernels

We can obtain interpolation properties for the (normalized) kernels like we did in the scalar case. This is related to polynomial kernels for varying measures in the sense that if $k_n^L(z,w)$ is a left reproducing kernel for the left inner product w.r.t. μ, then it can be written as $k_n^L(z,w) = k_{n,n}^L(z,w)/|\pi_n(z)|^2$ where $k_{n,n}^L(z,w)$ is the left polynomial reproducing kernel for the left inner product with respect to the measure $\mu/|\pi_n|^2$. We have

Theorem 10.1. *If $K_n^L(z,w)$ is the left normalized reproducing kernel for $\mathcal{L}_n$ with respect to μ, then for $d\mu_n^L(t) = [K_{n*}^L(t,0)K_n^L(t,0)]^{-1}d\lambda(t)$, we $\langle\langle f,g\rangle\rangle_{\ell,\mu} = \langle\langle f,g\rangle\rangle_{\ell,\mu_n^L}$ for all f and g in $\mathcal{L}_n$.*

Proof. It suffices to prove that $\langle\langle f,k_n^L(\cdot,w)\rangle\rangle_{\ell,\mu} = \langle\langle f,k_n^L(\cdot,w)\rangle\rangle_{\ell,\mu_n^L}$ for all $f\in\mathcal{L}_n$ because for a set of $n+1$ mutually different $\xi_k\in\mathbb{D}$, the $k_n^L(z,\xi_k)$ from a basis for $\mathcal{L}_n$. Therefore, it is sufficient to prove that for any polynomial p we have $\langle\langle p/\pi_n,k_n^L(\cdot,w)\rangle\rangle_{\ell,\mu_n^L} = p(w)/\pi_n(w)$. If $\phi_{k,n}^R$ is a set of right MOPVM for $\nu=\mu/|\pi_n|^2$, it follows that by using the polynomial Christoffel-Darboux relation that $K_n^L(z,0) = \phi_{nn}^{R*}(z)/\pi_n(z)$, and thus, $d\mu_n^L = |\pi_n|^2[\phi_{nn}^{R*}]_*\phi_{nn}^{R*}$. Therefore $\langle\langle p/\pi_n,k_n^L(\cdot,w)\rangle\rangle_{\ell,\mu_n^L} = I_1 - I_2$ with

$$I_1 = \frac{1}{\pi_n(w)}\left[\int p(t)\left[(\phi_{n,n}^{R*}(t))_*\phi_{n,n}^{R*}(t)\right]^{-1}(\phi_{n,n}^{R*}(t))_*\frac{t}{t-w}d\lambda(t)\right]\phi_{n,n}^{R*}(w)$$

and

$$I_2 = \frac{1}{\pi_n(w)}\left[\int p(t)\left[(\phi_{n,n}^{R*}(t))_*\phi_{n,n}^{R*}(t)\right]^{-1}(\phi_{n,n}^{L}(t))_*\frac{1}{t-w}d\lambda(t)\right]\phi_{n,n}^{L}(w).$$

Clearly $I_1 == \frac{p(w)}{\pi_n(w)}$ because the integral is by Cauchy's theorem equal to $p(w)[\phi_{n,n}^{R*}(w)]^{-1}$. The quantity I_2 is zero because $(\phi_{n,n}^{R*})_*\phi_{n,n}^{R*} = \phi_{n,n}^{L*}(\phi_{n,n}^{L*})_*$ which reduces the integral to the Hermitian conjugate of

$$\int[\phi_{n,n}^{L*}(t)]^{-1}p^*(t)\frac{t}{1-\overline{w}t}d\lambda(t),$$

and this is zero by Cauchy's theorem. This concludes the proof. □

Theorem 10.2. *With $\Omega_\mu(z) = \int D(t,z)d\mu(t)$ and $\Omega_n^L(z) = \int D(t,z)d\mu_n^L(t)$ we have $\Omega_n^L(z) = [K_n^L(z,0)]^{-1}L_n^L(z,0)$ and the left outer spectral factor of μ_n^L is $\sigma_n^L(z) = [K_n^L(z,0)]^{-1}$. Moreover we have the interpolation property*

$$\Omega_\mu(z) - \Omega_n^L(z) = zB_n(z)h(z)$$

with h analytic in $\mathbb{D}$.

Proof. We know that $[K_n^L(z,0)]^{-1}L_n^L(z,0)$ is a Carathéodory function and because $\frac{1}{2}[\Omega_n^L(t) + \Omega_{n*}^L(t)] = [K_{n*}^L(t,0)K_n^L(t,0)]^{-1}$, it is the Riesz-Herglotz transform of μ_n^L. Also, it is obvious that σ_n^L is a spectral factor. It is outer in $\mathbb{H}_2$ because $K_n^L(z,0)$ is a rational function in $\mathbb{H}_2$ and it does not vanish in $\mathbb{D} \cup \mathbb{T}$. For the interpolation property, note that with $\nu = \mu - \mu_n^L$

$$\frac{\Omega_\mu(z) - \Omega_n^L(z)}{B_n(z)} = \int \left[\frac{1}{B_n(z)} - \frac{1}{B_n(t)}\right] D(t,z)d\nu(t) + \int \frac{1}{B_n(t)} D(t,z)d\nu(t).$$

The first integral is zero because the term in square brackets is of the form f_* with $f = p/\pi_n$ where p is a scalar polynomial in t of degree less than $n+1$. Thus it equals $\langle\langle I, fI\rangle\rangle_{\ell,\nu}$. Because $fI \in \mathcal{L}_n$, this integral is 0. The integrand for the second term is analytic in $\mathbb{D}$ because $\int f(t)D(t,z)d\mu(t)$ is analytic in $\mathbb{D}$ for any $f \in \mathbb{L}_1(\nu)$ and any measure ν. Because $\langle\langle I, I\rangle\rangle_{\ell,\mu} = \langle\langle I, I\rangle\rangle_{\ell,\mu_n^L}$ we get the interpolation at the origin. □

Corollary 10.1.

$$\Omega_n^L(z) = [K_n^L(z,0)]^{-1}L_n^L(z,0) = L_n^R(0,z)[K_n^R(0,z)]^{-1} = \Omega_n^R(z).$$

Proof. Because $[K_{n*}^L(t,0)K_n^L(t,0)]^{-1} = [K_n^R(0,t)K_{n*}^R(0,t)]^{-1}$, we have by the Riesz-Herglotz theorem that $\Omega_n^L = \Omega_n^R$ in $\mathbb{D}$, but because these are just rational functions, equality also holds on $\mathbb{C}$.

10.2 The MORF

Similar properties can be derived for the MORF. Recall that $P(t,z) = \operatorname{Re} D(t,z)$ is the Poisson kernel. We have

Theorem 10.3. *Consider the measure*

$$d\hat{\mu}_n^L(t) = P(t,\alpha_n)[\phi_{n*}^L(t)\phi_n^L(t)]^{-1}d\lambda(t),$$

then in $\mathcal{L}_n$, the inner products $\langle\langle\cdot,\cdot\rangle\rangle_{\ell,\mu}$ and $\langle\langle\cdot,\cdot\rangle\rangle_{\ell,\hat{\mu}_n}$ are the same.

Proof. A proof will be given if it can be shown that the ϕ_n^L, which is a left orthonormal basis for μ, is also a left orthonormal basis for $\hat{\mu}_n^L$. This can be shown by a backward recursion. Obviously $\langle\langle\phi_n^L, \phi_n^L\rangle\rangle_{\ell,\hat{\mu}_n^L} = I$. Also for $k = 0, \ldots, n-1$:

$$\langle\langle\phi_n^L, \phi_k^L\rangle\rangle_{\ell,\hat{\mu}_n^L} = \int [\phi_n^{L*}(t)]^{-1}\phi_k^{L*}(t)P(t,\alpha_n)B_n(t)/B_k(t)d\lambda(t),$$

which is zero by the Poisson formula. □

Theorem 10.4. *With $\Omega_\mu(z) = \int D(t,z)d\mu(t)$ and $\hat{\Omega}_n^L(z) = \int D(t,z)d\hat{\mu}_n^L(t)$ where $\hat{\mu}_n^L$ is as in the previous theorem, then $\hat{\Omega}_n^L(z) = \psi_n^{L*}(z)[\phi_n^{L*}(z)]^{-1}$, and*

$\hat{\sigma}_n^R(z) = \frac{\sqrt{1-|\alpha_k|^2}}{1-\overline{\alpha}_n z}[\phi_n^{L*}(z)]^{-1}$ *is a right outer spectral factor of* $\hat{\mu}_n^L$. *Furthermore we have the interpolation property*

$$\Omega_\mu(z) - \hat{\Omega}_n^L(z) = zB_n(z)h(z)$$

with h *analytic in* $\mathbb{D}$.

Proof. The proof is completely parallel to the proof of Theorem 10.2. □

Corollary 10.2.

$$\hat{\Omega}_n^L(z) = \psi_n^{L*}(z)[\phi_n^{L*}(z)]^{-1} = [\phi_n^{R*}(z)]^{-1}\psi_n^{R*}(z) = \hat{\Omega}_n^R(z).$$

With these results, it is not difficult to derive a Favard type theorem.

Theorem 10.5. *Let* ϕ_n^L *and* ϕ_n^R *be generated by the recursion of Theorem 9.1. Then there is a positive definite measure on* $\mathbb{T}$ *for which they form a left, resp. a right MORF sequence.*

Proof. This is along the same lines as the scalar proof of Theorem 2.6. □

This measure will be unique if $A \in \mathrm{BD}$ and $\log\det\mu' \in L_1$ which follows from the completeness of the basis of Blaschke products in $\mathbb{H}_2$ with respect to the norm $|\cdot|_{\ell,\mu}$.

11 Minimalisation and Szegő's Problem

If $\Phi(s)$ is a function with Hermitian nonnegative definite values, then the problems $\inf_s \Phi(s)$ and $\inf_s \mathrm{tr}\Phi(s)$ have the same solutions. By a solution we mean an s such that $\Phi(s) \leq \Phi(t)$ for all t. We have the following.

Theorem 11.1. *The minimum of* $\langle\langle f, f\rangle\rangle_{\ell,\mu}$ *for all* $f \in \mathcal{L}_n$ *with* $f(w) = I$ *is given by* $[k_n^L(w.w)]^{-1}$ *and it is obtained for* $f = [k_n^L(w,w)]^{-1}k_n^L(\cdot,w)$, *provided that* $\det k_n^L(w,w) \neq 0$.

Proof. Similar to the proof of Theorem 6.1. □

We have formulated this theorem for $\mathcal{L}_n$ but it actually holds for any Hilbert module with reproducing kernel.

Define the left Szegő kernel $s^L(z,w) = [(1-\overline{w}z)\sigma^L(z)(\sigma^L(w))^H]^{-1}$ with $\sigma^L(0) > 0$ with $w \in \mathbb{D}$, then it is left reproducing in $\mathbb{H}_2$. We can conclude that, as in the scalar case, some vector valued stochastic process will be completely predictable if $\exp\{\int \log\det\mu'(t)d\lambda(t)\} = 0$, i.e., $\log\det\mu' \notin L_1$. If we assume that the Blaschke products diverge ($A \in \mathrm{BD}$) and that $\log\det\mu' \in L_1$ ($\mu \in \mathrm{SZ}$), then the set $\{B_n\}_{n\geq 0}$ is complete in the Banach space $\mathbb{H}_2(\mu)$ (thus with respect to the trace norm $|\cdot|_{\ell,\mu}$). So the analog of Theorem 6.2 is

Theorem 11.2. *Let $\mu \in \mathrm{SZ}$, $w \in \mathbb{D}$ a fixed number, and let $s_w^L(z) = s^L(z,w)$ be the left Szegő kernel. Consider the problem* $\min \|f - s_w^L\|_{\ell,\mu}^2$ *over all $f \in \mathcal{L}_n$, then the solution is $k_n^L(z,w)$ and the minimum is $s_w^L(w) - k_n^L(w,w)$. If $A \in \mathrm{BD}$ then $\lim_{n\to\infty} |k_n^L(t,w) - s_w^L(t)|_{\ell,\mu} = 0$.*

Proof. Clearly

$$\langle\langle s_w^L - f, s_w^L - f\rangle\rangle_{\ell,\mu} = \langle\langle s_w^L, s_w^L\rangle\rangle_{\ell,\mu} + \langle\langle f, f\rangle\rangle_{\ell,\mu} - 2\mathrm{Re}\, \langle\langle f, s_w^L\rangle\rangle_{\ell,\mu}$$

By Cauchy's theorem $\langle\langle s_w^L, s_w^L\rangle\rangle_{\ell,\mu} = s_w^L(w)$ and $\langle\langle f, s_w^L\rangle\rangle_{\ell,\mu} = f(w)$, so that we have to minimize $s_w^L(w) + \langle\langle f, f\rangle\rangle_{\ell,\mu} - 2\mathrm{Re}\, f(w)$. Like in the scalar case it can be seen that this minimum is obtained for f as claimed. Because $A \in \mathrm{BD}$ implies completeness in the Banach space $\mathbb{H}_2$, we have convergence in the norm of this space. □

This problem is immediately related to the prediction of vector valued stochastic processes, but it can also be interpreted in the context of inverse scattering theory or network synthesis.

12 What We did not Discuss

There are many things that were not discussed in this introduction.

First of all there is the scalar case with many things to be explored like further convergence results, characterization of the properties of the ORF in terms of the recurrence coefficients, moment problems, Nevanlinna-Pick and other classical interpolation problems. Some of these are included to some extend in [3].

There is a completely similar theory of ORF on the real line when all the poles are outside the real line. Interesting special cases are measures whose support is an interval or a half line. In the latter case results in the vein of what has been developped here can be obtained [29].

A slightly different theory emerges when poles are allowed to fall inside the support of the measure (see e.g. [3]).

In all those cases there is practically no result about an operator theoretical approach that generalizes the Jacobi matrices of classical polynomials.

For the matrix case, there is a huge literature on so-called Schur analysis. This involves matricial Nevanlinna-Pick interpolation problems and tangential or directional versions. A lot of energy is also spent on discussing generalizations of the Blaschke factors. Indeed instead of just using $B_k I$ as a basis function, one could consider products of factors of the form $\zeta_j U_j$ where U_j represents a rank one matrix. The recursion, which is in our treatment connected with J-lossless matrices, can be considerably generalized by replacing J by much more general matrices with a finer structure. And all this could be extended to the non-square matrix case.

Acknowledgments

The work of the first author is partially supported by the Fund for Scientific Research (FWO), projects "CORFU: Constructive study of orthogonal functions", grant #G.0184.02 and, "SMA: Structured matrices and their applications", grant G#0078.01, "ANCILA: Asymptotic analysis of the convergence behavior of iterative methods in numerical linear algebra", grant #G.0176.02, the K.U.Leuven research project "SLAP: Structured linear algebra package", grant OT-00-16, the Belgian Programme on Interuniversity Poles of Attraction, initiated by the Belgian State, Prime Minister's Office for Science, Technology and Culture. The scientific responsibility rests with the author.

References

1. N.I. Achieser, *Theory of Approximation*, Frederick Ungar Publ. Co., New York, 1956.
2. A. Bultheel, *Orthogonal matrix functions related to the multivariable Nevanlinna-Pick problem*, Bull. Soc. Math. Belg. Sér. B, **32** no. 2 (1980), 149–170.
3. A. Bultheel, P. González-Vera, E. Hendriksen, and O. Njåstad, *Orthogonal Rational Functions*, Cambridge Monographs on Applied and Computational Mathematics vol. 5, Cambridge University Press, 1999.
4. Ph. Delsarte, Y. Genin, and Y. Kamp, *Orthogonal polynomial matrices on the unit circle*, IEEE Trans. Circuits and Systems **25** (1978), 149–160.
5. P. Dewilde and H. Dym, *Lossless chain scattering matrices and optimum linear prediction: the vector case*, Internat. J. Circuit Th. Appl. **9** (1981), 135–175.
6. R.G. Douglas, *Banach Algebra Techniques in Operator Theory*, Academic Press, New York, 1972.
7. V.K. Dubovoj, B. Fritzsche, and B. Kirstein, *Matricial Version of the Classical Schur Problem*, Teubner-Texte zur Mathematik, vol. 129, Teubner Verlagsgesellschaft, Stuttgart, Leipzig, 1992.
8. P.L. Duren, *The Theory of H^p Spaces*, Pure and Applied Mathematics, vol. 38, Academic Press, New York, 1970.
9. H. Dym, *J-Contractive Matrix functions, Reproducing Kernel Hilbert Spaces and Interpolation*, CBMS Regional Conf. Ser. in Math., vol. 37, Amer. Math. Soc., Providence, Rhode Island, 1989.
10. B. Fritzsche and B. Kirstein and A. Lasarow, *Orthogonal rational matrix-valued functions on the unit circle*, Math. Nachr. **278**, no. 5 (2005), 525–553.
11. B. Fritzsche and B. Kirstein and A. Lasarow, *Further aspects of the theory of orthogonal rational matrix-valued functions on the unit circle*, Math. Nachr. (to appear).
12. B. Fritzsche and B. Kirstein and A. Lasarow, *On a moment problem for rational matrix-valued functions*, Linear Algebra Appl. **372** (2003), 1–31.
13. B. Fritzsche and B. Kirstein and A. Lasarow, *On rank invariance of moment matrices of nonnegative Hermitian-valued Borel measures on the unit circle*, Math. Nachr. **263–264** (2004), 103–132.
14. J. B. Garnett, *Bounded Analytic Functions*, Academic Press, New York, 1981.

15. K. Hoffman, *Banach Spaces of Analytic Functions*, Prentice-Hall, Englewood Cliffs, 1962.
16. P. Koosis, *Introduction to H_p Spaces*, Cambridge Tracts in Mathematics, vol. 115, Cambridge University Press, 2nd edition, 1998.
17. P. Kravanja, *Een Szegő theorie voor rationale matrixfuncties*, Master thesis, K.U.Leuven, Dept. Computerwetenschappen, 1994.
18. H. Langer and A. Lasarow, *Solution of a multiple Nevanlinna-Pick problem via orthogonal rational functions*, J. Comput. Appl. Math. (submitted).
19. A. Lasarow, *Aufbau einer Szegő-Theorie für rationale Matrixfunktionen*, Ph.D. thesis, Universität Leipzig, Fak. Mathematik Informatik, 2000.
20. G.L. Lopes [López-Lagomasino], *On the asymptotics of the ratio of orthogonal polynomials and convergence of multipoint Padé approximants*, Mat. Sb. (N.S.) **128** (170) (1985), no. 2, 216–229 (in Russian); Math. USSR-Sb. **56** (1987), 207–219.
21. G.L. López [López-Lagomasino], *Szegő's theorem for orthogonal polynomials with respect to varying measures*, in "Orthogonal polynomials and their applications" (M. Alfaro et al., editors), Lecture Notes in Mathematics **1329**, Springer, 1988, pp. 255–260.
22. G.L. López [López-Lagomasino], *Asymptotics of polynomials orthogonal with respect to varying measures*, Constr. Approx. **5** (1989), 199–219.
23. P. Masani, *The prediction theory of multivariate stochastic processes, III*, Acta Math. **104** (1960), 141–162.
24. M. Rosenblum and J. Rovnyak, *Hardy Classes and Operator Theory*, Oxford University Press, New York, 1985.
25. W. Rudin, *Real and Complex Analysis*, McGraw-Hill, New York, 3rd edition, 1987.
26. A. Ruhe, *Rational Krylov sequence methods for eigenvalue computation*, Linear Algebra Appl. **58** (1984), 391–405.
27. A. Ruhe, *Rational Krylov: A practical algorithm for large sparse nonsymmetric matrix pencils*, SIAM J. Sci. Comput. **19**, no. 5 (1998), 1535–1551.
28. A. Ruhe, *The rational Krylov algorithm for nonlinear matrix eigenvalue problems*, J. Math. Sci. **114** (2003), 176–180.
29. J. Van Deun, *Orthogonal rational functions: asymptotics and computational aspects*, Ph.D. thesis, K.U.Leuven, dept. computer science, 2004.
30. J.L. Walsh, *Interpolation and functions analytic interior to the unit circle*, Trans. Amer. Math. Soc. **34** (1932), 523–556.
31. J.L. Walsh, *Interpolation and Approximation*, Amer. Math. Soc. Colloq. Publ., vol. 20, Amer. Math. Soc., Providence, Rhode Island, 3rd edition, 1960. First edition 1935.
32. N. Wiener and P. Masani, *The prediction theory of multivariate stochastic processes, I. The regularity condition*, Acta Math. **98** (1957), 111–150.
33. N. Wiener and P. Masani, *The prediction theory of multivariate stochastic processes, II. The linear predictor*, Acta Math. **99** (1958), 93–139.

Orthogonal Polynomials and Separation of Variables

Vadim B. Kuznetsov

Department of Applied Mathematics, University of Leeds,
Leeds, LS2 9JT UK

Summary. These are the notes for an introductory course into the theory of quantum integrability and separability, a rather incomplete status of which dictates a special way of presentation. Therefore, I have chosen to work out a few very basic examples which shall help to exemplify the most important notions and the general methodology of the separation approach to solving quantum integrable systems. All bases considered (and factorized) in these notes appear to be special cases of the celebrated Jack polynomials which were the object of our study in the work [15], joint with Vladimir Mangazeev and Evgeny Sklyanin. These notes are designed to be a good preparation for a more dedicated reading into the subject.

1 Chebyshev Polynomials

1.1 Pafnuty Lvovich Chebyshev

Born: 16 May 1821 in Okatovo, Russia; died: 8 December 1894 in St. Petersburg, Russia (see, e.g., [2]).

"Chebyshev was probably the first mathematician to recognize the general concept of orthogonal polynomials. A few particular orthogonal polynomials were known before his work. Legendre and Laplace had encountered the Legendre polynomials in their work on celestial mechanics in the late eighteenth century. Laplace had found and studied the Hermite polynomials in the course of his discoveries in probability theory during the early nineteenth century. It was Chebyshev who saw the possibility of a general theory and its applications. His work arose out of the theory of least squares approximation and probability; he applied his results to interpolation, approximate quadrature and other areas. He discovered the discrete analogue of the Jacobi polynomials but their importance was not recognized until this century. They were rediscovered by Hahn and named after him upon their rediscovery. Geronimus has pointed out that in his first paper on orthogonal polynomials, Chebyshev already had the Christoffel-Darboux formula." [23]

1.2 Notation and Standard Formulae

We need a few basic facts from the theory of **hypergeometric series**, i.e., series $\sum_{n=0}^{\infty} c_n$ with $\frac{c_{n+1}}{c_n}$ a rational function of n. First, define the **Pochhammer symbol** $(a)_n$, for any $a \in \mathbb{C}$, $n = 0, 1, 2, \ldots$, as follows:

$$(a)_0 = 1, \qquad (a)_n = a(a+1)\cdots(a+n-1) = \frac{\Gamma(a+n)}{\Gamma(a)}. \tag{1.1}$$

Exercise 1. Check the following formulae:

$$(1)_n = n!, \qquad (2n)! = 2^{2n} n! (\tfrac{1}{2})_n, \qquad (a+n)_{k-n} = \frac{(a)_k}{(a)_n}, \tag{1.2}$$

$$(a)_{n+k} = (a)_n (a+n)_k, \qquad (a-n)_n = (-1)^n (1-a)_n. \tag{1.3}$$

A non-trivial (but still elementary function) example of a hypergeometric series is given by the **binomial theorem**:

$$(1-x)^{-a} = \sum_{n=0}^{\infty} \frac{(a)_n}{n!} x^n, \qquad |x| < 1, \tag{1.4}$$

which can be recognized as the Maclaurin expansion of the function on the left.

Let us expand the obvious identity $(1-x)^{-a}(1-x)^{-b} = (1-x)^{-a-b}$ using (1.4) and get

$$\sum_{n=0}^{\infty}\sum_{m=0}^{\infty} \frac{(a)_n}{n!}\frac{(b)_m}{m!}\, x^{n+m} = \sum_{k=0}^{\infty} \frac{(a+b)_k}{k!}\, x^k. \tag{1.5}$$

Comparing the coefficients of the powers of x on both sides we have

$$\sum_{n=0}^{k} \frac{(a)_n}{n!}\frac{(b)_{k-n}}{(k-n)!} = \frac{(a+b)_k}{k!}, \qquad k = 0, 1, 2, \ldots. \tag{1.6}$$

The Pochhammer symbol satisfies a number of useful identities, one of them is as follows:

$$(b)_{k-n} = \frac{(b)_k}{(b+k-n)\cdots(b+k-1)} = \frac{(b)_k}{(-1)^k(1-b-k)_n}. \tag{1.7}$$

Using this, we rewrite the equation (1.6) in the form

$$\sum_{n=0}^{k} \frac{(a)_n}{n!}\frac{(-k)_n}{(1-b-k)_n} = \frac{(a+b)_k}{(b)_k}, \tag{1.8}$$

and, putting $b = 1-c-k$, in an alternative form

$$\sum_{n=0}^{k} \frac{(-k)_n(a)_n}{(c)_n n!} = \frac{(c-a)_k}{(c)_k}, \tag{1.9}$$

since $(1-c-k)_k = (1-c-k)\cdots(-c) = (-1)^k(c)_k$. The terminating **Gauss hypergeometric series**

$${}_2F_1\left(\begin{matrix}-k, a\\ c\end{matrix}; x\right) = \sum_{n=0}^{k} \frac{(-k)_n(a)_n}{(c)_n n!}\, x^n, \qquad k = 0, 1, \ldots. \tag{1.10}$$

is well defined when $c \neq 0, -1, \ldots, -k+1$. As follows from (1.9), it takes the following factorized value at $x = 1$:

$${}_2F_1\left(\begin{matrix}-k, a\\ c\end{matrix}; 1\right) = \frac{(c-a)_k}{(c)_k}. \tag{1.11}$$

The above is called the **Chu-Vandermonde formula**. It can be put into a simpler (binomial) form

$$(a+b)_k = \sum_{n=0}^{k} \binom{k}{n} (a)_n (b)_{k-n}, \tag{1.12}$$

which is the same as (1.6).

1.3 Polynomials $T_n(x)$ and $U_n(x)$

Chebyshev polynomials were introduced in 1854 (see [3]). The **Chebyshev polynomials** of the first kind, $T_n(x)$, and of the second kind, $U_n(x)$, can be obtained by specifying the Jacobi polynomials $P_n^{(\alpha,\beta)}(x)$ [9]:

$$T_n(x) = \frac{P_n^{(-\frac{1}{2},-\frac{1}{2})}(x)}{P_n^{(-\frac{1}{2},-\frac{1}{2})}(1)} = {}_2F_1\left(\begin{matrix}-n, n\\ \frac{1}{2}\end{matrix}; \frac{1-x}{2}\right), \tag{1.13}$$

$$U_n(x) = (n+1)\,\frac{P_n^{(\frac{1}{2},\frac{1}{2})}(x)}{P_n^{(\frac{1}{2},\frac{1}{2})}(1)} = (n+1)\,{}_2F_1\left(\begin{matrix}-n, n+2\\ \frac{3}{2}\end{matrix}; \frac{1-x}{2}\right). \tag{1.14}$$

Although, in themselves, they are much simpler:

$$T_n(x) = \cos n\theta, \quad U_n(x) = \frac{\sin(n+1)\theta}{\sin\theta}, \qquad x = \cos\theta. \tag{1.15}$$

Now, of course, we assume that $-1 \le x \le 1$. Using the identity

$$\begin{aligned} e^{in\theta} &= \cos n\theta + i\sin n\theta = (\cos\theta + i\sin\theta)^n \\ &= \sum_{k=0}^{n} \binom{n}{k} \cos^{n-k}\theta\,(i\sin\theta)^k, \end{aligned} \tag{1.16}$$

we see that both sets are, indeed, polynomials of degree n in x and that they can be explicitly written as the following sums:

$$T_n(x) = \sum_{k=0}^{[n/2]} \binom{n}{2k} x^{n-2k}(x^2-1)^k, \tag{1.17a}$$

$$U_n(x) = \sum_{k=0}^{[n/2]} \binom{n+1}{2k+1} x^{n-2k}(x^2-1)^k. \tag{1.17b}$$

If instead of (1.16) we use a similar identity

$$\begin{aligned} e^{in\theta} &= \cos n\theta + i\sin n\theta = \left(\cos\tfrac{\theta}{2} + i\sin\tfrac{\theta}{2}\right)^{2n} \\ &= \sum_{k=0}^{2n} \binom{2n}{k} \cos^{2n-k}\tfrac{\theta}{2}\left(i\sin\tfrac{\theta}{2}\right)^k, \end{aligned} \tag{1.18}$$

then we obtain alternative expressions for the Chebyshev polynomials

$$\cos n\theta = \sum_{k=0}^{n} \binom{2n}{2k} \left(1-\sin^2\tfrac{\theta}{2}\right)^{n-k}\left(-\sin^2\tfrac{\theta}{2}\right)^k, \tag{1.19a}$$

$$\frac{\sin(n+1)\theta}{\sin\theta} = \frac{1}{2}\sum_{k=0}^{n} \binom{2n+2}{2k+1} \left(1-\sin^2\tfrac{\theta}{2}\right)^{n-k}\left(-\sin^2\tfrac{\theta}{2}\right)^k. \tag{1.19b}$$

Exercise 2. The last two formulae convert to the expressions (1.13)–(1.14) if one expands the binomial $\left(1 - \sin^2 \frac{\theta}{2}\right)^{n-k}$ and then evaluates one of the sums by the Chu-Vandermonde formula (1.11).

One can similarly convert the formulae (1.17b) into the following explicit expressions:

$$T_n(x) = \frac{n}{2} \sum_{k=0}^{[n/2]} \frac{(-1)^k (n-k-1)!}{k!(n-2k)!} (2x)^{n-2k}, \tag{1.20a}$$

$$U_n(x) = \sum_{k=0}^{[n/2]} \frac{(-1)^k (n-k)!}{k!(n-2k)!} (2x)^{n-2k}, \tag{1.20b}$$

which give expansions in terms of powers of x as opposed to the expansions (1.13)–(1.14) as powers of $1-x$.

All orthogonal polynomials satisfy a **three-term recurrence relation.** The polynomials $T_n(x)$ and $U_n(x)$ are generated by the same recurrence relation with different initial values:

$$T_n(x) = 2xT_{n-1}(x) - T_{n-2}(x), \qquad T_0(x) = 1, \quad T_1(x) = x, \tag{1.21a}$$

$$U_n(x) = 2xU_{n-1}(x) - U_{n-2}(x), \qquad U_0(x) = 1, \quad U_1(x) = 2x. \tag{1.21b}$$

Exercise 3. Check these recurrence relations for the Chebyshev polynomials.

The first few T-polynomials are:

$$\begin{aligned}
T_0(x) &= 1,\\
T_1(x) &= x,\\
T_2(x) &= 2x^2 - 1,\\
T_3(x) &= 4x^3 - 3x,\\
T_4(x) &= 8x^4 - 8x^2 + 1,\\
T_5(x) &= 16x^5 - 20x^3 + 5x,\\
T_6(x) &= 32x^6 - 48x^4 + 18x^2 - 1.
\end{aligned}$$

The first few U-polynomials are:

$$\begin{aligned}
U_0(x) &= 1,\\
U_1(x) &= 2x,\\
U_2(x) &= 4x^2 - 1,\\
U_3(x) &= 8x^3 - 4x,\\
U_4(x) &= 16x^4 - 12x^2 + 1,\\
U_5(x) &= 32x^5 - 32x^3 + 6x,\\
U_6(x) &= 64x^6 - 80x^4 + 24x^2 - 1.
\end{aligned}$$

1.4 Orthogonality

The **orthogonality relations** for $T_n(x)$ and $U_n(x)$,

$$\int_{-1}^{1} T_m(x)T_n(x)(1-x^2)^{-\frac{1}{2}}\,dx = \begin{cases} \frac{\pi}{2}\,\delta_{mn}, & n \neq 0 \\ \pi\delta_{mn}, & n = 0 \end{cases} \tag{1.22a}$$

$$\int_{-1}^{1} U_m(x)U_n(x)(1-x^2)^{\frac{1}{2}}\,dx = \frac{\pi}{2}\,\delta_{mn}, \tag{1.22b}$$

are the direct consequence of the well-known orthogonalities of the sine and cosine functions for integer m and n:

$$\int_{-\pi}^{\pi} \cos mx\,\cos nx\,dx = \begin{cases} 0, & m \neq n, \\ \pi, & m = n \neq 0, \\ 2\pi, & m = n = 0, \end{cases} \tag{1.23a}$$

$$\int_{-\pi}^{\pi} \sin mx\,\sin nx\,dx = \begin{cases} 0, & m \neq n, \\ \pi, & m = n \neq 0, \\ 0, & m = n = 0, \end{cases} \tag{1.23b}$$

which in turn are easily proved using the linearization formulae:

$$\cos a \cos b = \tfrac{1}{2}\left(\cos(a+b) + \cos(a-b)\right), \tag{1.24a}$$

$$\sin a \sin b = \tfrac{1}{2}\left(\cos(a-b) - \cos(a+b)\right). \tag{1.24b}$$

1.5 Other Results

Orthogonality relations such as (1.22a)–(1.22b), which are given by the weight function, the interval and the normalization, define the corresponding orthogonal polynomials uniquely, up to signs[1] (cf., e.g., [25]). Many other properties of the orthogonal polynomials, notably those for the classical ones, can be derived directly from the definition, including explicit forms, recurrence relation, **differential equation, differentiation formula** and **generating functions.** We cite here only the differential equations, differentiation formulae and simplest generating functions:

$$(1-x^2)T_n''(x) - xT_n'(x) + n^2T_n(x) = 0, \tag{1.25a}$$

$$(1-x^2)U_n''(x) - 3xU_n'(x) + n(n+2)U_n(x) = 0, \tag{1.25b}$$

$$\left((1-x^2)\,\tfrac{d}{dx} + nx\right)T_n(x) = nT_{n-1}(x), \tag{1.26a}$$

$$\left((1-x^2)\,\tfrac{d}{dx} + nx\right)U_n(x) = (n+1)U_{n-1}(x), \tag{1.26b}$$

[1] which are usually fixed by choosing the coefficients with the highest powers to be positive

$$\frac{1-xt}{1-2xt+t^2} = \sum_{n=0}^{\infty} T_n(x)t^n, \qquad \frac{1}{1-2xt+t^2} = \sum_{n=0}^{\infty} U_n(x)t^n. \tag{1.27}$$

Exercise 4. Prove (1.25)–(1.27).

Notice that the existence of differentiation (or shift) formulae such as (1.26) or second order differential equations such as (1.25) is characteristic for the so-called **'very-classical' orthogonal polynomials** [11].

1.6 Least Possible Deviation from Zero

Let us introduce the coefficients

$$L_n = \begin{cases} 1, & n = 0 \\ 2^{1-n}, & n = 1, 2, \ldots. \end{cases} \tag{1.28}$$

Then the polynomials $\tilde{T}_n(x) := L_n T_n(x)$ are monic

$$\tilde{T}_n(x) := L_n T_n(x) = x^n + \ldots. \tag{1.29}$$

The polynomials $\tilde{T}_n(x)$ have the following extremal and completely characterising property: they have **the least possible deviation from zero** on the interval $[-1, 1]$ among all monic polynomials $F_n(x)$ of degree n, i.e.,

$$\max_{x\in[-1,1]} |F_n(x)| \geq \max_{x\in[-1,1]} |\tilde{T}_n(x)| = L_n. \tag{1.30}$$

To prove this property, let $P_n(x) = x^n + \ldots$ be a monic polynomial whose deviation from zero in the interval $[-1, 1]$ is $\leq L_n$. Then, the polynomial $Q(x) := L_n T_n(x) - P_n(x)$ is of degree $\leq n-1$ and, at the points

$$x_k = \cos\tfrac{(n-k)\pi}{n}, \qquad k = 0, 1, \ldots, n, \tag{1.31}$$

has the following values

$$Q(x_k) = (-1)^{n-k} L_n - P_n(x_k) = (-1)^{n-k}[L_n \pm |P_n(x_k)|]. \tag{1.32}$$

The expression in the square bracket is not negative, meaning that the polynomial $Q(x)$ has at least n roots in the interval $[-1, 1]$, and hence $Q(x) \equiv 0$.

2 Gegenbauer Polynomials

2.1 Leopold Bernhard Gegenbauer

Born: 2 February 1849 in Asperhofen (E of Herzogenburg), Austria, died: 3 June 1903 in Vienna, Austria.

"Leopold Gegenbauer studied at the University of Vienna from 1869 until 1873. He then went to Berlin where he studied from 1873 to 1875 working under Weierstrass and Kronecker. After graduating from Berlin, Gegenbauer was appointed to a position at the University of Czernowitz (then in the Austrian Empire but now Chernovtsy, Ukraine) in 1875. He remained in Czernowitz for three years before moving to the University of Innsbruck where he worked with Stolz. After three years teaching in Innsbruck Gegenbauer was appointed full professor in 1881, then he was appointed full professor at the University of Vienna in 1893. He remained there until his death. Gegenbauer had many mathematical interests but was chiefly an algebraist. He is remembered for the Gegenbauer polynomials." [6]

2.2 Polynomials $C_n^{(g)}(x)$

The **Gegenbauer polynomials** $C_n^{(g)}(x)$ can be thought of as a g-deformation of the Chebyshev polynomials:

$$C_n^{(g)}(\cos\theta) = \frac{2g}{n}\left(\cos n\theta + O(g)\right), \qquad n = 1, 2, \ldots. \tag{2.1}$$

We shall derive their most important properties using the method of **shift operators** (cf [10, §4], [11, §2]).

The orthogonality relations with the standard normalization are ([9])

$$\int_{-1}^{1} (1-x^2)^{g-\frac{1}{2}} C_m^{(g)}(x) C_n^{(g)}(x) dx = \frac{\pi\Gamma(n+2g)2^{1-2g}}{\Gamma^2(g)(n+g)n!}\,\delta_{mn},$$
$$g > -\frac{1}{2}, \quad g \neq 0. \tag{2.2}$$

Let us also introduce the corresponding monic polynomials $p_n(x) \propto C_n^{(g)}(x)$,

$$\int_{-1}^{1} (1-x^2)^{g-\frac{1}{2}} p_m(x) p_n(x) dx = h_n(g)\,\delta_{mn}, \qquad p_n(x) = x^n + \ldots, \tag{2.3}$$

and the other set, $q_n(x) \propto C_n^{(g+1)}(x)$, corresponding to the shifted parameter, $g \to g+1$,

$$\int_{-1}^{1} (1-x^2)^{g+\frac{1}{2}} q_m(x) q_n(x) dx = h_n(g+1)\,\delta_{mn}, \qquad q_n(x) = x^n + \ldots, \tag{2.4}$$

with some normalization constants $h_n(g)$ to be determined later. Integration by parts gives

$$\int_{-1}^{1} p_n'(x) q_{m-1}(x) (1-x^2)^{g+\frac{1}{2}} dx$$
$$= -\int_{-1}^{1} p_n(x)(1-x^2)^{\frac{1}{2}-g} \tfrac{d}{dx}\Big((1-x^2)^{g+\frac{1}{2}} q_{m-1}(x)\Big)(1-x^2)^{g-\frac{1}{2}} dx. \tag{2.5}$$

Notice that

$$p'_n(x) = nq_{n-1}(x) + (\text{lower degree } q_k\text{'s})\,, \tag{2.6}$$

$$(1-x^2)^{\frac{1}{2}-g}\tfrac{d}{dx}\left((1-x^2)^{g+\frac{1}{2}}q_{m-1}(x)\right)$$
$$= (-m-2g)p_m(x) + (\text{lower degree } p_k\text{'s})\,. \tag{2.7}$$

Now, putting first $m = 1, \ldots, n-1$ and then $n = 0, \ldots, m-1$ in (2.5) and using the orthogonality, we conclude that there are no lower terms in the above equations, resulting in two **shift formulae** for the action of two shift operators $D_\pm$ on the monic Gegenbauer polynomials:

$$[D_- p_n^{(g)}](x) = \tfrac{d}{dx}\, p_n^{(g)}(x) = np_{n-1}^{(g+1)}(x), \tag{2.8}$$

$$[D_+^{(g)} p_{n-1}^{(g+1)}](x) = \left((1-x^2)\,\tfrac{d}{dx} - (2g+1)x\right)\, p_{n-1}^{(g+1)}(x)$$
$$= (-n-2g)p_n^{(g)}(x). \tag{2.9}$$

Since both operators act in opposite directions, composing them produces the second order differential equation for the Gegenbauer polynomials, $[(D_+^{(g)} \circ D_-)p_n^{(g)}](x) = -n(n+2g)p_n^{(g)}(x)$:

$$\left((1-x^2)\tfrac{d^2}{dx^2} - (2g+1)x\tfrac{d}{dx} + n(n+2g)\right)p_n^{(g)}(x) = 0. \tag{2.10}$$

Substituting $n = m$ into (2.5) yields the recurrence

$$\int_{-1}^{1} \left(p_n^{(g)}(x)\right)^2 (1-x^2)^{g-\frac{1}{2}}\,dx$$
$$= \frac{n}{n+2g}\int_{-1}^{1} \left(p_{n-1}^{(g+1)}(x)\right)^2 (1-x^2)^{g+\frac{1}{2}}\,dx, \tag{2.11}$$

iterating which one gets

$$h_n(g) = \int_{-1}^{1} \left(p_n^{(g)}(x)\right)^2 (1-x^2)^{g-\frac{1}{2}}\,dx = \frac{2^{2g+2n}\Gamma^2(g+n+\frac{1}{2})n!}{\Gamma(2g+2n+1)(n+2g)_n}, \tag{2.12}$$

where at the last step we have used a variant of the **beta integral**:

$$\int_{-1}^{1} (1-x)^\alpha(1+x)^\beta dx = \frac{2^{\alpha+\beta+1}\Gamma(\alpha+1)\Gamma(\beta+1)}{\Gamma(\alpha+\beta+2)}\,. \tag{2.13}$$

Finally, using Legendre's duplication formula

$$\Gamma(2a)\Gamma\left(\tfrac{1}{2}\right) = 2^{2a-1}\Gamma(a)\Gamma\left(a+\tfrac{1}{2}\right) \tag{2.14}$$

and the value $\Gamma\left(\frac{1}{2}\right) = \sqrt{\pi}$ we obtain

$$h_n(g) = \frac{\pi\Gamma(2g+n)n!}{2^{2g+2n-1}(g+n)\Gamma^2(g+n)}. \tag{2.15}$$

Exercise 5. Check this.

2.3 Beta Integral and Elliptic Coordinates

Let us start from a "better" form of the beta integral (2.13):

$$\int_0^\infty d\xi_1 \int_0^\infty d\xi_2 \;\delta(\xi_1+\xi_2-1)\xi_1^{\alpha_1-1}\xi_2^{\alpha_2-1} = \frac{\Gamma(\alpha_1)\Gamma(\alpha_2)}{\Gamma(\alpha_1+\alpha_2)} \tag{2.16}$$
$$= \int_0^1 d\xi_1\;\xi_1^{\alpha_1-1}(1-\xi_1)^{\alpha_2-1}.$$

Here, of course, we must assume that $\operatorname{Re}\alpha_i > 0$.

Theorem 2.1. *Let y_1 and y_2 be a fixed pair of real positive parameters such that*

$$0 < y_1 < y_2 < \infty. \tag{2.17}$$

Then for $\operatorname{Re}\alpha_i > 0$ *and* $z > 1$,

$$\int_{y_2}^\infty dx_2 \int_{y_1}^{y_2} dx_1 \delta(x_1x_2 - zy_1y_2)(x_2-x_1)\xi_1^{\alpha_1-1}\xi_2^{\alpha_2-1}$$
$$= (z-1)^{\alpha_1+\alpha_2-1}(y_2-y_1)\frac{y_1^{\alpha_1-1}y_2^{\alpha_2-1}\Gamma(\alpha_1)\Gamma(\alpha_2)}{\Gamma(\alpha_1+\alpha_2)}, \tag{2.18}$$

where

$$\xi_1 = \frac{(x_1-y_1)(x_2-y_1)}{y_2-y_1}, \qquad \xi_2 = \frac{(x_1-y_2)(x_2-y_2)}{y_1-y_2}. \tag{2.19}$$

Proof. The first step is to rescale the variables $\xi_k \mapsto \frac{\xi_k}{y_k(z-1)}$ in the beta integral (2.16) and therefore rewrite it in the form

$$\int_0^\infty d\xi_1 \int_0^\infty d\xi_2 \;\delta\left(\frac{\xi_1}{y_1}+\frac{\xi_2}{y_2}-(z-1)\right)\xi_1^{\alpha_1-1}\xi_2^{\alpha_2-1}$$
$$= (z-1)^{\alpha_1+\alpha_2-1}\frac{y_1^{\alpha_1}y_2^{\alpha_2}\Gamma(\alpha_1)\Gamma(\alpha_2)}{\Gamma(\alpha_1+\alpha_2)} \tag{2.20}$$

We consider (2.19) as the defining relations for the change of variables from (ξ_1, ξ_2) to (x_1, x_2), with (y_1, y_2) being the parameters, and perform this change of integration variables in the integral (2.20).

Notice first that the integration variables x_1 and x_2 in (2.18) satisfy the inequalities

$$y_1 < x_1 < y_2 < x_2 < \infty, \tag{2.21}$$

which guarantee the positivity of the functions ξ_1 and ξ_2, defined by (2.19). Conversely, every pair of the positive variables ξ_1, ξ_2 leads to a pair of variables x_1, x_2 satisfying (2.21). Indeed, the defining relations (2.19) can be rewritten in the equivalent form

$$1-\frac{\xi_1}{t-y_1}-\frac{\xi_2}{t-y_2}=\frac{(t-x_1)(t-x_2)}{(t-y_1)(t-y_2)},\qquad t\in\mathbb{C},\tag{2.22}$$

with the former formulae following by comparing the residues of both sides at $t = y_k$. When $\xi_k > 0$, the rational function on the left of (2.22) takes opposite signs near the ends of the intervals (y_1+, y_2-) and (y_2+, ∞), meaning that it has two zeros x_k interspersed with the poles y_k, as in (2.21).

Finally, calculating the Jacobian of the transformation and setting $t = 0$ in (2.22), we obtain

$$d\xi_1 d\xi_2=\frac{x_2-x_1}{y_2-y_1}\,dx_1dx_2,\qquad 1+\frac{\xi_1}{y_1}+\frac{\xi_2}{y_2}=\frac{x_1x_2}{y_1y_2},\tag{2.23}$$

which obviously imply the statement. □

Remark 2.1. When putting $\xi_k = z_k^2$, the change of variables $(z_1, z_2) \mapsto (x_1, x_2)$, defined by the relation (2.22), becomes the well-known transformation from the rectangular Cartesian coordinates in the plane, z_k, to the **elliptic coordinates**, x_k. The latter constitute a system of curvilinear orthogonal coordinates generated by the coordinate lines consisting of confocal ellipses and hyperbolas ([21, 4]). The value $\sqrt{y_2 - y_1}$ gives the distance of the foci from the coordinate origin.

2.4 Q-Operator

We use the results of the previous section to define an important integral operator, called a **Q-operator**, which depends on a positive parameter $g > 0$ and another parameter $z > 1$ and acts on any **symmetric polynomial** $p(x_1, x_2)$ of two variables as follows:

$$[Q_z p](y_1,y_2)=\frac{\Gamma(2g)}{\Gamma^2(g)}\,\frac{(z-1)^{1-2g}}{(y_2-y_1)y_1^{g-1}y_2^{g-1}}$$
$$\times\int_{y_2}^{\infty}dx_2\int_{y_1}^{y_2}dx_1\delta(x_1x_2-zy_1y_2)(x_2-x_1)\xi_1^{\alpha_1-1}\xi_2^{\alpha_2-1}p(x_1,x_2).\tag{2.24}$$

Notice that one should substitute into the above the values of the functions ξ_1 and ξ_2:

$$\xi_1=\frac{(x_1-y_1)(x_2-y_1)}{y_2-y_1},\qquad \xi_2=\frac{(x_1-y_2)(x_2-y_2)}{y_1-y_2}.\tag{2.25}$$

Taking into account Theorem 2.1, the above definition of the operator Q_z implies that it transforms the unit function into itself and provides a simple action on the monomials in ξ_j's:

$$Q_z[1]=1,\quad Q_z:\xi_1^{k_1}\xi_2^{k_2}\mapsto y_1^{k_1}y_2^{k_2}\,\frac{(z-1)^{k_1+k_2}(g)_{k_1}(g)_{k_2}}{(2g)_{k_1+k_2}},\tag{2.26}$$

for $k_j = 0, 1, 2, \ldots$

Introduce new variables η_j:

$$\xi_j = \eta_j y_j, \qquad j = 1, 2. \tag{2.27}$$

One can check that the x-, y- and η-variables are related by the following equalities (cf. (2.22)):

$$x_1 + x_2 = (1 + \eta_1)y_1 + (1 + \eta_2)y_2, \tag{2.28}$$

$$x_1 x_2 = (1 + \eta_1 + \eta_2)y_1 y_2. \tag{2.29}$$

The operator Q_z acts on the monomials in η_j's as follows

$$Q_z : \eta_1^{k_1}\eta_2^{k_2} \mapsto \frac{(z-1)^{k_1+k_2}(g)_{k_1}(g)_{k_2}}{(2g)_{k_1+k_2}}, \qquad k_j = 0, 1, 2, \ldots. \tag{2.30}$$

Formulae (2.28)–(2.29) and (2.30) determine how Q_z acts on any symmetric polynomial $p(x_1, x_2)$. This must be performed according to the algorithm:

- expand $p(x_1, x_2)$ in terms of elementary symmetric functions $e_1 = x_1 + x_2$ and $e_2 = x_1 x_2$,
- use substitution (2.28)–(2.29),
- replace all monomials in η_j's by numbers making use of (2.30).

The result clearly is

- a polynomial in y_1, y_2 and z,
- a rational function in g,
- a *symmetric* polynomial in y_1, y_2.

In what follows we adopt the following approach to the action of the operator Q_z, which seems natural because of the above: we shall look at it as a certain coordinate transformation (cf. (2.28)–(2.29)) from the variables x to the variables y, which depends on some parameters, the variables η, which in their turn are being specialized according to (2.30). The most interesting fact is that, recalling the analogy with the elliptic coordinates (cf. Remark 1), we are looking at the map sending the 'elliptic coordinates', x_j's, to the 'elliptic parameters', y_j's, while specializing the 'Cartesian variables', η_j's.

Regardless what viewpoint one chooses, this operator possesses many neat properties and, as a result, plays a fundamental role in the area of symmetric functions. The main motivation from now on will be to understand whether Q_z self-commutes,

$$Q_{z_1}Q_{z_2} = Q_{z_2}Q_{z_1}, \qquad \forall z_1, z_2 \in \mathbb{C}, \tag{2.31}$$

and therefore defines a **quantum integrable map**, that is whether there is a basis of good functions invariant under its action. The answer is yes and the basis is given by the Jack polynomials, which in fact can be uniquely characterized by this property.

3 Monomial and Elementary Symmetric Functions

3.1 Definitions

Let $\lambda = (\lambda_1, \lambda_2, \ldots, \lambda_n) \in \mathbb{N}^n$, $\lambda_1 \geq \lambda_2 \geq \ldots \geq \lambda_n \geq 0$, be a **partition** of a **weight** $|\lambda|$,

$$|\lambda| = \sum_{i=1}^{n} \lambda_i. \tag{3.1}$$

Hereafter, we shall use the weight notation $|\,.\,|$ as in (3.1) for any finite sum of indexed variables. The **dominance partial ordering** $\preceq$ for two partitions μ and λ is defined as follows:

$$\mu \preceq \lambda \quad \Longleftrightarrow \quad \left\{ |\mu| = |\lambda| \,; \quad \sum_{j=1}^{k} \mu_j \leq \sum_{j=1}^{k} \lambda_j \,, \quad k = 1, \ldots, n-1 \right\}. \tag{3.2}$$

Usually (cf. [22]), the **length** $l(\lambda)$ of a partition λ is the number of *non-zero* **parts** λ_i. We shall consider the partitions $\lambda = (\lambda_1, \lambda_2, \ldots, \lambda_n)$ that are finite sequences of exact length n, *including* any zero parts. For instance, $(6,6,2,0)$ and $(6,0,0,0)$ are both partitions of the length 4, while $(6,6,2)$ and (6) have lengths 3 and 1, respectively. For any partition $\lambda = (\lambda_1, \ldots, \lambda_n)$ of the length n, we shall often use the notation $\lambda_{i,j} = \lambda_i - \lambda_j$, in particular $\lambda_{i,i+1} = \lambda_i - \lambda_{i+1}$, $i = 1, \ldots, n$, $\lambda_{n+1} \equiv 0$.

Let $\mathbb{C}(a, b, c, \ldots)$ be the **field of rational functions** in indeterminates a, b, $c, \ldots$ over the field $\mathbb{C}$ of complex numbers, $\mathbb{C}(a, b, c, \ldots)[x, y, z, \ldots]$ be the **ring of polynomials** in the variables x, y, $z, \ldots$ with coefficients from $\mathbb{C}(a, b, c, \ldots)$, and $\mathbb{C}[x_1, \ldots, x_n]^{\mathrm{sym}}$ be the **subring of symmetric polynomials.**

We need two standard bases in the linear space of symmetric polynomials of n variables, $\mathbb{C}[\mathbf{x}]^{\mathrm{sym}}$, labelled by partitions λ. For each $\mathbf{a} = (a_1, \ldots, a_n) \in \mathbb{N}^n$ let us denote by $\mathbf{x}^{\mathbf{a}}$ the monomial

$$\mathbf{x}^{\mathbf{a}} = x_1^{a_1} \cdots x_n^{a_n}. \tag{3.3}$$

The **monomial symmetric functions** $m_\lambda(\mathbf{x})$ are defined by

$$m_\lambda(\mathbf{x}) = \sum \mathbf{x}^{\nu}, \tag{3.4}$$

where the sum is taken over all distinct permutations ν of λ. The $m_\lambda(\mathbf{x})$ form a basis in $\mathbb{C}[\mathbf{x}]^{\mathrm{sym}}$.

For each $r = 0, \ldots, n$ the rth **elementary symmetric function** $e_r(\mathbf{x})$ is the sum of all products of r distinct variables x_i, so that $e_0(\mathbf{x}) = 1$ and

$$e_r(\mathbf{x}) = \sum_{1 \leq i_1 < \ldots < i_r \leq n} x_{i_1} x_{i_2} \cdots x_{i_r} = m_{(1^r 0^{n-r})}(\mathbf{x}) \tag{3.5}$$

for $r = 1, \ldots, n$. Their generating function and explicit forms are

$$\prod_{j=1}^{n}(x+x_j)=\sum_{k=0}^{n}e_k(\mathbf{x})x^{n-k}, \tag{3.6}$$

$$e_1=x_1+\ldots, \quad e_2=x_1x_2+\ldots, \quad e_3=x_1x_2x_3+\ldots, \tag{3.7}$$

where '$+\ldots$' means adding corresponding permutations.

For each partition $\lambda=(\lambda_1,\ldots,\lambda_n)$ define the polynomials $E_\lambda(\mathbf{x})$ as

$$E_\lambda(\mathbf{x})=e_1^{\lambda_1-\lambda_2}(\mathbf{x})e_2^{\lambda_2-\lambda_3}(\mathbf{x})\cdots e_n^{\lambda_n}(\mathbf{x}). \tag{3.8}$$

The polynomials $E_\lambda(\mathbf{x})$ also form a basis in $\mathbb{C}[\mathbf{x}]^{\mathrm{sym}}$ and the transition matrix between E_λ and m_λ is triangular with respect to the dominance partial ordering (see [22, Chapter 1, Section 2]):

$$E_\lambda(\mathbf{x})=\sum_{\mu\preceq\lambda}a_{\lambda\mu}m_\mu(\mathbf{x}), \tag{3.9}$$

where $a_{\lambda\mu}\in\mathbb{N}$ and $a_{\lambda\lambda}=1$. For instance:

$$\begin{aligned}
&E_{000}=m_{000}, \quad E_{100}=m_{100}, \quad E_{110}=m_{110}, \quad E_{200}=m_{200}+2m_{110},\\
&E_{111}=m_{111}, \quad E_{210}=m_{210}+3m_{111}, \quad E_{300}=m_{300}+3m_{210}+6m_{111},\\
&E_{211}=m_{211}, \quad E_{220}=m_{220}+2m_{211}, \quad E_{221}=m_{221}, \quad E_{222}=m_{222}.
\end{aligned}$$

3.2 Factorizing Symmetric Polynomials $E_{\lambda_1,\lambda_2}(x_1,x_2)$

By definition

$$E_{\lambda_1,\lambda_2}(x_1,x_2)=(x_1+x_2)^{\lambda_{1,2}}(x_1x_2)^{\lambda_2}. \tag{3.10}$$

Consider the limit $g\to\infty$ in the construction of the Q-operator Q_z in Section 2.4, notably take this limit in the explicit action on the monomials in η_j's (2.30):

$$Q_z:\ \eta_1^{k_1}\eta_2^{k_2}\mapsto\left(\tfrac{z-1}{2}\right)^{k_1+k_2}. \tag{3.11}$$

It means that one can put $\eta_1=\eta_2=\frac{z-1}{2}$, so that in this case the action of the integral operator Q_z turns into a **pure change of coordinates** $(\mathbf{x})\to(\mathbf{y})$ defined by the equations:

$$x_1+x_2=\tfrac{z+1}{2}(y_1+y_2), \tag{3.12}$$

$$x_1x_2=zy_1y_2. \tag{3.13}$$

Obviously, the action of the Q-operator on the basis $E_{\lambda_1,\lambda_2}(x_1,x_2)$ is

$$Q_z:E_{\lambda_1,\lambda_2}(x_1,x_2)\mapsto\frac{E_{\lambda_1,\lambda_2}(z,1)}{E_{\lambda_1,\lambda_2}(1,1)}E_{\lambda_1,\lambda_2}(y_1,y_2). \tag{3.14}$$

This formula defines the operator Q_z for all $z\in\mathbb{C}$ on the space $\mathbb{C}[x_1,x_2]^{\mathrm{sym}}$. Moreover, the polynomials E_λ are its **eigenfunctions** with the **eigenvalues** $q_\lambda(z)$:

$$Q_z E_\lambda = q_\lambda(z) E_\lambda, \qquad q_\lambda(z) = \frac{E_{\lambda_1,\lambda_2}(z,1)}{E_{\lambda_1,\lambda_2}(1,1)} = \left(\frac{z+1}{2}\right)^{\lambda_{1,2}} z^{\lambda_2}. \tag{3.15}$$

The symmetric polynomials $E_{\lambda_1,\lambda_2}(x_1,x_2)$ can be defined as being the eigenfunctions,

$$D_1 E_\lambda = \lambda_{1,2} E_\lambda, \qquad D_2 E_\lambda = \lambda_2 E_\lambda, \tag{3.16}$$

of the **commuting differential operators** D_1 and D_2:

$$D_1 = \frac{x_1+x_2}{x_1-x_2}\left(x_1\frac{\partial}{\partial x_1} - x_2\frac{\partial}{\partial x_2}\right), \quad D_2 = -\frac{x_1 x_2}{x_1-x_2}\left(\frac{\partial}{\partial x_1} - \frac{\partial}{\partial x_2}\right), \tag{3.17}$$

$$[D_1, D_2] = 0 \tag{3.18}$$

Exercise 6. Check (3.16) and (3.18)

Therefore, the Q-operator commutes with the operators D_1 and D_2 as well as self-commutes:

$$[Q_z, D_j] = 0, \qquad [Q_{z_1}, Q_{z_2}] = 0, \qquad j = 1,2, \quad z, z_1, z_2 \in \mathbb{C}. \tag{3.19}$$

Setting $y_2 = 1$ and renaming y_1 and z into z_1 and z_2 one observes from (3.12)–(3.14) that the transform $\mathcal{S}_2 : (x_1,x_2) \to (z_1,z_2)$:

$$x_1 + x_2 = \tfrac{1}{2}(z_1+1)(z_2+1), \tag{3.20}$$

$$x_1 x_2 = z_1 z_2, \tag{3.21}$$

factorizes the basis functions into a symmetric product of two polynomials each in one variable:

$$\mathcal{S}_2 : E_\lambda(x_1,x_2) \mapsto E_\lambda(1,1)\,\frac{E_{\lambda_1,\lambda_2}(z_1,1)}{E_{\lambda_1,\lambda_2}(1,1)}\,\frac{E_\lambda(z_2,1)}{E_\lambda(1,1)}\,. \tag{3.22}$$

The inverse transform, $\mathcal{S}_2^{-1} : (z_1,z_2) \to (x_1,x_2)$:

$$z_1 + z_2 = 2(x_1+x_2) - x_1 x_2 - 1, \tag{3.23}$$

$$z_1 z_2 = x_1 x_2, \tag{3.24}$$

provides a representation for the basis functions in terms of the polynomials of one variable:

$$\mathcal{S}_2^{-1} : E_\lambda(1,1)\,\frac{E_{\lambda_1,\lambda_2}(z_1,1)}{E_{\lambda_1,\lambda_2}(1,1)}\,\frac{E_\lambda(z_2,1)}{E_\lambda(1,1)} \mapsto E_\lambda(x_1,x_2). \tag{3.25}$$

Such maps, $\mathcal{S}_2$ and $\mathcal{S}_2^{-1}$, we call the **separating and inverse separating maps**, respectively.

3.3 Factorizing the Basis $E_\lambda(x_1, \ldots, x_n)$

Let us study the change of coordinates $(\mathbf{x}) \leftrightarrow (\mathbf{e})$, defined by the elementary symmetric functions:

$$\prod_{j=1}^{n}(x + x_j) = x^n + \sum_{k=1}^{n} e_k x^{n-k}, \qquad x \in \mathbb{C}. \tag{3.26}$$

Taking the differential of the both sides and then dividing by the product one obtains

$$\sum_{j=1}^{n} \frac{dx_j}{x + x_j} = \sum_{k=1}^{n} \frac{x^{n-k}}{\prod_{m=1}^{n}(x + x_m)} de_k. \tag{3.27}$$

Comparing the residues at the poles one gets

$$dx_j = \sum_{k=1}^{n} \frac{(-x_j)^{n-k}}{\prod_{m\neq j}^{n}(x_m - x_j)} de_k, \qquad j = 1, \ldots, n, \tag{3.28}$$

and therefore

$$\frac{\partial}{\partial e_k} = \sum_{j=1}^{n} \frac{(-x_j)^{n-k}}{\prod_{m\neq j}^{n}(x_m - x_j)} \frac{\partial}{\partial x_j}, \qquad k = 1, \ldots, n. \tag{3.29}$$

Clearly, the operators $e_k \frac{\partial}{\partial e_k}$ are diagonal on $E_\lambda = \prod_{j=1}^{n} e_j^{\lambda_{j,j+1}}$:

$$e_k \frac{\partial}{\partial e_k} E_\lambda = \lambda_{k,k+1} E_\lambda, \qquad k = 1, \ldots, n. \tag{3.30}$$

Hence, the operators

$$D_k = \mathrm{e}_k(\mathbf{x}) \sum_{j=1}^{n} \frac{(-x_j)^{n-k}}{\prod_{m\neq j}^{n}(x_m - x_j)} \frac{\partial}{\partial x_j}, \qquad k = 1, \ldots, n, \tag{3.31}$$

mutually commute and have the functions $E_\lambda(\mathbf{x})$ as their eigenfunctions:

$$[D_j, D_k] = 0, \qquad D_k E_\lambda(\mathbf{x}) = \lambda_{k,k+1} E_\lambda(\mathbf{x}), \qquad k = 1, \ldots, n. \tag{3.32}$$

The change of coordinates $\mathcal{S}_n^{(0)} : (\mathbf{x}) \mapsto (\mathbf{e})$ is, by the above, a separating map for a **quantum integrable system** indexquantum integrable systemdefined by the **complete set of commuting operators** D_k[2]. It factorizes the corresponding eigenfunctions:

[2] completeness is understood here in the sense that the corresponding eigenfunctions form a basis in a Hilbert space, so that the set of commuting operators provides enough quantum numbers (parts λ_j) to label every eigenstate

$$\mathcal{S}_n^{(0)} : E_\lambda(\mathbf{x}) \mapsto \prod_{j=1}^{n} e_j^{\lambda_{j,j+1}}. \tag{3.33}$$

Under $\mathcal{S}_n^{(0)}$ the operators D_k are mapped into simpler operators

$$\mathcal{S}_n^{(0)} D_k = e_k \frac{\partial}{\partial e_k} \mathcal{S}_n^{(0)}, \qquad k = 1, \ldots, n, \tag{3.34}$$

and the initial multivariable spectral problem (3.32) is mapped into a set of one-variable **separation equations** for the factorized functions:

$$e_k \frac{\partial}{\partial e_k} e_k^{\lambda_{k,k+1}} = \lambda_{k,k+1} e_k^{\lambda_{k,k+1}}, \qquad k = 1, \ldots, n. \tag{3.35}$$

The main problem for any quantum integrable system given by a complete set of commuting operators $\{D_j\}_{j=1}^n$ is to explicitly construct the corresponding separating map which would factorize the multivariable eigenfunctions into functions of one variable. This is generally a rather complicated problem, but there are various approaches to tackle it, e.g., by making use of the Q-operator.

We shall now describe another purely coordinate separating map $\mathcal{S}_n$ for the basis $E_\lambda(\mathbf{x})$ which is similar to the map $\mathcal{S}_2$ from the previous subsection and which intrinsically comes from a n-variable generalization of the Q-operator.

The n-variable operator Q_z is defined as the following change of coordinates $(\mathbf{x}) \to (\mathbf{y})$ (cf (3.12)–(3.13)):

$$e_k(\mathbf{x}) = q_k(z) e_k(\mathbf{y}),$$
$$q_k(z) \equiv \frac{e_k(z, 1, \ldots, 1)}{e_k(1, \ldots, 1)} = 1 + \frac{(z-1)k}{n}, \qquad k = 1, \ldots, n. \tag{3.36}$$

Exercise 7. Show that

$$e_k(1, \ldots, 1) = \binom{n}{k}, \qquad q_k(z) = 1 + \frac{(z-1)k}{n}, \qquad k = 1, \ldots, n. \tag{3.37}$$

Obviously, Q_z has the diagonal action

$$Q_z : E_\lambda(\mathbf{x}) \mapsto q_\lambda(z) E_\lambda(\mathbf{y}),$$
$$q_\lambda(z) = \prod_{k=1}^{n} q_k^{\lambda_{k,k+1}} = \prod_{k=1}^{n} \left(1 + \tfrac{(z-1)k}{n}\right)^{\lambda_{k,k+1}}. \tag{3.38}$$

Iterate Q_z n times and restrict the out-variables by the evaluation homomorphism $\varrho_0 : f(y_1, \ldots, y_n) \mapsto f(1, \ldots, 1)$. In this way one obtains the separating operator $\mathcal{S}_n = \varrho_0 Q_{z_1} \cdots Q_{z_n}$ as the following change of coordinates $(\mathbf{x}) \to (\mathbf{z})$:

$$e_k(\mathbf{x}) = \binom{n}{k} \prod_{j=1}^{n} q_k(z_j), \qquad k = 1, \ldots, n. \tag{3.39}$$

Its action on the basis $E_\lambda(\mathbf{x})$ is as follows:

$$\mathcal{S}_n : E_\lambda(\mathbf{x}) \mapsto c_\lambda \prod_{j=1}^{n} q_\lambda(z_j), \qquad c_\lambda = \prod_{k=1}^{n} \binom{n}{k}^{\lambda_{k,k+1}}. \tag{3.40}$$

The transform $\mathcal{S}_n$ (3.39) can be rewritten in the form

$$e_k(\mathbf{x}) = \binom{n}{k} \left(\frac{k}{n}\right)^n \sum_{m=0}^{n} e_m(\mathbf{z}) \left(\frac{n-k}{k}\right)^{n-m}, \qquad k = 1, \ldots, n. \tag{3.41}$$

The inverse form of this change is complicated in general, e.g. for $n = 3$ one has

$$e_1(\mathbf{z}) = e_3(\mathbf{x}) - \tfrac{3}{2}\, e_2(\mathbf{x}) + 3e_1(\mathbf{x}) - \tfrac{5}{2}, \tag{3.42}$$

$$e_2(\mathbf{z}) = -\tfrac{5}{2}\, e_3(\mathbf{x}) + 3e_2(\mathbf{x}) - \tfrac{3}{2}\, e_1(\mathbf{x}) + 1, \tag{3.43}$$

$$e_3(\mathbf{z}) = e_3(\mathbf{x}). \tag{3.44}$$

The transform $\mathcal{S}_n$ (3.39) maps the differential operators D_k (3.31) in x-variables into the corresponding differential operators D_k^{sep} in the **separation variables** $\{z_j\}_{j=1}^n$:

$$\mathcal{S}_n D_k(\mathbf{x}) = D_k^{\text{sep}}(\mathbf{z}) \mathcal{S}_n. \tag{3.45}$$

The operators $D_k^{\text{sep}}(\mathbf{z})$ are in a special **separated form**, that is they admit the **separation of variables**. At the same time, the separating transform $\mathcal{S}_n$ (3.39), which factorizes their common eigenfunctions (3.40), sends the **original n-variable spectral problem** for the commuting operators $D_k(\mathbf{x})$,

$$D_k(\mathbf{x}) E_\lambda(\mathbf{x}) = \lambda_{k,k+1} E_\lambda(\mathbf{x}), \qquad k = 1, \ldots, n, \tag{3.46}$$

into a **multi-parameter spectral problem** [1, 24, 26] for the separated polynomials $q_\lambda(z_k)$:

$$\frac{dq_\lambda(z_k)}{dz_k} = \sum_{j=1}^{n} \frac{\lambda_{j,j+1}}{z_k + \frac{n-j}{j}}\, q_\lambda(z_k), \qquad k = 1, \ldots, n. \tag{3.47}$$

Let us derive the explicit form for the operators D_k^{sep}. Apply logarithmic differential to both sides of the defining equations (3.39) for the change of variables $(\mathbf{x}) \to (\mathbf{z})$ and get

$$\frac{de_j}{e_j} = \sum_{k=1}^{n} \frac{dz_k}{z_k + \frac{n-j}{j}}, \qquad j = 1, \ldots, n. \tag{3.48}$$

Therefore,

$$\frac{\partial}{\partial z_k} = \sum_{j=1}^{n} \frac{1}{z_k + \frac{n-j}{j}} D_j^{\text{sep}}(\mathbf{z}), \qquad k = 1, \ldots, n, \tag{3.49}$$

which is just the operator form of the separation equations (3.47). In order to derive the operators $D_j^{\text{sep}}(\mathbf{z})$, one must invert the above system of linear simultaneous equations.

Let us solve a more general problem formulated as follows. Suppose one has two sequences of distinct variables, $(z_1, \ldots, z_n)$ and $(w_1, \ldots, w_n)$, and the $n \times n$ **Cauchy matrix** A:

$$A_{jk} = \frac{1}{z_j - w_k}, \qquad j, k = 1, \ldots, n. \tag{3.50}$$

What is the inverse matrix A^{-1}?

Write down the following equality coming from the partial fraction decomposition of the left hand side:

$$\frac{\prod_{l=1}^{n}(z - z_l)}{(z - z_p)\prod_{l=1}^{n}(z - w_l)} = \sum_{j=1}^{n} \frac{1}{z - w_j} \frac{\prod_{l\neq p}(w_j - z_l)}{\prod_{l\neq j}(w_j - w_l)}, \quad p = 1, \ldots, n. \tag{3.51}$$

Substituting $u = z_q$, $q = 1, \ldots, n$, one derives

$$\delta_{qp} = \sum_{j=1}^{n} \frac{1}{z_q - w_j} \frac{\prod_{l=1}^{n}(z_p - w_l)}{\prod_{l\neq j}(w_j - w_l)} \prod_{l\neq p} \frac{w_j - z_l}{z_p - z_l}. \tag{3.52}$$

Hence, the matrix elements of the inverse of the Cauchy matrix are:

$$\left(A^{-1}\right)_{jk} = \frac{\prod_{l=1}^{n}(z_k - w_l)}{\prod_{l\neq j}(w_j - w_l)} \prod_{l\neq k} \frac{w_j - z_l}{z_k - z_l} \qquad j, k = 1, \ldots, n. \tag{3.53}$$

Applying this result to invert the problem (3.49) one derives the explicit representation for the n commuting operators $D_j^{\text{sep}}(\mathbf{z})$:

$$D_j^{\text{sep}}(\mathbf{z}) = \sum_{k=1}^{n} \frac{n^n}{j\prod_{l\neq j}(l - j)} \left(\prod_{l=1}^{n} q_l(z_k)\right) \left(\prod_{l\neq k} \frac{q_j(z_l)}{z_k - z_l}\right) \frac{\partial}{\partial z_k}. \tag{3.54}$$

Exercise 8. For $n = 2$, start from the two commuting operators $D_j^{\text{sep}}(z_1, z_2)$ (3.54) and separate the variables, thereby deriving the separation equations (3.47).

3.4 Liouville Integrable Systems

The quantum integrable system we have studied has a quasi-classical limit in which it turns into a **Liouville integrable system**, defined by the n **functionally independent** and **Poisson-commuting integrals of motion**:

$$d_k = e_k(\mathbf{x}) \sum_{j=1}^{n} \frac{(-x_j)^{n-k}}{\prod_{m\neq j}^{n}(x_m - x_j)}\, p_{x_k}, \qquad \{d_j, d_k\} = 0, \quad j,k = 1,\ldots,n, \tag{3.55}$$

where the canonical coordinates, x_j, and the conjugated momenta, p_j, have the standard Poisson brackets of the Darboux variables:

$$\{x_j, x_k\} = \{p_{x_j}, p_{x_k}\} = 0, \qquad \{p_{x_j}, x_k\} = \delta_{jk}, \quad j,k = 1,\ldots,n. \tag{3.56}$$

The same separating transform $\mathcal{S}_n$ defines now a separating canonical transformation $(p_{x_1},\ldots,p_{x_n};\mathbf{x}) \to (p_{z_1},\ldots,p_{z_n};\mathbf{z})$ such that the new (separation) variables satisfy the separation equations:

$$p_{z_k} = \sum_{j=1}^{n} \frac{d_j}{z_k + \frac{n-j}{j}}, \qquad k = 1,\ldots,n. \tag{3.57}$$

For the problem in question there is very little difference between separating the quantum or the classical system. In both cases it is the same purely coordinate[3] change of variables. In the classical case it amounts to a special canonical transformation which brings the system into a separated form when each new conjugated momentum depends only on its own coordinate. In the quantum case the map factorizes the eigenfunctions into some (separated) functions of one variable, thereby again reducing the n-variable problem to a set of one-variable problems, connected by the common eigenvalues of the commuting integrals.

Exercise 9. For $n = 2$ separate variables for the Liouville integrable system with the integrals d_1 and d_2 (3.55) and derive the corresponding separation equations (3.57).

3.5 Separation of Variables

The interested reader is guided to the book [7] and the papers [12, 13, 8, 5] for more information on separation of variables in the Hamilton-Jacobi equation and in the Laplace operator on Riemannian spaces of constant curvature. The last paper explains how to add separating potentials.

All these references though are about separation in curvilinear **orthogonal coordinate systems**. There are only few classification results about separability in **non-orthogonal coordinate systems**. Both separations, $\mathcal{S}_n^{(0)}$ and $\mathcal{S}_n$, we have considered are non-orthogonal. Moreover, in the next sections we encounter separations which are given by specific local operators and even by **integral operators** and corresponding canonical transformations (cf [17, 18, 19, 20, 15, 16]).

[3] it does not depend on the conjugated momenta

3.6 Factorizing Symmetric Monomials $m_{\lambda_1,\lambda_2}(x_1, x_2)$

Let us now come back to the Q-operator of Section 2.4 whose action on any symmetric polynomial in the variables x_1 and x_2 is defined by the following rules:

$$x_1 + x_2 = (1+\eta_1)y_1 + (1+\eta_2)y_2, \tag{3.58}$$

$$x_1 x_2 = (1+\eta_1+\eta_2)y_1 y_2. \tag{3.59}$$

$$Q_z \; : \; \eta_1^{k_1}\eta_2^{k_2} \mapsto \frac{(z-1)^{k_1+k_2}(g)_{k_1}(g)_{k_2}}{(2g)_{k_1+k_2}}, \qquad k_j = 0,1,2,\ldots. \tag{3.60}$$

Consider the case of $g = 0$. One immediately notices that the Q-operator annihilates most of the η-monomials:

$$Q_z[1] = 1, \qquad Q_z : \eta_j^k \mapsto \tfrac{(z-1)^k}{2}, \quad j = 1,2, \quad k = 1,2,\ldots, \tag{3.61}$$

$$Q_z \; : \; \eta_1^{k_1}\eta_2^{k_2} \mapsto 0 \qquad \text{if} \quad k_1 k_2 \neq 0. \tag{3.62}$$

Because of this property, the second substitution (3.59) can be replaced by $x_1 x_2 = (1+\eta_1)\, y_1(1+\eta_2)y_2$, thereby the whole change of variables involved in the action of the Q-operator becomes as simple as this

$$x_1 = (1+\eta_1)y_1, \qquad x_2 = (1+\eta_2)y_2. \tag{3.63}$$

A straightforward calculation gives

$$Q_z : (1+\eta_1)^{k_1}(1+\eta_2)^{k_2} \mapsto \tfrac{1}{2}\left(z^{k_1} + z^{k_2}\right), \qquad k_1, k_2 = 0,1,2,\ldots. \tag{3.64}$$

Therefore, if one identifies the out-variables, $y_{1,2}$, with the in-variables, $x_{1,2}$, one can realize the operator Q_z as the following (local) shift operator:

$$Q_z : f(x_1,x_2) \mapsto \tfrac{1}{2}\left(f(zx_1,x_2) + f(x_1,zx_2)\right), \quad \forall f \in \mathbb{C}[x_1,x_2]^{\rm sym}. \tag{3.65}$$

The monomial symmetric functions $m_{\lambda_1,\lambda_2}(x_1,x_2)$ are defined as follows:

$$m_{\lambda_1,\lambda_2}(x_1,x_2) = \begin{cases} x_1^{\lambda_1}x_2^{\lambda_2} + x_2^{\lambda_1}x_1^{\lambda_2}, & \lambda_1 \neq \lambda_2, \\ x_1^{\lambda_1}x_2^{\lambda_1}, & \lambda_1 = \lambda_2. \end{cases} \tag{3.66}$$

It is easy to check that the action of the Q-operator Q_z (3.65) on this basis is diagonal

$$Q_z : m_\lambda(x_1,x_2) \mapsto q_\lambda(z)\, m_\lambda(x_1,x_2), \qquad q_\lambda(z) = \tfrac{1}{2}\left(z^{\lambda_1} + z^{\lambda_2}\right). \tag{3.67}$$

Exercise 10. Prove that

$$q_\lambda(z) = \frac{m_{\lambda_1,\lambda_2}(z,1)}{m_{\lambda_1,\lambda_2}(1,1)}. \tag{3.68}$$

As before, the separating operator $\mathcal{S}_2$ is constructed as the restriction, by the evaluation homomorphism ϱ_0, of the product of two Q-operators:

$$\mathcal{S}_2 = \varrho_0 Q_{z_1} Q_{z_2}, \qquad \mathcal{S}_2 : m_\lambda(x_1, x_2) \mapsto m_\lambda(1,1) q_\lambda(x_1) q_\lambda(x_2), \tag{3.69}$$

where we have again identified the out- with the in-variables. The local operator $\mathcal{S}_2$ has the following explicit action:

$$(\mathcal{S}_2 f)(x_1, x_2) = \tfrac{1}{2}\left(f(x_1, x_2) + f(x_1 x_2, 1)\right), \qquad \forall f \in \mathbb{C}[x_1, x_2]^{\mathrm{sym}}, \tag{3.70}$$

which we prefer to rewrite in terms of the operator P_{12},

$$\mathcal{S}_2 = \tfrac{1}{2}\left(1 + P_{12}\right), \quad (P_{12} f)(x_1, x_2) = f(x_1 x_2, 1), \qquad f \in \mathbb{C}[x_1, x_2]^{\mathrm{sym}}. \tag{3.71}$$

Now the inverse separating map reads

$$\mathcal{S}_2^{-1} = 2 - P_{12} : m_\lambda(1,1) q_\lambda(x_1) q_\lambda(x_2) \mapsto m_\lambda(x_1, x_2). \tag{3.72}$$

Exercise 11. Prove that the operator P_{12} is a projector, $P_{12}^2 = P_{12}$, and then prove (3.72).

3.7 Factorizing the Basis $m_\lambda(x_1, \dots, x_n)$

Let us recall that the basis $m_\lambda(x_1, \dots, x_n)$ is defined as the sum over all distinct permutations $\nu = (\nu_1, \dots, \nu_n)$ of the partition $\lambda = (\lambda_1, \dots, \lambda_n)$

$$m_\lambda(\mathbf{x}) = \sum x_1^{\nu_1} \cdots x_n^{\nu_n}. \tag{3.73}$$

It can be characterized as the eigenbasis of the complete set of symmetric differential operators:

$$D_j(\mathbf{x}) = e_j\left(x_1 \frac{\partial}{\partial x_1}, \dots, x_n \frac{\partial}{\partial x_n}\right), \qquad [D_j, D_k] = 0, \quad j, k = 1, \dots, n, \tag{3.74}$$

$$D_j m_\lambda = e_j(\lambda) m_\lambda. \tag{3.75}$$

The Q-operator is defined as follows:

$$(Q_z f)(\mathbf{x}) = \frac{1}{n} \sum_{j=1}^{n} f(x_1, \dots, x_{j-1}, z x_j, x_{j+1}, \dots, x_n), \quad f \in \mathbb{C}[\mathbf{x}]^{\mathrm{sym}}. \tag{3.76}$$

This clearly gives a symmetric operator which can be simply represented in the operator form with the diagonal action on the monomial symmetric functions:

$$Q_z = \frac{1}{n} \sum_{j=1}^{n} z^{x_j \frac{\partial}{\partial x_j}}, \qquad Q_z m_\lambda = q_\lambda(z) m_\lambda, \qquad q_\lambda(z) = \frac{1}{n} \sum_{j=1}^{n} z^{\lambda_j}. \tag{3.77}$$

Exercise 12. Prove that

$$q_\lambda(z) = \frac{m_\lambda(z,1,\dots,1)}{m_\lambda(1,\dots,1)}. \tag{3.78}$$

The separating operator $\mathcal{S}_n$ is, by construction, given by the formula

$$\mathcal{S}_n = \varrho_0 Q_{z_1} \cdots Q_{z_n}, \tag{3.79}$$

where ϱ_0 is the evaluation homomorphism, $\varrho_0 : f(x_1,\dots,x_n) \mapsto f(1,\dots,1)$. Introduce a string of similar homomorphisms:

$$\varrho_j : f(x_1,\dots,x_n) \mapsto f(x_1,\dots,x_j,1,\dots,1), \qquad j = 0,\dots,n, \tag{3.80}$$

with $\varrho_n \equiv 1$. It is easy to check that the separating operator (3.79) can be factorized into a product of simpler operators $\mathcal{A}_k$, $k = 1,\dots,n$:

$$\mathcal{S}_n = \mathcal{A}_1 \cdots \mathcal{A}_n, \qquad \mathcal{A}_k = \frac{1}{n}\left(n - k + 1 + \sum_{j=1}^{k-1} P_{jk}\right), \qquad \mathcal{A}_n \equiv 1, \tag{3.81}$$

where the operators P_{jk}, $j < k$, are similar to the operator P_{12} from the previous section:

$$(P_{jk}f)(x_1,\dots,x_n) = f(\dots,\overset{j\text{th}}{x_j x_k},\dots,\overset{k\text{th}}{1},\dots). \tag{3.82}$$

Indeed, the required factorization follows from the easily verified identities which arise from pulling ϱ_0 through the string of Q's in the definition (3.79):

$$\varrho_{k-1} Q_{z_k} = \mathcal{A}_k \varrho_k, \qquad k = 1,\dots,n. \tag{3.83}$$

Notice that in the formula (3.81) above we have identified the out, z-variables, with the corresponding in, x-variables, to obtain local operators.

For instance, for $n = 3$ the separating operator $\mathcal{S}_3 = \mathcal{A}_2\mathcal{A}_3$ is the following operator:

$$\begin{aligned}(\mathcal{S}_3 f)(x_1,x_2,x_3) = \tfrac{1}{9}\,&(2f(x_1,x_2,x_3) + 2f(x_1x_3,x_2,1) + 2f(x_1x_2,1,x_3)\\ &+ 2f(x_1,x_2x_3,1) + f(x_1x_2x_3,1,1)),\end{aligned} \tag{3.84}$$

for $f \in \mathbb{C}[\mathbf{x}]^{\mathrm{sym}}$. As an example, its factors move the polynomial m_{300} as follows:

$$\mathcal{A}_3 : x_1^3 + x_2^3 + x_3^3 \mapsto \tfrac{1}{3}\left(2 + x_3^3\right)\left(x_1^3 + x_2^3 + 1\right), \tag{3.85}$$

$$\mathcal{A}_2 : \tfrac{1}{3}\left(2 + x_3^3\right)\left(x_1^3 + x_2^3 + 1\right) \mapsto \tfrac{1}{9}\left(2 + x_1^3\right)\left(2 + x_2^3\right)\left(2 + x_3^3\right). \tag{3.86}$$

The general map $\mathcal{S}_n$, by construction, factorizes the basis m_λ,

$$\mathcal{S}_n = \mathcal{A}_2 \cdots \mathcal{A}_n : \; m_\lambda(\mathbf{x}) \mapsto m_\lambda(1,\dots,1) \prod_{k=1}^{n} q_\lambda(x_k). \tag{3.87}$$

Exercise 13. Show that

$$\mathcal{A}_k : m_\lambda(x_1,\ldots,x_k,1,\ldots,1) \mapsto m_\lambda(x_1,\ldots,x_{k-1},1,1,\ldots,1)q_\lambda(x_k). \quad (3.88)$$

The above action means that the operators $\mathcal{A}_k$ possess the **dimension reduction** property. They successively factorize out separation variables one-by-one and each time those with a smaller index act on a smaller number of variables.

Exercise 14. Show that

$$P_{jk}P_{lk} = P_{jk}, \qquad j,l = 1,\ldots,k-1, \quad (3.89)$$

and use it to prove the formula for the inverse operators,

$$\mathcal{A}_k^{-1} = \frac{1}{n-k+1}\left(n - \sum_{j=1}^{k-1} P_{jk}\right), \qquad k = 1,\ldots,n. \quad (3.90)$$

As a result, one derives an explicit representation for the symmetric monomial functions in terms of the separated polynomials $q_\lambda(x)$:

$$\mathcal{S}_n^{-1} = \mathcal{A}_n^{-1}\cdots\mathcal{A}_2^{-1} : \ m_\lambda(1,\ldots,1)\prod_{k=1}^{n} q_\lambda(x_k) \mapsto m_\lambda(\mathbf{x}). \quad (3.91)$$

The observed factorization of the separating and inverse separating maps and the corresponding structure of the dimension reduction action of the factor-operators has a general name **factorized separation chain**. It was introduced in [14] for a wide class of integrable systems with a 2×2 Lax matrix.

References

1. F. V. Atkinson, *Multiparameter Eigenvalue Problems: Matrices and Compact Operators*, Academic Press, New York and London, 1972.
2. P.L. Chebyshev: `http://www-groups.dcs.st-and.ac.uk/~history/Mathematicians/Chebyshev.html`
3. P.L. Chebyshev, *Collected Works*, 2, Moscow-Leningrad (1947), pp. 23–51 (In Russian).
4. G. Darboux, *Leçons sur la théorie générale des surfaces et ses applications géométriques du calcul infinitésimal*, 1, Gauthier-Villars (1887), pp. 1–18.
5. J.C. Eilbeck, V.Z. Enol'skii, V.B. Kuznetsov and A.V. Tsiganov, *Linear r-matrix algebra for classical separable systems*, J. Phys. A: Math. Gen. **27** (1994), 567–578.
   ArXiv: `http://arxiv.org/abs/hep-th/9306155`

6. L.B. Gegenbauer: `http://www-groups.dcs.st-and.ac.uk/~history/Mathematicians/Gegenbauer.html`
7. E. G. Kalnins, *Separation of Variables for Riemannian Spaces of Constant Curvature*, Longman Scientific & Technical, Harlow, 1986.
8. E.G. Kalnins, V.B. Kuznetsov and W. Miller, Jr, *Quadrics on complex Riemannian spaces of constant curvature, separation of variables and the Gaudin magnet*, J. Math. Phys. **35** (1994), 1710–1731.
ArXiv: `http://arxiv.org/abs/hep-th/9308109`
9. R. Koekoek and R.F. Swarttouw, *The Askey-scheme of hypergeometric orthogonal polynomials and its q-analogue*, Delft University of Technology, Faculty of Information Technology and Systems, Department of Technical Mathematics and Informatics, Report no. 98–17, 1998.
`http://aw.twi.tudelft.nl/~koekoek/askey/index.html`
10. T.H. Koornwinder, *Compact quantum groups and q-special functions*, in: Representations of Lie groups and quantum groups, V. Baldoni & M. A. Picardello (eds.), Pitman Research Notes in Mathematics Series 311, Longman Scientific & Technical, 1994, pp. 46–128.
11. T.H. Koornwinder, *q-Special functions, a tutorial*, Pitman Research Notes in Mathematics Series **311**, Longman Scientific & Technical, Harlow, 1994, pp. 46–128.
ArXiv: `http://arxiv.org/abs/math.CA/9403216`
12. V. B. Kuznetsov, *Quadrics on real Riemannian spaces of constant curvature: separation of variables and connection with Gaudin magnet.* J. Math. Phys. **33** (9) (1992), 3240–3254.
13. V. B. Kuznetsov, *Equivalence of two graphical calculi.* J. Phys. A **25** (22) (1992), 6005–6026.
14. V.B. Kuznetsov, *Inverse problem for $sl(2)$ lattices*, In: Symmetry and Perturbation Theory, World Scientific, Singapore (2003); Proceedings of the International Conference SPT 2002, Cala Gonone, Sardinia, 19–26 May 2002 (S. Abenda, G. Gaeta, S. Walcher, eds.); pp. 136–152.
Arxiv: `http://arxiv.org/abs/nlin.SI/0207025`
15. V.B. Kuznetsov, V.V. Mangazeev and E.K. Sklyanin, *Q-operator and factorized separation chain for Jack polynomials.* Indag. Mathem. **14** (3,4) (2003), 451–482.
ArXiv: `http://arxiv.org/abs/math.CA/0306242`
16. V.B. Kuznetsov, F.W. Nijhoff and E.K. Sklyanin, *Separation of variables for the Ruijsenaars system*, Commun. Math. Phys. **189** (1997), 855–877.
ArXiv: `http://arxiv.org/abs/solv-int/9701004`
17. V.B. Kuznetsov and E.K. Sklyanin, *Separation of variables in the A_2 type Jack polynomials,* RIMS Kokyuroku, **919** (1995), 27–34.
ArXiv: `http://arxiv.org/abs/solv-int/9508002`
18. V.B. Kuznetsov and E.K. Sklyanin, *Separation of variables for the A_2 Ruijsenaars model and a new integral representation for the A_2 Macdonald polynomials*, J. Phys. A: Math. Gen. **29** (1996), 2779–2804.
ArXiv: `http://arxiv.org/abs/q-alg/9602023`
19. V.B. Kuznetsov and E.K. Sklyanin, *Factorization of Macdonald polynomials*, In: Symmetries and integrability of difference equations (Canterbury, 1996), pp. 370–384, London Math. Soc. Lecture Note Ser. **255**, Cambridge Univ. Press, Cambridge, 1999.
ArXiv: `http://arxiv.org/abs/q-alg/9703013`

20. V.B. Kuznetsov and E.K. Sklyanin, *Separation of variables and integral relations for special functions*, The Ramanujan J. **3** (1999), 5–35. ArXiV: `http://arxiv.org/abs/q-alg/9705006`
21. G. Lamé, *Leçons sur les coordonnées curvilignés et leurs diverses applications*, Paris, 1859.
22. I.G. Macdonald, *Symmetric Functions and Hall Polynomials*, Oxford Univ. Press, Oxford, Second Edition, 1995.
23. R. Roy, *The work of Chebyshev on orthogonal polynomials*, in Topics in polynomials of one and several variables and their applications, River Edge, NJ, 1993, pp. 495–512.
24. B. D. Sleeman, *Multiparameter Spectral Theory in Hilbert Space*, Pitman Research Notes in Mathematics **22**, 1978.
25. G. Szegő, *Orthogonal Polynomials*, AMS, Colloquium Publications **23**, Providence RI, 4th ed., 1975.
26. H. Volkmer, *Multiparameter Eigenvalue Problems and Expansion Theorems*, Springer Lecture Notes in Mathematics **1356** (1988).

An Algebraic Approach to the Askey Scheme of Orthogonal Polynomials

Paul Terwilliger

Department of Mathematics, University of Wisconsin,
Madison, Wisconsin 53706, USA
e-mail: terwilli@math.wisc.edu

Summary. Let $\mathbb{K}$ denote a field, and let V denote a vector space over $\mathbb{K}$ with finite positive dimension. We consider a pair of linear transformations $A : V \to V$ and $A^* : V \to V$ that satisfy the following two conditions:

1. There exists a basis for V with respect to which the matrix representing A is irreducible tridiagonal and the matrix representing A^* is diagonal.
2. There exists a basis for V with respect to which the matrix representing A^* is irreducible tridiagonal and the matrix representing A is diagonal.

We call such a pair a *Leonard pair* on V. We give a correspondence between Leonard pairs and a class of orthogonal polynomials. This class coincides with the terminating branch of the Askey scheme and consists of the q-Racah, q-Hahn, dual q-Hahn, q-Krawtchouk, dual q-Krawtchouk, quantum q-Krawtchouk, affine q-Krawtchouk, Racah, Hahn, dual Hahn, Krawtchouk, Bannai/Ito, and orphan polynomials. We describe the above correspondence in detail. We show how, for the listed polynomials, the 3-term recurrence, difference equation, Askey-Wilson duality, and orthogonality can be expressed in a uniform and attractive manner using the corresponding Leonard pair. We give some examples that indicate how Leonard pairs arise in representation theory and algebraic combinatorics. We discuss a mild generalization of a Leonard pair called a tridiagonal pair. At the end we list some open problems. Throughout these notes our argument is elementary and uses only linear algebra. No prior exposure to the topic is assumed.

1 Leonard Pairs and Leonard Systems

1.1 Leonard Pairs

These notes are based on the papers [45], [46], [47], [79], [80], [81], [82], [83], [84], [85], [86], [87], [88]. We begin by recalling the notion of a *Leonard pair*. We will use the following terms. Let X denote a square matrix. Then X is called *tridiagonal* whenever each nonzero entry lies on either the diagonal, the subdiagonal, or the superdiagonal. Assume X is tridiagonal. Then X is called *irreducible* whenever each entry on the subdiagonal is nonzero and each entry on the superdiagonal is nonzero.

We now define a Leonard pair. For the rest of this paper $\mathbb{K}$ will denote a field.

Definition 1.1. [79] *Let V denote a vector space over $\mathbb{K}$ with finite positive dimension. By a* Leonard pair *on V, we mean an ordered pair of linear transformations $A : V \to V$ and $A^* : V \to V$ that satisfy both (i), (ii) below.*

(i) There exists a basis for V with respect to which the matrix representing A is irreducible tridiagonal and the matrix representing A^* is diagonal.
(ii) There exists a basis for V with respect to which the matrix representing A^* is irreducible tridiagonal and the matrix representing A is diagonal.

Note 1.1. According to a common notational convention A^* denotes the conjugate-transpose of A. We are not using this convention. In a Leonard pair A, A^* the linear transformations A and A^* are arbitrary subject to (i), (ii) above.

Note 1.2. Our use of the name "Leonard pair" is motivated by a connection to a theorem of Doug Leonard [62], [11, p. 260] that involves the q-Racah and related polynomials of the Askey scheme.

1.2 An Example

Here is an example of a Leonard pair. Set $V = \mathbb{K}^4$ (column vectors), set

$$A = \begin{pmatrix} 0&3&0&0\\ 1&0&2&0\\ 0&2&0&1\\ 0&0&3&0 \end{pmatrix}, \qquad A^* = \begin{pmatrix} 3&0&0&0\\ 0&1&0&0\\ 0&0&-1&0\\ 0&0&0&-3 \end{pmatrix},$$

and view A and A^* as linear transformations from V to V. We assume the characteristic of $\mathbb{K}$ is not 2 or 3, to ensure A is irreducible. Then A, A^* is a Leonard pair on V. Indeed, condition (i) in Definition 1.1 is satisfied by the basis for V consisting of the columns of the 4 by 4 identity matrix. To verify condition (ii), we display an invertible matrix P such that $P^{-1}AP$ is diagonal and $P^{-1}A^*P$ is irreducible tridiagonal. Set

$$P = \begin{pmatrix} 1&3&3&1\\ 1&1&-1&-1\\ 1&-1&-1&1\\ 1&-3&3&-1 \end{pmatrix}.$$

By matrix multiplication $P^2 = 8I$, where I denotes the identity, so P^{-1} exists. Also by matrix multiplication,

$$AP = PA^*. \tag{1.1}$$

Apparently $P^{-1}AP$ is equal to A^* and is therefore diagonal. By (1.1) and since P^{-1} is a scalar multiple of P, we find $P^{-1}A^*P$ is equal to A and is therefore irreducible tridiagonal. Now condition (ii) of Definition 1.1 is satisfied by the basis for V consisting of the columns of P.

The above example is a member of the following infinite family of Leonard pairs. For any nonnegative integer d the pair

$$A = \begin{pmatrix} 0 & d & & & \mathbf{0} \\ 1 & 0 & d-1 & & \\ & 2 & \ddots & \ddots & \\ & & \ddots & \ddots & 1 \\ \mathbf{0} & & & d & 0 \end{pmatrix}, \qquad A^* = \mathrm{diag}(d, d-2, d-4, \ldots, -d) \tag{1.2}$$

is a Leonard pair on the vector space $\mathbb{K}^{d+1}$, provided the characteristic of $\mathbb{K}$ is zero or an odd prime greater than d. This can be proved by modifying the proof for $d = 3$ given above. One shows $P^2 = 2^d I$ and $AP = PA^*$, where P denotes the matrix with ij entry

$$P_{ij} = \binom{d}{j} {}_2F_1\left(\begin{matrix} -i, -j \\ -d \end{matrix} \,\middle|\, 2 \right) \qquad (0 \le i, j \le d). \tag{1.3}$$

We follow the standard notation for hypergeometric series [28, p. 3]. The details of the above calculations are given in Section 6 below.

1.3 Leonard Systems

When working with a Leonard pair, it is often convenient to consider a closely related and somewhat more abstract object called a *Leonard system*. In order to define this we first make an observation about Leonard pairs.

Lemma 1.1. *Let V denote a vector space over $\mathbb{K}$ with finite positive dimension and let A, A^* denote a Leonard pair on V. Then the eigenvalues of A are mutually distinct and contained in $\mathbb{K}$. Moreover, the eigenvalues of A^* are mutually distinct and contained in $\mathbb{K}$.*

Proof. Concerning A, recall by Definition 1.1(ii) that there exists a basis for V consisting of eigenvectors for A. Consequently the eigenvalues of A are all in $\mathbb{K}$, and the minimal polynomial of A has no repeated roots. To show the eigenvalues of A are distinct, we show that the minimal polynomial of A has degree equal to $\dim V$. By Definition 1.1(i), there exists a basis for V with respect to which the matrix representing A is irreducible tridiagonal. Denote this matrix by B. On one hand, A and B have the same minimal polynomial. On the other hand, using the tridiagonal shape of B, we find $I, B, B^2, \ldots, B^d$

are linearly independent, where $d = \dim V - 1$, so the minimal polynomial of B has degree $d + 1 = \dim V$. We conclude that the mininimal polynomial of A has degree equal to $\dim V$, so the eigenvalues of A are distinct. We have now obtained our assertions about A, and the case of A^* is similar. □

To prepare for our definition of a Leonard system, we recall a few concepts from linear algebra. Let d denote a nonnegative integer and let $\mathrm{Mat}_{d+1}(\mathbb{K})$ denote the $\mathbb{K}$-algebra consisting of all $d+1$ by $d+1$ matrices that have entries in $\mathbb{K}$. We index the rows and columns by $0, 1, \ldots, d$. We let $\mathbb{K}^{d+1}$ denote the $\mathbb{K}$-vector space consisting of all $d + 1$ by 1 matrices that have entries in $\mathbb{K}$. We index the rows by $0, 1, \ldots, d$. We view $\mathbb{K}^{d+1}$ as a left module for $\mathrm{Mat}_{d+1}(\mathbb{K})$. We observe this module is irreducible. For the rest of this paper we let $\mathcal{A}$ denote a $\mathbb{K}$-algebra isomorphic to $\mathrm{Mat}_{d+1}(\mathbb{K})$. When we refer to an $\mathcal{A}$-module we mean a left $\mathcal{A}$-module. Let V denote an irreducible $\mathcal{A}$-module. We remark that V is unique up to isomorphism of $\mathcal{A}$-modules, and that V has dimension $d + 1$. Let $v_0, v_1, \ldots, v_d$ denote a basis for V. For $X \in \mathcal{A}$ and $Y \in \mathrm{Mat}_{d+1}(\mathbb{K})$, we say Y *represents* X *with respect to* $v_0, v_1, \ldots, v_d$ whenever $Xv_j = \sum_{i=0}^{d} Y_{ij} v_i$ for $0 \le j \le d$. Let A denote an element of $\mathcal{A}$. We say A is *multiplicity-free* whenever it has $d+1$ mutually distinct eigenvalues in $\mathbb{K}$. Let A denote a multiplicity-free element of $\mathcal{A}$. Let $\theta_0, \theta_1, \ldots, \theta_d$ denote an ordering of the eigenvalues of A, and for $0 \le i \le d$ put

$$E_i = \prod_{\substack{0 \le j \le d \\ j \ne i}} \frac{A - \theta_j I}{\theta_i - \theta_j}, \tag{1.4}$$

where I denotes the identity of $\mathcal{A}$. We observe

(i) $AE_i = \theta_i E_i \ (0 \le i \le d)$;
(ii) $E_i E_j = \delta_{ij} E_i \ (0 \le i, j \le d)$;
(iii) $\sum_{i=0}^{d} E_i = I$;
(iv) $A = \sum_{i=0}^{d} \theta_i E_i$.

Let $\mathcal{D}$ denote the subalgebra of $\mathcal{A}$ generated by A. Using (i)–(iv) we find the sequence $E_0, E_1, \ldots, E_d$ is a basis for the $\mathbb{K}$-vector space $\mathcal{D}$. We call E_i the *primitive idempotent* of A associated with θ_i. It is helpful to think of these primitive idempotents as follows. Observe

$$V = E_0 V + E_1 V + \cdots + E_d V \qquad \text{(direct sum)}. \tag{1.5}$$

For $0 \le i \le d$, $E_i V$ is the (one dimensional) eigenspace of A in V associated with the eigenvalue θ_i, and E_i acts on V as the projection onto this eigenspace. We remark that $\{A^i | 0 \le i \le d\}$ is a basis for the $\mathbb{K}$-vector space $\mathcal{D}$ and that $\prod_{i=0}^{d}(A - \theta_i I) = 0$. By a *Leonard pair in* $\mathcal{A}$ we mean an ordered pair of elements taken from $\mathcal{A}$ that act on V as a Leonard pair in the sense of Definition 1.1. We call $\mathcal{A}$ the *ambient algebra* of the pair and say the pair is *over* $\mathbb{K}$. We refer to d as the *diameter* of the pair. We now define a Leonard system.

Definition 1.2. [79, Definition 1.4] *By a* Leonard system *in $\mathcal{A}$ we mean a sequence $\Phi := (A; A^*; \{E_i\}_{i=0}^d; \{E_i^*\}_{i=0}^d)$ that satisfies (i)–(v) below.*

(i) Each of A, A^* is a multiplicity-free element in $\mathcal{A}$.
(ii) $E_0, E_1, \ldots, E_d$ is an ordering of the primitive idempotents of A.
(iii) $E_0^*, E_1^*, \ldots, E_d^*$ is an ordering of the primitive idempotents of A^*.
(iv) $E_i A^* E_j = \begin{cases} 0, & \text{if } |i-j|>1, \\ \neq 0, & \text{if } |i-j|=1, \end{cases} \qquad (0 \le i,j \le d).$
(v) $E_i^* A E_j^* = \begin{cases} 0, & \text{if } |i-j|>1, \\ \neq 0, & \text{if } |i-j|=1, \end{cases} \qquad (0 \le i,j \le d).$

We refer to d as the diameter *of Φ and say Φ is* over $\mathbb{K}$. *We call $\mathcal{A}$ the* ambient algebra *of Φ.*

We comment on how Leonard pairs and Leonard systems are related. In the following discussion V denotes an irreducible $\mathcal{A}$-module. Let $(A; A^*; \{E_i\}_{i=0}^d; \{E_i^*\}_{i=0}^d)$ denote a Leonard system in $\mathcal{A}$. For $0 \le i \le d$ let v_i denote a nonzero vector in E_iV. Then the sequence $v_0, v_1, \ldots, v_d$ is a basis for V that satisfies Definition 1.1(ii). For $0 \le i \le d$ let v_i^* denote a nonzero vector in E_i^*V. Then the sequence $v_0^*, v_1^*, \ldots, v_d^*$ is a basis for V that satisfies Definition 1.1(i). By these comments the pair A, A^* is a Leonard pair in $\mathcal{A}$. Conversely let A, A^* denote a Leonard pair in $\mathcal{A}$. Then each of A, A^* is multiplicity-free by Lemma 1.1. Let $v_0, v_1, \ldots, v_d$ denote a basis for V that satisfies Definition 1.1(ii). For $0 \le i \le d$ the vector v_i is an eigenvector for A; let E_i denote the corresponding primitive idempotent. Let $v_0^*, v_1^*, \ldots, v_d^*$ denote a basis for V that satisfies Definition 1.1(i). For $0 \le i \le d$ the vector v_i^* is an eigenvector for A^*; let E_i^* denote the corresponding primitive idempotent. Then $(A; A^*; \{E_i\}_{i=0}^d; \{E_i^*\}_{i=0}^d)$ is a Leonard system in $\mathcal{A}$. In summary we have the following.

Lemma 1.2. *Let A and A^* denote elements of $\mathcal{A}$. Then the pair A, A^* is a Leonard pair in $\mathcal{A}$ if and only if the following (i), (ii) hold.*

(i) Each of A, A^* is multiplicity-free.
(ii) There exists an ordering $E_0, E_1, \ldots, E_d$ of the primitive idempotents of A and there exists an ordering $E_0^*, E_1^*, \ldots, E_d^*$ of the primitive idempotents of A^* such that $(A; A^*; \{E_i\}_{i=0}^d; \{E_i^*\}_{i=0}^d)$ is a Leonard system in $\mathcal{A}$.

We recall the notion of *isomorphism* for Leonard pairs and Leonard systems.

Definition 1.3. *Let A, A^* and B, B^* denote Leonard pairs over $\mathbb{K}$. By an* isomorphism of Leonard pairs from A, A^* to B, B^* *we mean an isomorphism of $\mathbb{K}$-algebras from the ambient algebra of A, A^* to the ambient algebra of B, B^* that sends A to B and A^* to B^*. The Leonard pairs A, A^* and B, B^* are said to be* isomorphic *whenever there exists an isomorphism of Leonard pairs from A, A^* to B, B^*.*

Let Φ denote the Leonard system from Definition 1.2 and let $\sigma : \mathcal{A} \to \mathcal{A}'$ denote an isomorphism of $\mathbb{K}$-algebras. We write

$$\Phi^\sigma := (A^\sigma; A^{*\sigma}; \{E_i^\sigma\}_{i=0}^d; \{E_i^{*\sigma}\}_{i=0}^d)$$

and observe Φ^σ is a Leonard system in $\mathcal{A}'$.

Definition 1.4. *Let Φ and Φ' denote Leonard systems over $\mathbb{K}$. By an* isomorphism of Leonard systems from Φ to Φ' *we mean an isomorphism of $\mathbb{K}$-algebras σ from the ambient algebra of Φ to the ambient algebra of Φ' such that $\Phi^\sigma = \Phi'$. The Leonard systems Φ, Φ' are said to be* isomorphic *whenever there exists an isomorphism of Leonard systems from Φ to Φ'.*

We have a remark. Let $\sigma : \mathcal{A} \to \mathcal{A}$ denote any map. By the Skolem-Noether theorem [75, Corollary 9.122], σ is an isomorphism of $\mathbb{K}$-algebras if and only if there exists an invertible $S \in \mathcal{A}$ such that $X^\sigma = SXS^{-1}$ for all $X \in \mathcal{A}$.

1.4 The D_4 Action

A given Leonard system can be modified in several ways to get a new Leonard system. For instance, let Φ denote the Leonard system from Definition 1.2, and let $\alpha, \alpha^*, \beta, \beta^*$ denote scalars in $\mathbb{K}$ such that $\alpha \neq 0$, $\alpha^* \neq 0$. Then the sequence

$$(\alpha A + \beta I; \alpha^* A^* + \beta^* I; \{E_i\}_{i=0}^d; \{E_i^*\}_{i=0}^d)$$

is a Leonard system in $\mathcal{A}$. Also, each of the following three sequences is a Leonard system in $\mathcal{A}$.

$$\Phi^* := (A^*; A; \{E_i^*\}_{i=0}^d; \{E_i\}_{i=0}^d),$$

$$\Phi^\downarrow := (A; A^*; \{E_i\}_{i=0}^d; \{E_{d-i}^*\}_{i=0}^d),$$

$$\Phi^\Downarrow := (A; A^*; \{E_{d-i}\}_{i=0}^d; \{E_i^*\}_{i=0}^d).$$

Viewing $*, \downarrow, \Downarrow$ as permutations on the set of all Leonard systems,

$$*^2 = \downarrow^2 = \Downarrow^2 = 1, \tag{1.6}$$

$$\Downarrow * = * \downarrow, \qquad \downarrow * = * \Downarrow, \qquad \downarrow\Downarrow = \Downarrow\downarrow . \tag{1.7}$$

The group generated by symbols $*, \downarrow, \Downarrow$ subject to the relations (1.6), (1.7) is the dihedral group D_4. We recall D_4 is the group of symmetries of a square, and has 8 elements. Apparently $*, \downarrow, \Downarrow$ induce an action of D_4 on the set of all Leonard systems. Two Leonard systems will be called *relatives* whenever they are in the same orbit of this D_4 action. The relatives of Φ are as follows:

name	relative
Φ	$(A; A^*; \{E_i\}_{i=0}^d; \{E_i^*\}_{i=0}^d)$
$\Phi^{\downarrow}$	$(A; A^*; \{E_i\}_{i=0}^d; \{E_{d-i}^*\}_{i=0}^d)$
$\Phi^{\Downarrow}$	$(A; A^*; \{E_{d-i}\}_{i=0}^d; \{E_i^*\}_{i=0}^d)$
$\Phi^{\downarrow\Downarrow}$	$(A; A^*; \{E_{d-i}\}_{i=0}^d; \{E_{d-i}^*\}_{i=0}^d)$
Φ^*	$(A^*; A; \{E_i^*\}_{i=0}^d; \{E_i\}_{i=0}^d)$
$\Phi^{\downarrow *}$	$(A^*; A; \{E_{d-i}^*\}_{i=0}^d; \{E_i\}_{i=0}^d)$
$\Phi^{\Downarrow *}$	$(A^*; A; \{E_i^*\}_{i=0}^d; \{E_{d-i}\}_{i=0}^d)$
$\Phi^{\downarrow\Downarrow *}$	$(A^*; A; \{E_{d-i}^*\}_{i=0}^d; \{E_{d-i}\}_{i=0}^d)$

There may be some isomorphisms among the above Leonard systems.

For the rest of this paper we will use the following notational convention.

Definition 1.5. *Let Φ denote a Leonard system. For any element g in the group D_4 and for any object f that we associate with Φ, we let f^g denote the corresponding object for the Leonard system $\Phi^{g^{-1}}$. We have been using this convention all along; an example is $E_i^*(\Phi) = E_i(\Phi^*)$.*

2 The Structure of a Leonard System

In this section we establish a few basic facts concerning Leonard systems. We begin with a definition and two routine lemmas.

Definition 2.1. *Let Φ denote the Leonard system from Definition 1.2. For $0 \leq i \leq d$, we let θ_i (resp. θ_i^*) denote the eigenvalue of A (resp. A^*) associated with E_i (resp. E_i^*). We refer to $\theta_0, \theta_1, \ldots, \theta_d$ as the* eigenvalue sequence *of Φ. We refer to $\theta_0^*, \theta_1^*, \ldots, \theta_d^*$ as the* dual eigenvalue sequence *of Φ. We observe $\theta_0, \theta_1, \ldots, \theta_d$ are mutually distinct and contained in $\mathbb{K}$. Similarly $\theta_0^*, \theta_1^*, \ldots, \theta_d^*$ are mutually distinct and contained in $\mathbb{K}$.*

Lemma 2.1. *Let Φ denote the Leonard system from Definition 1.2 and let V denote an irreducible $\mathcal{A}$-module. For $0 \leq i \leq d$ let v_i denote a nonzero vector in E_i^*V and observe $v_0, v_1, \ldots, v_d$ is a basis for V. Then (i), (ii) hold below.*

(i) For $0 \leq i \leq d$ the matrix in $\mathrm{Mat}_{d+1}(\mathbb{K})$ that represents E_i^* with respect to $v_0, v_1, \ldots, v_d$ has ii entry 1 and all other entries 0.
(ii) The matrix in $\mathrm{Mat}_{d+1}(\mathbb{K})$ that represents A^* with respect to $v_0, v_1, \ldots, v_d$ is equal to $diag(\theta_0^*, \theta_1^*, \ldots, \theta_d^*)$.

Lemma 2.2. *Let A denote an irreducible tridiagonal matrix in $\mathrm{Mat}_{d+1}(\mathbb{K})$. Pick any integers i, j $(0 \leq i, j \leq d)$. Then (i)–(iii) hold below.*

(i) The entry $(A^r)_{ij} = 0$ if $r < |i - j|$, $\qquad (0 \leq r \leq d)$.
(ii) Suppose $i \leq j$. Then the entry $(A^{j-i})_{ij} = \prod_{h=i}^{j-1} A_{h,h+1}$. Moreover $(A^{j-i})_{ij} \neq 0$.

(iii) Suppose $i \geq j$. *Then the entry* $(A^{i-j})_{ij} = \prod_{h=j}^{i-1} A_{h+1,h}$. *Moreover* $(A^{i-j})_{ij} \neq 0$.

Theorem 2.1. *Let* Φ *denote the Leonard system from Definition 1.2. Then the elements*

$$A^r E_0^* A^s \qquad (0 \leq r, s \leq d) \tag{2.1}$$

form a basis for the $\mathbb{K}$*-vector space* $\mathcal{A}$.

Proof. The number of elements in (2.1) is equal to $(d+1)^2$, and this number is the dimension of $\mathcal{A}$. Therefore it suffices to show the elements in (2.1) are linearly independent. To do this, we represent the elements in (2.1) by matrices. Let V denote an irreducible $\mathcal{A}$-module. For $0 \leq i \leq d$ let v_i denote a nonzero vector in $E_i^* V$, and observe $v_0, v_1, \ldots, v_d$ is a basis for V. For the purpose of this proof, let us identify each element of $\mathcal{A}$ with the matrix in $\mathrm{Mat}_{d+1}(\mathbb{K})$ that represents it with respect to the basis $v_0, v_1, \ldots, v_d$. Adopting this point of view we find A is irreducible tridiagonal and A^* is diagonal. For $0 \leq r, s \leq d$ we show the entries of $A^r E_0^* A^s$ satisfy

$$(A^r E_0^* A^s)_{ij} = \begin{cases} 0, & \text{if } i > r \text{ or } j > s, \\ \neq 0, & \text{if } i = r \text{ and } j = s. \end{cases} \qquad (0 \leq i, j \leq d). \tag{2.2}$$

By Lemma 2.1(i) the matrix E_0^* has 00 entry 1 and all other entries 0. Therefore

$$(A^r E_0^* A^s)_{ij} = (A^r)_{i0} (A^s)_{0j} \qquad (0 \leq i, j \leq d). \tag{2.3}$$

We mentioned A is irreducible tridiagonal. Applying Lemma 2.2 we find that for $0 \leq i \leq d$ the entry $(A^r)_{i0}$ is zero if $i > r$, and nonzero if $i = r$. Similarly for $0 \leq j \leq d$ the entry $(A^s)_{0j}$ is zero if $j > s$, and nonzero if $j = s$. Combining these facts with (2.3) we routinely obtain (2.2) and it follows the elements (2.1) are linearly independent. Apparently the elements (2.1) form a basis for $\mathcal{A}$, as desired. □

Corollary 2.1. *Let* Φ *denote the Leonard system from Definition 1.2. Then the elements* A, E_0^* *together generate* $\mathcal{A}$. *Moreover the elements* A, A^* *together generate* $\mathcal{A}$.

Proof. The first assertion is immediate from Theorem 2.1. The second assertion follows from the first assertion and the observation that E_0^* is a polynomial in A^*. □

The following is immediate from Corollary 2.1.

Corollary 2.2. *Let* A, A^* *denote a Leonard pair in* $\mathcal{A}$. *Then the elements* A, A^* *together generate* $\mathcal{A}$.

We mention a few implications of Theorem 2.1 that will be useful later in the paper.

Lemma 2.3. *Let Φ denote the Leonard system from Definition 1.2. Let $\mathcal{D}$ denote the subalgebra of $\mathcal{A}$ generated by A. Let $X_0, X_1, \ldots, X_d$ denote a basis for the $\mathbb{K}$-vector space $\mathcal{D}$. Then the elements*

$$X_r E_0^* X_s \qquad (0 \leq r, s \leq d) \tag{2.4}$$

form a basis for the $\mathbb{K}$-vector space $\mathcal{A}$.

Proof. The number of elements in (2.4) is equal to $(d+1)^2$, and this number is the dimension of $\mathcal{A}$. Therefore it suffices to show the elements (2.4) span $\mathcal{A}$. But this is immediate from Theorem 2.1, and since each element in (2.1) is contained in the span of the elements (2.4). □

Corollary 2.3. *Let Φ denote the Leonard system from Definition 1.2. Then the elements*

$$E_r E_0^* E_s \qquad (0 \leq r, s \leq d) \tag{2.5}$$

form a basis for the $\mathbb{K}$-vector space $\mathcal{A}$.

Proof. Immediate from Lemma 2.3, with $X_i = E_i$ for $0 \leq i \leq d$. □

Lemma 2.4. *Let Φ denote the Leonard system from Definition 1.2. Let $\mathcal{D}$ denote the subalgebra of $\mathcal{A}$ generated by A. Let X and Y denote elements in $\mathcal{D}$ and assume $XE_0^*Y = 0$. Then $X = 0$ or $Y = 0$.*

Proof. Let $X_0, X_1, \ldots, X_d$ denote a basis for the $\mathbb{K}$-vector space $\mathcal{D}$. Since $X \in \mathcal{D}$ there exists $\alpha_i \in \mathbb{K}$ $(0 \leq i \leq d)$ such that $X = \sum_{i=0}^{d} \alpha_i X_i$. Similarly there exists $\beta_i \in \mathbb{K}$ $(0 \leq i \leq d)$ such that $Y = \sum_{i=0}^{d} \beta_i X_i$. Evaluating $0 = XE_0^*Y$ using these equations we get $0 = \sum_{i=0}^{d} \sum_{j=0}^{d} \alpha_i \beta_j X_i E_0^* X_j$. From this and Lemma 2.3 we find $\alpha_i \beta_j = 0$ for $0 \leq i, j \leq d$. We assume $X \neq 0$ and show $Y = 0$. Since $X \neq 0$ there exists an integer i $(0 \leq i \leq d)$ such that $\alpha_i \neq 0$. Now for $0 \leq j \leq d$ we have $\alpha_i \beta_j = 0$ so $\beta_j = 0$. It follows $Y = 0$. □

We finish this section with a comment.

Lemma 2.5. *Let Φ denote the Leonard system from Definition 1.2. Pick any integers i, j $(0 \leq i, j \leq d)$. Then (i)–(iii) hold below.*

(i) $E_i^* A^r E_j^* = 0$ *if* $r < |i-j|$, $\qquad (0 \leq r \leq d)$.
(ii) Suppose $i \leq j$. Then

$$E_i^* A^{j-i} E_j^* = E_i^* A E_{i+1}^* A \cdots E_{j-1}^* A E_j^*. \tag{2.6}$$

Moreover $E_i^* A^{j-i} E_j^* \neq 0$.
(iii) Suppose $i \geq j$. Then

$$E_i^* A^{i-j} E_j^* = E_i^* A E_{i-1}^* A \cdots E_{j+1}^* A E_j^*. \tag{2.7}$$

Moreover $E_i^* A^{i-j} E_j^* \neq 0$.

Proof. Represent the elements of Φ by matrices as in the proof of Theorem 2.1, and use Lemma 2.2. □

2.1 The Antiautomorphism †

We recall the notion of an *antiautomorphism* of $\mathcal{A}$. Let $\gamma : \mathcal{A} \to \mathcal{A}$ denote any map. We call γ an *antiautomorphism* of $\mathcal{A}$ whenever γ is an isomorphism of $\mathbb{K}$-vector spaces and $(XY)^\gamma = Y^\gamma X^\gamma$ for all $X, Y \in \mathcal{A}$. For example assume $\mathcal{A} = \mathrm{Mat}_{d+1}(\mathbb{K})$. Then γ is an antiautomorphism of $\mathcal{A}$ if and only if there exists an invertible element R in $\mathcal{A}$ such that $X^\gamma = R^{-1}X^tR$ for all $X \in \mathcal{A}$, where t denotes transpose. This follows from the Skolem-Noether theorem [75, Corollary 9.122].

Theorem 2.2. *Let A, A^* denote a Leonard pair in $\mathcal{A}$. Then there exists a unique antiautomorphism † of $\mathcal{A}$ such that $A^\dagger = A$ and $A^{*\dagger} = A^*$. Moreover $X^{\dagger\dagger} = X$ for all $X \in \mathcal{A}$.*

Proof. Concerning existence, let V denote an irreducible $\mathcal{A}$-module. By Definition 1.1(i) there exists a basis for V with respect to which the matrix representing A is irreducible tridiagonal and the matrix representing A^* is diagonal. Let us denote this basis by $v_0, v_1, \ldots, v_d$. For $X \in \mathcal{A}$ let X^σ denote the matrix in $\mathrm{Mat}_{d+1}(\mathbb{K})$ that represents X with respect to the basis $v_0, v_1, \ldots, v_d$. We observe $\sigma : \mathcal{A} \to \mathrm{Mat}_{d+1}(\mathbb{K})$ is an isomorphism of $\mathbb{K}$-algebras. We abbreviate $B = A^\sigma$ and observe B is irreducible tridiagonal. We abbreviate $B^* = A^{*\sigma}$ and observe B^* is diagonal. Let D denote the diagonal matrix in $\mathrm{Mat}_{d+1}(\mathbb{K})$ that has ii entry

$$D_{ii} = \frac{B_{01}B_{12}\cdots B_{i-1,i}}{B_{10}B_{21}\cdots B_{i,i-1}} \qquad (0 \le i \le d).$$

It is routine to verify $D^{-1}B^tD = B$. Each of D, B^* is diagonal so $DB^* = B^*D$; also $B^{*t} = B^*$ so $D^{-1}B^{*t}D = B^*$. Let $\gamma : \mathrm{Mat}_{d+1}(\mathbb{K}) \to \mathrm{Mat}_{d+1}(\mathbb{K})$ denote the map that satisfies $X^\gamma = D^{-1}X^tD$ for all $X \in \mathrm{Mat}_{d+1}(\mathbb{K})$. We observe γ is an antiautomorphism of $\mathrm{Mat}_{d+1}(\mathbb{K})$ such that $B^\gamma = B$ and $B^{*\gamma} = B^*$. We define the map $\dagger : \mathcal{A} \to \mathcal{A}$ to be the composition $\dagger = \sigma\gamma\sigma^{-1}$. We observe † is an antiautomorphism of $\mathcal{A}$ such that $A^\dagger = A$ and $A^{*\dagger} = A^*$. We have now shown there exists an antiautomorphism † of $\mathcal{A}$ such that $A^\dagger = A$ and $A^{*\dagger} = A^*$. This antiautomorphism is unique since A, A^* together generate $\mathcal{A}$. The map $X \to X^{\dagger\dagger}$ is an isomorphism of $\mathbb{K}$-algebras from $\mathcal{A}$ to itself. This isomorphism is the identity since $A^{\dagger\dagger} = A$, $A^{*\dagger\dagger} = A^*$, and since A, A^* together generate $\mathcal{A}$. □

Definition 2.2. *Let A, A^* denote a Leonard pair in $\mathcal{A}$. By the* antiautomorphism which corresponds to A, A^* *we mean the map $\dagger : \mathcal{A} \to \mathcal{A}$ from Theorem 2.2. Let $\Phi = (A; A^*; \{E_i\}_{i=0}^d; \{E_i^*\}_{i=0}^d)$ denote a Leonard system in $\mathcal{A}$. By the* antiautomorphism which corresponds to Φ *we mean the antiautomorphism which corresponds to the Leonard pair A, A^*.*

Lemma 2.6. *Let Φ denote the Leonard system from Definition 1.2 and let † denote the corresponding antiautomorphism. Then the following (i), (ii) hold.*

(i) Let $\mathcal{D}$ denote the subalgebra of $\mathcal{A}$ generated by A. Then $X^\dagger = X$ for all $X \in \mathcal{D}$; in particular $E_i^\dagger = E_i$ for $0 \leq i \leq d$.
(ii) Let $\mathcal{D}^*$ denote the subalgebra of $\mathcal{A}$ generated by A^*. Then $X^\dagger = X$ for all $X \in \mathcal{D}^*$; in particular $E_i^{*\dagger} = E_i^*$ for $0 \leq i \leq d$.

Proof. (i) The sequence A^i $(0 \leq i \leq d)$ is a basis for the $\mathbb{K}$-vector space $\mathcal{D}$. Observe $\dagger$ stabilizes A^i for $0 \leq i \leq d$. The result follows.
(ii) Similar to the proof of (i) above. □

2.2 The Scalars a_i, x_i

In this section we introduce some scalars that will help us describe Leonard systems.

Definition 2.3. *Let Φ denote the Leonard system from Definition 1.2. We define*

$$a_i = \mathrm{tr}(E_i^* A) \qquad (0 \leq i \leq d), \tag{2.8}$$
$$x_i = \mathrm{tr}(E_i^* A E_{i-1}^* A) \qquad (1 \leq i \leq d), \tag{2.9}$$

where tr *denotes trace. For notational convenience we define $x_0 = 0$.*

We have a comment.

Lemma 2.7. *Let Φ denote the Leonard system from Definition 1.2 and let V denote an irreducible $\mathcal{A}$-module. For $0 \leq i \leq d$ let v_i denote a nonzero vector in $E_i^* V$ and observe $v_0, v_1, \ldots, v_d$ is a basis for V. Let B denote the matrix in $\mathrm{Mat}_{d+1}(\mathbb{K})$ that represents A with respect to $v_0, v_1, \ldots, v_d$. We observe B is irreducible tridiagonal. The following (i)–(iii) hold.*

(i) $B_{ii} = a_i$ $(0 \leq i \leq d)$.
(ii) $B_{i,i-1} B_{i-1,i} = x_i$ $(1 \leq i \leq d)$.
(iii) $x_i \neq 0$ $(1 \leq i \leq d)$.

Proof. (i), (ii) For $0 \leq i \leq d$ the matrix in $\mathrm{Mat}_{d+1}(\mathbb{K})$ that represents E_i^* with respect to $v_0, v_1, \ldots, v_d$ has ii entry 1 and all other entries 0. The result follows in view of Definition 2.3.
(iii) Immediate from (ii) and since B is irreducible. □

Theorem 2.3. *Let Φ denote the Leonard system from Definition 1.2. Let V denote an irreducible $\mathcal{A}$-module and let v denote a nonzero vector in $E_0^* V$. Then for $0 \leq i \leq d$ the vector $E_i^* A^i v$ is nonzero and hence a basis for $E_i^* V$. Moreover the sequence*

$$E_i^* A^i v \qquad (0 \leq i \leq d) \tag{2.10}$$

is a basis for V.

Proof. We show $E_i^* A^i v \neq 0$ for $0 \leq i \leq d$. Let i be given. Setting $j = 0$ in Lemma 2.5(iii) we find $E_i^* A^i E_0^* \neq 0$. Therefore $E_i^* A^i E_0^* V \neq 0$. The space $E_0^* V$ is spanned by v so $E_i^* A^i v \neq 0$ as desired. The remaining claims follow. □

Theorem 2.4. *Let Φ denote the Leonard system from Definition 1.2 and let the scalars a_i, x_i be as in Definition 2.3. Let V denote an irreducible $\mathcal{A}$-module. With respect to the basis for V given in (2.10) the matrix that represents A is equal to*

$$\begin{pmatrix} a_0 & x_1 & & & & \mathbf{0} \\ 1 & a_1 & x_2 & & & \\ & 1 & \cdot & \cdot & & \\ & & \cdot & \cdot & \cdot & \\ & & & \cdot & \cdot & x_d \\ \mathbf{0} & & & & 1 & a_d \end{pmatrix}. \tag{2.11}$$

Proof. With reference to (2.10) abbreviate $v_i = E_i^* A^i v$ for $0 \leq i \leq d$. Let B denote the matrix in $\mathrm{Mat}_{d+1}(\mathbb{K})$ that represents A with respect to $v_0, v_1, \ldots, v_d$. We show B is equal to (2.11). In view of Lemma 2.7 it suffices to show $B_{i,i-1} = 1$ for $1 \leq i \leq d$. For $0 \leq i \leq d$ the matrix B^i represents A^i with respect to $v_0, v_1, \ldots, v_d$; therefore $A^i v_0 = \sum_{j=0}^{d} (B^i)_{j0} v_j$. Applying E_i^* and using $v_0 = v$ we find $v_i = (B^i)_{i0} v_i$ so $(B^i)_{i0} = 1$, forcing $B_{i,i-1} \cdots B_{21} B_{10} = 1$ by Lemma 2.2. We have shown $B_{i,i-1} \cdots B_{21} B_{10} = 1$ for $1 \leq i \leq d$ so $B_{i,i-1} = 1$ for $1 \leq i \leq d$. We now see B is equal to (2.11). □

We finish this section with a few comments.

Lemma 2.8. *Let Φ denote the Leonard system from Definition 1.2 and let the scalars a_i, x_i be as in Definition 2.3. Then the following (i)–(iii) hold.*

(i) $E_i^* A E_i^* = a_i E_i^* \qquad (0 \leq i \leq d)$.
(ii) $E_i^* A E_{i-1}^* A E_i^* = x_i E_i^* \qquad (1 \leq i \leq d)$.
(iii) $E_{i-1}^* A E_i^* A E_{i-1}^* = x_i E_{i-1}^* \qquad (1 \leq i \leq d)$.

Proof. (i) Observe E_i^* is a basis for $E_i^* \mathcal{A} E_i^*$. By this and since $E_i^* A E_i^*$ is contained in $E_i^* \mathcal{A} E_i^*$ we find there exists $\alpha_i \in \mathbb{K}$ such that $E_i^* A E_i^* = \alpha_i E_i^*$. Taking the trace of both sides and using $\mathrm{tr}(XY) = \mathrm{tr}(YX)$, $\mathrm{tr}(E_i^*) = 1$ we find $a_i = \alpha_i$.
(ii) We mentioned above that E_i^* is a basis for $E_i^* \mathcal{A} E_i^*$. By this and since $E_i^* A E_{i-1}^* A E_i^*$ is contained in $E_i^* \mathcal{A} E_i^*$ we find there exists $\beta_i \in \mathbb{K}$ such that $E_i^* A E_{i-1}^* A E_i^* = \beta_i E_i^*$. Taking the trace of both sides we find $x_i = \beta_i$.
(iii) Similar to the proof of (ii) above. □

Lemma 2.9. *Let Φ denote the Leonard system from Definition 1.2 and let the scalars x_i be as in Definition 2.3. Then the following (i), (ii) hold.*

(i) $E_j^* A^{j-i} E_i^* A^{j-i} E_j^* = x_{i+1} x_{i+2} \cdots x_j E_j^* \qquad (0 \leq i \leq j \leq d)$.

(ii) $E^*_i A^{j-i} E^*_j A^{j-i} E^*_i = x_{i+1} x_{i+2} \cdots x_j E^*_i \qquad (0 \le i \le j \le d)$.

Proof. (i) Evaluate the expression on the left using Lemma 2.5(ii), (iii) and Lemma 2.8(ii).
(ii) Evaluate the expression on the left using Lemma 2.5(ii), (iii) and Lemma 2.8(iii). □

2.3 The Polynomials p_i

In this section we begin our discussion of polynomials. We will use the following notation. Let λ denote an indeterminate. We let $\mathbb{K}[\lambda]$ denote the $\mathbb{K}$-algebra consisting of all polynomials in λ that have coefficients in $\mathbb{K}$. For the rest of this paper all polynomials that we discuss are assumed to lie in $\mathbb{K}[\lambda]$.

Definition 2.4. *Let Φ denote the Leonard system from Definition 1.2 and let the scalars a_i, x_i be as in Definition 2.3. We define a sequence of polynomials $p_0, p_1, \ldots, p_{d+1}$ by*

$$p_0 = 1, \tag{2.12}$$

$$\lambda p_i = p_{i+1} + a_i p_i + x_i p_{i-1} \qquad (0 \le i \le d), \tag{2.13}$$

where $p_{-1} = 0$. We observe p_i is monic with degree exactly i for $0 \le i \le d+1$.

Lemma 2.10. *Let Φ denote the Leonard system from Definition 1.2 and let the polynomials p_i be as in Definition 2.4. Let V denote an irreducible $\mathcal{A}$-module and let v denote a nonzero vector in $E^*_0 V$. Then $p_i(A)v = E^*_i A^i v$ for $0 \le i \le d$ and $p_{d+1}(A)v = 0$.*

Proof. We abbreviate $v_i = p_i(A)v$ for $0 \le i \le d+1$. We define $v'_i = E^*_i A^i v$ for $0 \le i \le d$ and $v'_{d+1} = 0$. We show $v_i = v'_i$ for $0 \le i \le d+1$. From the construction $v_0 = v$ and $v'_0 = v$ so $v_0 = v'_0$. From (2.13) we obtain

$$Av_i = v_{i+1} + a_i v_i + x_i v_{i-1} \qquad (0 \le i \le d) \tag{2.14}$$

where $v_{-1} = 0$. From Theorem 2.4 we find

$$Av'_i = v'_{i+1} + a_i v'_i + x_i v'_{i-1} \qquad (0 \le i \le d) \tag{2.15}$$

where $v'_{-1} = 0$. Comparing (2.14), (2.15) and using $v_0 = v'_0$ we find $v_i = v'_i$ for $0 \le i \le d+1$. The result follows. □

We mention a few consequences of Lemma 2.10.

Theorem 2.5. *Let Φ denote the Leonard system from Definition 1.2 and let the polynomials p_i be as in Definition 2.4. Let V denote an irreducible $\mathcal{A}$-module. Then*

$$p_i(A) E^*_0 V = E^*_i V \quad (0 \le i \le d).$$

Proof. Let v denote a nonzero vector in E_0^*V. Then $p_i(A)v = E_i^*A^iv$ by Lemma 2.10. Observe v is a basis for E_0^*V. By Theorem 2.3 we find $E_i^*A^iv$ is a basis for E_i^*V. Combining these facts we find $p_i(A)E_0^*V = E_i^*V$. □

Theorem 2.6. *Let Φ denote the Leonard system from Definition 1.2 and let the polynomials p_i be as in Definition 2.4. Then*

$$p_i(A)E_0^* = E_i^*A^iE_0^* \qquad (0 \le i \le d). \tag{2.16}$$

Proof. Let the integer i be given and abbreviate $\Delta = p_i(A) - E_i^*A^i$. We show $\Delta E_0^* = 0$. In order to do this we show $\Delta E_0^*V = 0$, where V denotes an irreducible $\mathcal{A}$-module. Let v denote a nonzero vector in E_0^*V and recall v is a basis for E_0^*V. By Lemma 2.10 we have $\Delta v = 0$ so $\Delta E_0^*V = 0$. Now $\Delta E_0^* = 0$ so $p_i(A)E_0^* = E_i^*A^iE_0^*$. □

Theorem 2.7. *Let Φ denote the Leonard system from Definition 1.2 and let the polynomial p_{d+1} be as in Definition 2.4. Then the following (i), (ii) hold.*

(i) p_{d+1} is both the minimal polynomial and the characteristic polynomial of A.

(ii) $p_{d+1} = \prod_{i=0}^d(\lambda - \theta_i)$.

Proof. (i) We first show p_{d+1} is equal to the minimal polynomial of A. Recall $I, A, \ldots, A^d$ are linearly independent and that p_{d+1} is monic with degree $d+1$. We show $p_{d+1}(A) = 0$. Let V denote an irreducible $\mathcal{A}$-module. Let v denote a nonzero vector in E_0^*V and recall v is a basis for E_0^*V. From Lemma 2.10 we find $p_{d+1}(A)v = 0$. It follows $p_{d+1}(A)E_0^*V = 0$ so $p_{d+1}(A)E_0^* = 0$. Applying Lemma 2.4 (with $X = p_{d+1}(A)$ and $Y = I$) we find $p_{d+1}(A) = 0$. We have now shown p_{d+1} is the minimal polynomial of A. By definition the characteristic polynomial of A is equal to $\det(\lambda I - A)$. This polynomial is monic with degree $d+1$ and has p_{d+1} as a factor; therefore it is equal to p_{d+1}.
(ii) For $0 \le i \le d$ the scalar θ_i is an eigenvalue of A and therefore a root of the characteristic polynomial of A. □

The following result will be useful.

Lemma 2.11. *Let Φ denote the Leonard system from Definition 1.2 and let the polynomials p_i be as in Definition 2.4. Let the scalars x_i be as in Definition 2.3. Then*

$$E_i^* = \frac{p_i(A)E_0^*p_i(A)}{x_1x_2\cdots x_i} \qquad (0 \le i \le d). \tag{2.17}$$

Proof. Let $\dagger : \mathcal{A} \to \mathcal{A}$ denote the antiautomorphism which corresponds to Φ. From Theorem 2.6 we have $p_i(A)E_0^* = E_i^*A^iE_0^*$. Applying $\dagger$ we find $E_0^*p_i(A) = E_0^*A^iE_i^*$. From these comments we find

$$\begin{aligned} p_i(A)E_0^*p_i(A) &= E_i^*A^iE_0^*A^iE_i^* \\ &= x_1x_2\cdots x_iE_i^* \end{aligned}$$

in view of Lemma 2.9(i). The result follows. □

2.4 The scalars ν, m_i

In this section we introduce some more scalars that will help us describe Leonard systems.

Definition 2.5. *Let Φ denote the Leonard system from Definition 1.2. We define*

$$m_i = \mathrm{tr}(E_i E_0^*) \qquad (0 \le i \le d). \tag{2.18}$$

Lemma 2.12. *Let Φ denote the Leonard system from Definition 1.2. Then (i)–(v) hold below.*

(i) $E_i E_0^* E_i = m_i E_i \qquad (0 \le i \le d)$.
(ii) $E_0^* E_i E_0^* = m_i E_0^* \qquad (0 \le i \le d)$.
(iii) $m_i \neq 0 \qquad (0 \le i \le d)$.
(iv) $\sum_{i=0}^{d} m_i = 1$.
(v) $m_0 = m_0^*$.

Proof. (i) Observe E_i is a basis for $E_i \mathcal{A} E_i$. By this and since $E_i E_0^* E_i$ is contained in $E_i \mathcal{A} E_i$, there exists $\alpha_i \in \mathbb{K}$ such that $E_i E_0^* E_i = \alpha_i E_i$. Taking the trace of both sides in this equation and using $\mathrm{tr}(XY) = \mathrm{tr}(YX)$, $\mathrm{tr}(E_i) = 1$ we find $\alpha_i = m_i$.
(ii) Similar to the proof of (i).
(iii) Observe $m_i E_i$ is equal to $E_i E_0^* E_i$ by part (i) above and $E_i E_0^* E_i$ is nonzero by Corollary 2.3. It follows $m_i E_i \neq 0$ so $m_i \neq 0$.
(iv) Multiply each term in the equation $\sum_{i=0}^{d} E_i = I$ on the right by E_0^*, and then take the trace. Evaluate the result using Definition 2.5.
(v) The elements $E_0 E_0^*$ and $E_0^* E_0$ have the same trace. □

Definition 2.6. *Let Φ denote the Leonard system from Definition 1.2. Recall $m_0 = m_0^*$ by Lemma 2.12(v); we let ν denote the multiplicative inverse of this common value. We observe $\nu = \nu^*$. We emphasize*

$$\mathrm{tr}(E_0 E_0^*) = \nu^{-1}. \tag{2.19}$$

Lemma 2.13. *Let Φ denote the Leonard system from Definition 1.2 and let the scalar ν be as in Definition 2.6. Then the following (i), (ii) hold.*

(i) $\nu E_0 E_0^* E_0 = E_0$.
(ii) $\nu E_0^* E_0 E_0^* = E_0^*$.

Proof. (i) Set $i = 0$ in Lemma 2.12(i) and recall $m_0 = \nu^{-1}$.
(ii) Set $i = 0$ in Lemma 2.12(ii) and recall $m_0 = \nu^{-1}$. □

3 The Standard Basis

In this section we discuss the notion of a *standard basis.* We begin with a comment.

Lemma 3.1. *Let Φ denote the Leonard system from Definition 1.2 and let V denote an irreducible $\mathcal{A}$-module. Then*

$$E_i^* V = E_i^* E_0 V \qquad (0 \le i \le d). \tag{3.1}$$

Proof. The space $E_i^* V$ has dimension 1 and contains $E_i^* E_0 V$. We show $E_i^* E_0 V \neq 0$. Applying Corollary 2.3 to Φ^* we find $E_i^* E_0 \neq 0$. It follows $E_i^* E_0 V \neq 0$. We conclude $E_i^* V = E_i^* E_0 V$. □

Lemma 3.2. *Let Φ denote the Leonard system from Definition 1.2 and let V denote an irreducible $\mathcal{A}$-module. Let u denote a nonzero vector in $E_0 V$. Then for $0 \le i \le d$ the vector $E_i^* u$ is nonzero and hence a basis for $E_i^* V$. Moreover the sequence*

$$E_0^* u, E_1^* u, \ldots, E_d^* u \tag{3.2}$$

is a basis for V.

Proof. Let the integer i be given. We show $E_i^* u \neq 0$. Recall $E_0 V$ has dimension 1 and u is a nonzero vector in $E_0 V$ so u spans $E_0 V$. Applying E_i^* we find $E_i^* u$ spans $E_i^* E_0 V$. The space $E_i^* E_0 V$ is nonzero by Lemma 3.1 so $E_i^* u$ is nonzero. The remaining assertions are clear. □

Definition 3.1. *Let Φ denote the Leonard system from Definition 1.2 and let V denote an irreducible $\mathcal{A}$-module. By a Φ-*standard basis *for V, we mean a sequence*

$$E_0^* u, E_1^* u, \ldots, E_d^* u,$$

where u is a nonzero vector in $E_0 V$.

We give a few characterizations of the standard basis.

Lemma 3.3. *Let Φ denote the Leonard system from Definition 1.2 and let V denote an irreducible $\mathcal{A}$-module. Let $v_0, v_1, \ldots, v_d$ denote a sequence of vectors in V, not all 0. Then this sequence is a Φ-standard basis for V if and only if both (i), (ii) hold below.*

(i) $v_i \in E_i^* V$ for $0 \le i \le d$.
(ii) $\sum_{i=0}^d v_i \in E_0 V$.

Proof. To prove the lemma in one direction, assume $v_0, v_1, \ldots, v_d$ is a Φ-standard basis for V. By Definition 3.1 there exists a nonzero $u \in E_0 V$ such that $v_i = E_i^* u$ for $0 \le i \le d$. Apparently $v_i \in E_i^* V$ for $0 \le i \le d$ so (i) holds. Let I denote the identity element of $\mathcal{A}$ and recall $I = \sum_{i=0}^d E_i^*$. Applying this to u we find $u = \sum_{i=0}^d v_i$ and (ii) follows. We have now proved the lemma in one direction. To prove the lemma in the other direction, assume $v_0, v_1, \ldots, v_d$ satisfy (i), (ii) above. We define $u = \sum_{i=0}^{d} v_i$ and observe $u \in E_0V$. Using (i) we find $E_i^* v_j = \delta_{ij} v_j$ for $0 \leq i, j \leq d$; it follows $v_i = E_i^* u$ for $0 \leq i \leq d$. Observe $u \neq 0$ since at least one of $v_0, v_1, \ldots, v_d$ is nonzero. Now $v_0, v_1, \ldots, v_d$ is a Φ-standard basis for V by Definition 3.1. □

We recall some notation. Let d denote a nonnegative integer and let B denote a matrix in $\mathrm{Mat}_{d+1}(\mathbb{K})$. Let α denote a scalar in $\mathbb{K}$. Then B is said to have *constant row sum* α whenever $B_{i0} + B_{i1} + \cdots + B_{id} = \alpha$ for $0 \leq i \leq d$.

Lemma 3.4. *Let Φ denote the Leonard system from Definition 1.2 and let the scalars θ_i, θ_i^* be as in Definition 2.1. Let V denote an irreducible $\mathcal{A}$-module and let $v_0, v_1, \ldots, v_d$ denote a basis for V. Let B (resp. B^*) denote the matrix in $\mathrm{Mat}_{d+1}(\mathbb{K})$ that represents A (resp. A^*) with respect to this basis. Then $v_0, v_1, \ldots, v_d$ is a Φ-standard basis for V if and only if both (i), (ii) hold below.*

(i) B has constant row sum θ_0.
(ii) $B^* = diag(\theta_0^*, \theta_1^*, \ldots, \theta_d^*)$.

Proof. Observe $A \sum_{j=0}^{d} v_j = \sum_{i=0}^{d} v_i (B_{i0} + B_{i1} + \cdots B_{id})$. Recall E_0V is the eigenspace for A and eigenvalue θ_0. Apparently B has constant row sum θ_0 if and only if $\sum_{i=0}^{d} v_i \in E_0V$. Recall that for $0 \leq i \leq d$, E_i^*V is the eigenspace for A^* and eigenvalue θ_i^*. Apparently $B^* = \mathrm{diag}(\theta_0^*, \theta_1^*, \ldots, \theta_d^*)$ if and only if $v_i \in E_i^*V$ for $0 \leq i \leq d$. The result follows in view of Lemma 3.3. □

Definition 3.2. *Let Φ denote the Leonard system from Definition 1.2. We define a map $\flat : \mathcal{A} \to \mathrm{Mat}_{d+1}(\mathbb{K})$ as follows. Let V denote an irreducible $\mathcal{A}$-module. For all $X \in \mathcal{A}$ we let $X^\flat$ denote the matrix in $Mat_{d+1}(\mathbb{K})$ that represents X with respect to a Φ-standard basis for V. We observe $\flat : \mathcal{A} \to \mathrm{Mat}_{d+1}(\mathbb{K})$ is an isomorphism of $\mathbb{K}$-algebras.*

Lemma 3.5. *Let Φ denote the Leonard system from Definition 1.2 and let the scalars θ_i, θ_i^* be as in Definition 2.1. Let the map $\flat : \mathcal{A} \to Mat_{d+1}(\mathbb{K})$ be as in Definition 3.2. Then (i)–(iii) hold below.*

(i) $A^\flat$ has constant row sum θ_0.
(ii) $A^{*\flat} = diag(\theta_0^*, \theta_1^*, \ldots, \theta_d^*)$.
(iii) For $0 \leq i \leq d$ the matrix $E_i^{*\flat}$ has ii entry 1 and all other entries 0.

Proof. (i), (ii) Combine Lemma 3.4 and Definition 3.2.
(iii) Immediate from Lemma 2.1(i). □

3.1 The Scalars b_i, c_i

In this section we consider some scalars that arise naturally in the context of the standard basis.

Definition 3.3. *Let Φ denote the Leonard system from Definition 1.2 and let the map $\flat : \mathcal{A} \to \mathrm{Mat}_{d+1}(\mathbb{K})$ be as in Definition 3.2. For $0 \le i \le d-1$ we let b_i denote the $i, i+1$ entry of $A^\flat$. For $1 \le i \le d$ we let c_i denote the $i, i-1$ entry of $A^\flat$. We observe*

$$A^\flat = \begin{pmatrix} a_0 & b_0 & & & & \mathbf{0} \\ c_1 & a_1 & b_1 & & & \\ & c_2 & \cdot & \cdot & & \\ & & \cdot & \cdot & \cdot & \\ & & & \cdot & \cdot & b_{d-1} \\ \mathbf{0} & & & & c_d & a_d \end{pmatrix}, \tag{3.3}$$

where the a_i are from Definition 2.3. For notational convenience we define $b_d = 0$ and $c_0 = 0$.

Lemma 3.6. *Let Φ denote the Leonard system from Definition 1.2 and let the scalars b_i, c_i be as in Definition 3.3. Then with reference to Definition 2.1 and Definition 2.3 the following (i), (ii) hold.*

(i) $b_{i-1}c_i = x_i \qquad (1 \le i \le d)$.
(ii) $c_i + a_i + b_i = \theta_0 \qquad (0 \le i \le d)$.

Proof. (i) Apply Lemma 2.7(ii) with $B = A^\flat$.
(ii) Combine (3.3) and Lemma 3.5(i). □

Lemma 3.7. *Let Φ denote the Leonard system from Definition 1.2 and let the scalars b_i, c_i be as in Definition 3.3. Let the polynomials p_i be as in Definition 2.4 and let the scalar θ_0 be as in Definition 2.1. Then the following (i)–(iii) hold.*

(i) $b_i \ne 0 \qquad (0 \le i \le d-1)$.
(ii) $c_i \ne 0 \qquad (1 \le i \le d)$.
(iii) $b_0 b_1 \cdots b_{i-1} = p_i(\theta_0) \qquad (0 \le i \le d+1)$.

Proof. (i), (ii) Immediate from Lemma 3.6(i) and since each of $x_1, x_2, \ldots, x_d$ is nonzero.
(iii) Assume $0 \le i \le d$; otherwise each side is zero. Let $\dagger : \mathcal{A} \to \mathcal{A}$ denote the antiautomorphism which corresponds to Φ. Applying $\dagger$ to both sides of (2.16) we get $E_0^* p_i(A) = E_0^* A^i E_i^*$. Let u denote a nonzero vector in $E_0 V$ and observe $Au = \theta_0 u$. Recall $E_0^* u, E_1^* u, \ldots, E_d^* u$ is a Φ-standard basis for V, and that $A^\flat$ represents A with respect to this basis. From (3.3) we find $b_0 b_1 \cdots b_{i-1}$ is the $0i$ entry of $A^{i\flat}$. Now

$$\begin{aligned} b_0 b_1 \cdots b_{i-1} E_0^* u &= E_0^* A^i E_i^* u \\ &= E_0^* p_i(A) u \\ &= p_i(\theta_0) E_0^* u \end{aligned}$$

and it follows $b_0 b_1 \cdots b_{i-1} = p_i(\theta_0)$. □

Theorem 3.1. *Let Φ denote the Leonard system from Definition 1.2 and let the polynomials p_i be as in Definition 2.4. Let the scalar θ_0 be as in Definition 2.1. Then $p_i(\theta_0) \neq 0$ for $0 \leq i \leq d$. Let the scalars b_i, c_i be as in Definition 3.3. Then*

$$b_i = \frac{p_{i+1}(\theta_0)}{p_i(\theta_0)} \qquad (0 \leq i \leq d) \tag{3.4}$$

and

$$c_i = \frac{x_i p_{i-1}(\theta_0)}{p_i(\theta_0)} \qquad (1 \leq i \leq d). \tag{3.5}$$

Proof. Observe $p_i(\theta_0) \neq 0$ for $0 \leq i \leq d$ by Lemma 3.7(i), (iii). Line (3.4) is immediate from Lemma 3.7(iii). To get (3.5) combine (3.4) and Lemma 3.6(i). □

Theorem 3.2. *Let Φ denote the Leonard system from Definition 1.2 and let the scalars c_i be as in Definition 3.3. Let the scalars θ_i be as in Definition 2.1 and let the scalar ν be as in Definition 2.6. Then*

$$(\theta_0 - \theta_1)(\theta_0 - \theta_2) \cdots (\theta_0 - \theta_d) = \nu c_1 c_2 \cdots c_d. \tag{3.6}$$

Proof. Let δ denote the expression on the left-hand side of (3.6). Setting $i = 0$ in (1.4) we find $\delta E_0 = \prod_{j=1}^{d}(A - \theta_j I)$. We multiply both sides of this equation on the left by E^*_d and on the right by E^*_0. We evaluate the resulting equation using Lemma 2.5(i) to obtain $\delta E^*_d E_0 E^*_0 = E^*_d A^d E^*_0$. We multiply both sides of this equation on the right by E_0 and use Lemma 2.13(i) to obtain

$$\delta \nu^{-1} E^*_d E_0 = E^*_d A^d E^*_0 E_0. \tag{3.7}$$

Let u denote a nonzero vector in $E_0 V$ and observe that $E_0 u = u$. Recall that $E^*_0 u, E^*_1 u, \ldots, E^*_d u$ is a Φ-standard basis for V, and that $A^\flat$ represents A with respect to this basis. From (3.3) we find $c_1 c_2 \cdots c_d$ is the $d0$ entry of $A^{d\flat}$. Now

$$\begin{aligned} c_1 c_2 \cdots c_d E^*_d u &= E^*_d A^d E^*_0 u \\ &= E^*_d A^d E^*_0 E_0 u \\ &= \delta \nu^{-1} E^*_d u \end{aligned}$$

so $c_1 c_2 \cdots c_d = \delta \nu^{-1}$. The result follows. □

3.2 The Scalars k_i

In this section we consider some scalars that are closely related to the scalars from Definition 2.5.

Definition 3.4. *Let Φ denote the Leonard system from Definition 1.2. We define*

$$k_i = m_i^* \nu \qquad (0 \leq i \leq d), \tag{3.8}$$

where the m_i^* are from Definition 2.5 and ν is from Definition 2.6.

Lemma 3.8. *Let Φ denote the Leonard system from Definition 1.2 and let the scalars k_i be as in Definition 3.4. Then*

(i) $k_0 = 1$;
(ii) $k_i \neq 0$ *for* $0 \leq i \leq d$;
(iii) $\sum_{i=0}^d k_i = \nu$.

Proof. (i) Set $i = 0$ in (3.8) and recall $m_0^* = \nu^{-1}$.
(ii) Applying Lemma 2.12(iii) to Φ^* we find $m_i^* \neq 0$ for $0 \leq i \leq d$. We have $\nu \neq 0$ by Definition 2.6. The result follows in view of (3.8).
(iii) Applying Lemma 2.12(iv) to Φ^* we find $\sum_{i=0}^d m_i^* = 1$. The result follows in view of (3.8). □

Lemma 3.9. *Let Φ denote the Leonard system from Definition 1.2 and let the scalars k_i be as in Definition 3.4. Then with reference to Definition 2.1, Definition 2.3, and Definition 2.4,*

$$k_i = \frac{p_i(\theta_0)^2}{x_1 x_2 \cdots x_i} \qquad (0 \leq i \leq d). \tag{3.9}$$

Proof. We show that each side of (3.9) is equal to $\nu \mathrm{tr}(E_i^* E_0)$. Using (2.18) and (3.8) we find $\nu \mathrm{tr}(E_i^* E_0)$ is equal to the left-hand side of (3.9). Using Lemma 2.11 we find $\nu \mathrm{tr}(E_i^* E_0)$ is equal to the right-hand side of (3.9). □

Theorem 3.3. *Let Φ denote the Leonard system from Definition 1.2 and let the scalars k_i be as in Definition 3.4. Let the scalars b_i, c_i be as in Definition 3.3. Then*

$$k_i = \frac{b_0 b_1 \cdots b_{i-1}}{c_1 c_2 \cdots c_i} \qquad (0 \leq i \leq d). \tag{3.10}$$

Proof. Evaluate the expression on the right in (3.9) using Lemma 3.6(i) and Lemma 3.7(iii). □

3.3 The Polynomials v_i

Let Φ denote the Leonard system from Definition 1.2 and let the polynomials p_i be as in Definition 2.4. The p_i have two normalizations of interest; we call these the u_i and the v_i. In this section we discuss the v_i. In the next section we will discuss the u_i.

Definition 3.5. *Let Φ denote the Leonard system from Definition 1.2 and let the polynomials p_i be as in Definition 2.4. For $0 \leq i \leq d$ we define the polynomial v_i by*

$$v_i = \frac{p_i}{c_1 c_2 \cdots c_i}, \tag{3.11}$$

where the c_j are from Definition 3.3. We observe $v_0 = 1$.

Lemma 3.10. *Let Φ denote the Leonard system from Definition 1.2 and let the polynomials v_i be as in Definition 3.5. Let the scalar θ_0 be as in Definition 2.1 and let the scalars k_i be as in Definition 3.4. Then*

$$v_i(\theta_0) = k_i \qquad (0 \leq i \leq d). \tag{3.12}$$

Proof. Use Lemma 3.7(iii), Theorem 3.3, and (3.11). □

Lemma 3.11. *Let Φ denote the Leonard system from Definition 1.2 and let the polynomials v_i be as in Definition 3.5. Let the scalars a_i, b_i, c_i be as in Definition 2.3 and Definition 3.3. Then*

$$\lambda v_i = c_{i+1} v_{i+1} + a_i v_i + b_{i-1} v_{i-1} \qquad (0 \leq i \leq d-1), \tag{3.13}$$

where $b_{-1} = 0$ and $v_{-1} = 0$. Moreover

$$\lambda v_d - a_d v_d - b_{d-1} v_{d-1} = (c_1 c_2 \cdots c_d)^{-1} p_{d+1}. \tag{3.14}$$

Proof. In (2.13), divide both sides by $c_1 c_2 \cdots c_i$. Evaluate the result using Lemma 3.6(i) and (3.11). □

Theorem 3.4. *Let Φ denote the Leonard system from Definition 1.2 and let the polynomials v_i be as in Definition 3.5. Let V denote an irreducible $\mathcal{A}$-module and let u denote a nonzero vector in $E_0 V$. Then*

$$v_i(A) E_0^* u = E_i^* u \qquad (0 \leq i \leq d). \tag{3.15}$$

Proof. For $0 \leq i \leq d$ we define $w_i = v_i(A) E_0^* u$ and $w_i' = E_i^* u$. We show $w_i = w_i'$. Each of w_0, w_0' is equal to $E_0^* u$ so $w_0 = w_0'$. Using Lemma 3.11 we obtain

$$A w_i = c_{i+1} w_{i+1} + a_i w_i + b_{i-1} w_{i-1} \qquad (0 \leq i \leq d-1) \tag{3.16}$$

where $w_{-1} = 0$ and $b_{-1} = 0$. By Definition 3.2, Definition 3.3, and since $w_0', w_1', \ldots, w_d'$ is a Φ-standard basis,

$$A w_i' = c_{i+1} w_{i+1}' + a_i w_i' + b_{i-1} w_{i-1}' \qquad (0 \leq i \leq d-1) \tag{3.17}$$

where $w_{-1}' = 0$. Comparing (3.16), (3.17) and using $w_0 = w_0'$ we find $w_i = w_i'$ for $0 \leq i \leq d$. The result follows. □

3.4 The Polynomials u_i

Let Φ denote the Leonard system from Definition 1.2 and let the polynomials p_i be as in Definition 2.4. In the previous section we gave a normalization of the p_i that we called the v_i. In this section we give a normalization for the p_i that we call the u_i.

Definition 3.6. *Let Φ denote the Leonard system from Definition 1.2 and let the polynomials p_i be as in Definition 2.4. For $0 \leq i \leq d$ we define the polynomial u_i by*

$$u_i = \frac{p_i}{p_i(\theta_0)}, \tag{3.18}$$

where θ_0 is from Definition 2.1. We observe $u_0 = 1$. Moreover

$$u_i(\theta_0) = 1 \qquad (0 \leq i \leq d). \tag{3.19}$$

Lemma 3.12. *Let Φ denote the Leonard system from Definition 1.2 and let the polynomials u_i be as in Definition 3.6. Let the scalars a_i, b_i, c_i be as in Definition 2.3 and Definition 3.3. Then*

$$\lambda u_i = b_i u_{i+1} + a_i u_i + c_i u_{i-1} \qquad (0 \leq i \leq d-1), \tag{3.20}$$

where $u_{-1} = 0$. Moreover

$$\lambda u_d - c_d u_{d-1} - a_d u_d = p_d(\theta_0)^{-1} p_{d+1}, \tag{3.21}$$

where θ_0 is from Definition 2.1.

Proof. In (2.13), divide both sides by $p_i(\theta_0)$ and evaluate the result using Lemma 3.6(i), Theorem 3.1, and (3.18). □

The u_i and v_i are related as follows.

Lemma 3.13. *Let Φ denote the Leonard system from Definition 1.2. Let the polynomials u_i, v_i be as in Definition 3.6 and Definition 3.5 respectively. Then*

$$v_i = k_i u_i \qquad (0 \leq i \leq d), \tag{3.22}$$

where the k_i are from Definition 3.4.

Proof. Compare (3.11) and (3.18) in light of Lemma 3.7(iii) and Theorem 3.3.
□

3.5 A Bilinear Form

In this section we associate with each Leonard pair a certain bilinear form. To prepare for this we recall a few concepts from linear algebra.

Let V denote a finite dimensional vector space over $\mathbb{K}$. By a *bilinear form on V* we mean a map $\langle \cdot, \cdot \rangle : V \times V \to \mathbb{K}$ that satisfies the following four conditions for all $u, v, w \in V$ and for all $\alpha \in \mathbb{K}$:

(i) $\langle u+v, w\rangle = \langle u, w\rangle + \langle v, w\rangle$;
(ii) $\langle \alpha u, v\rangle = \alpha\langle u, v\rangle$;
(iii) $\langle u, v+w\rangle = \langle u, v\rangle + \langle u, w\rangle$;
(iv) $\langle u, \alpha v\rangle = \alpha\langle u, v\rangle$.

We observe that a scalar multiple of a bilinear form on V is a bilinear form on V. Let $\langle \cdot,\cdot \rangle$ denote a bilinear form on V. This form is said to be *symmetric* whenever $\langle u, v\rangle = \langle v, u\rangle$ for all $u, v \in V$. Also, the following are equivalent:

(i) there exists a nonzero $u \in V$ such that $\langle u, v\rangle = 0$ for all $v \in V$;
(ii) there exists a nonzero $v \in V$ such that $\langle u, v\rangle = 0$ for all $u \in V$.

The form $\langle \cdot,\cdot \rangle$ is said to be *degenerate* whenever (i), (ii) hold and *nondegenerate* otherwise. Let $\gamma : \mathcal{A} \to \mathcal{A}$ denote an antiautomorphism and let V denote an irreducible $\mathcal{A}$-module. Then there exists a nonzero bilinear form $\langle \cdot,\cdot \rangle$ on V such that $\langle Xu, v\rangle = \langle u, X^\gamma v\rangle$ for all $u, v \in V$ and for all $X \in \mathcal{A}$. The form is unique up to multiplication by a nonzero scalar in $\mathbb{K}$. The form is nondegenerate. We refer to this form as the *bilinear form on V associated with γ*. This form is not symmetric in general.

We now return our attention to Leonard pairs.

Definition 3.7. *Let $\Phi = (A; A^*; \{E_i\}_{i=0}^d; \{E_i^*\}_{i=0}^d)$ denote a Leonard system in $\mathcal{A}$. Let $\dagger : \mathcal{A} \to \mathcal{A}$ denote the corresponding antiautomorphism from Definition 2.2. Let V denote an irreducible $\mathcal{A}$-module. For the rest of this paper we let $\langle \cdot,\cdot \rangle$ denote the bilinear form on V associated with $\dagger$. We abbreviate $\|u\|^2 = \langle u, u\rangle$ for all $u \in V$. By the construction, for $X \in \mathcal{A}$ we have*

$$\langle Xu, v\rangle = \langle u, X^\dagger v\rangle \qquad (\forall u \in V, \forall v \in V). \tag{3.23}$$

We make an observation.

Lemma 3.14. *With reference to Definition 3.7, let $\mathcal{D}$ (resp. $\mathcal{D}^*$) denote the subalgebra of $\mathcal{A}$ generated by A (resp. A^*). Then for $X \in \mathcal{D} \cup \mathcal{D}^*$ we have*

$$\langle Xu, v\rangle = \langle u, Xv\rangle \qquad (\forall u \in V, \forall v \in V). \tag{3.24}$$

Proof. Combine (3.23) and Lemma 2.6. □

Theorem 3.5. *With reference to Definition 3.7, let u denote a nonzero vector in E_0V and recall $E_0^*u, E_1^*u, \dots, E_d^*u$ is a Φ-standard basis for V. We have*

$$\langle E_i^*u, E_j^*u\rangle = \delta_{ij} k_i \nu^{-1} \|u\|^2 \qquad (0 \le i, j \le d), \tag{3.25}$$

where the k_i are from Definition 3.4 and ν is from Definition 2.6.

Proof. By (3.24) and since $E_0u = u$ we find $\langle E_i^*u, E_j^*u\rangle = \langle u, E_0E_i^*E_j^*E_0u\rangle$. Using Lemma 2.12(ii) and (3.8) we find $\langle u, E_0E_i^*E_j^*E_0u\rangle = \delta_{ij}k_i\nu^{-1}\|u\|^2$. □

Corollary 3.1. *With reference to Definition 3.7, the bilinear form* $\langle \cdot, \cdot \rangle$ *is symmetric.*

Proof. Let u denote a nonzero vector in E_0V and abbreviate $v_i = E_i^* u$ for $0 \le i \le d$. From Theorem 3.5 we find $\langle v_i, v_j \rangle = \langle v_j, v_i \rangle$ for $0 \le i, j \le d$. The result follows since $v_0, v_1, \ldots, v_d$ is a basis for V. □

We have a comment.

Lemma 3.15. *With reference to Definition 3.7, let* u *denote a nonzero vector in* E_0V *and let* v *denote a nonzero vector in* E_0^*V. *Then the following (i)–(iv) hold.*

(i) Each of $\|u\|^2, \|v\|^2, \langle u, v \rangle$ *is nonzero.*
(ii) $E_0^* u = \langle u, v \rangle \|v\|^{-2} v$.
(iii) $E_0 v = \langle u, v \rangle \|u\|^{-2} u$.
(iv) $\nu \langle u, v \rangle^2 = \|u\|^2 \|v\|^2$.

Proof. (i) Observe $\|u\|^2 \ne 0$ by Theorem 3.5 and since $\langle \cdot, \cdot \rangle$ is not 0. Similarly $\|v\|^2 \ne 0$. To see that $\langle u, v \rangle \ne 0$, observe that v is a basis for E_0^*V so there exists $\alpha \in \mathbb{K}$ such that $E_0^* u = \alpha v$. Recall $E_0^* u \ne 0$ by Lemma 3.2 so $\alpha \ne 0$. Using (3.24) and $E_0^* v = v$ we routinely find $\langle u, v \rangle = \alpha \|v\|^2$ and it follows $\langle u, v \rangle \ne 0$.
(ii) In the proof of part (i) we found $E_0^* u = \alpha v$ where $\langle u, v \rangle = \alpha \|v\|^2$. The result follows.
(iii) Similar to the proof of (ii) above.
(iv) Using $u = E_0 u$ and $\nu E_0 E_0^* E_0 = E_0$ we find $\nu^{-1} u = E_0 E_0^* u$. To finish the proof, evaluate $E_0 E_0^* u$ using (ii) above and then (iii) above. □

4 Askey-Wilson Duality

In this section we show the polynomials u_i, v_i, p_i satisfy a relation known as *Askey-Wilson duality.* We begin with a lemma.

Lemma 4.1. *With reference to Definition 3.7, let* u *denote a nonzero vector in* E_0V *and let* v *denote a nonzero vector in* E_0^*V. *Then*

$$\langle E_i^* u, E_j v \rangle = \nu^{-1} k_i k_j^* u_i(\theta_j) \langle u, v \rangle \qquad (0 \le i, j \le d). \tag{4.1}$$

Proof. Using Theorem 3.4 we find

$$\begin{aligned} \langle E_i^* u, E_j v \rangle &= \langle v_i(A) E_0^* u, E_j v \rangle \\ &= \langle E_0^* u, v_i(A) E_j v \rangle \\ &= v_i(\theta_j) \langle E_0^* u, E_j v \rangle \\ &= v_i(\theta_j) \langle E_0^* u, v_j^*(A^*) E_0 v \rangle \\ &= v_i(\theta_j) \langle v_j^*(A^*) E_0^* u, E_0 v \rangle \\ &= v_i(\theta_j) v_j^*(\theta_0^*) \langle E_0^* u, E_0 v \rangle. \end{aligned} \tag{4.2}$$

Using Lemma 3.15(ii)–(iv) we find $\langle E_0^* u, E_0 v\rangle = \nu^{-1}\langle u, v\rangle$. Observe $v_i(\theta_j) = u_i(\theta_j)k_i$ by (3.22). Applying Lemma 3.10 to Φ^* we find $v_j^*(\theta_0^*) = k_j^*$. Evaluating (4.2) using these comments we obtain (4.1). □

Theorem 4.1. *Let Φ denote the Leonard system from Definition 1.2. Let the polynomials u_i be as in Definition 3.6 and recall the u_i^* are the corresponding polynomials for Φ^*. Let the scalars θ_i, θ_i^* be as in Definition 2.1. Then*

$$u_i(\theta_j) = u_j^*(\theta_i^*) \qquad (0 \le i, j \le d). \tag{4.3}$$

Proof. Applying Lemma 4.1 to Φ^* we find

$$\langle E_j v, E_i^* u\rangle = \nu^{-1} k_j^* k_i u_j^*(\theta_i^*)\langle u, v\rangle \qquad (0 \le i, j \le d). \tag{4.4}$$

To finish the proof, compare (4.1), (4.4), and recall $\langle \cdot, \cdot \rangle$ is symmetric. □

In the following two theorems we show how (4.3) looks in terms of the polynomials v_i and p_i.

Theorem 4.2. *Let Φ denote the Leonard system from Definition 1.2. With reference to Definition 1.5, Definition 2.1, and Definition 3.5,*

$$v_i(\theta_j)/k_i = v_j^*(\theta_i^*)/k_j^* \qquad (0 \le i, j \le d). \tag{4.5}$$

Proof. Evaluate (4.3) using Lemma 3.13. □

Theorem 4.3. *Let Φ denote the Leonard system from Definition 1.2. With reference to Definition 1.5, Definition 2.1, and Definition 2.4,*

$$\frac{p_i(\theta_j)}{p_i(\theta_0)} = \frac{p_j^*(\theta_i^*)}{p_j^*(\theta_0^*)} \qquad (0 \le i, j \le d). \tag{4.6}$$

Proof. Evaluate (4.3) using Definition 3.6. □

The equations (4.3), (4.5), (4.6) are often referred to as *Askey-Wilson duality*.

4.1 The Three-Term Recurrence and the Difference Equation

In Lemma 3.12 we gave a three-term recurrence for the polynomials u_i. This recurrence is often expressed as follows.

Theorem 4.4. *Let Φ denote the Leonard system from Definition 1.2 and let the polynomials u_i be as in Definition 3.6. Let the scalars θ_i be as in Definition 2.1. Then for $0 \le i, j \le d$ we have*

$$\theta_j u_i(\theta_j) = b_i u_{i+1}(\theta_j) + a_i u_i(\theta_j) + c_i u_{i-1}(\theta_j), \tag{4.7}$$

where $u_{-1} = 0$ and $u_{d+1} = 0$.

Proof. Apply Lemma 3.12 (with $\lambda = \theta_j$) and observe $p_{d+1}(\theta_j) = 0$ by Theorem 2.7(ii). □

Applying Theorem 4.4 to Φ^* and using Theorem 4.1 we routinely obtain the following.

Theorem 4.5. *Let Φ denote the Leonard system from Definition 1.2 and let the polynomials u_i be as in Definition 3.6. Then for $0 \le i, j \le d$ we have*

$$\theta_i^* u_i(\theta_j) = b_j^* u_i(\theta_{j+1}) + a_j^* u_i(\theta_j) + c_j^* u_i(\theta_{j-1}), \tag{4.8}$$

where θ_{-1}, θ_{d+1} denote indeterminates.

We refer to (4.8) as the *difference equation* satisfied by the u_i.

4.2 The Orthogonality Relations

In this section we show that each of the polynomials p_i, u_i, v_i satisfy an orthogonality relation. We begin with a lemma.

Lemma 4.2. *With reference to Definition 3.7, let u denote a nonzero vector in E_0V and let v denote a nonzero vector in E_0^*V. Then for $0 \le i \le d$, both*

$$E_i^* u = \frac{\langle u, v\rangle}{\|v\|^2} \sum_{j=0}^{d} v_i(\theta_j) E_j v, \tag{4.9}$$

$$E_i v = \frac{\langle u, v\rangle}{\|u\|^2} \sum_{j=0}^{d} v_i^*(\theta_j^*) E_j^* u. \tag{4.10}$$

Proof. We first show (4.9). To do this we show each side of (4.9) is equal to $v_i(A)E_0^*u$. By Theorem 3.4 we find $v_i(A)E_0^*u$ is equal to the left-hand side of (4.9). To see that $v_i(A)E_0^*u$ is equal to the right-hand side of (4.9), multiply $v_i(A)E_0^*u$ on the left by the identity I, expand using $I = \sum_{j=0}^{d} E_j$, and simplify the result using $E_jA = \theta_j E_j$ $(0 \le j \le d)$ and Lemma 3.15(ii). We have now proved (4.9). Applying (4.9) to Φ^* we obtain (4.10). □

We now display the orthogonality relations for the polynomials v_i.

Theorem 4.6. *Let Φ denote the Leonard system from Definition 1.2 and let the polynomials v_i be as in Definition 3.5. Then both*

$$\sum_{r=0}^{d} v_i(\theta_r) v_j(\theta_r) k_r^* = \delta_{ij} \nu k_i \qquad (0 \le i, j \le d), \tag{4.11}$$

$$\sum_{i=0}^{d} v_i(\theta_r) v_i(\theta_s) k_i^{-1} = \delta_{rs} \nu k_r^{*-1} \qquad (0 \le r, s \le d). \tag{4.12}$$

Proof. Let V denote an irreducible $\mathcal{A}$-module and let v denote a nonzero vector in E_0^*V. Applying Theorem 3.5 to Φ^* we find $\langle E_r v, E_s v\rangle = \delta_{rs} k_r^* \nu^{-1} \|v\|^2$ for $0 \le r, s \le d$. To obtain (4.11), in equation (3.25), eliminate each of $E_i^* u$, $E_j^* u$ using (4.9), and simplify the result using our preliminary comment and Lemma 3.15. To obtain (4.12), apply (4.11) to Φ^* and use Askey-Wilson duality. □

We now turn to the polynomials u_i.

Theorem 4.7. *Let Φ denote the Leonard system from Definition 1.2 and let the polynomials u_i be as in Definition 3.6. Then both*

$$\sum_{r=0}^{d} u_i(\theta_r) u_j(\theta_r) k_r^* = \delta_{ij} \nu k_i^{-1} \qquad (0 \le i, j \le d),$$

$$\sum_{i=0}^{d} u_i(\theta_r) u_i(\theta_s) k_i = \delta_{rs} \nu k_r^{*-1} \qquad (0 \le r, s \le d).$$

Proof. Evaluate each of (4.11), (4.12) using Lemma 3.13. □

We now turn to the polynomials p_i.

Theorem 4.8. *Let Φ denote the Leonard system from Definition 1.2 and let the polynomials p_i be as in Definition 2.4. Then both*

$$\sum_{r=0}^{d} p_i(\theta_r) p_j(\theta_r) m_r = \delta_{ij} x_1 x_2 \cdots x_i \qquad (0 \le i, j \le d),$$

$$\sum_{i=0}^{d} \frac{p_i(\theta_r) p_i(\theta_s)}{x_1 x_2 \cdots x_i} = \delta_{rs} m_r^{-1} \qquad (0 \le r, s \le d).$$

Proof. Applying Definition 3.4 to Φ^* we find $k_r^* = m_r \nu$ for $0 \le r \le d$. Evaluate each of (4.11), (4.12) using this and Definition 3.5, Lemma 3.6(i), (3.10). □

4.3 The Matrix P

In this section we express Lemma 4.2 in matrix form and consider the consequences.

Definition 4.1. *Let Φ denote the Leonard system from Definition 1.2. We define a matrix $P \in \mathrm{Mat}_{d+1}(\mathbb{K})$ as follows. For $0 \le i, j \le d$ the entry $P_{ij} = v_j(\theta_i)$, where θ_i is from Definition 2.1 and v_j is from Definition 3.5.*

Theorem 4.9. *Let Φ denote the Leonard system from Definition 1.2. Let the matrix P be as in Definition 4.1 and recall P^* is the corresponding matrix for Φ^*. Then $P^* P = \nu I$, where ν is from Definition 2.6.*

Proof. Compare (4.9), (4.10) and use Lemma 3.15(iv). □

Theorem 4.10. *Let Φ denote the Leonard system from Definition 1.2 and let the matrix P be as in Definition 4.1. Let the map $\flat : \mathcal{A} \to \mathrm{Mat}_{d+1}(\mathbb{K})$ be as in Definition 3.2 and let $\sharp : \mathcal{A} \to \mathrm{Mat}_{d+1}(\mathbb{K})$ denote the corresponding map for Φ^*. Then for all $X \in \mathcal{A}$ we have*

$$X^\sharp P = PX^\flat. \tag{4.13}$$

Proof. Let V denote an irreducible $\mathcal{A}$-module. Let u denote a nonzero vector in E_0V and recall $E_0^*u, E_1^*u, \ldots, E_d^*u$ is a Φ-standard basis for V. By Definition 3.2, $X^\flat$ is the matrix in $\mathrm{Mat}_{d+1}(\mathbb{K})$ that represents X with respect to $E_0^*u, E_1^*u, \ldots, E_d^*u$. Similarly for a nonzero $v \in E_0^*V$, $X^\sharp$ is the matrix in $\mathrm{Mat}_{d+1}(\mathbb{K})$ that represents X with respect to $E_0v, E_1v, \ldots, E_dv$. In view of (4.9), the transition matrix from $E_0v, E_1v, \ldots, E_dv$ to $E_0^*u, E_1^*u, \ldots, E_d^*u$ is a scalar multiple of P. The result follows from these comments and elementary linear algebra. □

5 The Split Decomposition

Notation 5.1 Throughout this section we let $\Phi = (A; A^*; \{E_i\}_{i=0}^d; \{E_i^*\}_{i=0}^d)$ denote a Leonard system in $\mathcal{A}$, with eigenvalue sequence $\theta_0, \theta_1, \ldots, \theta_d$ and dual eigenvalue sequence $\theta_0^*, \theta_1^*, \ldots, \theta_d^*$. We let V denote an irreducible $\mathcal{A}$-module.

With reference to Notation 5.1, by a *decomposition* of V we mean a sequence $U_0, U_1, \ldots, U_d$ consisting of 1-dimensional subspaces of V such that

$$V = U_0 + U_1 + \cdots + U_d \qquad \text{(direct sum)}.$$

In this section we are concerned with the following type of decomposition.

Definition 5.2. *With reference to Notation 5.1, a decomposition $U_0, \ldots, U_d$ of V is said to be Φ-*split *whenever both*

$$(A - \theta_i I)U_i \subseteq U_{i+1} \quad (0 \le i \le d-1), \qquad (A - \theta_d I)U_d = 0, \tag{5.1}$$

$$(A^* - \theta_i^* I)U_i \subseteq U_{i-1} \quad (1 \le i \le d), \qquad (A^* - \theta_0^* I)U_0 = 0. \tag{5.2}$$

Our goal in this section is to show there exists a unique Φ-split decomposition of V. The following definition will be useful.

Definition 5.3. *With reference to Notation 5.1, we set*

$$V_{ij} = \left(\sum_{h=0}^{i} E_h^* V\right) \cap \left(\sum_{k=j}^{d} E_k V\right) \tag{5.3}$$

for all integers i, j. We interpret the sum on the left in (5.3) to be 0 (resp. V) if $i < 0$ (resp. $i > d$). Similarily, we interpret the sum on the right in (5.3) to be V (resp. 0) if $j < 0$ (resp. $j > d$).

Lemma 5.1. *With reference to Notation 5.1 and Definition 5.3, we have*

(i) $V_{i0} = E_0^*V + E_1^*V + \cdots + E_i^*V \qquad (0 \le i \le d)$,
(ii) $V_{dj} = E_jV + E_{j+1}V + \cdots + E_dV \qquad (0 \le j \le d)$.

Proof. To get (i), set $j = 0$ in (5.3), and apply (1.5). Line (ii) is similarily obtained. □

Lemma 5.2. *With reference to Notation 5.1 and Definition 5.3, the following (i)–(iv) hold for* $0 \le i, j \le d$.

(i) $(A - \theta_j I)V_{ij} \subseteq V_{i+1,j+1}$,
(ii) $AV_{ij} \subseteq V_{ij} + V_{i+1,j+1}$,
(iii) $(A^* - \theta_i^* I)V_{ij} \subseteq V_{i-1,j-1}$,
(iv) $A^*V_{ij} \subseteq V_{ij} + V_{i-1,j-1}$.

Proof. (i) Using Definition 1.2(v) we find

$$(A - \theta_j I)\sum_{h=0}^{i} E_h^*V \subseteq \sum_{h=0}^{i+1} E_h^*V. \tag{5.4}$$

Also observe

$$(A - \theta_j I)\sum_{k=j}^{d} E_kV = \sum_{k=j+1}^{d} E_kV. \tag{5.5}$$

Evaluating $(A-\theta_j I)V_{ij}$ using (5.3), (5.4), (5.5) we routinely find it is contained in $V_{i+1,j+1}$.
(ii) Immediate from (i) above.
(iii) Similar to the proof of (i) above.
(iv) Immediate from (iii) above. □

Lemma 5.3. *With reference to Definition 5.3, we have*

$$V_{ij} = 0 \quad \text{if} \quad i < j, \qquad (0 \le i, j \le d). \tag{5.6}$$

Proof. We show the sum

$$V_{0r} + V_{1,r+1} + \cdots + V_{d-r,d} \tag{5.7}$$

is zero for $0 < r \le d$. Let r be given, and let W denote the sum in (5.7). Applying Lemma 5.2(ii),(iv), we find $AW \subseteq W$ and $A^*W \subseteq W$. Now $W = 0$ or $W = V$ in view of Corollary 2.2. By Definition 5.3, each term in (5.7) is contained in

$$E_rV + E_{r+1}V + \cdots + E_dV, \tag{5.8}$$

so W is contained in (5.8). The sum (5.8) is properly contained in V by (1.5), and since $r > 0$. Apparently $W \ne V$, so $W = 0$. We have now shown (5.7) is zero for $0 < r \le d$, and (5.6) follows. □

Theorem 5.1. *With reference to Notation 5.1, let $U_0, U_1, \ldots, U_d$ denote subspaces of V. Then the following (i)–(iii) are equivalent.*

(i) $U_i = (E_0^*V+E_1^*V+\cdots+E_i^*V)\cap(E_iV+E_{i+1}V+\cdots+E_dV) \qquad (0 \le i \le d).$
(ii) The sequence $U_0, U_1, \ldots, U_d$ is a Φ-split decomposition of V.
(iii) For $0 \le i \le d$, both

$$U_i + U_{i+1} + \cdots + U_d = E_iV + E_{i+1}V + \cdots + E_dV, \tag{5.9}$$
$$U_0 + U_1 + \cdots + U_i = E_0^*V + E_1^*V + \cdots + E_i^*V. \tag{5.10}$$

Proof. (*i*) → (*ii*) To get (5.1) and (5.2), set $j = i$ in Lemma 5.2(i),(iii), and observe $U_i = V_{ii}$. We now show the sequence $U_0, U_1, \ldots, U_d$ is a decomposition of V. Define $W = \sum_{i=0}^{d} U_i$. Then $AW \subseteq W$ by (5.1) and $A^*W \subseteq W$ by (5.2). Now $W = 0$ or $W = V$ in view of Corollary 2.2. However W contains U_0, and $U_0 = E_0^*V$ is nonzero, so $W \neq 0$. It follows $W = V$, and in other words

$$V = U_0 + U_1 + \cdots + U_d. \tag{5.11}$$

We show the sum (5.11) is direct. To do this, we show

$$(U_0 + U_1 + \cdots + U_{i-1}) \cap U_i = 0$$

for $1 \le i \le d$. Let the integer i be given. From the construction

$$U_j \subseteq E_0^*V + E_1^*V + \cdots + E_{i-1}^*V$$

for $0 \le j \le i-1$, and

$$U_i \subseteq E_iV + E_{i+1}V + \cdots + E_dV.$$

It follows

$$\begin{aligned}
&(U_0 + U_1 + \cdots + U_{i-1}) \cap U_i \\
&\quad \subseteq (E_0^*V + E_1^*V + \cdots + E_{i-1}^*V) \cap (E_iV + E_{i+1}V + \cdots + E_dV) \\
&\quad = V_{i-1,i} \\
&\quad = 0
\end{aligned}$$

in view of Lemma 5.3. We have now shown the sum (5.11) is direct. We now show U_i has dimension 1 for $0 \le i \le d$. Since the sum (5.11) is direct, this will follow if we can show $U_i \neq 0$ for $0 \le i \le d$. Suppose there exists an integer i $(0 \le i \le d)$ such that $U_i = 0$. We observe $i \neq 0$, since $U_0 = E_0^*V$ is nonzero, and $i \neq d$, since $U_d = E_dV$ is nonzero. Set

$$U = U_0 + U_1 + \cdots + U_{i-1},$$

and observe $U \neq 0$ and $U \neq V$ by our remarks above. By Lemma 5.2(ii) and since $U_i = 0$ we find $AU \subseteq U$. By Lemma 5.2(iv) we find $A^*U \subseteq U$. Now

$U = 0$ or $U = V$ in view of Corollary 2.2, for a contradiction. We conclude $U_i \neq 0$ for $0 \leq i \leq d$ and it follows $U_0, U_1, \ldots, U_d$ is a decomposition of V.

(ii) $\rightarrow$ *(iii)* First consider (5.9). Let i be given, and abbreviate

$$Z = E_iV + E_{i+1}V + \cdots + E_dV, \qquad W = U_i + U_{i+1} + \cdots + U_d.$$

We show $Z = W$. To obtain $Z \subseteq W$, set $X = \prod_{h=0}^{i-1}(A - \theta_h I)$, and observe $Z = XV$. Using (5.1) we find $XU_j \subseteq W$ for $0 \leq j \leq d$. By this and since $U_0, U_1, \ldots, U_d$ is a decomposition of V we find $XV \subseteq W$. We now have $Z \subseteq W$. Each of Z, W has dimension $d - i + 1$ so $Z = W$. We now have (5.9). Line (5.10) is similarily obtained.

(iii) $\rightarrow$ *(i)* We first show the sum $U_0 + \cdots + U_d$ is direct. To do this, we show

$$(U_0 + U_1 + \cdots + U_{i-1}) \cap U_i \tag{5.12}$$

is zero for $1 \leq i \leq d$. Let i be given. From (5.9), (5.10), we find (5.12) is contained in

$$(E_0^*V + E_1^*V + \cdots + E_{i-1}^*V) \cap (E_iV + E_{i+1}V + \cdots + E_dV). \tag{5.13}$$

The expression (5.13) equals $V_{i-1,i}$, and is hence zero by Lemma 5.3. It follows (5.12) is zero, and we have now shown the sum $U_0 + \cdots + U_d$ is direct. Combining this with (5.9), (5.10), we find

$$\begin{aligned} U_i &= (U_0 + U_1 + \cdots + U_i) \cap (U_i + U_{i+1} + \cdots + U_d) \\ &= (E_0^*V + E_1^*V + \cdots + E_i^*V) \cap (E_iV + E_{i+1}V + \cdots + E_dV), \end{aligned}$$

as desired. □

Corollary 5.1. *With reference to Notation 5.1, there exists a unique Φ-split decomposition of V.*

Proof. Immediate from Theorem 5.1(i),(ii). □

We finish this section with a comment.

Lemma 5.4. *With reference to Notation 5.1, let $U_0, U_1, \ldots, U_d$ denote the Φ-split decomposition of V. Then the following (i), (ii) hold.*

(i) $(A - \theta_i I)U_i = U_{i+1} \qquad (0 \leq i \leq d - 1)$.
(ii) $(A^* - \theta_i^* I)U_i = U_{i-1} \qquad (1 \leq i \leq d)$.

Proof. (i) Let i be given. Recall $(A - \theta_i I)U_i$ is contained in U_{i+1} by (5.1) and U_{i+1} has dimension 1, so it suffices to show

$$(A - \theta_i I)U_i \neq 0. \tag{5.14}$$

Assume $(A - \theta_i I)U_i = 0$, and set $W = \sum_{h=0}^{i} U_h$. Since $U_0, U_1, \ldots, U_d$ is a decomposition of V, and since $0 \leq i \leq d - 1$ we find $W \neq 0$ and $W \neq V$.

Observe $AU_i \subseteq U_i$ by our above assumption; combining this with (5.1) we find $AW \subseteq W$. By (5.2) we find $A^*W \subseteq W$. Now $W = 0$ or $W = V$ in view of Corollary 2.2, for a contradiction. We conclude (5.14) holds and the result follows.
(ii) Similar to the proof of (i) above. □

5.1 The Split Basis

Let $\Phi = (A; A^*; \{E_i\}_{i=0}^d; \{E_i^*\}_{i=0}^d)$ denote a Leonard system in $\mathcal{A}$, with eigenvalue sequence $\theta_0, \theta_1, \ldots, \theta_d$ and dual eigenvalue sequence $\theta_0^*, \theta_1^*, \ldots, \theta_d^*$. Let V denote an irreducible $\mathcal{A}$-module and let $U_0, U_1, \ldots, U_d$ denote the Φ-split decomposition of V from Definition 5.2. Pick any integer i $(1 \leq i \leq d)$. By Lemma 5.4 we have $(A^* - \theta_i^* I)U_i = U_{i-1}$ and $(A - \theta_{i-1} I)U_{i-1} = U_i$. Apparently U_i is an eigenspace for $(A - \theta_{i-1}I)(A^* - \theta_i^* I)$, and the corresponding eigenvalue is a nonzero scalar in $\mathbb{K}$. We denote this eigenvalue by φ_i. We display a basis for V that illuminates the significance of φ_i. Setting $i = 0$ in Theorem 5.1(i) we find $U_0 = E_0^* V$. Combining this with Lemma 5.4(i) we find

$$U_i = (A - \theta_{i-1}I) \cdots (A - \theta_1 I)(A - \theta_0 I)E_0^* V \qquad (0 \leq i \leq d). \tag{5.15}$$

Let v denote a nonzero vector in $E_0^* V$. From (5.15) we find that for $0 \leq i \leq d$ the vector $(A - \theta_{i-1}I) \cdots (A - \theta_0 I)v$ is a basis for U_i. By this and since $U_0, U_1, \ldots, U_d$ is a decomposition of V we find the sequence

$$(A - \theta_{i-1}I) \cdots (A - \theta_1 I)(A - \theta_0 I)v \qquad (0 \leq i \leq d) \tag{5.16}$$

is a basis for V. With respect to this basis the matrices representing A and A^* are

$$\begin{pmatrix} \theta_0 & & & & \mathbf{0} \\ 1 & \theta_1 & & & \\ & 1 & \theta_2 & & \\ & & \ddots & \ddots & \\ \mathbf{0} & & & 1 & \theta_d \end{pmatrix}, \qquad \begin{pmatrix} \theta_0^* & \varphi_1 & & & \mathbf{0} \\ & \theta_1^* & \varphi_2 & & \\ & & \theta_2^* & \ddots & \\ & & & \ddots & \varphi_d \\ \mathbf{0} & & & & \theta_d^* \end{pmatrix} \tag{5.17}$$

respectively. By a *Φ-split basis* for V we mean a sequence of the form (5.16), where v is a nonzero vector in $E_0^* V$. We call $\varphi_1, \varphi_2, \ldots, \varphi_d$ the *first split sequence* of Φ. We let $\phi_1, \phi_2, \ldots, \phi_d$ denote the first split sequence of $\Phi^{\Downarrow}$ and call this the *second split sequence* of Φ. For notational convenience we define $\varphi_0 = 0$, $\varphi_{d+1} = 0$, $\phi_0 = 0$, $\phi_{d+1} = 0$.

5.2 The Parameter Array and the Classifying Space

Our next goal is to describe the relationship between the eigenvalue sequence, the dual eigenvalue sequence, the first split sequence, and the second split sequence. We will use the following concept.

Definition 5.4. *Let d denote a nonnegative integer. By a* parameter array *over $\mathbb{K}$ of diameter d we mean a sequence of scalars $(\theta_i, \theta^*_i, i = 0, \dots, d; \varphi_j, \phi_j, j = 1, \dots, d)$ taken from $\mathbb{K}$ that satisfy the following conditions (PA1)–(PA5).*

(PA1) $\theta_i \neq \theta_j, \qquad \theta^*_i \neq \theta^*_j \qquad \text{if} \quad i \neq j, \qquad (0 \leq i, j \leq d).$

(PA2) $\varphi_i \neq 0, \qquad \phi_i \neq 0 \qquad (1 \leq i \leq d).$

(PA3) $\varphi_i = \phi_1 \sum_{h=0}^{i-1} \frac{\theta_h - \theta_{d-h}}{\theta_0 - \theta_d} + (\theta^*_i - \theta^*_0)(\theta_{i-1} - \theta_d) \qquad (1 \leq i \leq d).$

(PA4) $\phi_i = \varphi_1 \sum_{h=0}^{i-1} \frac{\theta_h - \theta_{d-h}}{\theta_0 - \theta_d} + (\theta^*_i - \theta^*_0)(\theta_{d-i+1} - \theta_0) \qquad (1 \leq i \leq d).$

(PA5) The expressions

$$\frac{\theta_{i-2} - \theta_{i+1}}{\theta_{i-1} - \theta_i}, \qquad \frac{\theta^*_{i-2} - \theta^*_{i+1}}{\theta^*_{i-1} - \theta^*_i} \tag{5.18}$$

are equal and independent of i for $2 \leq i \leq d - 1$.

Theorem 5.2. [79, Theorem 1.9] *Let d denote a nonnegative integer and let $(\theta_i, \theta^*_i, i = 0, \dots, d; \varphi_j, \phi_j, j = 1, \dots, d)$ denote a sequence of scalars taken from $\mathbb{K}$. Then the following (i), (ii) are equivalent.*

(i) The sequence $(\theta_i, \theta^*_i, i = 0, \dots, d; \varphi_j, \phi_j, j = 1, \dots, d)$ is a parameter array over $\mathbb{K}$.

(ii) There exists a Leonard system Φ over $\mathbb{K}$ that has eigenvalue sequence $\theta_0, \theta_1, \dots, \theta_d$, dual eigenvalue sequence $\theta^*_0, \theta^*_1, \dots, \theta^*_d$, first split sequence $\varphi_1, \varphi_2, \dots, \varphi_d$ and second split sequence $\phi_1, \phi_2, \dots, \phi_d$.

Suppose (i), (ii) hold. Then Φ is unique up to isomorphism of Leonard systems.

Our proof of Theorem 5.2 is too long to be included in these notes. A complete proof can be found in [79].

Definition 5.5. *Let Φ denote the Leonard system from Definition 1.2. By the* parameter array *of Φ we mean the sequence $(\theta_i, \theta^*_i, i = 0, \dots, d; \varphi_j, \phi_j, j = 1, \dots, d)$, where $\theta_0, \theta_1, \dots, \theta_d$ (resp. $\theta^*_0, \theta^*_1, \dots, \theta^*_d$) is the eigenvalue sequence (resp. dual eigenvalue sequence) of Φ and $\varphi_1, \varphi_2, \dots, \varphi_d$ (resp. $\phi_1, \phi_2, \dots, \phi_d$) is the first split sequence (resp. second split sequence) of Φ.*

By Theorem 5.2 the map which sends a given Leonard system to its parameter array induces a bijection from the set of isomorphism classes of Leonard systems over $\mathbb{K}$ to the set of parameter arrays over $\mathbb{K}$. Consequently we view the set of parameter arrays over $\mathbb{K}$ as a "classifying space" for the Leonard systems over $\mathbb{K}$.

In the appendix to these notes we display all the parameter arrays over $\mathbb{K}$.

We now cite a result that shows how the parameter arrays behave with respect to the D_4 action given in Section 4.

Theorem 5.3. [79, Theorem 1.11] *Let Φ denote a Leonard system with parameter array $(\theta_i, \theta^*_i, i = 0, \ldots, d;\ \varphi_j, \phi_j, j = 1, \ldots, d)$. Then (i)–(iii) hold below.*

(i) The parameter array of Φ^* is $(\theta^*_i, \theta_i, i = 0, \ldots, d; \varphi_j, \phi_{d-j+1}, j = 1, \ldots, d)$.
(ii) The parameter array of $\Phi^{\downarrow}$ is $(\theta_i, \theta^*_{d-i}, i = 0, \ldots, d; \phi_{d-j+1}, \varphi_{d-j+1}, j = 1, \ldots, d)$.
(iii) The parameter array of $\Phi^{\Downarrow}$ is $(\theta_{d-i}, \theta^*_i, i = 0, \ldots, d; \phi_j, \varphi_j, j = 1, \ldots, d)$.

5.3 Everything in Terms of the Parameter Array

In this section we express all the polynomials and scalars that came up so far in the paper, in terms the parameter array. We will use the following notation.

Definition 5.6. *Suppose we are given an integer $d \geq 0$ and two sequences of scalars*

$$\theta_0, \theta_1, \ldots, \theta_d; \quad \theta^*_0, \theta^*_1, \ldots, \theta^*_d$$

taken from $\mathbb{K}$. Then for $0 \leq i \leq d+1$ we let τ_i, τ^*_i, η_i, η^*_i denote the following polynomials in $\mathbb{K}[\lambda]$.

$$\tau_i = \prod_{h=0}^{i-1}(\lambda - \theta_h), \qquad \tau^*_i = \prod_{h=0}^{i-1}(\lambda - \theta^*_h), \tag{5.19}$$

$$\eta_i = \prod_{h=0}^{i-1}(\lambda - \theta_{d-h}), \qquad \eta^*_i = \prod_{h=0}^{i-1}(\lambda - \theta^*_{d-h}). \tag{5.20}$$

We observe that each of τ_i, τ^*_i, η_i, η^*_i is monic with degree i.

Theorem 5.4. *Let Φ denote the Leonard system from Definition 1.2 and let $(\theta_i, \theta^*_i, i = 0, \ldots, d; \varphi_j, \phi_j, j = 1, \ldots, d)$ denote the corresponding parameter array. Let the polynomials u_i be as in Definition 3.6. Then*

$$u_i = \sum_{h=0}^{i} \frac{\tau^*_h(\theta^*_i)}{\varphi_1\varphi_2\cdots\varphi_h}\tau_h \qquad (0 \leq i \leq d). \tag{5.21}$$

We are using the notation (5.19).

Proof. Let the integer i be given. The polynomial u_i has degree i so there exists scalars $\alpha_0, \alpha_1, \ldots, \alpha_i$ in $\mathbb{K}$ such that

$$u_i = \sum_{h=0}^{i} \alpha_h \tau_h. \tag{5.22}$$

We show

$$\alpha_h = \frac{\tau_h^*(\theta_i^*)}{\varphi_1\varphi_2\cdots\varphi_h} \qquad (0 \le h \le i). \tag{5.23}$$

In order to do this we show $\alpha_0 = 1$ and $\alpha_{h+1}\varphi_{h+1} = \alpha_h(\theta_i^* - \theta_h^*)$ for $0 \le h \le i-1$. We now show $\alpha_0 = 1$. We evaluate (5.22) at $\lambda = \theta_0$ and find $u_i(\theta_0) = \sum_{h=0}^{i} \alpha_h \tau_h(\theta_0)$. Recall $u_i(\theta_0) = 1$ by (3.19). Using (5.19) we find $\tau_h(\theta_0) = 1$ for $h = 0$ and $\tau_h(\theta_0) = 0$ for $1 \le h \le i$. From these comments we find $\alpha_0 = 1$. We now show $\alpha_{h+1}\varphi_{h+1} = \alpha_h(\theta_i^* - \theta_h^*)$ for $0 \le h \le i-1$. Let V denote an irreducible $\mathcal{A}$-module. Let v denote a nonzero vector in E_0^*V and define $e_i = \tau_i(A)v$ for $0 \le i \le d$. Observe that the sequence $e_0, e_1, \ldots, e_d$ is the basis for V from (5.16). Using (5.17) we find $(A^* - \theta_j^* I)e_j = \varphi_j e_{j-1}$ for $1 \le j \le d$ and $(A^* - \theta_0^* I)e_0 = 0$. By Theorem 2.5 and (3.18) we find $u_i(A)E_0^*V = E_i^*V$. By this and since $v \in E_0^*V$ we find $u_i(A)v \in E_i^*V$. Apparently $u_i(A)v$ is an eigenvector for A^* with eigenvalue θ_i^*. We may now argue

$$\begin{aligned}
0 &= (A^* - \theta_i^* I)u_i(A)v \\
&= (A^* - \theta_i^* I)\sum_{h=0}^{i} \alpha_h \tau_h(A)v \\
&= (A^* - \theta_i^* I)\sum_{h=0}^{i} \alpha_h e_h \\
&= \sum_{h=0}^{i-1} e_h(\alpha_{h+1}\varphi_{h+1} - \alpha_h(\theta_i^* - \theta_h^*)).
\end{aligned}$$

By this and since $e_0, e_1, \ldots, e_d$ are linearly independent we find $\alpha_{h+1}\varphi_{h+1} = \alpha_h(\theta_i^* - \theta_h^*)$ for $0 \le h \le i-1$. Line (5.23) follows and the theorem is proved. □

Lemma 5.5. *Let Φ denote the Leonard system from Definition 1.2 and let $(\theta_i, \theta_i^*, i = 0, \ldots, d; \varphi_j, \phi_j, j = 1, \ldots, d)$ denote the corresponding parameter array. Let the polynomials p_i be as in Definition 2.4. With reference to Definition 5.6 we have*

$$p_i(\theta_0) = \frac{\varphi_1\varphi_2\cdots\varphi_i}{\tau_i^*(\theta_i^*)} \qquad (0 \le i \le d). \tag{5.24}$$

Proof. In equation (5.21), each side is a polynomial of degree i in λ. For the polynomial on the left in (5.21) the coefficient of λ^i is $p_i(\theta_0)^{-1}$ by (3.18) and since p_i is monic. For the polynomial on the right in (5.21) the coefficient of λ^i is $\tau_i^*(\theta_i^*)(\varphi_1\varphi_2\cdots\varphi_i)^{-1}$. Comparing these coefficients we obtain the result. □

Theorem 5.5. *Let Φ denote the Leonard system from Definition 1.2 and let $(\theta_i, \theta_i^*, i = 0, \ldots, d; \varphi_j, \phi_j, j = 1, \ldots, d)$ denote the corresponding parameter array. Let the polynomials p_i be as in Definition 2.4. Then with reference to Definition 5.6,*

$$p_i = \sum_{h=0}^{i} \frac{\varphi_1\varphi_2\cdots\varphi_i}{\varphi_1\varphi_2\cdots\varphi_h} \frac{\tau_h^*(\theta_i^*)}{\tau_i^*(\theta_i^*)}\tau_h \qquad (0 \le i \le d).$$

Proof. Observe $p_i = p_i(\theta_0)u_i$ by (3.18). In this equation we evaluate $p_i(\theta_0)$ using (5.24) and we evaluate u_i using (5.21). The result follows. □

Theorem 5.6. *Let Φ denote the Leonard system from Definition 1.2 and let $(\theta_i, \theta_i^*, i = 0, \ldots, d; \varphi_j, \phi_j, j = 1, \ldots, d)$ denote the corresponding parameter array. Let the scalars b_i, c_i be as in Definition 3.3. Then with reference to Definition 5.6 the following (i), (ii) hold.*

(i) $b_i = \varphi_{i+1}\dfrac{\tau_i^*(\theta_i^*)}{\tau_{i+1}^*(\theta_{i+1}^*)} \qquad (0 \le i \le d-1).$

(ii) $c_i = \phi_i\dfrac{\eta_{d-i}^*(\theta_i^*)}{\eta_{d-i+1}^*(\theta_{i-1}^*)} \qquad (1 \le i \le d).$

Proof. (i) Evaluate (3.4) using Lemma 5.5.
(ii) Using Definition 3.3 we find, with reference to Definition 1.5, that $c_i = b_{d-i}^{\downarrow}$. Applying part (i) above to $\Phi^{\downarrow}$ and using Theorem 5.3(ii) we routinely obtain the result. □

Let Φ denote the Leonard system from Definition 1.2 and let the scalars a_i be as in Definition 2.3. We mention two formulae that give a_i in terms of the parameter array of Φ. The first formula is obtained using Lemma 3.6(ii) and Theorem 5.6. The second formula is given in the following theorem.

Theorem 5.7. *Let Φ denote the Leonard system from Definition 1.2 and let $(\theta_i, \theta_i^*, i = 0, \ldots, d; \varphi_j, \phi_j, j = 1, \ldots, d)$ denote the corresponding parameter array. Let the scalars a_i be as in Definition 2.3. Then*

$$a_i = \theta_i + \frac{\varphi_i}{\theta_i^* - \theta_{i-1}^*} + \frac{\varphi_{i+1}}{\theta_i^* - \theta_{i+1}^*} \qquad (0 \le i \le d), \tag{5.25}$$

where we recall $\varphi_0 = 0$, $\varphi_{d+1} = 0$, and where $\theta_{-1}^*, \theta_{d+1}^*$ denote indeterminates.

Proof. Let the polynomials $p_0, p_1, \ldots, p_{d+1}$ be as in Definition 2.4 and recall these polynomials are monic. Let i be given and consider the polynomial

$$\lambda p_i - p_{i+1}. \tag{5.26}$$

From (2.13) we find the polynomial (5.26) is equal to $a_i p_i + x_i p_{i-1}$. Therefore the polynomial (5.26) has degree i and leading coefficient a_i. In order to compute this leading coefficient, in (5.26) we evaluate each of p_i, p_{i+1} using Theorem 2.7(ii) and Theorem 5.5. By this method we routinely obtain (5.25). □

Theorem 5.8. *Let Φ denote the Leonard system from Definition 1.2 and let $(\theta_i, \theta^*_i, i = 0, \dots, d; \varphi_j, \phi_j, j = 1, \dots, d)$ denote the corresponding parameter array. Let the scalars x_i be as in Definition 2.3. Then with reference to Definition 5.6,*

$$x_i = \varphi_i \phi_i \frac{\tau^*_{i-1}(\theta^*_{i-1})\eta^*_{d-i}(\theta^*_i)}{\tau^*_i(\theta^*_i)\eta^*_{d-i+1}(\theta^*_{i-1})} \qquad (1 \le i \le d). \tag{5.27}$$

Proof. Use $x_i = b_{i-1}c_i$ and Theorem 5.6. □

Theorem 5.9. *Let Φ denote the Leonard system from Definition 1.2 and let $(\theta_i, \theta^*_i, i = 0, \dots, d; \varphi_j, \phi_j, j = 1, \dots, d)$ denote the corresponding parameter array. Let the scalar ν be as in Definition 2.6. Then with reference to Definition 5.6,*

$$\nu = \frac{\eta_d(\theta_0)\eta^*_d(\theta^*_0)}{\phi_1\phi_2\cdots\phi_d}. \tag{5.28}$$

Proof. Evaluate (3.6) using Theorem 5.6(ii). □

Theorem 5.10. *Let Φ denote the Leonard system from Definition 1.2 and let $(\theta_i, \theta^*_i, i = 0, \dots, d; \varphi_j, \phi_j, j = 1, \dots, d)$ denote the corresponding parameter array. Let the scalars k_i be as in Definition 3.4. Then with reference to Definition 5.6,*

$$k_i = \frac{\varphi_1\varphi_2\cdots\varphi_i}{\phi_1\phi_2\cdots\phi_i} \frac{\eta^*_d(\theta^*_0)}{\tau^*_i(\theta^*_i)\eta^*_{d-i}(\theta^*_i)} \qquad (0 \le i \le d). \tag{5.29}$$

Proof. Evaluate (3.10) using Theorem 5.6. □

Theorem 5.11. *Let Φ denote the Leonard system from Definition 1.2 and let $(\theta_i, \theta^*_i, i = 0, \dots, d; \varphi_j, \phi_j, j = 1, \dots, d)$ denote the corresponding parameter array. Let the scalars m_i be as in Definition 2.5. Then with reference to Definition 5.6,*

$$m_i = \frac{\varphi_1\varphi_2\cdots\varphi_i\phi_1\phi_2\cdots\phi_{d-i}}{\eta^*_d(\theta^*_0)\tau_i(\theta_i)\eta_{d-i}(\theta_i)} \qquad (0 \le i \le d). \tag{5.30}$$

Proof. Applying Definition 3.4 to Φ^* we find $m_i = k^*_i \nu^{-1}$. We compute k^*_i using Theorem 5.10 and Theorem 5.3(i). We compute ν using Theorem 5.9. The result follows. □

6 The Terminating Branch of the Askey Scheme

Let Φ denote the Leonard system from Definition 1.2 and let the polynomials u_i be as in Definition 3.6. In this section we discuss how the u_i fit into the Askey scheme [53], [11, p. 260]. Our argument is summarized as follows.

In the appendix to these notes we display all the parameter arrays over $\mathbb{K}$. These parameter arrays fall into 13 families. In (5.21) the u_i are expressed as a sum involving the parameter array of Φ. For each of the 13 families of parameter arrays we evaluate this sum. We find the corresponding u_i form a class consisting of the q-Racah, q-Hahn, dual q-Hahn, q-Krawtchouk, dual q-Krawtchouk, quantum q-Krawtchouk, affine q-Krawtchouk, Racah, Hahn, dual Hahn, Krawtchouk, Bannai/Ito, and orphan polynomials. This class coincides with the terminating branch of the Askey scheme. See the appendix for the details. We remark the Bannai/Ito polynomials can be obtained from the q-Racah polynomials by letting q tend to -1 [11, p. 260]. The orphan polynomials exist for diameter $d = 3$ and $\mathrm{Char}(\mathbb{K}) = 2$ only.

In this section we illustrate what is going on with some examples. We will consider two families of parameter arrays. For the first family the corresponding u_i will turn out to be some Krawtchouk polynomials. For the second family the corresponding u_i will turn out to be the q-Racah polynomials.

Our first example is associated with the Leonard pair (1.2). Let d denote a nonnegative integer and consider the following elements of $\mathbb{K}$.

$$\theta_i = d - 2i, \quad \theta^*_i = d - 2i \qquad (0 \le i \le d), \tag{6.1}$$

$$\varphi_i = -2i(d - i + 1), \quad \phi_i = 2i(d - i + 1) \qquad (1 \le i \le d). \tag{6.2}$$

In order to avoid degenerate situations we assume the characteristic of $\mathbb{K}$ is zero or an odd prime greater than d. It is routine to show (6.1), (6.2) satisfy the conditions PA1–PA5 of Definition 5.4, so $(\theta_i, \theta^*_i, i = 0, \ldots, d; \varphi_j, \phi_j, j = 1, \ldots, d)$ is a parameter array over $\mathbb{K}$. By Theorem 5.2 there exists a Leonard system Φ over $\mathbb{K}$ with this parameter array. Let the scalars a_i for Φ be as in (2.8). Applying Theorem 5.7 to Φ we find

$$a_i = 0 \qquad (0 \le i \le d). \tag{6.3}$$

Let the scalars b_i, c_i for Φ be as in Definition 3.3. Applying Theorem 5.6 to Φ we find

$$b_i = d - i, \quad c_i = i \qquad (0 \le i \le d). \tag{6.4}$$

Pick any integers i, j $(0 \le i, j \le d)$. Applying Theorem 5.4 to Φ we find

$$u_i(\theta_j) = \sum_{n=0}^{d} \frac{(-i)_n(-j)_n 2^n}{(-d)_n n!}, \tag{6.5}$$

where

$$(a)_n := a(a+1)(a+2)\cdots(a+n-1) \qquad n = 0, 1, 2, \ldots$$

Hypergeometric series are defined in [28, p. 3]. From this definition we find the sum on the right in (6.5) is the hypergeometric series

$$ {}_2F_1\left(\begin{matrix} -i, -j \\ -d \end{matrix} \,\middle|\, 2 \right). \tag{6.6} $$

A definition of the Krawtchouk polynomials can be found in [4] or [53]. Comparing this definition with (6.5), (6.6) we find the u_i are Krawtchouk polynomials but not the most general ones. Let the scalar ν for Φ be as in Definition 2.6. Applying Theorem 5.9 to Φ we find $\nu = 2^d$. Let the scalars k_i for Φ be as in Definition 3.4. Applying Theorem 5.10 to Φ we obtain a binomial coefficient

$$ k_i = \binom{d}{i} \qquad (0 \le i \le d). \tag{6.7} $$

Let the scalars m_i for Φ be as in Definition 2.5. Applying Theorem 5.11 to Φ we find

$$ m_i = \binom{d}{i} 2^{-d} \qquad (0 \le i \le d). \tag{6.8} $$

We now give our second example. For this example the polynomials u_i will turn out to be the q-Racah polynomials. To begin, let d denote a nonnegative integer and consider the following elements in $\mathbb{K}$.

$$ \theta_i = \theta_0 + h(1-q^i)(1-sq^{i+1})/q^i, \tag{6.9} $$

$$ \theta^*_i = \theta^*_0 + h^*(1-q^i)(1-s^*q^{i+1})/q^i \tag{6.10} $$

for $0 \le i \le d$, and

$$ \varphi_i = hh^*q^{1-2i}(1-q^i)(1-q^{i-d-1})(1-r_1q^i)(1-r_2q^i), \tag{6.11} $$

$$ \phi_i = hh^*q^{1-2i}(1-q^i)(1-q^{i-d-1})(r_1-s^*q^i)(r_2-s^*q^i)/s^* \tag{6.12} $$

for $1 \le i \le d$. We assume $q, h, h^*, s, s^*, r_1, r_2$ are nonzero scalars in the algebraic closure of $\mathbb{K}$, and that $r_1r_2 = ss^*q^{d+1}$. To avoid degenerate situations we assume none of $q^i, r_1q^i, r_2q^i, s^*q^i/r_1, s^*q^i/r_2$ is equal to 1 for $1 \le i \le d$ and neither of sq^i, s^*q^i is equal to 1 for $2 \le i \le 2d$. It is routine to show (6.9)–(6.12) satisfy the conditions PA1–PA5 of Definition 5.4, so $(\theta_i, \theta^*_i, i = 0, \ldots, d; \varphi_j, \phi_j, j = 1, \ldots, d)$ is a parameter array over $\mathbb{K}$. By Theorem 5.2 there exists a Leonard system Φ over $\mathbb{K}$ with this parameter array. Let the scalars b_i, c_i for Φ be as in Definition 3.3. Applying Theorem 5.6 to Φ we find

$$ b_0 = \frac{h(1-q^{-d})(1-r_1q)(1-r_2q)}{1-s^*q^2}, $$

$$ b_i = \frac{h(1-q^{i-d})(1-s^*q^{i+1})(1-r_1q^{i+1})(1-r_2q^{i+1})}{(1-s^*q^{2i+1})(1-s^*q^{2i+2})} \qquad (1 \le i \le d-1), $$

$$ c_i = \frac{h(1-q^i)(1-s^*q^{i+d+1})(r_1-s^*q^i)(r_2-s^*q^i)}{s^*q^d(1-s^*q^{2i})(1-s^*q^{2i+1})} \qquad (1 \le i \le d-1), $$

$$ c_d = \frac{h(1-q^d)(r_1-s^*q^d)(r_2-s^*q^d)}{s^*q^d(1-s^*q^{2d})}. $$

Pick integers i, j $(0 \leq i, j \leq d)$. Applying Theorem 5.4 to Φ we find

$$u_i(\theta_j) = \sum_{n=0}^{d} \frac{(q^{-i};q)_n(s^*q^{i+1};q)_n(q^{-j};q)_n(sq^{j+1};q)_n q^n}{(r_1q;q)_n(r_2q;q)_n(q^{-d};q)_n(q;q)_n}, \tag{6.13}$$

where

$$(a;q)_n := (1-a)(1-aq)(1-aq^2)\cdots(1-aq^{n-1}) \qquad n = 0,1,2,\ldots$$

Basic hypergeometric series are defined in [28, p. 4]. From that definition we find the sum on the right in (6.13) is the basic hypergeometric series

$${}_4\phi_3\left(\begin{matrix} q^{-i},\ s^*q^{i+1},\ q^{-j},\ sq^{j+1} \\ r_1q,\ r_2q,\ q^{-d} \end{matrix} \,\middle|\, q,\ q\right). \tag{6.14}$$

A definition of the q-Racah polynomials can be found in [5] or [53]. Comparing this definition with (6.13), (6.14) and recalling $r_1r_2 = ss^*q^{d+1}$, we find the u_i are the q-Racah polynomials. Let the scalar ν for Φ be as in Definition 2.6. Applying Theorem 5.9 to Φ we find

$$\nu = \frac{(sq^2;q)_d(s^*q^2;q)_d}{r_1^d q^d (sq/r_1;q)_d(s^*q/r_1;q)_d}.$$

Let the scalars k_i for Φ be as in Definition 3.4. Applying Theorem 5.10 to Φ we obtain

$$k_i = \frac{(r_1q;q)_i(r_2q;q)_i(q^{-d};q)_i(s^*q;q)_i(1-s^*q^{2i+1})}{s^iq^i(q;q)_i(s^*q/r_1;q)_i(s^*q/r_2;q)_i(s^*q^{d+2};q)_i(1-s^*q)} \qquad (0 \leq i \leq d).$$

Let the scalars m_i for Φ be as in Definition 2.5. Applying Theorem 5.11 to Φ we find

$$m_i = \frac{(r_1q;q)_i(r_2q;q)_i(q^{-d};q)_i(sq;q)_i(1-sq^{2i+1})}{s^{*i}q^i(q;q)_i(sq/r_1;q)_i(sq/r_2;q)_i(sq^{d+2};q)_i(1-sq)\nu} \qquad (0 \leq i \leq d).$$

7 Applications and Related Topics

We are done describing the correspondence between Leonard pairs and the terminating branch of the Askey scheme. For the remainder of these notes we discuss applications and related topics. We begin with a characterization of Leonard systems.

7.1 A Characterization of Leonard Systems

We recall some results from earlier in the paper. Let Φ denote the Leonard system from Definition 1.2. Let the polynomials $p_0, p_1, \ldots, p_{d+1}$ be as in Definition 2.4 and recall $p_0^*, p_1^*, \ldots, p_{d+1}^*$ are the corresponding polynomials for

Φ^*. For the purpose of this section, we call $p_0, p_1, \ldots, p_{d+1}$ the *monic polynomial sequence* (or *MPS*) of Φ. We call $p_0^*, p_1^*, \ldots, p_{d+1}^*$ the *dual MPS* of Φ. By Definition 2.4 we have

$$p_0 = 1, \qquad p_0^* = 1, \tag{7.1}$$

$$\lambda p_i = p_{i+1} + a_i p_i + x_i p_{i-1} \qquad (0 \leq i \leq d), \tag{7.2}$$

$$\lambda p_i^* = p_{i+1}^* + a_i^* p_i^* + x_i^* p_{i-1}^* \qquad (0 \leq i \leq d), \tag{7.3}$$

where x_0, x_0^*, p_{-1}, p_{-1}^* are all zero, and where

$$a_i = \operatorname{tr}(E_i^* A), \qquad a_i^* = \operatorname{tr}(E_i A^*) \qquad (0 \leq i \leq d),$$
$$x_i = \operatorname{tr}(E_i^* A E_{i-1}^* A), \qquad x_i^* = \operatorname{tr}(E_i A^* E_{i-1} A^*) \qquad (1 \leq i \leq d).$$

By Lemma 2.7(iii) we have

$$x_i \neq 0, \quad x_i^* \neq 0 \qquad (1 \leq i \leq d). \tag{7.4}$$

Let $\theta_0, \theta_1, \ldots, \theta_d$ (resp. $\theta_0^*, \theta_1^*, \ldots, \theta_d^*$) denote the eigenvalue sequence (resp. dual eigenvalue sequence) of Φ, and recall

$$\theta_i \neq \theta_j, \quad \theta_i^* \neq \theta_j^* \qquad \text{if} \quad i \neq j, \qquad (0 \leq i, j \leq d). \tag{7.5}$$

By Theorem 2.7(ii) we have

$$p_{d+1}(\theta_i) = 0, \quad p_{d+1}^*(\theta_i^*) = 0 \qquad (0 \leq i \leq d). \tag{7.6}$$

By Theorem 3.1 we have

$$p_i(\theta_0) \neq 0, \quad p_i^*(\theta_0^*) \neq 0 \qquad (0 \leq i \leq d). \tag{7.7}$$

By Theorem 4.3 we have

$$\frac{p_i(\theta_j)}{p_i(\theta_0)} = \frac{p_j^*(\theta_i^*)}{p_j^*(\theta_0^*)} \qquad (0 \leq i, j \leq d). \tag{7.8}$$

In the following theorem we show the equations (7.1)–(7.8) characterize the Leonard systems.

Theorem 7.1. *Let d denote a nonnegative integer. Given polynomials*

$$p_0, p_1, \ldots, p_{d+1}, \tag{7.9}$$

$$p_0^*, p_1^*, \ldots, p_{d+1}^* \tag{7.10}$$

in $\mathbb{K}[\lambda]$ satisfying (7.1)–(7.4) and given scalars

$$\theta_0, \theta_1, \ldots, \theta_d, \tag{7.11}$$

$$\theta_0^*, \theta_1^*, \ldots, \theta_d^* \tag{7.12}$$

in $\mathbb{K}$ satisfying (7.5)–(7.8), there exists a Leonard system Φ over $\mathbb{K}$ that has MPS (7.9), dual MPS (7.10), eigenvalue sequence (7.11) and dual eigenvalue sequence (7.12). The system Φ is unique up to isomorphism of Leonard systems.

Proof. We abbreviate $V = \mathbb{K}^{d+1}$. Let A and A^* denote the following matrices in $\mathrm{Mat}_{d+1}(\mathbb{K})$:

$$A := \begin{pmatrix} a_0 & x_1 & & & & \mathbf{0} \\ 1 & a_1 & x_2 & & & \\ & 1 & \cdot & \cdot & & \\ & & \cdot & \cdot & \cdot & \\ & & & \cdot & \cdot & x_d \\ \mathbf{0} & & & & 1 & a_d \end{pmatrix}, \qquad A^* := \mathrm{diag}(\theta_0^*, \theta_1^*, \ldots, \theta_d^*).$$

We show the pair A, A^* is a Leonard pair on V. To do this we apply Definition 1.1. Observe that A is irreducible tridiagonal and A^* is diagonal. Therefore condition (i) of Definition 1.1 is satisfied by the basis for V consisting of the columns of I, where I denotes the identity matrix in $\mathrm{Mat}_{d+1}(\mathbb{K})$. To verify condition (ii) of Definition 1.1, we display an invertible matrix X such that $X^{-1}AX$ is diagonal and $X^{-1}A^*X$ is irreducible tridiagonal. Let X denote the matrix in $\mathrm{Mat}_{d+1}(\mathbb{K})$ that has entries

$$X_{ij} = \frac{p_i(\theta_j)p_j^*(\theta_0^*)}{x_1x_2\cdots x_i} \tag{7.13}$$

$$= \frac{p_j^*(\theta_i^*)p_i(\theta_0)}{x_1x_2\cdots x_i} \tag{7.14}$$

$0 \leq i, j \leq d$. The matrix X is invertible since it is essentially Vandermonde. Using (7.2) and (7.13) we find $AX = XH$ where $H = \mathrm{diag}(\theta_0, \theta_1, \ldots, \theta_d)$. Apparently $X^{-1}AX$ is equal to H and is therefore diagonal. Using (7.3) and (7.14) we find $A^*X = XH^*$ where

$$H^* := \begin{pmatrix} a_0^* & x_1^* & & & & \mathbf{0} \\ 1 & a_1^* & x_2^* & & & \\ & 1 & \cdot & \cdot & & \\ & & \cdot & \cdot & \cdot & \\ & & & \cdot & \cdot & x_d^* \\ \mathbf{0} & & & & 1 & a_d^* \end{pmatrix}.$$

Apparently $X^{-1}A^*X$ is equal to H^* and is therefore irreducible tridiagonal. Now condition (ii) of Definition 1.1 is satisfied by the basis for V consisting of the columns of X. We have now shown the pair A, A^* is a Leonard pair on V. Pick an integer j $(0 \leq j \leq d)$. Using $X^{-1}AX = H$ we find θ_j is the eigenvalue of A associated with column j of X. From the definition of A^* we find θ_j^* is the eigenvalue of A^* associated with column j of I. Let E_j (resp. E_j^*) denote the primitive idempotent of A (resp. A^*) for θ_j (resp. θ_j^*). From our above comments the sequence $\Phi := (A; A^*; \{E_i\}_{i=0}^d; \{E_i^*\}_{i=0}^d)$ is a Leonard system. From the construction Φ is over $\mathbb{K}$. We show (7.9) is the *MPS* of Φ. To do this is suffices to show $a_i = \mathrm{tr}(E_i^*A)$ for $0 \leq i \leq d$ and $x_i = \mathrm{tr}(E_i^*AE_{i-1}^*A)$ for $1 \leq i \leq d$. Applying Lemma 2.7(i),(ii) to Φ

(with v_i = column i of I, $B = A$) we find $a_i = \mathrm{tr}(E_i^* A)$ for $0 \le i \le d$ and $x_i = \mathrm{tr}(E_i^* A E_{i-1}^* A)$ for $1 \le i \le d$. Therefore (7.9) is the MPS of Φ. We show (7.10) is the dual MPS of Φ. Applying Lemma 2.7(i),(ii) to Φ^* (with v_i = column i of X, $B = H^*$) we find $a_i^* = \mathrm{tr}(E_i A^*)$ for $0 \le i \le d$ and $x_i^* = \mathrm{tr}(E_i A^* E_{i-1} A^*)$ for $1 \le i \le d$. Therefore (7.10) is the dual MPS of Φ. From the construction we find (7.11) (resp. (7.12)) is the eigenvalue sequence (resp. dual eigenvalue sequence) of Φ. We show Φ is uniquely determined by (7.9)–(7.12) up to isomorphism of Leonard systems. Recall that Φ is determined up to isomorphism of Leonard systems by its own parameter array. We show the parameter array of Φ is determined by (7.9)–(7.12). Recall the parameter array consists of the eigenvalue sequence, the dual eigenvalue sequence, the first split sequence and the second split sequence. We mentioned earlier that the eigenvalue sequence of Φ is (7.11) and the dual eigenvalue sequence of Φ is (7.12). By Lemma 5.5 the first split sequence of Φ is determined by (7.9)–(7.12). By this and Theorem 5.8 we find the second split sequence of Φ is determined by (7.9)–(7.12). We have now shown the parameter array of Φ is determined by (7.9)–(7.12). We now see that Φ is uniquely determined by (7.9)–(7.12) up to isomorphism of Leonard systems. □

7.2 Leonard Pairs A, A^* with A Lower Bidiagonal and A^* Upper Bidiagonal

Let A, A^* denote matrices in $\mathrm{Mat}_{d+1}(\mathbb{K})$. Let us assume A is lower bidiagonal and A^* is upper bidiagonal. We cite a necessary and sufficient condition for A, A^* to be a Leonard pair.

Theorem 7.2. [82, Theorem 17.1] *Let d denote a nonnegative integer and let A, A^* denote matrices in $\mathrm{Mat}_{d+1}(\mathbb{K})$. Assume A lower bidiagonal and A^* is upper bidiagonal. Then the following (i), (ii) are equivalent.*

(i) The pair A, A^* is a Leonard pair.
(ii) There exists a parameter array $(\theta_i, \theta_i^*, i = 0, \ldots, d; \varphi_j, \phi_j, j = 1, \ldots, d)$ over $\mathbb{K}$ such that

$$A_{ii} = \theta_i, \qquad A_{ii}^* = \theta_i^* \qquad (0 \le i \le d),$$
$$A_{i,i-1} A_{i-1,i}^* = \varphi_i \qquad (1 \le i \le d).$$

7.3 Leonard Pairs A, A^* with A Tridiagonal and A^* Diagonal

Let A, A^* denote matrices in $\mathrm{Mat}_{d+1}(\mathbb{K})$. Let us assume A is tridiagonal and A^* is diagonal. We cite a necessary and sufficient condition for A, A^* to be a Leonard pair.

Theorem 7.3. [82, Theorem 25.1] *Let d denote a nonnegative integer and let A, A^* denote matrices in $\mathrm{Mat}_{d+1}(\mathbb{K})$. Assume A is tridiagonal and A^* is diagonal. Then the following (i), (ii) are equivalent.*

(i) The pair A, A^* is a Leonard pair.
(ii) There exists a parameter array $(\theta_i, \theta^*_i, i = 0, \ldots, d; \varphi_j, \phi_j, j = 1, \ldots, d)$ over $\mathbb{K}$ such that

$$A_{ii} = \theta_i + \frac{\varphi_i}{\theta^*_i - \theta^*_{i-1}} + \frac{\varphi_{i+1}}{\theta^*_i - \theta^*_{i+1}} \qquad (0 \le i \le d),$$

$$A_{i,i-1}A_{i-1,i} = \varphi_i \phi_i \frac{\prod_{h=0}^{i-2}(\theta^*_{i-1} - \theta^*_h)}{\prod_{h=0}^{i-1}(\theta^*_i - \theta^*_h)} \frac{\prod_{h=i+1}^{d}(\theta^*_i - \theta^*_h)}{\prod_{h=i}^{d}(\theta^*_{i-1} - \theta^*_h)} \qquad (1 \le i \le d),$$

$$A^*_{ii} = \theta^*_i \qquad (0 \le i \le d).$$

7.4 Characterizations of the Parameter Arrays

In this section we cite a characterization of the parameter arrays in terms of bidiagonal matrices. We will refer to the following set-up.

Definition 7.1. *Let d denote a nonnegative integer and let $(\theta_i, \theta^*_i, i = 0, \ldots, d; \varphi_j, \phi_j, j = 1, \ldots, d)$ denote a sequence of scalars taken from $\mathbb{K}$. We assume this sequence satisfies PA1 and PA2.*

Theorem 7.4. [86, Theorem 3.2] *With reference to Definition 7.1, the following (i), (ii) are equivalent.*

(i) The sequence $(\theta_i, \theta^*_i, i = 0, \ldots, d; \varphi_j, \phi_j, j = 1, \ldots, d)$ satisfies PA3–PA5.
(ii) There exists an invertible matrix $G \in \mathrm{Mat}_{d+1}(\mathbb{K})$ such that both

$$G^{-1}\begin{pmatrix} \theta_0 & & & & \mathbf{0} \\ 1 & \theta_1 & & & \\ & 1 & \theta_2 & & \\ & & \ddots & \ddots & \\ \mathbf{0} & & & 1 & \theta_d \end{pmatrix} G = \begin{pmatrix} \theta_d & & & & \mathbf{0} \\ 1 & \theta_{d-1} & & & \\ & 1 & \theta_{d-2} & & \\ & & \ddots & \ddots & \\ \mathbf{0} & & & 1 & \theta_0 \end{pmatrix},$$

$$G^{-1}\begin{pmatrix} \theta^*_0 & \varphi_1 & & & \mathbf{0} \\ & \theta^*_1 & \varphi_2 & & \\ & & \theta^*_2 & \ddots & \\ & & & \ddots & \varphi_d \\ \mathbf{0} & & & & \theta^*_d \end{pmatrix} G = \begin{pmatrix} \theta^*_0 & \phi_1 & & & \mathbf{0} \\ & \theta^*_1 & \phi_2 & & \\ & & \theta^*_2 & \ddots & \\ & & & \ddots & \phi_d \\ \mathbf{0} & & & & \theta^*_d \end{pmatrix}.$$

Next we cite a characterization of the parameter arrays in terms of polynomials.

Theorem 7.5. [86, Theorem 4.1] *With reference to Definition 7.1, the following (i), (ii) are equivalent.*

(i) The sequence $(\theta_i, \theta^*_i, i = 0, \ldots, d; \varphi_j, \phi_j, j = 1, \ldots, d)$ satisfies PA3–PA5.

(ii) For $0 \le i \le d$ *the polynomial*

$$\sum_{n=0}^{i} \frac{(\lambda-\theta_0)(\lambda-\theta_1)\cdots(\lambda-\theta_{n-1})(\theta_i^*-\theta_0^*)(\theta_i^*-\theta_1^*)\cdots(\theta_i^*-\theta_{n-1}^*)}{\varphi_1\varphi_2\cdots\varphi_n} \tag{7.15}$$

is a scalar multiple of the polynomial

$$\sum_{n=0}^{i} \frac{(\lambda-\theta_d)(\lambda-\theta_{d-1})\cdots(\lambda-\theta_{d-n+1})(\theta_i^*-\theta_0^*)(\theta_i^*-\theta_1^*)\cdots(\theta_i^*-\theta_{n-1}^*)}{\phi_1\phi_2\cdots\phi_n}.$$

8 The Askey-Wilson Relations

We turn our attention to the representation theoretic aspects of Leonard pairs.

Theorem 8.1. [88, Theorem 1.5] *Let* V *denote a vector space over* $\mathbb{K}$ *with finite positive dimension. Let* A, A^* *denote a Leonard pair on* V. *Then there exists a sequence of scalars* $\beta, \gamma, \gamma^*, \varrho, \varrho^*, \omega, \eta, \eta^*$ *taken from* $\mathbb{K}$ *such that both*

$$A^2A^* - \beta AA^*A + A^*A^2 - \gamma(AA^*+A^*A) - \varrho A^* = \gamma^*A^2 + \omega A + \eta I, \tag{8.1}$$
$$A^{*2}A - \beta A^*AA^* + AA^{*2} - \gamma^*(A^*A+AA^*) - \varrho^*A = \gamma A^{*2} + \omega A^* + \eta^* I. \tag{8.2}$$

The sequence is uniquely determined by the pair A, A^* *provided the diameter* $d \ge 3$.

We refer to (8.1), (8.2) as the *Askey-Wilson relations.* As far as we know these relations first appeared in [90].

Our next result is a kind of converse to Theorem 8.1.

Theorem 8.2. [88, Theorem 6.2] *Given a sequence of scalars* $\beta, \gamma, \gamma^*, \varrho, \varrho^*, \omega, \eta, \eta^*$ *taken from* $\mathbb{K}$, *we let* A_w *denote the unital associative* $\mathbb{K}$*-algebra generated by two symbols* A, A^* *subject to the relations (8.1), (8.2). Let* V *denote a finite dimensional irreducible* A_w*-module and assume each of* A, A^* *is multiplicity-free on* V. *Then* A, A^* *act on* V *as a Leonard pair provided* q *is not a root of unity, where* $q + q^{-1} = \beta$.

The algebra A_w in Theorem 8.2 is called the *Askey-Wilson algebra* [90].

We finish this section with an open problem.

Problem 8.1. Let A_w denote the Askey-Wilson algebra from Theorem 8.2. Let V denote an irreducible A_w-module with either finite or countably infinite dimension. We say V has *polynomial type* whenever there exists a basis $v_0, v_1, \ldots$ for V with respect to which the matrix representing A is irreducible tridiagonal and the matrix representing A^* is diagonal. Determine up to isomorphism the irreducible A_w-modules of polynomial type. We expect that the solutions correspond to the entire Askey scheme of orthogonal polynomials.

Remark 8.1. The papers [31], [32], [33], [34], [35], [80], [90], [91], [92] contain some results related to Problem 8.1, but a complete and rigorous treatment has yet to be carried out. See also the work of Grünbaum and Haine on the "bispectral problem" [37], [38], [39], [40], [41], [42], [43] as well as [6], [7], [8], [11, p. 263], [63], [65], [66], [67], [68].

Remark 8.2. Referring to Theorem 8.2, for the special case $\beta = q + q^{-1}$, $\gamma = \gamma^* = 0$, $\omega = 0$, $\eta = \eta^* = 0$ the Askey-Wilson algebra is related to the quantum groups $U_q(\mathrm{su}_2)$, $U_q(\mathrm{so}_3)$ [16, Theorem 8.10], [27], [44], [69] as well as the bipartite 2-homogeneous distance-regular graphs [17, Lemma 3.3], [18], [30, p. 427].

8.1 Leonard Pairs and the Lie Algebra sl_2

In this section we assume the field $\mathbb{K}$ is algebraically closed with characteristic zero.

We recall the Lie algebra $sl_2 = sl_2(\mathbb{K})$. This algebra has a basis e, f, h satisfying

$$[h, e] = 2e, \qquad [h, f] = -2f, \qquad [e, f] = h,$$

where $[\cdot, \cdot]$ denotes the Lie bracket.

We recall the irreducible finite dimensional modules for sl_2.

Lemma 8.1. [49, p. 102] *There exists a family*

$$V_d \qquad d = 0, 1, 2, \ldots \tag{8.3}$$

of irreducible finite dimensional sl_2-modules with the following properties. The module V_d has a basis $v_0, v_1, \ldots, v_d$ satisfying $hv_i = (d - 2i)v_i$ for $0 \le i \le d$, $fv_i = (i+1)v_{i+1}$ for $0 \le i \le d-1$, $fv_d = 0$, $ev_i = (d - i + 1)v_{i-1}$ for $1 \le i \le d$, $ev_0 = 0$. Every irreducible finite dimensional sl_2-module is isomorphic to exactly one of the modules in line (8.3).

Example 8.1. Let A and A^* denote the following elements of sl_2.

$$A = e + f, \qquad A^* = h.$$

Let d denote a nonnegative integer and consider the action of A, A^* on the module V_d. With respect to the basis $v_0, v_1, \ldots, v_d$ from Lemma 8.1, the matrices representing A and A^* are

$$A : \begin{pmatrix} 0 & d & & & & \mathbf{0} \\ 1 & 0 & d-1 & & & \\ & 2 & \cdot & \cdot & & \\ & & \cdot & \cdot & \cdot & \\ & & & \cdot & \cdot & 1 \\ \mathbf{0} & & & & d & 0 \end{pmatrix}, \qquad A^* : \ \mathrm{diag}(d, d-2, d-4, \ldots, -d).$$

The pair A, A^* acts on V_d as a Leonard pair. The resulting Leonard pair is isomorphic to the one in (1.2).

The Leonard pairs in Example 8.1 are not the only ones associated with sl_2. To get more Leonard pairs we replace A and A^* by more general elements in sl_2. Our result is the following.

Theorem 8.3. [45, Ex. 1.5] *Let A and A^* denote semi-simple elements in sl_2 and assume sl_2 is generated by these elements. Let V denote an irreducible finite dimensional module for sl_2. Then the pair A, A^* acts on V as a Leonard pair.*

We remark the Leonard pairs in Theorem 8.3 correspond to the Krawtchouk polynomials [53].

8.2 Leonard Pairs and the Quantum Algebra $U_q(sl_2)$

In this section we assume $\mathbb{K}$ is algebraically closed. We fix a nonzero scalar $q \in \mathbb{K}$ that is not a root of unity. We recall the quantum algebra $U_q(sl_2)$.

Definition 8.1. [49, p. 122] *Let $U_q(sl_2)$ denote the unital associative $\mathbb{K}$-algebra with generators e, f, k, k^{-1} and relations*

$$kk^{-1} = k^{-1}k = 1,$$

$$ke = q^2 ek, \qquad kf = q^{-2} fk,$$

$$ef - fe = \frac{k - k^{-1}}{q - q^{-1}}.$$

We recall the irreducible finite dimensional modules for $U_q(sl_2)$. We use the following notation.

$$[n]_q = \frac{q^n - q^{-n}}{q - q^{-1}} \qquad n \in \mathbb{Z}.$$

Lemma 8.2. [49, p. 128] *With reference to Definition 8.1, there exists a family*

$$V_{\varepsilon,d} \qquad \varepsilon \in \{1, -1\}, \qquad d = 0, 1, 2, \ldots \tag{8.4}$$

of irreducible finite dimensional $U_q(sl_2)$-modules with the following properties. The module $V_{\varepsilon,d}$ has a basis $u_0, u_1, \ldots, u_d$ satisfying $ku_i = \varepsilon q^{d-2i} u_i$ for $0 \leq i \leq d$, $fu_i = [i+1]_q u_{i+1}$ for $0 \leq i \leq d-1$, $fu_d = 0$, $eu_i = \varepsilon [d-i+1]_q u_{i-1}$ for $1 \leq i \leq d$, $eu_0 = 0$. Every irreducible finite dimensional $U_q(sl_2)$-module is isomorphic to exactly one of the modules $V_{\varepsilon,d}$. (Referring to line (8.4), if $\mathbb{K}$ has characteristic 2 we interpret the set $\{1, -1\}$ as having a single element.)

Theorem 8.4. [56], [57], [84] *Referring to Definition 8.1 and Lemma 8.2, let* α, β *denote nonzero scalars in* $\mathbb{K}$ *and define* A, A^* *as follows.*

$$A = \alpha f + \frac{k}{q - q^{-1}}, \qquad A^* = \beta e + \frac{k^{-1}}{q - q^{-1}}.$$

Let d *denote a nonnegative integer and choose* $\varepsilon \in \{1, -1\}$. *Then the pair* A, A^* *acts on* $V_{\varepsilon,d}$ *as a Leonard pair provided* $\varepsilon\alpha\beta$ *is not among* $q^{d-1}, q^{d-3}, \ldots, q^{1-d}$.

We remark the Leonard pairs in Theorem 8.4 correspond to the quantum q-Krawtchouk polynomials [53], [55].

8.3 Leonard Pairs in Combinatorics

Leonard pairs arise in many branches of combinatorics. For instance they arise in the theory of partially ordered sets (posets). We illustrate this with a poset called the subspace lattice $L_n(q)$.

In this section we assume our field $\mathbb{K}$ is the field $\mathbb{C}$ of complex numbers.

To define the subspace lattice we introduce a second field. Let $GF(q)$ denote a finite field of order q. Let n denote a positive integer and let W denote an n-dimensional vector space over $GF(q)$. Let P denote the set consisting of all subspaces of W. The set P, together with the containment relation, is a poset called $L_n(q)$.

Using $L_n(q)$ we obtain a family of Leonard pairs as follows. Let $\mathbb{C}P$ denote the vector space over $\mathbb{C}$ consisting of all formal $\mathbb{C}$-linear combinations of elements of P. We observe P is a basis for $\mathbb{C}P$ so the dimension of $\mathbb{C}P$ is equal to the cardinality of P.

We define three linear transformations on $\mathbb{C}P$. We call these K, R (for "raising"), L (for "lowering").

We begin with K. For all $x \in P$,

$$Kx = q^{n/2 - \dim x} x.$$

Apparently each element of P is an eigenvector for K.

To define R and L we use the following notation. For $x, y \in P$ we say y *covers* x whenever (i) $x \subseteq y$ and (ii) $\dim y = 1 + \dim x$.

The maps R and L are defined as follows. For all $x \in P$,

$$Rx = \sum_{y \text{ covers } x} y.$$

Similarly

$$Lx = q^{(1-n)/2} \sum_{x \text{ covers } y} y.$$

(The scalar $q^{(1-n)/2}$ is included for aesthetic reasons.)

We consider the properties of K, R, L. From the construction we find K^{-1} exists. By combinatorial counting we verify

$$KL = qLK, \qquad KR = q^{-1}RK,$$
$$LR - RL = \frac{K - K^{-1}}{q^{1/2} - q^{-1/2}}.$$

We recognize these equations. They are the defining relations for $U_{q^{1/2}}(sl_2)$. Apparently K, R, L turn $\mathbb{C}P$ into a module for $U_{q^{1/2}}(sl_2)$.

We now see how to get Leonard pairs from $L_n(q)$. Let α, β denote nonzero complex scalars and define A, A^* as follows.

$$A = \alpha R + \frac{K}{q^{1/2} - q^{-1/2}}, \qquad A^* = \beta L + \frac{K^{-1}}{q^{1/2} - q^{-1/2}}.$$

To avoid degenerate situations we assume $\alpha\beta$ is not among $q^{(n-1)/2}, q^{(n-3)/2}, \ldots, q^{(1-n)/2}$.

The $U_{q^{1/2}}(sl_2)$-module $\mathbb{C}P$ is completely reducible [49, p. 144]. In other words $\mathbb{C}P$ is a direct sum of irreducible $U_{q^{1/2}}(sl_2)$-modules. On each irreducible module in this sum the pair A, A^* acts as a Leonard pair. This follows from Theorem 8.4.

We just saw how the subspace lattice gives Leonard pairs. We expect that some other classical posets, such as the polar spaces and attenuated spaces [76], give Leonard pairs in a similar fashion. However the details remain to be worked out. See [76] for more information on this topic.

Another combinatorial object that gives Leonard pairs is a P- and Q-polynomial association scheme [11], [13], [77]. Leonard pairs have been used to describe certain irreducible modules for the subconstituent algebra of these schemes [14], [18], [19], [45], [77].

9 Tridiagonal Pairs

There is a mild generalization of a Leonard pair called a *tridiagonal pair* [45], [46], [47], [80]. In order to define this, we use the following terms. Let V denote a vector space over $\mathbb{K}$ with finite positive dimension. Let $A : V \to V$ denote a linear transformation and let W denote a subspace of V. We call W an *eigenspace* of A whenever $W \neq 0$ and there exists $\theta \in \mathbb{K}$ such that

$$W = \{v \in V \mid Av = \theta v\}.$$

We say A is *diagonalizable* whenever V is spanned by the eigenspaces of A.

Definition 9.1. [45, Definition 1.1] *Let V denote a vector space over $\mathbb{K}$ with finite positive dimension. By a* tridiagonal pair *on V, we mean an ordered pair of linear transformations $A : V \to V$ and $A^* : V \to V$ that satisfy the following four conditions.*

(i) Each of A, A^* is diagonalizable.
(ii) There exists an ordering $V_0, V_1, \ldots, V_d$ of the eigenspaces of A such that

$$A^*V_i \subseteq V_{i-1} + V_i + V_{i+1} \qquad (0 \le i \le d), \tag{9.1}$$

where $V_{-1} = 0$, $V_{d+1} = 0$.
(iii) There exists an ordering $V_0^*, V_1^*, \ldots, V_\delta^*$ of the eigenspaces of A^* such that

$$AV_i^* \subseteq V_{i-1}^* + V_i^* + V_{i+1}^* \qquad (0 \le i \le \delta), \tag{9.2}$$

where $V_{-1}^* = 0$, $V_{\delta+1}^* = 0$.
(iv) There does not exist a subspace W of V such that $AW \subseteq W$, $A^*W \subseteq W$, $W \neq 0$, $W \neq V$.

The following problem is open.

Problem 9.1. Classify the tridiagonal pairs.

For the rest of this section we discuss what is known about tridiagonal pairs, and give some conjectures.

Let A, A^* denote a tridiagonal pair on V and let the integers d, δ be as in Definition 9.1(ii), (iii) respectively. By [45, Lemma 4.5] we have $d = \delta$; we call this common value the *diameter* of the pair. An ordering of the eigenspaces of A (resp. A^*) will be called *standard* whenever it satisfies (9.1) (resp. (9.2)). We comment on the uniqueness of the standard ordering. Let $V_0, V_1, \ldots, V_d$ denote a standard ordering of the eigenspaces of A. Then the ordering $V_d, V_{d-1}, \ldots, V_0$ is standard and no other ordering is standard. A similar result holds for the eigenspaces of A^*. Let $V_0, V_1, \ldots, V_d$ (resp. $V_0^*, V_1^*, \ldots, V_d^*$) denote a standard ordering of the eigenspaces of A (resp. A^*). By [45, Corollary 5.7], for $0 \le i \le d$ the spaces V_i, V_i^* have the same dimension; we denote this common dimension by ρ_i. By the construction $\rho_i \neq 0$. By [45, Corollary 5.7] and [45, Corollary 6.6], the sequence $\rho_0, \rho_1, \ldots, \rho_d$ is symmetric and unimodal; that is $\rho_i = \rho_{d-i}$ for $0 \le i \le d$ and $\rho_{i-1} \le \rho_i$ for $1 \le i \le d/2$. We refer to the sequence $(\rho_0, \rho_1, \ldots, \rho_d)$ as the *shape vector* of A, A^*. A Leonard pair is the same thing as a tridiagonal pair that has shape vector $(1, 1, \ldots, 1)$.

Conjecture 9.1. [45, Conjecture 13.5] Referring to Definition 9.1, assume $\mathbb{K}$ is algebraically closed and let $(\rho_0, \rho_1, \ldots, \rho_d)$ denote the shape vector for A, A^*. Then the entries in this shape vector are bounded above by binomial coefficients as follows:

$$\rho_i \le \binom{d}{i} \qquad (0 \le i \le d).$$

See [46] for some partial results on Conjecture 9.1. We now give some examples of tridiagonal pairs.

Example 9.1. [45, Example 1.6] Assume $\mathbb{K}$ is algebraically closed with characteristic 0. Let b, b^* denote nonzero scalars in $\mathbb{K}$. Let O denote the Lie algebra over $\mathbb{K}$ generated by symbols A, A^* subject to the relations

$$[A,[A,[A,A^*]]] = b^2[A,A^*], \tag{9.3}$$
$$[A^*,[A^*,[A^*,A]]] = b^{*2}[A^*,A]. \tag{9.4}$$

Let V denote a finite dimensional irreducible O-module. Then A, A^* act on V as a tridiagonal pair.

Remark 9.1. The algebra O from Example 9.1 is called the *Onsager algebra.* It first appeared in the seminal paper by Onsager [70] in which the free energy of the two dimensional Ising model was computed exactly. Onsager presented his algebra by displaying a basis; the above presentation using generators and relations (9.3), (9.4) was established by Perk [71]. The relations themselves first appeared in work of Dolan and Grady [26]. A few years later they were used by von Gehlen and Rittenberg [29] to describe the superintegrable chiral Potts model. In [72] Roan observed that O is isomorphic to the invariant subalgebra of the loop algebra $\mathbb{K}[t,t^{-1}] \otimes sl_2$ by an involution. Of course this last result was not available to Onsager since his discovery predates the invention of Kac-Moody algebras by some 25 years. See [1], [2], [3], [9], [10], [12], [23], [25], [64], [89] for recent work involving the Onsager algebra and integrable lattice models. The equations (9.3), (9.4) are called the *Dolan-Grady* relations [22], [24], [50], [51], [52], [61].

Example 9.2. [45, Example 1.7] Assume $\mathbb{K}$ is algebraically closed, and let q denote a nonzero scalar in $\mathbb{K}$ that is not a root of unity. Let $U_q(\widehat{sl}_2)^{>0}$ denote the unital associative $\mathbb{K}$-algebra generated by symbols A, A^* subject to the relations

$$0 = A^3A^* - [3]_q A^2A^*A + [3]_q AA^*A^2 - A^*A^3, \tag{9.5}$$
$$0 = A^{*3}A - [3]_q A^{*2}AA^* + [3]_q A^*AA^{*2} - AA^{*3}. \tag{9.6}$$

Let V denote a finite dimensional irreducible $U_q(\widehat{sl}_2)^{>0}$-module and assume neither of A, A^* is nilpotent on V. Then A, A^* act on V as a tridiagonal pair.

Remark 9.2. The equations (9.5), (9.6) are known as the *q-Serre relations,* and are among the defining relations for the quantum affine algebra $U_q(\widehat{sl}_2)$ [15], [48]. The algebra $U_q(\widehat{sl}_2)^{>0}$ is called the *positive part* of $U_q(\widehat{sl}_2)$. The tridiagonal pairs from Example 9.2 are said to have *q-geometric type.*

In order to get the most general tridiagonal pairs, we consider a pair of relations that generalize both the Dolan-Grady relations and the q-Serre relations. We call these the *tridiagonal relations.* These relations are given as follows.

Theorem 9.1. [45, Theorem 10.1] *Let V denote a vector space over $\mathbb{K}$ with finite positive dimension and let A, A^* denote a tridiagonal pair on V. Then there exists a sequence of scalars $\beta, \gamma, \gamma^*, \varrho, \varrho^*$ taken from $\mathbb{K}$ such that both*

$$0 = [A, A^2A^* - \beta AA^*A + A^*A^2 - \gamma(AA^* + A^*A) - \varrho A^*] \tag{9.7}$$

$$0 = [A^*, A^{*2}A - \beta A^*AA^* + AA^{*2} - \gamma^*(AA^* + A^*A) - \varrho^* A], \tag{9.8}$$

where $[r, s]$ means $rs - sr$. The sequence is unique if the diameter $d \geq 3$.

We call (9.7), (9.8) the *tridiagonal relations* [80]. As far as we know these relations first appeared in [78, Lemma 5.4].

Remark 9.3. The Dolan-Grady relations (9.3), (9.4) are the tridiagonal relations with parameters $\beta = 2$, $\gamma = \gamma^* = 0$, $\varrho = b^2$, $\varrho^* = b^{*2}$, if we interpret the bracket in (9.3), (9.4) as $[r, s] = rs - sr$. The q-Serre relations (9.5), (9.6) are the tridiagonal relations with parameters $\beta = q^2 + q^{-2}$, $\gamma = \gamma^* = 0$, $\varrho = \varrho^* = 0$.

Our next result is a kind of converse to Theorem 9.1.

Theorem 9.2. [80, Theorem 3.10] *Let $\beta, \gamma, \gamma^*, \varrho, \varrho^*$ denote a sequence of scalars taken from $\mathbb{K}$. Let T denote the unital associative $\mathbb{K}$-algebra generated by symbols A, A^* subject to the tridiagonal relations (9.7), (9.8). Let V denote an irreducible finite dimensional T-module and assume each of A, A^* is diagonalizable on V. Then A, A^* act on V as a tridiagonal pair provided q is not a root of unity, where $q + q^{-1} = \beta$.*

Remark 9.4. The algebra T in Theorem 9.2 is called the *tridiagonal algebra* [45], [79], [80].

So far in our research on tridiagonal pairs, our strongest result concerns the case of q-geometric type. In order to describe this result we define one more algebra.

Definition 9.2. *Assume $\mathbb{K}$ is algebraically closed. Let q denote a nonzero scalar in $\mathbb{K}$ that is not a root of unity. We let $\square_q$ denote the unital associative $\mathbb{K}$-algebra with generators x_i, k_i $(i = 0, 1, 2, 3)$ and relations*

$$\frac{qx_ik_i - q^{-1}k_ix_i}{q - q^{-1}} = 1, \qquad \frac{qx_ix_{i+1} - q^{-1}x_{i+1}x_i}{q - q^{-1}} = 1,$$

$$\frac{qk_{i+1}x_i - q^{-1}x_ik_{i+1}}{q - q^{-1}} = 1, \qquad k_ik_{i+2} = 1,$$

$$x_i^3x_{i+2} - [3]_qx_i^2x_{i+2}x_i + [3]_qx_ix_{i+2}x_i^2 - x_{i+2}x_i^3 = 0$$

for $0 \leq i \leq 3$, where all subscripts are computed modulo 4.

Remark 9.5. The algebra $\square_q$ is closely related to the quantum affine algebra $U_q(\widehat{sl}_2)$. Indeed there exists a homomorphism of $\mathbb{K}$-algebras from $U_q(\widehat{sl}_2)$ into $\square_q$ [47, Theorem 2.1]. This homomorphism induces on each $\square_q$-module the structure of a $U_q(\widehat{sl}_2)$-module. The $\square_q$-module structure is irreducible if and only if the $U_q(\widehat{sl}_2)$-module structure is irreducible.

Theorem 9.3. [47, Theorem 3.3] *Assume $\mathbb{K}$ is algebraically closed. Let q denote a nonzero scalar in $\mathbb{K}$ that is not a root of unity. Let V denote a vector space over $\mathbb{K}$ with finite positive dimension. Let A, A^* denote a tridiagonal pair on V that has q-geometric type. Then there exists an irreducible $\square_q$-module structure on V such that A acts as a scalar multiple of x_0 and A^* acts as a scalar multiple of x_2. Conversely, let V denote a finite dimensional irreducible $\square_q$-module. Then the elements x_0, x_2 act on V as a tridiagonal pair of q-geometric type.*

Problem 9.2. Determine up to isomorphism the tridiagonal pairs of q-geometric type. This can probably be done by using Remark 9.5 and the classification of finite dimensional irreducible $U_q(\widehat{sl}_2)$-modules [15].

We end this section with a conjecture.

Conjecture 9.2. Assume $\mathbb{K}$ is algebraically closed. Let V denote a vector space over $\mathbb{K}$ with finite positive dimension and let A, A^* denote a tridiagonal pair on V. To avoid degenerate situations we assume q is not a root of unity, where $\beta = q^2 + q^{-2}$, and where β is from Theorem 9.1. Then referring to Definition 9.2, there exists an irreducible $\square_q$-module structure on V such that A acts as a linear combination of x_0, x_1, I and A^* acts as a linear combination of x_2, x_3, I.

10 Appendix: List of Parameter Arrays

In this section we display all the parameter arrays over $\mathbb{K}$. We will use the following notation.

Definition 10.1. *Let $p = (\theta_i, \theta^*_i, i = 0, \ldots, d; \varphi_j, \phi_j, j = 1, \ldots, d)$ denote a parameter array over $\mathbb{K}$. For $0 \leq i \leq d$ we let u_i denote the following polynomial in $\mathbb{K}[\lambda]$.*

$$u_i = \sum_{n=0}^{i} \frac{(\lambda - \theta_0)\cdots(\lambda - \theta_{n-1})(\theta^*_i - \theta^*_0)\cdots(\theta^*_i - \theta^*_{n-1})}{\varphi_1\varphi_2\cdots\varphi_n}. \tag{10.1}$$

We call $u_0, u_1, \ldots, u_d$ the polynomials that correspond to p.

We now display all the parameter arrays over $\mathbb{K}$. For each displayed array $(\theta_i, \theta^*_i, i = 0, \ldots, d; \varphi_j, \phi_j, j = 1, \ldots, d)$ we present $u_i(\theta_j)$ for $0 \leq i, j \leq d$, where $u_0, u_1, \ldots, u_d$ are the corresponding polynomials. Our presentation is organized as follows. In each of Example 10.1–10.13 below we give a family of parameter arrays over $\mathbb{K}$. In Theorem 10.1 we show every parameter array over $\mathbb{K}$ is contained in at least one of these families.

In each of Example 10.1–10.13 below the following implicit assumptions apply: d denotes a nonnegative integer, the scalars $(\theta_i, \theta^*_i, i = 0, \ldots, d; \varphi_j, \phi_j, j = 1, \ldots, d)$ are contained in $\mathbb{K}$, and the scalars $q, h, h^*, \ldots$ are contained in the algebraic closure of $\mathbb{K}$.

Example 10.1. (q-Racah) Assume

$$\theta_i = \theta_0 + h(1-q^i)(1-sq^{i+1})q^{-i}, \tag{10.2}$$
$$\theta^*_i = \theta^*_0 + h^*(1-q^i)(1-s^*q^{i+1})q^{-i} \tag{10.3}$$

for $0 \le i \le d$ and

$$\varphi_i = hh^*q^{1-2i}(1-q^i)(1-q^{i-d-1})(1-r_1q^i)(1-r_2q^i), \tag{10.4}$$
$$\phi_i = hh^*q^{1-2i}(1-q^i)(1-q^{i-d-1})(r_1-s^*q^i)(r_2-s^*q^i)/s^* \tag{10.5}$$

for $1 \le i \le d$. Assume $h, h^*, q, s, s^*, r_1, r_2$ are nonzero and $r_1r_2 = ss^*q^{d+1}$. Assume none of $q^i, r_1q^i, r_2q^i, s^*q^i/r_1, s^*q^i/r_2$ is equal to 1 for $1 \le i \le d$ and that neither of sq^i, s^*q^i is equal to 1 for $2 \le i \le 2d$. Then $(\theta_i, \theta^*_i, i = 0, \ldots, d; \varphi_j, \phi_j, j = 1, \ldots, d)$ is a parameter array over $\mathbb{K}$. The corresponding polynomials u_i satisfy

$$u_i(\theta_j) = {}_4\phi_3\left(\begin{matrix} q^{-i},\ s^*q^{i+1},\ q^{-j},\ sq^{j+1} \\ r_1q,\ r_2q,\ q^{-d} \end{matrix} \,\middle|\, q, q\right)$$

for $0 \le i, j \le d$. These u_i are the q-Racah polynomials.

Example 10.2. (q-Hahn) Assume

$$\theta_i = \theta_0 + h(1-q^i)q^{-i},$$
$$\theta^*_i = \theta^*_0 + h^*(1-q^i)(1-s^*q^{i+1})q^{-i}$$

for $0 \le i \le d$ and

$$\varphi_i = hh^*q^{1-2i}(1-q^i)(1-q^{i-d-1})(1-rq^i),$$
$$\phi_i = -hh^*q^{1-i}(1-q^i)(1-q^{i-d-1})(r-s^*q^i)$$

for $1 \le i \le d$. Assume h, h^*, q, s^*, r are nonzero. Assume none of $q^i, rq^i, s^*q^i/r$ is equal to 1 for $1 \le i \le d$ and that $s^*q^i \ne 1$ for $2 \le i \le 2d$. Then the sequence $(\theta_i, \theta^*_i, i = 0, \ldots, d; \varphi_j, \phi_j, j = 1, \ldots, d)$ is a parameter array over $\mathbb{K}$. The corresponding polynomials u_i satisfy

$$u_i(\theta_j) = {}_3\phi_2\left(\begin{matrix} q^{-i},\ s^*q^{i+1},\ q^{-j} \\ rq,\ q^{-d} \end{matrix} \,\middle|\, q, q\right)$$

for $0 \le i, j \le d$. These u_i are the q-Hahn polynomials.

Example 10.3. (Dual q-Hahn) Assume

$$\begin{aligned}\theta_i &= \theta_0 + h(1-q^i)(1-sq^{i+1})q^{-i},\\ \theta^*_i &= \theta^*_0 + h^*(1-q^i)q^{-i}\end{aligned}$$

for $0 \le i \le d$ and

$$\begin{aligned}\varphi_i &= hh^*q^{1-2i}(1-q^i)(1-q^{i-d-1})(1-rq^i),\\ \phi_i &= hh^*q^{d+2-2i}(1-q^i)(1-q^{i-d-1})(s-rq^{i-d-1})\end{aligned}$$

for $1 \le i \le d$. Assume h, h^*, q, r, s are nonzero. Assume none of $q^i, rq^i, sq^i/r$ is equal to 1 for $1 \le i \le d$ and that $sq^i \ne 1$ for $2 \le i \le 2d$. Then the sequence $(\theta_i, \theta^*_i, i = 0, \ldots, d; \varphi_j, \phi_j, j = 1, \ldots, d)$ is a parameter array over $\mathbb{K}$. The corresponding polynomials u_i satisfy

$$u_i(\theta_j) = {}_3\phi_2\left(\begin{matrix} q^{-i},\ q^{-j},\ sq^{j+1} \\ rq,\ q^{-d} \end{matrix} \,\middle|\, q,\ q\right)$$

for $0 \le i, j \le d$. These u_i are the dual q-Hahn polynomials.

Example 10.4. (Quantum q-Krawtchouk) Assume

$$\begin{aligned}\theta_i &= \theta_0 - sq(1-q^i),\\ \theta^*_i &= \theta^*_0 + h^*(1-q^i)q^{-i}\end{aligned}$$

for $0 \le i \le d$ and

$$\begin{aligned}\varphi_i &= -rh^*q^{1-i}(1-q^i)(1-q^{i-d-1}),\\ \phi_i &= h^*q^{d+2-2i}(1-q^i)(1-q^{i-d-1})(s-rq^{i-d-1})\end{aligned}$$

for $1 \le i \le d$. Assume h^*, q, r, s are nonzero. Assume neither of $q^i, sq^i/r$ is equal to 1 for $1 \le i \le d$. Then the sequence $(\theta_i, \theta^*_i, i = 0, \ldots, d; \varphi_j, \phi_j, j = 1, \ldots, d)$ is a parameter array over $\mathbb{K}$. The corresponding polynomials u_i satisfy

$$u_i(\theta_j) = {}_2\phi_1\left(\begin{matrix} q^{-i},\ q^{-j} \\ q^{-d} \end{matrix} \,\middle|\, q,\ sr^{-1}q^{j+1}\right)$$

for $0 \le i, j \le d$. These u_i are the quantum q-Krawtchouk polynomials.

Example 10.5. (q-Krawtchouk) Assume

$$\begin{aligned}\theta_i &= \theta_0 + h(1-q^i)q^{-i},\\ \theta^*_i &= \theta^*_0 + h^*(1-q^i)(1-s^*q^{i+1})q^{-i}\end{aligned}$$

for $0 \le i \le d$ and

$$\begin{aligned}\varphi_i &= hh^*q^{1-2i}(1-q^i)(1-q^{i-d-1}),\\ \phi_i &= hh^*s^*q(1-q^i)(1-q^{i-d-1})\end{aligned}$$

for $1 \le i \le d$. Assume h, h^*, q, s^* are nonzero. Assume $q^i \neq 1$ for $1 \le i \le d$ and that $s^*q^i \neq 1$ for $2 \le i \le 2d$. Then the sequence $(\theta_i, \theta^*_i, i = 0, \ldots, d; \varphi_j, \phi_j, j = 1, \ldots, d)$ is a parameter array over $\mathbb{K}$. The corresponding polynomials u_i satisfy

$$u_i(\theta_j) = {}_3\phi_2\left(\begin{matrix} q^{-i},\ s^*q^{i+1},\ q^{-j} \\ 0,\ q^{-d} \end{matrix}\ \middle|\ q,\ q\right)$$

for $0 \le i, j \le d$. These u_i are the q-Krawtchouk polynomials.

Example 10.6. (Affine q-Krawtchouk) Assume

$$\theta_i = \theta_0 + h(1 - q^i)q^{-i},$$
$$\theta^*_i = \theta^*_0 + h^*(1 - q^i)q^{-i}$$

for $0 \le i \le d$ and

$$\varphi_i = hh^*q^{1-2i}(1 - q^i)(1 - q^{i-d-1})(1 - rq^i),$$
$$\phi_i = -hh^*rq^{1-i}(1 - q^i)(1 - q^{i-d-1})$$

for $1 \le i \le d$. Assume h, h^*, q, r are nonzero. Assume neither of q^i, rq^i is equal to 1 for $1 \le i \le d$. Then the sequence $(\theta_i, \theta^*_i, i = 0, \ldots, d; \varphi_j, \phi_j, j = 1, \ldots, d)$ is a parameter array over $\mathbb{K}$. The corresponding polynomials u_i satisfy

$$u_i(\theta_j) = {}_3\phi_2\left(\begin{matrix} q^{-i},\ 0,\ q^{-j} \\ rq,\ q^{-d} \end{matrix}\ \middle|\ q,\ q\right)$$

for $0 \le i, j \le d$. These u_i are the affine q-Krawtchouk polynomials.

Example 10.7. (Dual q-Krawtchouk) Assume

$$\theta_i = \theta_0 + h(1 - q^i)(1 - sq^{i+1})q^{-i},$$
$$\theta^*_i = \theta^*_0 + h^*(1 - q^i)q^{-i}$$

for $0 \le i \le d$ and

$$\varphi_i = hh^*q^{1-2i}(1 - q^i)(1 - q^{i-d-1}),$$
$$\phi_i = hh^*sq^{d+2-2i}(1 - q^i)(1 - q^{i-d-1})$$

for $1 \le i \le d$. Assume h, h^*, q, s are nonzero. Assume $q^i \neq 1$ for $1 \le i \le d$ and $sq^i \neq 1$ for $2 \le i \le 2d$. Then the sequence $(\theta_i, \theta^*_i, i = 0, \ldots, d; \varphi_j, \phi_j, j = 1, \ldots, d)$ is a parameter array over $\mathbb{K}$. The corresponding polynomials u_i satisfy

$$u_i(\theta_j) = {}_3\phi_2\left(\begin{matrix} q^{-i},\ q^{-j},\ sq^{j+1} \\ 0,\ q^{-d} \end{matrix}\ \middle|\ q,\ q\right)$$

for $0 \le i, j \le d$. These u_i are the dual q-Krawtchouk polynomials.

Example 10.8. (Racah) Assume

$$\theta_i = \theta_0 + hi(i+1+s), \tag{10.6}$$
$$\theta^*_i = \theta^*_0 + h^*i(i+1+s^*) \tag{10.7}$$

for $0 \le i \le d$ and

$$\varphi_i = hh^*i(i-d-1)(i+r_1)(i+r_2), \tag{10.8}$$
$$\phi_i = hh^*i(i-d-1)(i+s^*-r_1)(i+s^*-r_2) \tag{10.9}$$

for $1 \le i \le d$. Assume h, h^* are nonzero and that $r_1 + r_2 = s + s^* + d + 1$. Assume the characteristic of $\mathbb{K}$ is 0 or a prime greater than d. Assume none of $r_1, r_2, s^* - r_1, s^* - r_2$ is equal to $-i$ for $1 \le i \le d$ and that neither of s, s^* is equal to $-i$ for $2 \le i \le 2d$. Then the sequence $(\theta_i, \theta^*_i, i = 0, \ldots, d; \varphi_j, \phi_j, j = 1, \ldots, d)$ is a parameter array over $\mathbb{K}$. The corresponding polynomials u_i satisfy

$$u_i(\theta_j) = {}_4F_3\left(\begin{matrix} -i,\ i+1+s^*,\ -j,\ j+1+s \\ r_1+1,\ r_2+1,\ -d \end{matrix} \middle| \, 1\right)$$

for $0 \le i, j \le d$. These u_i are the Racah polynomials.

Example 10.9. (Hahn) Assume

$$\theta_i = \theta_0 + si,$$
$$\theta^*_i = \theta^*_0 + h^*i(i+1+s^*)$$

for $0 \le i \le d$ and

$$\varphi_i = h^*si(i-d-1)(i+r),$$
$$\phi_i = -h^*si(i-d-1)(i+s^*-r)$$

for $1 \le i \le d$. Assume h^*, s are nonzero. Assume the characteristic of $\mathbb{K}$ is 0 or a prime greater than d. Assume neither of $r, s^* - r$ is equal to $-i$ for $1 \le i \le d$ and that $s^* \ne -i$ for $2 \le i \le 2d$. Then the sequence $(\theta_i, \theta^*_i, i = 0, \ldots, d; \varphi_j, \phi_j, j = 1, \ldots, d)$ is a parameter array over $\mathbb{K}$. The corresponding polynomials u_i satisfy

$$u_i(\theta_j) = {}_3F_2\left(\begin{matrix} -i,\ i+1+s^*,\ -j \\ r+1,\ -d \end{matrix} \middle| \, 1\right)$$

for $0 \le i, j \le d$. These u_i are the Hahn polynomials.

Example 10.10. (Dual Hahn) Assume

$$\theta_i = \theta_0 + hi(i+1+s),$$
$$\theta^*_i = \theta^*_0 + s^*i$$

for $0 \le i \le d$ and

$$\varphi_i = hs^* i(i-d-1)(i+r),$$

$$\phi_i = hs^* i(i-d-1)(i+r-s-d-1)$$

for $1 \le i \le d$. Assume h, s^* are nonzero. Assume the characteristic of $\mathbb{K}$ is 0 or a prime greater than d. Assume neither of $r, s-r$ is equal to $-i$ for $1 \le i \le d$ and that $s \ne -i$ for $2 \le i \le 2d$. Then the sequence $(\theta_i, \theta^*_i, i = 0, \ldots, d; \varphi_j, \phi_j, j = 1, \ldots, d)$ is a parameter array over $\mathbb{K}$. The corresponding polynomials u_i satisfy

$$u_i(\theta_j) = {}_3F_2\left(\begin{matrix} -i,\ -j,\ j+1+s \\ r+1,\ -d \end{matrix}\ \middle|\ 1\right)$$

for $0 \le i, j \le d$. These u_i are the dual Hahn polynomials.

Example 10.11. (Krawtchouk) Assume

$$\theta_i = \theta_0 + si,$$

$$\theta^*_i = \theta^*_0 + s^* i$$

for $0 \le i \le d$ and

$$\varphi_i = ri(i-d-1)$$

$$\phi_i = (r - ss^*)i(i-d-1)$$

for $1 \le i \le d$. Assume r, s, s^* are nonzero. Assume the characteristic of $\mathbb{K}$ is 0 or a prime greater than d. Assume $r \ne ss^*$. Then the sequence $(\theta_i, \theta^*_i, i = 0, \ldots, d; \varphi_j, \phi_j, j = 1, \ldots, d)$ is a parameter array over $\mathbb{K}$. The corresponding polynomials u_i satisfy

$$u_i(\theta_j) = {}_2F_1\left(\begin{matrix} -i,\ -j \\ -d \end{matrix}\ \middle|\ r^{-1}ss^*\right)$$

for $0 \le i, j \le d$. These u_i are the Krawtchouk polynomials.

Example 10.12. (Bannai/Ito) Assume

$$\theta_i = \theta_0 + h(s - 1 + (1 - s + 2i)(-1)^i), \tag{10.10}$$

$$\theta^*_i = \theta^*_0 + h^*(s^* - 1 + (1 - s^* + 2i)(-1)^i) \tag{10.11}$$

for $0 \le i \le d$ and

$$\varphi_i = \begin{cases} -4hh^*i(i+r_1), & \text{if } i \text{ even, } d \text{ even,} \\ -4hh^*(i-d-1)(i+r_2), & \text{if } i \text{ odd, } d \text{ even,} \\ -4hh^*i(i-d-1), & \text{if } i \text{ even, } d \text{ odd,} \\ -4hh^*(i+r_1)(i+r_2), & \text{if } i \text{ odd, } d \text{ odd,} \end{cases} \tag{10.12}$$

$$\phi_i = \begin{cases} 4hh^*i(i-s^*-r_1), & \text{if } i \text{ even, } d \text{ even,} \\ 4hh^*(i-d-1)(i-s^*-r_2), & \text{if } i \text{ odd, } d \text{ even,} \\ -4hh^*i(i-d-1), & \text{if } i \text{ even, } d \text{ odd,} \\ -4hh^*(i-s^*-r_1)(i-s^*-r_2), & \text{if } i \text{ odd, } d \text{ odd,} \end{cases} \tag{10.13}$$

for $1 \leq i \leq d$. Assume h, h^* are nonzero and that $r_1 + r_2 = -s - s^* + d + 1$. Assume the characteristic of $\mathbb{K}$ is either 0 or an odd prime greater than $d/2$. Assume neither of $r_1, -s^* - r_1$ is equal to $-i$ for $1 \leq i \leq d$, $d - i$ even. Assume neither of $r_2, -s^* - r_2$ is equal to $-i$ for $1 \leq i \leq d$, i odd. Assume neither of s, s^* is equal to $2i$ for $1 \leq i \leq d$. Then the sequence $(\theta_i, \theta^*_i, i = 0, \ldots, d; \varphi_j, \phi_j, j = 1, \ldots, d)$ is a parameter array over $\mathbb{K}$. We call the corresponding polynomials from Definition 10.1 the Bannai/Ito polynomials [11, p. 260].

Example 10.13. (Orphan) For this example assume $\mathbb{K}$ has characteristic 2. For notational convenience we define some scalars $\gamma_0, \gamma_1, \gamma_2, \gamma_3$ in $\mathbb{K}$. We define $\gamma_i = 0$ for $i \in \{0, 3\}$ and $\gamma_i = 1$ for $i \in \{1, 2\}$. Assume

$$\theta_i = \theta_0 + h(si + \gamma_i), \tag{10.14}$$
$$\theta^*_i = \theta^*_0 + h^*(s^*i + \gamma_i) \tag{10.15}$$

for $0 \leq i \leq 3$. Assume $\varphi_1 = hh^*r$, $\varphi_2 = hh^*$, $\varphi_3 = hh^*(r + s + s^*)$ and $\phi_1 = hh^*(r + s(1 + s^*))$, $\phi_2 = hh^*$, $\phi_3 = hh^*(r + s^*(1 + s))$. Assume each of h, h^*, s, s^*, r is nonzero. Assume neither of s, s^* is equal to 1 and that r is equal to none of $s + s^*$, $s(1 + s^*)$, $s^*(1 + s)$. Then the sequence $(\theta_i, \theta^*_i, i = 0, \ldots, 3; \varphi_j, \phi_j, j = 1, \ldots, 3)$ is a parameter array over $\mathbb{K}$ which has diameter 3. We call the corresponding polynomials from Definition 10.1 the orphan polynomials.

Theorem 10.1. *Every parameter array over $\mathbb{K}$ is listed in at least one of the Examples 10.1–10.13.*

Proof. Let $p := (\theta_i, \theta^*_i, i = 0, \ldots, d; \varphi_j, \phi_j, j = 1, \ldots, d)$ denote a parameter array over $\mathbb{K}$. We show this array is given in at least one of the Examples 10.1–10.13. We assume $d \geq 1$; otherwise the result is trivial. For notational convenience let $\tilde{\mathbb{K}}$ denote the algebraic closure of $\mathbb{K}$. We define a scalar $q \in \tilde{\mathbb{K}}$ as follows. For $d \geq 3$, we let q denote a nonzero scalar in $\tilde{\mathbb{K}}$ such that $q + q^{-1} + 1$ is equal to the common value of (5.18). For $d < 3$ we let q denote a nonzero scalar in $\tilde{\mathbb{K}}$ such that $q \neq 1$ and $q \neq -1$. By PA5, both

$$\theta_{i-2} - \xi\theta_{i-1} + \xi\theta_i - \theta_{i+1} = 0, \tag{10.16}$$
$$\theta^*_{i-2} - \xi\theta^*_{i-1} + \xi\theta^*_i - \theta^*_{i+1} = 0 \tag{10.17}$$

for $2 \le i \le d-1$, where $\xi = q + q^{-1} + 1$. We divide the argument into the following four cases. (I) $q \ne 1$, $q \ne -1$; (II) $q = 1$ and $\mathrm{char}(\mathbb{K}) \ne 2$; (III) $q = -1$ and $\mathrm{char}(\mathbb{K}) \ne 2$; (IV) $q = 1$ and $\mathrm{char}(\mathbb{K}) = 2$.

Case I: $q \ne 1, q \ne -1$.
By (10.16) there exist scalars η, μ, h in $\tilde{\mathbb{K}}$ such that

$$\theta_i = \eta + \mu q^i + h q^{-i} \qquad (0 \le i \le d). \tag{10.18}$$

By (10.17) there exist scalars η^*, μ^*, h^* in $\tilde{\mathbb{K}}$ such that

$$\theta^*_i = \eta^* + \mu^* q^i + h^* q^{-i} \qquad (0 \le i \le d). \tag{10.19}$$

Observe μ, h are not both 0; otherwise $\theta_1 = \theta_0$ by (10.18). Similarly μ^*, h^* are not both 0. For $1 \le i \le d$ we have $q^i \ne 1$; otherwise $\theta_i = \theta_0$ by (10.18). Setting $i = 0$ in (10.18), (10.19) we obtain

$$\theta_0 = \eta + \mu + h, \tag{10.20}$$
$$\theta^*_0 = \eta^* + \mu^* + h^*. \tag{10.21}$$

We claim there exists $\tau \in \tilde{\mathbb{K}}$ such that both

$$\varphi_i = (q^i - 1)(q^{d-i+1} - 1)(\tau - \mu\mu^* q^{i-1} - hh^* q^{-i-d}), \tag{10.22}$$
$$\phi_i = (q^i - 1)(q^{d-i+1} - 1)(\tau - h\mu^* q^{i-d-1} - \mu h^* q^{-i}) \tag{10.23}$$

for $1 \le i \le d$. Since $q \ne 1$ and $q^d \ne 1$ there exists $\tau \in \tilde{\mathbb{K}}$ such that (10.22) holds for $i = 1$. In the equation of PA4, we eliminate φ_1 using (10.22) at $i = 1$, and evaluate the result using (10.18), (10.19) in order to obtain (10.23) for $1 \le i \le d$. In the equation of PA3, we eliminate ϕ_1 using (10.23) at $i = 1$, and evaluate the result using (10.18), (10.19) in order to obtain (10.22) for $1 \le i \le d$. We have now proved the claim. We now break the argument into subcases. For each subcase our argument is similar. We will discuss the first subcase in detail in order to give the idea; for the remaining subcases we give the essentials only.

Subcase q-Racah: $\mu \ne 0, \mu^* \ne 0, h \ne 0, h^* \ne 0$. We show p is listed in Example 10.1. Define

$$s := \mu h^{-1} q^{-1}, \qquad s^* := \mu^* h^{*-1} q^{-1}. \tag{10.24}$$

Eliminating η in (10.18) using (10.20) and eliminating μ in the result using the equation on the left in (10.24), we obtain (10.2) for $0 \le i \le d$. Similarly we obtain (10.3) for $0 \le i \le d$. Since $\tilde{\mathbb{K}}$ is algebraically closed it contains scalars r_1, r_2 such that both

$$r_1 r_2 = s s^* q^{d+1}, \qquad r_1 + r_2 = \tau h^{-1} h^{*-1} q^d. \tag{10.25}$$

Eliminating μ, μ^*, τ in (10.22), (10.23) using (10.24) and the equation on the right in (10.25), and evaluating the result using the equation on the left in

(10.25), we obtain (10.4), (10.5) for $1 \le i \le d$. By the construction each of h, h^*, q, s, s^* is nonzero. Each of r_1, r_2 is nonzero by the equation on the left in (10.25). The remaining inequalities mentioned below (10.5) follow from PA1, PA2 and (10.2)–(10.5). We have now shown p is listed in Example 10.1.

We now give the remaining subcases of Case I. We list the essentials only.

Subcase q-Hahn: $\mu = 0, \mu^* \neq 0, h \neq 0, h^* \neq 0, \tau \neq 0$. Definitions:

$$s^* := \mu^* h^{*-1} q^{-1}, \qquad r := \tau h^{-1} h^{*-1} q^d.$$

Subcase dual q-Hahn: $\mu \neq 0, \mu^* = 0, h \neq 0, h^* \neq 0, \tau \neq 0$. Definitions:

$$s := \mu h^{-1} q^{-1}, \qquad r := \tau h^{-1} h^{*-1} q^d.$$

Subcase quantum q-Krawtchouk: $\mu \neq 0, \mu^* = 0, h = 0, h^* \neq 0, \tau \neq 0$. Definitions:

$$s := \mu q^{-1}, \qquad r := \tau h^{*-1} q^d.$$

Subcase q-Krawtchouk: $\mu = 0, \mu^* \neq 0, h \neq 0, h^* \neq 0, \tau = 0$. Definition:

$$s^* := \mu^* h^{*-1} q^{-1}.$$

Subcase affine q-Krawtchouk: $\mu = 0, \mu^* = 0, h \neq 0, h^* \neq 0, \tau \neq 0$. Definition:

$$r := \tau h^{-1} h^{*-1} q^d.$$

Subcase dual q-Krawtchouk: $\mu \neq 0, \mu^* = 0, h \neq 0, h^* \neq 0, \tau = 0$. Definition:

$$s := \mu h^{-1} q^{-1}.$$

We have a few more comments concerning Case I. Earlier we mentioned that μ, h are not both 0 and that μ^*, h^* are not both 0. Suppose one of μ, h is 0 and one of μ^*, h^* is 0. Then $\tau \neq 0$; otherwise $\varphi_1 = 0$ by (10.22) or $\phi_1 = 0$ by (10.23). Suppose $\mu^* \neq 0$, $h^* = 0$. Replacing q by q^{-1} we obtain $\mu^* = 0$, $h^* \neq 0$. Suppose $\mu^* \neq 0$, $h^* \neq 0$, $\mu \neq 0$, $h = 0$. Replacing q by q^{-1} we obtain $\mu^* \neq 0$, $h^* \neq 0$, $\mu = 0$, $h \neq 0$. By these comments we find that after replacing q by q^{-1} if necessary, one of the above subcases holds. This completes our argument for Case I.

Case II: $q = 1$ and $\mathrm{char}(\mathbb{K}) \neq 2$.
By (10.16) and since $\mathrm{char}(\mathbb{K}) \neq 2$, there exist scalars η, μ, h in $\tilde{\mathbb{K}}$ such that

$$\theta_i = \eta + (\mu + h)i + hi^2 \qquad (0 \le i \le d). \tag{10.26}$$

Similarly there exist scalars η^*, μ^*, h^* in $\tilde{\mathbb{K}}$ such that

$$\theta^*_i = \eta^* + (\mu^* + h^*)i + h^* i^2 \qquad (0 \le i \le d). \tag{10.27}$$

Observe μ, h are not both 0; otherwise $\theta_1 = \theta_0$. Similarly μ^*, h^* are not both 0. For any prime i such that $i \leq d$ we have $\text{char}(\mathbb{K}) \neq i$; otherwise $\theta_i = \theta_0$ by (10.26). Therefore $\text{char}(\mathbb{K})$ is 0 or a prime greater than d. Setting $i = 0$ in (10.26), (10.27) we obtain

$$\theta_0 = \eta, \qquad \theta_0^* = \eta^*. \tag{10.28}$$

We claim there exists $\tau \in \tilde{\mathbb{K}}$ such that both

$$\varphi_i = i(d-i+1)(\tau - (\mu h^* + h\mu^*)i - hh^*i(i+d+1)), \tag{10.29}$$

$$\phi_i = i(d-i+1)(\tau + \mu\mu^* + h\mu^*(1+d) + (\mu h^* - h\mu^*)i + hh^*i(d-i+1)) \tag{10.30}$$

for $1 \leq i \leq d$. There exists $\tau \in \tilde{\mathbb{K}}$ such that (10.29) holds for $i = 1$. In the equation of PA4, we eliminate φ_1 using (10.29) at $i = 1$, and evaluate the result using (10.26), (10.27) in order to obtain (10.30) for $1 \leq i \leq d$. In the equation of PA3, we eliminate ϕ_1 using (10.30) at $i = 1$, and evaluate the result using (10.26), (10.27) in order to obtain (10.29) for $1 \leq i \leq d$. We have now proved the claim. We now break the argument into subcases.

Subcase Racah: $h \neq 0$, $h^* \neq 0$. We show p is listed in Example 10.8. Define

$$s := \mu h^{-1}, \qquad s^* := \mu^* h^{*-1}. \tag{10.31}$$

Eliminating η, μ in (10.26) using (10.28), (10.31) we obtain (10.6) for $0 \leq i \leq d$. Eliminating η^*, μ^* in (10.27) using (10.28), (10.31) we obtain (10.7) for $0 \leq i \leq d$. Since $\tilde{\mathbb{K}}$ is algebraically closed it contains scalars r_1, r_2 such that both

$$r_1 r_2 = -\tau h^{-1} h^{*-1}, \qquad r_1 + r_2 = s + s^* + d + 1. \tag{10.32}$$

Eliminating μ, μ^*, τ in (10.29), (10.30) using (10.31) and the equation on the left in (10.32) we obtain (10.8), (10.9) for $1 \leq i \leq d$. By the construction each of h, h^* is nonzero. The remaining inequalities mentioned below (10.9) follow from PA1, PA2 and (10.6)–(10.9). We have now shown p is listed in Example 10.8.

We now give the remaining subcases of Case II. We list the essentials only.

Subcase Hahn: $h = 0$, $h^* \neq 0$. Definitions:

$$s = \mu, \qquad s^* := \mu^* h^{*-1}, \qquad r := -\tau\mu^{-1} h^{*-1}.$$

Subcase dual Hahn: $h \neq 0, h^* = 0$. Definitions:

$$s := \mu h^{-1}, \qquad s^* = \mu^*, \qquad r := -\tau h^{-1} \mu^{*-1}.$$

Subcase Krawtchouk: $h = 0, h^* = 0$. Definitions:

$$s := \mu, \qquad s^* := \mu^*, \qquad r := -\tau.$$

Case III: $q = -1$ and $\mathrm{char}(\mathbb{K}) \neq 2$.
We show p is listed in Example 10.12. By (10.16) and since $\mathrm{char}(\mathbb{K}) \neq 2$, there exist scalars η, μ, h in $\tilde{\mathbb{K}}$ such that

$$\theta_i = \eta + \mu(-1)^i + 2hi(-1)^i \qquad (0 \leq i \leq d). \tag{10.33}$$

Similarly there exist scalars η^*, μ^*, h^* in $\tilde{\mathbb{K}}$ such that

$$\theta_i^* = \eta^* + \mu^*(-1)^i + 2h^*i(-1)^i \qquad (0 \leq i \leq d). \tag{10.34}$$

Observe $h \neq 0$; otherwise $\theta_2 = \theta_0$ by (10.33). Similarly $h^* \neq 0$. For any prime i such that $i \leq d/2$ we have $\mathrm{char}(\mathbb{K}) \neq i$; otherwise $\theta_{2i} = \theta_0$ by (10.33). By this and since $\mathrm{char}(\mathbb{K}) \neq 2$ we find $\mathrm{char}(\mathbb{K})$ is either 0 or an odd prime greater than $d/2$. Setting $i = 0$ in (10.33), (10.34) we obtain

$$\theta_0 = \eta + \mu, \qquad \theta_0^* = \eta^* + \mu^*. \tag{10.35}$$

We define

$$s := 1 - \mu h^{-1}, \qquad s^* = 1 - \mu^* h^{*-1}. \tag{10.36}$$

Eliminating η in (10.33) using (10.35) and eliminating μ in the result using (10.36) we find (10.10) holds for $0 \leq i \leq d$. Similarly we find (10.11) holds for $0 \leq i \leq d$. We now define r_1, r_2. First assume d is odd. Since $\tilde{\mathbb{K}}$ is algebraically closed it contains r_1, r_2 such that

$$r_1 + r_2 = -s - s^* + d + 1 \tag{10.37}$$

and such that

$$4hh^*(1 + r_1)(1 + r_2) = -\varphi_1. \tag{10.38}$$

Next assume d is even. Define

$$r_2 := -1 + \frac{\varphi_1}{4hh^*d} \tag{10.39}$$

and define r_1 so that (10.37) holds. We have now defined r_1, r_2 for either parity of d. In the equation of PA4, we eliminate φ_1 using (10.38) or (10.39), and evaluate the result using (10.10), (10.11) in order to obtain (10.13) for $1 \leq i \leq d$. In the equation of PA3, we eliminate ϕ_1 using (10.13) at $i = 1$, and evaluate the result using (10.10), (10.11) in order to obtain (10.12) for $1 \leq i \leq d$. We mentioned each of h, h^* is nonzero. The remaining inequalities mentioned below (10.13) follow from PA1, PA2 and (10.10)–(10.13). We have now shown p is listed in Example 10.12.

Case IV: $q = 1$ and $\mathrm{char}(\mathbb{K}) = 2$.
We show p is listed in Example 10.13. We first show $d = 3$. Recall $d \geq 3$ since

$q = 1$. Suppose $d \geq 4$. By (10.16) we have $\sum_{j=0}^{3} \theta_j = 0$ and $\sum_{j=1}^{4} \theta_j = 0$. Adding these sums we find $\theta_0 = \theta_4$ which contradicts PA1. Therefore $d = 3$. We claim there exist nonzero scalars h, s in $\mathbb{K}$ such that (10.14) holds for $0 \leq i \leq 3$. Define $h = \theta_0 + \theta_2$. Observe $h \neq 0$; otherwise $\theta_0 = \theta_2$. Define $s = (\theta_0 + \theta_3)h^{-1}$. Observe $s \neq 0$; otherwise $\theta_0 = \theta_3$. Using these values for h, s we find (10.14) holds for $i = 0, 2, 3$. By this and $\sum_{j=0}^{3} \theta_j = 0$ we find (10.14) holds for $i = 1$. We have now proved our claim. Similarly there exist nonzero scalars h^*, s^* in $\mathbb{K}$ such that (10.15) holds for $0 \leq i \leq 3$. Define $r := \varphi_1 h^{-1} h^{*-1}$. Observe $r \neq 0$ and that $\varphi_1 = hh^* r$. In the equation of PA4, we eliminate φ_1 using $\varphi_1 = hh^* r$ and evaluate the result using (10.14), (10.15) in order to obtain $\phi_1 = hh^*(r + s(1+s^*))$, $\phi_2 = hh^*$, $\phi_3 = hh^*(r + s^*(1+s))$. In the equation of PA3, we eliminate ϕ_1 using $\phi_1 = hh^*(r + s(1 + s^*))$ and evaluate the result using (10.14), (10.15) in order to obtain $\varphi_2 = hh^*$, $\varphi_3 = hh^*(r+s+s^*)$. We mentioned each of h, h^*, s, s^*, r is nonzero. Observe $s \neq 1$; otherwise $\theta_1 = \theta_0$. Similarly $s^* \neq 1$. Observe $r \neq s + s^*$; otherwise $\varphi_3 = 0$. Observe $r \neq s(1 + s^*)$; otherwise $\phi_1 = 0$. Observe $r \neq s^*(1 + s)$; otherwise $\phi_3 = 0$. We have now shown p is listed in Example 10.13. We are done with Case IV and the proof is complete. □

11 Suggestions for Further Research

In this section we give some suggestions for further research.

Problem 11.1. Let $\Phi = (A; A^*; \{E_i\}_{i=0}^d; \{E_i^*\}_{i=0}^d)$ denote a Leonard system in $\mathcal{A}$. Let $\alpha, \alpha^*, \beta, \beta^*$ denote scalars in $\mathbb{K}$ such that $\alpha \neq 0$ and $\alpha^* \neq 0$. We mentioned in Section 4 that the sequence $(\alpha A + \beta I; \alpha^* A^* + \beta^* I; \{E_i\}_{i=0}^d; \{E_i^*\}_{i=0}^d)$ is a Leonard system in $\mathcal{A}$. In some cases this system is isomorphic to a relative of Φ; determine all the cases where this occurs.

Problem 11.2. Assume $\mathbb{K} = \mathbb{R}$. With reference to Definition 3.7, find a necessary and sufficient condition on the parameter array of Φ, for the bilinear form $\langle \cdot, \cdot \rangle$ to be positive definite. By definition the form $\langle \cdot, \cdot \rangle$ is positive definite whenever $\|u\|^2 > 0$ for all nonzero $u \in V$.

Problem 11.3. Assume $\mathbb{K} = \mathbb{R}$ and let Φ denote the Leonard system from Definition 1.2. For $0 \leq i \leq d$ we define $A_i = v_i(A)$, where the polynomial v_i is from Definition 3.5. Observe there exist real scalars p_{ij}^h $(0 \leq h, i, j \leq d)$ such that

$$A_i A_j = \sum_{h=0}^{d} p_{ij}^h A_h \qquad (0 \leq i, j \leq d).$$

Determine those Φ for which $p_{ij}^h \geq 0$ for $0 \leq h, i, j \leq d$.

Problem 11.4. Assume $\mathbb{K} = \mathbb{R}$. Let Φ denote the Leonard system from Definition 1.2 and let $\theta_0, \theta_1, \ldots, \theta_d$ denote the corresponding eigenvalue sequence. Consider the permutation σ of $0, 1, \ldots, d$ such that $\theta_{\sigma(0)} > \theta_{\sigma(1)} > \cdots > \theta_{\sigma(d)}$. What are the possibilities for σ?

Problem 11.5. Assume $\mathbb{K} = \mathbb{R}$. Let Φ denote the Leonard system from Definition 1.2 and let $\theta_0, \theta_1, \ldots, \theta_d$ denote the corresponding eigenvalue sequence. Let the polynomials u_i be as in Definition 3.6. Find a necessary and sufficient condition on the parameter array of Φ, so that the absolute value $|u_i(\theta_j)| \leq 1$ for $0 \leq i, j \leq d$. See [60, Conjecture 2] for an application.

Conjecture 11.1. Let Φ denote the Leonard system from Definition 1.2 and let the scalars a_i be as in Definition 2.3. Then for $0 \leq i \leq d$ the following are equivalent:

(i) $a_h = a_{d-h}$ for $0 \leq h \leq i$;
(ii) $a^*_h = a^*_{d-h}$ for $0 \leq h \leq i$.

Problem 11.6. Let Φ denote the Leonard system from Definition 1.2 and let $\varphi_1, \varphi_2, \ldots, \varphi_d$ denote the corresponding first split sequence. Let the scalar ν be as in Definition 2.6. For $0 \leq i, j \leq d$ we consider the expression

$$\frac{\nu \tau_i(A) E^*_0 E_0 \tau^*_j(A^*)}{\varphi_1 \varphi_2 \cdots \varphi_j}, \tag{11.1}$$

where the τ_i, τ^*_j are from Definition 5.6. Show that with respect to a Φ-split basis the matrix representing (11.1) has ij entry 1 and all other entries 0.

Problem 11.7. Let Φ denote the Leonard system from Definition 1.2. Define $G = u_d(A)$ where the polynomial u_d is from Definition 3.6. Show G^{-1} exists and is equal to $G^{\downarrow}$. Show that with respect to a Φ-split basis the matrix representing $G^* G^{-1} G^{*-1} G$ is diagonal. Find the diagonal entries of this matrix.

Problem 11.8. Let $(A; A^*; \{E_i\}_{i=0}^d; \{E^*_i\}_{i=0}^d)$ denote a Leonard system and let $(\theta_i, \theta^*_i, i = 0, \ldots, d; \varphi_j, \phi_j, j = 1, \ldots, d)$ denote the corresponding parameter array. Show that

$$\varphi_i = (\theta^*_0 - \theta^*_i) \frac{\mathrm{tr}(\tau_i(A) E^*_0)}{\mathrm{tr}(\tau_{i-1}(A) E^*_0)} \qquad (1 \leq i \leq d),$$

where tr means trace and the τ_j are from (5.19).

Problem 11.9. Find a short direct proof of Theorem 7.5. Such a proof is likely to lead to an improved proof of Theorem 5.2. The current proof of Theorem 5.2 is in [79].

Problem 11.10. Let Φ denote the Leonard system from Definition 1.2 and let V denote an irreducible $\mathcal{A}$-module. Let $\langle \cdot, \cdot \rangle$ denote the bilinear form on V from Definition 3.7. We recall $\langle \cdot, \cdot \rangle$ is nondegenerate. What is the Witt index of $\langle \cdot, \cdot \rangle$? The definition of the Witt index is given in [36].

Problem 11.11. Let A_w denote the Askey-Wilson algebra from Theorem 8.2. An element of A_w is called *central* whenever it commutes with every element of A_w. By definition the *center* of A_w is the $\mathbb{K}$-subalgebra of A_w consisting of the central elements of A_w. Describe the center of A_w. Find a generating set for this center. By [32, p. 6] the following element of A_w is central:

$$\begin{aligned} &AA^*AA^* - \beta AA^{*2}A + A^*AA^*A - \gamma^*AA^*A - \gamma(1+\beta)A^*AA^* - \beta\varrho A^{*2} \\ &-(\omega + \gamma\gamma^*)(AA^* + A^*A) - (\eta^* + \gamma\varrho^*)A - (\eta(1+\beta) + \gamma^*\varrho)A^*. \end{aligned}$$

This can be verified using the Askey-Wilson relations. Does this element generate the center of A_w?

Problem 11.12. Let d denote a nonnegative integer. Find all Leonard pairs A, A^* in $\mathrm{Mat}_{d+1}(\mathbb{K})$ that satisfy the following two conditions:

(i) A is irreducible tridiagonal;
(ii) A^* is lower bidiagonal with $A^*_{i,i-1} = 1$ for $1 \le i \le d$.

Problem 11.13. Let d denote a nonnegative integer. Find all Leonard pairs A, A^* in $\mathrm{Mat}_{d+1}(\mathbb{K})$ such that each of A, A^* is irreducible tridiagonal.

Problem 11.14. Let V denote a vector space over $\mathbb{K}$ with finite positive dimension. By a *Leonard triple* on V, we mean a three-tuple of linear transformations $A : V \to V$, $A^* : V \to V$, $A^\varepsilon : V \to V$ that satisfy conditions (i)–(iii) below.

(i) There exists a basis for V with respect to which the matrix representing A is diagonal and the matrices representing A^* and A^ε are each irreducible tridiagonal.
(ii) There exists a basis for V with respect to which the matrix representing A^* is diagonal and the matrices representing A^ε and A are each irreducible tridiagonal.
(iii) There exists a basis for V with respect to which the matrix representing A^ε is diagonal and the matrices representing A and A^* are each irreducible tridiagonal.

Find all the Leonard triples. See [20] for a connection between Leonard triples and spin models.

Problem 11.15. Let V denote a vector space over $\mathbb{K}$ with finite positive dimension. Let $\mathrm{End}(V)$ denote the $\mathbb{K}$-algebra consisting of all linear transformations from V to V. Let A, A^*, A^ε denote a Leonard triple on V. Each of the pairs A, A^*; A, A^ε; A^*, A^ε is a Leonard pair on V; let r, s, t denote the corresponding antiautomorphisms of End(V) from Definition 2.2. Determine the subgroup of GL(End(V)) generated by r, s, t. Since $r^2 = s^2 = t^2 = 1$, it is conceivable that this subgroup is a Coxeter group. For which Leonard triples is this the case?

Problem 11.16. Let V denote a vector space over $\mathbb{K}$ with finite positive dimension and let A, A^*, A^ε denote a Leonard triple on V. Show that for any permutation x, y, z of A, A^*, A^ε there exists an antiautomorphism σ of End(V) such that $x^\sigma = x$ and each of $[x, y]^\sigma, [x, z]^\sigma$ is a scalar multiple of the other. Here $[r, s]$ means $rs - sr$.

Problem 11.17. Assume $\mathbb{K}$ is algebraically closed with characteristic 0. Let d denote a nonnegative integer and let A, A^* denote the Leonard pair on $\mathbb{K}^{d+1}$ given in (1.2). Find all the matrices A^ε such that A, A^*, A^ε is a Leonard triple on $\mathbb{K}^{d+1}$. Given a solution A^ε, show that each of

$$[A, A^*], \qquad [A^*, A^\varepsilon], \qquad [A^\varepsilon, A]$$

is contained in the $\mathbb{K}$-linear span of I, A, A^*, A^ε. Here $[r, s]$ means $rs - sr$.

Problem 11.18. Let V denote a vector space over $\mathbb{K}$ with finite positive dimension and let A, A^*, A^ε denote a Leonard triple on V. Show that there exists a nonzero scalar $q \in \mathbb{K}$ such that each of

$$AA^* - qA^*A, \qquad A^*A^\varepsilon - qA^\varepsilon A^*, \qquad A^\varepsilon A - qAA^\varepsilon$$

is contained in the $\mathbb{K}$-linear span of I, A, A^*, A^ε.

Problem 11.19. Assume $\mathbb{K}$ is algebraically closed. Let q denote a nonzero scalar in $\mathbb{K}$ that is not a root of unity. Let B denote the unital associative $\mathbb{K}$-algebra with generators x, y, z and the following relations. The relations are that each of

$$\frac{q^{-1}xy - qyx}{q^2 - q^{-2}} - z, \qquad \frac{q^{-1}yz - qzy}{q^2 - q^{-2}} - x, \qquad \frac{q^{-1}zx - qxz}{q^2 - q^{-2}} - y$$

is central in B. Let V denote a finite dimensional irreducible B-module on which each of x, y, z is multiplicity-free. Show that x, y, z act on V as a Leonard triple. Determine all the B-modules of this type, up to isomorphism.

Problem 11.20. Classify up to isomorphism the finite dimensional irreducible B-modules, where the algebra B is from Problem 11.19. This problem is closely related to Problem 8.1.

Conjecture 11.2. Let Φ denote the Leonard system from Definition 1.2 and let I denote the identity element of $\mathcal{A}$. Then for all $X \in \mathcal{A}$ the following are equivalent:

(i) both

$$E_i X E_j = 0 \quad \text{if} \quad |i-j| > 1, \qquad (0 \le i, j \le d),$$
$$E_i^* X E_j^* = 0 \quad \text{if} \quad |i-j| > 1, \qquad (0 \le i, j \le d);$$

(ii) X is a $\mathbb{K}$-linear combination of I, A, A^*, AA^*, A^*A.

Conjecture 11.3. Let Φ denote the Leonard system from Definition 1.2. Then for $0 \le r \le d$ the elements

$$E_0^*, E_1^*, \ldots, E_r^*, E_r, E_{r+1}, \ldots, E_d$$

together generate $\mathcal{A}$.

Remark 11.1. Conjecture 11.3 holds for $r = 0$ by Corollary 2.1, and since A is a linear combination of $E_0, E_1, \ldots, E_d$. Similarly Conjecture 11.3 holds for $r = d$.

Problem 11.21. Recall the algebra $U_q(\widehat{sl}_2)^{>0}$ from Example 9.2. Describe the center of $U_q(\widehat{sl}_2)^{>0}$. Find a generating set for this center. We remark that $U_q(\widehat{sl}_2)^{>0}$ has infinite dimension as a vector space over $\mathbb{K}$. A basis for this vector space is given in [46, Theorem 2.29].

Problem 11.22. Assume $\mathbb{K}$ is algebraically closed. Let V denote a vector space over $\mathbb{K}$ with finite positive dimension and let A, A^* denote a tridiagonal pair on V. Compute the Jordan canonical form for $q^{-1}AA^* - qA^*A$, where $q^2 + q^{-2} = \beta$ and β is from Theorem 9.1.

Conjecture 11.4. Let $\Phi = (A; A^*; \{E_i\}_{i=0}^d; \{E_i^*\}_{i=0}^d)$ denote a Leonard system and let $(\theta_i, \theta_i^*, i = 0, \ldots, d; \varphi_j, \phi_j, j = 1, \ldots, d)$ denote the corresponding parameter array. Assume $q \ne 1$, $q \ne -1$, where $q + q^{-1} + 1$ is the common value of (5.18). For d odd,

$$\det(AA^* - A^*A) = \prod_{\substack{1 \le i \le d \\ i \text{ odd}}} \varphi_i \phi_i \frac{q-1}{1-q^i} \frac{1-q^{-1}}{1-q^{-i}}.$$

Remark 11.2. Referring to Conjecture 11.4, for d even we have $\det(AA^* - A^*A) = 0$. This is because $\det(X^\dagger) = \det(X)$ and $X^\dagger = -X$, where $X = AA^* - A^*A$ and where $\dagger$ is the antiautomorphism from Definition 2.2.

Problem 11.23. Referring to Conjecture 11.4 and Remark 11.2, show that for d even the null space of $AA^* - A^*A$ has dimension 1. Find a basis vector for this null space. Express this basis vector in terms of a Φ-standard basis and a Φ^*-standard basis.

Problem 11.24. Let V denote a vector space over $\mathbb{K}$ with finite positive dimension. By an *inverting pair* on V we mean an ordered pair of invertible linear transformations $K : V \to V$ and $K^* : V \to V$ that satisfy both (i), (ii) below.

(i) There exists a basis for V with respect to which the matrix representing K has all entries 0 above the superdiagonal, the matrix representing K^{-1} has all entries 0 below the subdiagonal, and the matrix representing K^* is diagonal.
(ii) There exists a basis for V with respect to which the matrix representing K^* has all entries 0 above the superdiagonal, the matrix representing K^{*-1} has all entries 0 below the subdiagonal, and the matrix representing K is diagonal.

Find all the inverting pairs. See [47, Section 11] for a connection between inverting pairs and Leonard pairs of q-geometric type.

Problem 11.25. For an integer $d \geq 0$, find all the invertible matrices $K \in \mathrm{Mat}_{d+1}(\mathbb{K})$ such that

(i) K has all entries 0 above the superdiagonal,
(ii) K^{-1} has all entries 0 below the subdiagonal.

Problem 11.26. Let d denote a nonnegative integer and let $\theta_0, \theta_1, \ldots, \theta_d$ denote a sequence of mutually distinct scalars in $\mathbb{K}$. Let λ denote an indeterminate and let V denote the vector space over $\mathbb{K}$ consisting of all polynomials in λ that have degree at most d. Define a polynomial $\tau_i = \prod_{h=0}^{i-1}(\lambda - \theta_h)$ for $0 \leq i \leq d$ and observe $\tau_0, \tau_1, \ldots, \tau_d$ is a basis for V. Define $\rho_i = \prod_{h=0}^{i-1}(\lambda - \theta_{d-h})$ for $0 \leq i \leq d$ and observe $\rho_0, \rho_1, \ldots, \rho_d$ is a basis for V. By a *lowering* map on V we mean a linear transformation $\Psi : V \to V$ that satisfies both

$$\begin{aligned} &\Psi\tau_i \in \mathrm{span}(\tau_{i-1}) \qquad (1 \leq i \leq d), \qquad \Psi\tau_0 = 0, \\ &\Psi\rho_i \in \mathrm{span}(\rho_{i-1}) \qquad (1 \leq i \leq d), \qquad \Psi\rho_0 = 0. \end{aligned}$$

Show that there exists a nonzero lowering map on V if and only if $(\theta_{i-2} - \theta_{i+1})(\theta_{i-1} - \theta_i)^{-1}$ is independent of i for $2 \leq i \leq d-1$.

Problem 11.27. By a *generalized Leonard system* in $\mathcal{A}$ we mean a sequence $(A; A^*; \{\theta_i\}_{i=0}^d; \{\theta_i^*\}_{i=0}^d)$ that satisifies (i)–(v) below.

(i) $A, A^* \in \mathcal{A}$ and $\theta_i, \theta_i^* \in \mathbb{K}$ for $0 \leq i \leq d$.
(ii) $\theta_0, \theta_1, \ldots, \theta_d$ is an ordering of the roots of the characteristic polynomial of A.
(iii) $\theta_0^*, \theta_1^*, \ldots, \theta_d^*$ is an ordering of the roots of the characteristic polynomial of A^*.
(iv) For $0 \leq i, j \leq d$,

$$\tau_i(A)A^*\eta_{d-j}(A) = \begin{cases} 0, & \text{if } i - j > 1, \\ \neq 0, & \text{if } i - j = 1, \end{cases}$$

$$\eta_{d-i}(A)A^*\tau_j(A) = \begin{cases} 0, & \text{if } j - i > 1, \\ \neq 0, & \text{if } j - i = 1. \end{cases}$$

(v) For $0 \leq i, j \leq d$,

$$\tau_i^*(A^*)A\eta_{d-j}^*(A^*) = \begin{cases} 0, & \text{if } i-j>1, \\ \neq 0, & \text{if } i-j=1, \end{cases}$$

$$\eta_{d-i}^*(A^*)A\tau_j^*(A^*) = \begin{cases} 0, & \text{if } j-i>1, \\ \neq 0, & \text{if } j-i=1. \end{cases}$$

(We are using the notation (5.19), (5.20)). We are not assuming $\theta_0, \theta_1, \ldots, \theta_d$ are mutually distinct or that $\theta_0^*, \theta_1^*, \ldots, \theta_d^*$ are mutually distinct. Classify the generalized Leonard systems. Extend the theory of Leonard systems to the level of generalized Leonard systems.

Problem 11.28. For an integer $d \geq 0$ and for $X \in \mathrm{Mat}_{d+1}(\mathbb{K})$, we define X to be *north Vandermonde* whenever the entries $X_{ij} = X_{0j}f_i(\theta_j)$ for $0 \leq i, j \leq d$, where $\theta_0, \theta_1, \ldots, \theta_d$ are mutually distinct scalars in $\mathbb{K}$ and $f_i \in \mathbb{K}[\lambda]$ has degree i for $0 \leq i \leq d$. Let $X' \in \mathrm{Mat}_{d+1}(\mathbb{K})$ denote the matrix obtained by rotating X counterclockwise 90 degrees. We define X to be *east Vandermonde* (resp. *south Vandermonde*) (resp. *west Vandermonde*) whenever X' (resp. X'') (resp. X''') is north Vandermonde. Find all the matrices in $\mathrm{Mat}_{d+1}(\mathbb{K})$ that are simultaneously north, south, east, and west Vandermonde.

Problem 11.29. Let V denote a vector space over $\mathbb{K}$ with finite positive dimension n. By a *cyclic Leonard pair* on V, we mean an ordered pair of linear transformations $A : V \to V$ and $A^* : V \to V$ that satisfy (i)–(iv) below.

(i) Each of A, A^* is multiplicity free.
(ii) There exists a bijection $i \to V_i$ from the cyclic group $\mathbb{Z}/n\mathbb{Z}$ to the set of eigenspaces of A such that

$$A^*V_i \subseteq V_{i-1} + V_i + V_{i+1} \qquad (\forall i \in \mathbb{Z}/n\mathbb{Z}).$$

(iii) There exists a bijection $i \to V_i^*$ from $\mathbb{Z}/n\mathbb{Z}$ to the set of eigenspaces of A^* such that

$$AV_i^* \subseteq V_{i-1}^* + V_i^* + V_{i+1}^* \qquad (\forall i \in \mathbb{Z}/n\mathbb{Z}).$$

(iv) There does not exist a subspace W of V such that $AW \subseteq W$, $A^*W \subseteq W$, $W \neq 0$, $W \neq V$.

Classify the cyclic Leonard pairs. Extend the theory of Leonard pairs to the level of cyclic Leonard pairs.

Problem 11.30. Referring to the tridiagonal pair A, A^* in Definition 9.1, for $0 \leq i \leq d$ let θ_i (resp. θ_i^*) denote the eigenvalue of A (resp. A^*) associated with the eigenspace V_i (resp. V_i^*). Assume V_0^* has dimension 1. Observe that for $0 \leq i \leq d$ the space V_0^* is an eigenspace for

$$(A^* - \theta_1^* I)(A^* - \theta_2^* I) \cdots (A^* - \theta_i^* I)(A - \theta_{i-1} I) \cdots (A - \theta_1 I)(A - \theta_0 I);$$

let ζ_i denote the corresponding eigenvalue. Show that the tridiagonal pair A, A^* is determined up to isomorphism by the array $\{\theta_i, \theta_i^*, \zeta_i, i = 0, \ldots, d\}$.

Problem 11.31. Classify the tridiagonal pairs that have shape vector $(\rho, \rho, \ldots, \rho)$, where ρ is an integer at least 2. See Section 34 for the definition of the shape vector.

References

1. C. Ahn and K. Shigemoto, *Onsager algebra and integrable lattice models*, Modern Phys. Lett. A **6 (38)** (1991), 3509–3515.
2. G. Albertini, B. McCoy, J. Perk, *Eigenvalue spectrum of the superintegrable chiral Potts model*, in “Integrable systems in quantum field theory and statistical mechanics”, Adv. Stud. Pure Math. **19**, Academic Press, Boston, MA, 1989, pp. 1–55.
3. G. Albertini, B. McCoy, J. H. H. Perk, S. Tang, *Excitation spectrum and order parameter for the integrable N-state chiral Potts model*, Nuclear Phys. B **314** (1989), 741–763.
4. G. Andrews, R. Askey, and R. Roy, *Special Functions*, Cambridge University Press, Cambridge, 1999.
5. R. Askey and J.A. Wilson, *A set of orthogonal polynomials that generalize the Racah coefficients or* $6-j$ *symbols*, SIAM J. Math. Anal. **10** (1979), 1008–1016.
6. N. Atakishiyev and A. Klimyk, *On q-orthogonal polynomials, dual to little and big q-Jacobi polynomials*, J. Math. Anal. Appl. **294** (2004), 246–254; arXiv:math.CA/0307250.
7. M. N. Atakishiyev and V. Groza, *The quantum algebra $U_q(\mathrm{su}_2)$ and q-Krawtchouk families of polynomials*, J. Phys. A: Math. Gen. **37** (2004), 2625–2635.
8. M. N. Atakishiyev, N. M. Atakishiyev, and A. Klimyk, *Big q-Laguerre and q-Meixner polynomials and representations of the quantum algebra $U_q(\mathrm{su}_{1,1})$*, J. Phys. A: Math. Gen. **36** (2003), 10335–10347.
9. H. Au-Yang and J. H. H. Perk, *Onsager's star-triangle equation: master key to integrability*, in “Integrable systems in quantum field theory and statistical mechanics”, Adv. Stud. Pure Math. **19**, Academic Press, Boston, MA, 1989, pp. 57–94.
10. H. Au-Yang, B. McCoy, J. H. H. Perk, and S. Tang, *Solvable models in statistical mechanics and Riemann surfaces of genus greater than one*, in “Algebraic analysis”, Vol. I, Academic Press, Boston, MA, 1988, pp. 29–39.
11. E. Bannai and T. Ito, *Algebraic Combinatorics I: Association Schemes*, Benjamin/Cummings, London, 1984.
12. V. V. Bazhanov and Y. G. Stroganov, *Chiral Potts model as a descendant of the six-vertex model*, J. Statist. Phys. **59** (1990), 799–817.
13. A. E. Brouwer, A. M. Cohen, and A. Neumaier, *Distance-Regular Graphs*, Springer-Verlag, Berlin, 1989.
14. J. S. Caughman IV, *The Terwilliger algebras of bipartite P- and Q-polynomial schemes*, Discrete Math. **196** (1999), 65–95.

15. V. Chari and A. Pressley, *Quantum affine algebras*, Commun. Math. Phys. **142** (1991), 261–283.
16. N. Ciccoli and F. Gavarini, *A quantum duality principle for subgroups and homogeneous spaces*, preprint arXiv:math.QA/0312289.
17. B. Curtin, *The Terwilliger algebra of a 2-homogeneous bipartite distance-regular graph*, J. Combin. Theory Ser. B **81** (2001), 125–141.
18. B. Curtin and K. Nomura, *Distance-regular graphs related to the quantum enveloping algebra of sl(2)*, J. Algebraic Combin. **12** (2000), 25–36.
19. B. Curtin, *Distance-regular graphs which support a spin model are thin*, in "Proc. 16th British Combinatorial Conference" (London, 1997), Discrete Math. **197/198** (1999), 205–216.
20. B. Curtin, *Modular Leonard triples*, preprint.
21. C. Daskaloyannis, *Quadratic Poisson algebras of two-dimensional classical superintegrable systems and quadratic associative algebras of quantum superintegrable systems*, J. Math. Phys. **42** (2001), 100–1119.
22. E. Date and S.S. Roan, *The structure of quotients of the Onsager algebra by closed ideals*, J. Phys. A: Math. Gen. **33** (2000), 3275–3296.
23. B. Davies, *Onsager's algebra and superintegrability*, J. Phys. A: Math. Gen. **23** (1990), 2245–2261.
24. B. Davies, *Onsager's algebra and the Dolan-Grady condition in the non-self-dual case*, J. Math. Phys. **32** (1991), 2945–2950.
25. T. Deguchi, K. Fabricius, and B. McCoy, *The* sl_2 *loop algebra symmetry of the six-vertex model at roots of unity*, in "Proceedings of the Baxter Revolution in Mathematical Physics" (Canberra, 2000), J. Statist. Phys. **102** (2001), 701–736.
26. L. Dolan and M. Grady, *Conserved charges from self-duality*, Phys. Rev. D (3) **25** (1982), 1587–1604.
27. D. B. Fairlie, *Quantum deformations of SU(2)*. J. Phys. A: Math. Gen. **23** (1990), L183–L187.
28. G. Gasper and M. Rahman, *Basic Hypergeometric Series*, Encyclopedia of Mathematics and its Applications 35, Cambridge University Press, Cambridge, 1990.
29. G. von Gehlen and V. Rittenberg, *Z_n-symmetric quantum chains with infinite set of conserved charges and Z_n zero modes*, Nucl. Phys. B **257** (1985), 351–370.
30. J. T. Go, *The Terwilliger algebra of the hypercube*, European J. Combin. **23** (2002), 399–429.
31. Ya. A. Granovskiĭ and A. S. Zhedanov, *Nature of the symmetry group of the 6j-symbol*, Zh. Èksper. Teoret. Fiz. **94** (1988), 49–54.
32. Ya. I. Granovskiĭ, I. M. Lutzenko, and A. S. Zhedanov, *Mutual integrability, quadratic algebras, and dynamical symmetry*, Ann. Physics **217** (1992), 1–20.
33. Ya. I. Granovskiĭ and A. S. Zhedanov, *"Twisted" Clebsch-Gordan coefficients for* $\mathrm{su}_q(2)$, J. Phys. A: Math. Gen. **25** (1992), L1029–L1032.
34. Ya. I. Granovskiĭ and A. S. Zhedanov, *Linear covariance algebra for* $\mathrm{sl}_q(2)$, J. Phys. A: Math. Gen. **26** (1993), L357–L359.
35. Ya. I. Granovskiĭ and A. S. Zhedanov, *Spherical q-functions*, J. Phys. A **26** (1993), 4331–4338.
36. L. C. Grove, *Classical Groups and Geometric Algebra*, American Mathematical Society, Providence RI, 2002.
37. F. A. Grünbaum, *Some bispectral musings*, in "The bispectral problem" (Montreal, PQ, 1997), Amer. Math. Soc., Providence, RI, 1998, pp. 31–45.
38. F. A. Grünbaum and L. Haine, *Bispectral Darboux transformations: an extension of the Krall polynomials*, Internat. Math. Res. Notices **8** (1997), 359–392.

39. F. A. Grünbaum and L. Haine, *The q-version of a theorem of Bochner*, J. Comput. Appl. Math. **68** (1996), 103–114.
40. F. A. Grünbaum and L. Haine, *Some functions that generalize the Askey-Wilson polynomials*, Comm. Math. Phys. **184** (1997), 173–202.
41. F. A. Grünbaum and L. Haine, *On a q-analogue of the string equation and a generalization of the classical orthogonal polynomials*, in "Algebraic methods and q-special functions" (Montréal, QC, 1996), Amer. Math. Soc., Providence, RI, 1999, pp. 171–181.
42. F. A. Grünbaum and L. Haine, *The Wilson bispectral involution: some elementary examples*, in "Symmetries and integrability of difference equations" (Canterbury, 1996), Cambridge Univ. Press, Cambridge, 1999, pp. 353–369.
43. F. A. Grünbaum, L. Haine, and E. Horozov, *Some functions that generalize the Krall-Laguerre polynomials*, J. Comput. Appl. Math. **106** (1999), 271–297.
44. M. Havliček, A. U. Klimyk, and S. Pošta, *Representations of the cyclically symmetric q-deformed algebra* $\mathrm{so}_q(3)$, J. Math. Phys. **40** (1999), 2135–2161.
45. T. Ito, K. Tanabe, and P. Terwilliger, *Some algebra related to P- and Q-polynomial association schemes*, in "Codes and Association Schemes" (Piscataway NJ, 1999), Amer. Math. Soc., Providence RI, 2001, pp. 167–192; `arXiv:math.CO/0406556`.
46. T. Ito and P. Terwilliger, *The shape of a tridiagonal pair*, J. Pure Appl. Algebra **188** (2004), 145–160; `arXiv:math.QA/0304244`.
47. T. Ito and P. Terwilliger, *Tridiagonal pairs and the quantum affine algebra* $U_q(\widehat{sl}_2)$, Ramanujan J., to appear; `arXiv:math.QA/0310042`.
48. I. T. Ivanov and D. B. Uglov, *R-matrices for the semicyclic representations of* $U_q\widehat{\mathrm{sl}}(2)$, Phys. Lett. A, **167** (1992), 459–464.
49. C. Kassel, *Quantum Groups*, Springer-Verlag, New York, 1995.
50. S. Klishevich and M. Plyushchay, *Dolan-Grady relations and noncommutative quasi-exactly solvable systems*, J. Phys. A: Math. Gen. **36** (2003), 11299–11319.
51. S. Klishevich and M. Plyushchay, *Nonlinear holomorphic supersymmetry on Riemann surfaces*, Nuclear Phys. B **640** (2002), 481–503.
52. S. Klishevich and M. Plyushchay, *Nonlinear holomorphic supersymmetry, Dolan-Grady relations and Onsager algebra*, Nuclear Phys. B **628** (2002), 217–233.
53. R. Koekoek and R. F. Swarttouw, *The Askey scheme of hypergeometric orthogonal polyomials and its q-analog*, report 98-17, Delft University of Technology, The Netherlands, 1998,
Available at `http://aw.twi.tudelft.nl/~koekoek/research_en.html`
54. H. T. Koelink, *Askey-Wilson polynomials and the quantum* su(2) *group: survey and applications*, Acta Appl. Math. **44** (1996), 295–352.
55. H. T. Koelink, *q-Krawtchouk polynomials as spherical functions on the Hecke algebra of type B*, Trans. Amer. Math. Soc. **352** (2000), 4789–4813.
56. H. T. Koelink and J. Van der Jeugt, *Convolutions for orthogonal polynomials from Lie and quantum algebra representations*, SIAM J. Math. Anal. **29** (1998), 794–822.
57. H. T. Koelink and J. Van der Jeugt, *Bilinear generating functions for orthogonal polynomials*, Constr. Approx. **15** (1999), 481–497.
58. T. H. Koornwinder, *Askey-Wilson polynomials as zonal spherical functions on the* $SU(2)$ *quantum group*, SIAM J. Math. Anal. **24** (1993), 795–813.

59. T. H. Koornwinder and U. Onn, *Lower-upper triangular decompositions, $q = 0$ limits, and p-adic interpretations of some q-hypergeometric orthogonal polynomials*, preprint `arXiv:math.CA/0405309`.
60. A. Kresch and H. Tamvakis, *Standard conjectures for the arithmetic grassmannian $G(2, N)$ and Racah polynomials*, Duke Math. J. **110** (2001), 359–376.
61. C. W. H. Lee and S. G. Rajeev, *A Lie algebra for closed strings, spin chains, and gauge theories*, J. Math. Phys. **39** (1998), 5199–5230.
62. D. Leonard, *Orthogonal polynomials, duality, and association schemes*, SIAM J. Math. Anal. **13** (1982), 656–663.
63. J. Marco and J. Parcet, *A new approach to the theory of classical hypergeometric polynomials*, Trans. Amer. Math. Soc. **358** (2006), 183–214 (electronic).
64. B. McCoy, *Integrable models in statistical mechanics: the hidden field with unsolved problems*, Internat. J. Modern Phys. A **14** (1999), 3921–3933.
65. M. Noumi and K. Mimachi, *Askey-Wilson polynomials as spherical functions on* $SU_q(2)$, in "Quantum groups" (Leningrad, 1990), Lecture Notes in Math. **1510**, Springer, Berlin, 1992, pp. 98–103.
66. M. Noumi and K. Mimachi, *Spherical functions on a family of quantum 3-spheres*, Compositio Math. **83** (1992), 19–42.
67. M. Noumi, *Quantum groups and q-orthogonal polynomials—towards a realization of Askey-Wilson polynomials on* $SU_q(2)$, in "Special functions" (Okayama, 1990) 260–288. ICM-90 Satell. Conf. Proc., Springer, Tokyo, 1991, pp. 260–288.
68. M. Noumi and K. Mimachi, *Askey-Wilson polynomials and the quantum group* $SU_q(2)$, Proc. Japan Acad. Ser. A Math. Sci. **66** (1990), 146–149.
69. M. Odesskii, *An analog of the Sklyanin algebra*, Funct. Anal. Appl. **20** (1986), 78–79.
70. L. Onsager, *Crystal statistics. I. A two-dimensional model with an order-disorder transition*, Phys. Rev. (2) **65** (1944), 117–149.
71. J. H. H. Perk, *Star-triangle relations, quantum Lax pairs, and higher genus curves*, Proceedings of Symposia in Pure Mathematics **49**, Amer. Math. Soc., Providence, RI, 1989, pp. 341–354.
72. S. S. Roan, *Onsager's algebra, loop algebra and chiral Potts model*, Preprint MPI 91–70, Max Plank Institute for Mathematics, Bonn, 1991.
73. H. Rosengren, *Multivariable orthogonal polynomials as coupling coefficients for Lie and quantum algebra representations*, Ph.D. Thesis, Centre for Mathematical Sciences, Lund University, Sweden, 1999.
74. H. Rosengren, *An elementary approach to the 6j-symbols (classical, quantum, rational, trigonometric, and elliptic)*, preprint `arXiv:math.CA/0312310`.
75. J. J. Rotman, *Advanced Modern Algebra*, Prentice Hall, Saddle River NJ, 2002.
76. P. Terwilliger, *The incidence algebra of a uniform poset*, Math and its Applications **20** (1990), 193–212.
77. P. Terwilliger, *The subconstituent algebra of an association scheme I*, J. Algebraic Combin. **1** (1992), 363–388.
78. P. Terwilliger, *The subconstituent algebra of an association scheme III*, J. Algebraic Combin. **2** (1993), 177–210.
79. P. Terwilliger, *Two linear transformations each tridiagonal with respect to an eigenbasis of the other*, Linear Algebra Appl. **330** (2001), 149–203; `arXiv:math.RA/0406555`.
80. P. Terwilliger, *Two relations that generalize the q-Serre relations and the Dolan-Grady relations*, in "Physics and Combinatorics 1999" (Nagoya), World Scientific Publishing, River Edge, NJ, 2001, pp. 377–398; `arXiv:math.QA/0307016`.

81. P. Terwilliger, *Leonard pairs from 24 points of view*, Rocky Mountain J. Math. **32** (2) (2002), 827–888; `arXiv:math.RA/0406577`.
82. P. Terwilliger, *Two linear transformations each tridiagonal with respect to an eigenbasis of the other; the TD-D and the LB-UB canonical form*, J. Algebra **291** (2005), 1–45; `arXiv:math.RA/0304077`.
83. P. Terwilliger, *Introduction to Leonard pairs*, OPSFA Rome 2001, J. Comput. Appl. Math. **153** (2) (2003), 463–475.
84. P. Terwilliger, *Introduction to Leonard pairs and Leonard systems*, Sūrikaisekikenkyūsho Kōkyūroku **1109** (1999), 67–79; Algebraic combinatorics (Kyoto, 1999).
85. P. Terwilliger, *Two linear transformations each tridiagonal with respect to an eigenbasis of the other; comments on the split decomposition*, OPSFA7, J. Comput. Appl. Math. **178** (1–2) (2005), 437–452.
86. P. Terwilliger, *Two linear transformations each tridiagonal with respect to an eigenbasis of the other; comments on the parameter array.* Geometric and Algebraic Combinatorics 2, Oisterwijk, The Netherlands 2002, Des. Codes Cryptogr. **34** (2005), 307–332; `arXiv:math.RA/0306291`.
87. P. Terwilliger, *Leonard pairs and the q-Racah polynomials*, Linear Algebra Appl. **387** (2004), 235–276.
88. P. Terwilliger and R. Vidunas, *Leonard pairs and the Askey-Wilson relations*, J. Algebra Appl. **3**, no. 4 (2004), 411–426.
89. D. B. Uglov and I. T. Ivanov, sl(N) *Onsager's algebra and integrability*, J. Statist. Phys. **82** (1996), 87–113.
90. A. S. Zhedanov, *"Hidden symmetry" of Askey-Wilson polynomials*, Teoret. Mat. Fiz. **89** (1991), 190–204.
91. A. S. Zhedanov, *Quantum* $\mathrm{su}_q(2)$ *algebra: "Cartesian" version and overlaps*, Modern Phys. Lett. A **7** (1992), 1589–1593.
92. A. S. Zhedanov, *Hidden symmetry algebra and overlap coefficients for two ring-shaped potentials*, J. Phys. A: Math. Gen. **26** (1993), 4633–4641.
93. A. S. Zhedanov, and A. Korovnichenko, *"Leonard pairs" in classical mechanics*, J. Phys. A: Math. Gen. **5** (2002), 5767–5780.

Painlevé Equations — Nonlinear Special Functions

Peter A. Clarkson

Institute of Mathematics, Statistics and Actuarial Science, University of Kent, Canterbury, CT2 7NF UK
e-mail: P.A.Clarkson@kent.ac.uk

Summary. The six Painlevé equations (P_I–P_{VI}) were first discovered around the beginning of the twentieth century by Painlevé, Gambier and their colleagues in an investigation of nonlinear second-order ordinary differential equations. Recently there has been considerable interest in the Painlevé equations primarily due to the fact that they arise as reductions of the soliton equations which are solvable by inverse scattering. Consequently the Painlevé equations can be regarded as completely integrable equations and possess solutions which can be expressed in terms of solutions of linear integral equations, despite being nonlinear equations. Although first discovered from strictly mathematical considerations, the Painlevé equations have arisen in a variety of important physical applications including statistical mechanics, random matrices, plasma physics, nonlinear waves, quantum gravity, quantum field theory, general relativity, nonlinear optics and fibre optics.

The Painlevé equations may be thought of as nonlinear analogues of the classical special functions. They possess hierarchies of rational solutions and one-parameter families of solutions expressible in terms of the classical special functions, for special values of the parameters. Furthermore the Painlevé equations admit symmetries under affine Weyl groups which are related to the associated Bäcklund transformations.

In this paper many of the remarkable properties which the Painlevé equations possess are reviewed including connection formulae, Bäcklund transformations associated discrete equations, and hierarchies of exact solutions. In particular, the second Painlevé equation P_{II} is primarily used to illustrate these properties and some of the applications of P_{II} are also discussed.

1 Introduction

In this paper our interest is in the six Painlevé equations ($\mathrm{P_I}$–$\mathrm{P_{VI}}$)

$$w'' = 6w^2 + z, \tag{1.1}$$

$$w'' = 2w^3 + zw + \alpha, \tag{1.2}$$

$$w'' = \frac{(w')^2}{w} - \frac{w'}{z} + \frac{\alpha w^2 + \beta}{z} + \gamma w^3 + \frac{\delta}{w}, \tag{1.3}$$

$$w'' = \frac{(w')^2}{2w} + \frac{3}{2}w^3 + 4zw^2 + 2(z^2 - \alpha)w + \frac{\beta}{w}, \tag{1.4}$$

$$w'' = \left(\frac{1}{2w} + \frac{1}{w-1}\right)(w')^2 - \frac{w'}{z} + \frac{(w-1)^2}{z^2}\left(\alpha w + \frac{\beta}{w}\right) + \frac{\gamma w}{z} + \frac{\delta w(w+1)}{w-1}, \tag{1.5}$$

$$w'' = \frac{1}{2}\left(\frac{1}{w} + \frac{1}{w-1} + \frac{1}{w-z}\right)(w')^2 - \left(\frac{1}{z} + \frac{1}{z-1} + \frac{1}{w-z}\right)w' + \frac{w(w-1)(w-z)}{z^2(z-1)^2}\left\{\alpha + \frac{\beta z}{w^2} + \frac{\gamma(z-1)}{(w-1)^2} + \frac{\delta z(z-1)}{(w-z)^2}\right\}, \tag{1.6}$$

where $' \equiv \mathrm{d}/\mathrm{d}z$ and α, β, γ and δ are arbitrary constants. The solutions of $\mathrm{P_I}$–$\mathrm{P_{VI}}$ are called the *Painlevé transcendents*. The Painlevé equations $\mathrm{P_I}$–$\mathrm{P_{VI}}$ were discovered about a hundred years ago by Painlevé, Gambier and their colleagues whilst studying a problem posed by Picard [190]. Picard asked which second order ordinary differential equations of the form

$$w'' = F(z; w, w'), \tag{1.7}$$

where F is rational in w' and w and analytic in z, have the property that the solutions have no movable branch points, i.e., the locations of multi-valued singularities of any of the solutions are independent of the particular solution chosen and so are dependent only on the equation; this is now known as the *Painlevé property*. Painlevé, Gambier *et al.* showed that there were fifty canonical equations of the form (1.7) with this property, up to a Möbius (bilinear rational) transformation

$$W(\zeta) = \frac{a(z)w + b(z)}{c(z)w + d(z)}, \qquad \zeta = \phi(z), \tag{1.8}$$

where $a(z)$, $b(z)$, $c(z)$, $d(z)$ and $\phi(z)$ are locally analytic functions. Further they showed that of these fifty equations, forty-four are either integrable in terms of previously known functions (such as elliptic functions or are equivalent to linear equations) or reducible to one of six new nonlinear ordinary differential equations, which define new transcendental functions (cf. [101]).

Although first discovered from mathematical considerations, the Painlevé equations have subsequently arisen in a variety of applications including statistical mechanics (correlation functions of the XY model and the Ising model), random matrix theory, topological field theory, plasma physics, nonlinear waves (resonant oscillations in shallow water, convective flows with viscous dissipation, Görtler vortices in boundary layers and Hele-shaw problems), quantum gravity, quantum field theory, general relativity, nonlinear and fibre optics, polyelectrolytes, Bose-Einstein condensation and stimulated Raman scattering. Further the Painlevé equations have attracted much interest since they arise in many physical situations and as reductions of the soliton equations which are solvable by inverse scattering, as discussed in §10.1; see, e.g., [2, 8], and the references therein.

The Painlevé equations can be thought of as nonlinear analogues of the classical special functions (cf. [30, 102]). Iwasaki, Kimura, Shimomura and Yoshida [109] characterize the Painlevé equations as "the most important nonlinear ordinary differential equations" and also state that "many specialists believe that during the twenty-first century the Painlevé functions will become new members of the community of special functions". Further Umemura [224] states that "Kazuo Okamoto and his circle predict that in the 21st century a new chapter on Painlevé equations will be added to Whittaker and Watson's book".

The general solutions of the Painlevé equations are transcendental, i.e., irreducible in the sense that they cannot be expressed in terms of previously known functions, such as rational functions, elliptic functions or the classical special functions (cf., [224]). Essentially, generic solutions of the $\mathrm{P_I}$–$\mathrm{P_V}$ are meromorphic functions in the complex plane; solutions of $\mathrm{P_{VI}}$ are rather different since the equation has three singular points (cf., [97]). Hinkkanen and Laine [94, 95, 96], Shimomura [199, 200, 201, 202, 203] and Steinmetz [205, 206, 207, 208] have rigorously proved the following theorem.

Theorem 1.1. *All local solutions of* $\mathrm{P_I}$, $\mathrm{P_{II}}$ *and* $\mathrm{P_{IV}}$, *and all local solutions of* $\mathrm{P_{III}}$ *and* $\mathrm{P_V}$ *after making the transformation* $z = \mathrm{e}^{\zeta}$, *i.e.,*

$$\frac{\mathrm{d}^2 w}{\mathrm{d}\zeta^2} = \frac{1}{w}\left(\frac{\mathrm{d}w}{\mathrm{d}\zeta}\right)^2 + \alpha w^2 + \beta \mathrm{e}^{\zeta} + \gamma w^3 + \frac{\delta \mathrm{e}^{2\zeta}}{w}, \tag{1.9}$$

$$\begin{aligned}\frac{\mathrm{d}^2 w}{\mathrm{d}\zeta^2} &= \left(\frac{1}{2w} + \frac{1}{w-1}\right)\left(\frac{\mathrm{d}w}{\mathrm{d}\zeta}\right)^2 + (w-1)^2\left(\alpha w + \frac{\beta}{w}\right)\\ &\quad + \gamma \mathrm{e}^{\zeta} w + \frac{\delta \mathrm{e}^{2\zeta} w(w+1)}{w-1},\end{aligned} \tag{1.10}$$

respectively, can be analytically continued to single-valued meromorphic functions in the complex plane.

The Painlevé equations, like other integrable equations such as soliton equations, have a plethora of interesting properties. In the following sections we discuss some of these properties.

- The isomonodromy method has been developed for the study of the Painlevé equations and in this sense they are said to be integrable.

- The equations $\mathrm{P_{II}}$–$\mathrm{P_{VI}}$ also possess Bäcklund transformations which relate one solution to another solution either of the same equation, with different values of the parameters, or another equation which are discussed in §4.
- The equations $\mathrm{P_{II}}$–$\mathrm{P_{VI}}$ possess many rational solutions, algebraic solutions and solutions expressible in terms of special functions for certain values of the parameters, which we discuss in §§5–7, respectively; these special solutions are called "classical solutions" [223].
- Other properties include that they can be written in Hirota bilinear form, which we discuss in §8.1 below) and have a coalescence cascade, which we discuss in §8.2.

2 Inverse Problems for the Painlevé Equations

In this section we discuss inverse problems associated with the Painlevé equations. There are two approaches in order to develop inverse problems for the Painlevé equations, either through a Gel'fand-Levitan-Marchenko integral equation, or through the isomonodromic deformation method which can be viewed as the nonlinear analog of Laplace transform method.

The Painlevé equations arise as similarity reductions of partial differential equations solvable by inverse scattering, the so-called *soliton* equations, as we discuss in §10.1. We illustrate this in the following example given by Ablowitz and Segur [7].

Example 2.1. If we make the scaling reduction

$$u(x,t) = (3t)^{-1/3}w(z), \qquad z = x/(3t)^{1/3}, \tag{2.1}$$

in the modified Korteweg-de Vries (mKdV) equation

$$u_t - 6u^2u_x + u_{xxx} = 0, \tag{2.2}$$

then after integrating once, $w(z)$ satisfies $\mathrm{P_{II}}$ with α the arbitrary constant of integration.

Both the integral equation and isomonodromic deformation method approaches for the inverse problem for $\mathrm{P_{II}}$ can be derived from the inverse scattering method for the mKdV equation (2.2) as we shall illustrate. For details of the inverse scattering method for the mKdV equation see, for example, Ablowitz and Clarkson [2, §7.3], Ablowitz, Kaup, Newell and Segur [4].

2.1 Integral Equations

The inverse scattering method for nonlinear partial differential equations expresses the solution in terms of solutions of *linear* integral equations (cf. [2, 4, 8]). Hence, due to the relationship between Painlevé and soliton equations, then certain solutions of the Painlevé equations can be expressed in terms of solutions of linear integral equations.

Theorem 2.1. *Consider the integral equation*

$$K(z,\zeta) = k\,\mathrm{Ai}\left(\frac{z+\zeta}{2}\right) + \tfrac{1}{4}k^2 \int_z^\infty \int_z^\infty K(z,s)\,\mathrm{Ai}\left(\frac{s+t}{2}\right)\mathrm{Ai}\left(\frac{t+\zeta}{2}\right)\mathrm{d}s\,\mathrm{d}t, \tag{2.3}$$

with $\mathrm{Ai}(z)$ *the Airy function, then*

$$w_k(z) = K(z,z), \tag{2.4}$$

satisfies (9.16), i.e., $\mathrm{P_{II}}$ *with* $\alpha = 0$*, with the boundary condition*

$$w_k(z) \sim k\,\mathrm{Ai}(z), \qquad z \to \infty. \tag{2.5}$$

Proof. See Ablowitz, Ramani and Segur [5] and Ablowitz and Segur [7]. □

The integral equation (2.3) is derived from the Gel'fand-Levitan-Marchenko integral equation

$$\mathcal{K}(x,y;t) = \mathcal{F}\left(\frac{x+y}{2};t\right) + \tfrac{1}{4} \int_x^\infty \int_x^\infty \mathcal{K}(x,z;t)\,\mathcal{F}\left(\frac{z+s}{2};t\right)\mathcal{F}\left(\frac{z+y}{2};t\right)\mathrm{d}z\,\mathrm{d}s, \tag{2.6}$$

where $\mathcal{F}(x;t)$ is expressed in terms of the initial data and satisfies the linear equation

$$\mathcal{F}_t + \mathcal{F}_{xxx} = 0, \tag{2.7}$$

and $u(x,t)$ is obtained through

$$u(x,t) = \mathcal{K}(x,x;t), \tag{2.8}$$

for solving the mKdV equation (2.2) by inverse scattering [4] through the scaling reduction

$$\mathcal{K}(x,y;t) = (3t)^{-1/3}K(z,\zeta), \qquad \mathcal{F}\left(\frac{x+y}{2};t\right) = (3t)^{-1/3}F\left(\frac{z+\zeta}{2}\right), \tag{2.9}$$

with $z = x/(3t)^{1/3}$, $\zeta = y/(3t)^{1/3}$; see also [2, 5, 7, 8] for further details.

The construction of the one-parameter transcendental solution of $\mathrm{P_{II}}$ with $\alpha = 0$ satisfying the boundary condition (2.5) by Ablowitz and Segur [7] through the linear integral equation (2.3) is the first such construction for a Painlevé equation. Such integral equations associated with Painlevé equations have been used to prove properties such as existence and uniqueness of solutions and connection formulae; see, for example, §9 and [5, 18, 40, 41, 91]).

We remark that the method of studying the Painlevé equations using integral equations has now been superseded by the isomonodromy method (cf.,

[49, 53, 55, 60, 103, 104, 105, 106, 107]), which is introduced in §2.2 below. Indeed, the integral equations discussed here represent a reduction of the Riemann-Hilbert approach for a specific, quasilinear family of the solutions of Painlevé equations. For example, Theorem 9.3 was derived in [104] and proved in [49, 103], using the isomonodromy method; the integral equations given above do not apply in this case.

2.2 Isomonodromy Problems

Each of the Painlevé equations P_{I}–P_{VI} can be expressed as the compatibility condition of the linear system

$$\frac{\partial \boldsymbol{\Psi}}{\partial \lambda} = \boldsymbol{A}(z;\lambda)\boldsymbol{\Psi}, \qquad \frac{\partial \boldsymbol{\Psi}}{\partial z} = \boldsymbol{B}(z;\lambda)\boldsymbol{\Psi}, \tag{2.10}$$

where $\boldsymbol{A}$ and $\boldsymbol{B}$ are matrices whose entries depend on the solution $w(z)$ of the Painlevé equation. The equation

$$\frac{\partial^2 \boldsymbol{\Psi}}{\partial z \partial \lambda} = \frac{\partial^2 \boldsymbol{\Psi}}{\partial \lambda \partial z},$$

is satisfied provided that

$$\frac{\partial \boldsymbol{A}}{\partial z} - \frac{\partial \boldsymbol{B}}{\partial \lambda} + \boldsymbol{AB} - \boldsymbol{BA} = 0, \tag{2.11}$$

which is the compatibility condition of (2.10). Matrices $\boldsymbol{A}$ and $\boldsymbol{B}$ for P_{I}–P_{VI} satisfying (2.12) are given by Jimbo and Miwa [111], see also [107], though these are not unique as illustrated below.

Theorem 2.2. *Consider the matrices* $\boldsymbol{A}$ *and* $\boldsymbol{B}$ *given by*

$$\begin{aligned} \boldsymbol{A}(z;\lambda) &= \begin{pmatrix} -\mathrm{i}(4\lambda^2+2w^2+z) & 4\lambda w + 2\mathrm{i}w' + \alpha/\lambda \\ 4\lambda w - 2\mathrm{i}w' + \alpha/\lambda & \mathrm{i}(4\lambda^2+2w^2+z) \end{pmatrix}, \\ \boldsymbol{B}(z;\lambda) &= \begin{pmatrix} -\mathrm{i}\lambda & w \\ w & \mathrm{i}\lambda \end{pmatrix}. \end{aligned} \tag{2.12}$$

Then these satisfy the compatibility condition (2.11) if and only if $w(z)$ *satisfies* P_{II}.

Proof. This is easily verified by substituting the matrices (2.12) into (2.11); see Flaschka and Newell [53], also Fokas and Ablowitz [55], Fokas and Zhou [60], for further details of the application of the isomonodromy deformation method to P_{II}. □

The matrices $\boldsymbol{A}$ and $\boldsymbol{B}$ given by (2.12) are derived from the Lax pair of the mKdV equation (2.2)

$$\frac{\partial \psi}{\partial t} = \begin{pmatrix} -4ik^3 - 2iku^2 & 4k^2u + 2iku_x - u_{xx} + 2u^3 \\ 4k^2u - 2iku_x - u_{xx} + 2u^3 & 4ik^3 + 2iku^2 \end{pmatrix} \psi, \qquad \frac{\partial \psi}{\partial x} = \begin{pmatrix} -ik & u \\ u & ik \end{pmatrix} \psi, \tag{2.13}$$

[4] through the scaling reduction

$$\psi(x,t) = \boldsymbol{\Psi}(z;\lambda), \qquad u(x,t) = (3t)^{-1/3} w(z), \tag{2.14}$$

where $z = x/(3t)^{1/3}$ and $\lambda = k(3t)^{1/3}$, and using the fact that $w(z)$ satisfies $\mathrm{P_{II}}$; see Ablowitz and Clarkson [2, §7.3], Flaschka and Newell [53] for further details.

As remarked above, the isomonodromy problem is not unique. Another isomonodromy problem for $\mathrm{P_{II}}$ given by Jimbo and Miwa [111] arises for the matrices

$$\boldsymbol{A}_{\mathrm{JM}}(z,\lambda) = \begin{pmatrix} \lambda^2 + v + \frac{1}{2}z & u(\lambda - w) \\ -\dfrac{2(\lambda v + \theta + wu)}{u} & -(\lambda^2 + v + \frac{1}{2}z) \end{pmatrix}, \qquad \boldsymbol{B}_{\mathrm{JM}}(z,\lambda) = \begin{pmatrix} \frac{1}{2}\lambda & \frac{1}{2}u \\ -v/u & -\frac{1}{2}\lambda \end{pmatrix}. \tag{2.15}$$

Substituting these into (2.11) yields

$$u' = -uw, \qquad v' = -2vw - \theta, \qquad w' = v + w^2 + \tfrac{1}{2}z. \tag{2.16}$$

Eliminating u and v yields $\mathrm{P_{II}}$, with $\alpha = \frac{1}{2} - \theta$. Note that eliminating u and w yields $\mathrm{P_{34}}$

$$v'' = \frac{(v')^2}{2v} - 2v^2 - zv - \frac{\theta^2}{2v}, \tag{2.17}$$

so-called since it is equivalent to equation XXXIV of Chapter 14 in [101], which itself is equivalent to $\mathrm{P_{II}}$ since there is a one-to-one relationship between solutions of (2.17) and those of $\mathrm{P_{II}}$. The matrices (2.15) are derived from the Garnier systems of second-order scalar equations.

3 Hamiltonian Structure

The Hamiltonian structure associated with the Painlevé equations $\mathrm{P_I}$–$\mathrm{P_{VI}}$ is $\mathcal{H}_{\mathrm{J}} = (q, p, \mathrm{H_J}, z)$, where $\mathrm{H_J}$, the Hamiltonian function associated with $\mathrm{H_J}$

a polynomial in q, p and z. Each of the Painlevé equations $\mathrm{P_I}$–$\mathrm{P_{VI}}$ can be written as a Hamiltonian system

$$\frac{\mathrm{D}q}{\mathrm{D}z} = \frac{\partial \mathrm{H_J}}{\partial p}, \qquad \frac{\mathrm{D}p}{\mathrm{D}z} = -\frac{\partial \mathrm{H_J}}{\partial q}, \tag{3.1}$$

where the derivative $\mathrm{D}/\mathrm{D}z$ is given by

$$\begin{aligned} \frac{\mathrm{D}}{\mathrm{D}z} &\equiv \frac{\mathrm{d}}{\mathrm{d}z}, && \text{for} \quad \mathrm{J} = \mathrm{I, II, V}, \\ \frac{\mathrm{D}}{\mathrm{D}z} &\equiv z\frac{\mathrm{d}}{\mathrm{d}z}, && \text{for} \quad \mathrm{J} = \mathrm{III, V}, \\ \frac{\mathrm{D}}{\mathrm{D}z} &\equiv z(z-1)\frac{\mathrm{d}}{\mathrm{d}z}, && \text{for} \quad \mathrm{J} = \mathrm{VI}, \end{aligned} \tag{3.2}$$

for a suitable Hamiltonian function $\mathrm{H_J}(q,p,z)$ (cf. Okamoto [188]). Further the function $\sigma(z) \equiv \mathrm{H_J}(q,p,z)$ satisfies a second-order, second-degree ordinary differential equation, whose solution is expressible in terms of the solution of the associated Painlevé equation.

Example 3.1. The Hamiltonian for $\mathrm{P_I}$ is

$$\mathrm{H_I}(q,p,z) = \tfrac{1}{2}p^2 - 2q^3 - zq, \tag{3.3}$$

and so

$$q' = p, \qquad p' = 6q^2 + z. \tag{3.4}$$

Then $q = w$ satisfies $\mathrm{P_I}$. The function $\sigma = \mathrm{H_I}(q,p,z)$, defined by (3.3) satisfies

$$(\sigma'')^2 + 4(\sigma')^3 + 2z\sigma' - 2\sigma = 0 \tag{3.5}$$

(Jimbo and Miwa [110], Okamoto [182]). Conversely, if σ is a solution of (3.5), then

$$q = -\sigma', \qquad p = -\sigma'', \tag{3.6}$$

are solutions of (3.4).

Example 3.2. The Hamiltonian for $\mathrm{P_{II}}$ is

$$\mathrm{H_{II}}(q,p,z;\alpha) = \tfrac{1}{2}p^2 - (q^2 + \tfrac{1}{2}z)p - (\alpha + \tfrac{1}{2})q \tag{3.7}$$

and so

$$q' = p - q^2 - \tfrac{1}{2}z, \qquad p' = 2qp + \alpha + \tfrac{1}{2} \tag{3.8}$$

(Jimbo and Miwa [110], Okamoto [184]). Eliminating p in (3.8) then $q = w$ satisfies $\mathrm{P_{II}}$ whilst eliminating q yields

$$pp'' = \tfrac{1}{2}\left(\frac{\mathrm{d}p}{\mathrm{d}z}\right)^2 = \tfrac{1}{2}(p')^2 + 2p^3 - zp^2 - \tfrac{1}{2}(\alpha + \tfrac{1}{2})^2, \tag{3.9}$$

which is known as P_{34}, since it is equivalent to equation XXXIV of Chapter 14 in [101]. Further if q satisfies P_{II} then $p = q' + q^2 + \frac{1}{2}z$ satisfies (3.9) and conversely if p satisfies (3.9) then $q = (p' - \alpha - \frac{1}{2})/(2p)$ satisfies P_{II}. Thus there is a one-to-one correspondence between solutions of P_{II} and those of P_{34} (3.9). The function $\sigma(z;\alpha) = H_{II}(q,p,z;\alpha)$ defined by (3.7) satisfies the second-order, second-degree equation

$$(\sigma'')^2 + 4(\sigma')^3 + 2\sigma'(z\sigma' - \sigma) = \tfrac{1}{4}(\alpha + \tfrac{1}{2})^2. \tag{3.10}$$

Conversely if $\sigma(z;\alpha)$ is a solution of (3.10), then

$$q(z;\alpha) = \frac{4\sigma''(z;\alpha) + 2\alpha + 1}{8\sigma'(z;\alpha)}, \qquad p(z;\alpha) = -2\sigma'(z;\alpha), \tag{3.11}$$

are solutions of (3.8).

We remark that equation (3.10) is equation SD-I.d in the classification of second-order, second-degree equations which have the Painlevé property by Cosgrove and Scoufis [45], an equation first derived by Chazy [27]. Further frequently in applications it is an associated second-order, second-degree equation such as (3.10) which arises rather than the Painlevé equation.

For Hamiltonian structure for P_{III}–P_{VI} see Jimbo and Miwa [110], Okamoto [184, 185, 186, 187]; see also Forrester and Witte [61, 62, 63].

Remarks

1. Each Hamiltonian function $\sigma = H_J$ satisfies a second-order second-degree ordinary differential equation whose solutions are in a 1-1 correspondence with solutions of the associated Painlevé equation through (3.1) since
$$q = F_J(\sigma, \sigma', \sigma'', z), \qquad p = G_J(\sigma, \sigma', \sigma'', z),$$
for suitable functions $F_J(\sigma, \sigma', \sigma'', z)$ and $G_J(\sigma, \sigma', \sigma'', z)$. Thus given q and p one can determine σ and conversely, given σ one can determine q and p.
2. The ordinary differential equations which the σ functions satisfy are part of the classification of second-order, second-degree equations of Painlevé type by Cosgrove and Scoufis [45]. They were first derived by Chazy [27] and subsequently rederived by Bureau [25].
3. The Hamiltonian functions $\sigma = H_J$ frequently arise in applications, e.g, random matrix theory (Tracy and Widom [214, 215, 216, 217, 218, 219]; see also [61, 62, 63]) and statistical physics (Jimbo, Miwa, Mori and Sato [112]).

4 Bäcklund Transformations

The Painlevé equations P_{II}–P_{VI} possess *Bäcklund transformations* which relate one solution to another solution either of the same equation, with different values of the parameters, or another equation, as we illustrate here.

4.1 Bäcklund Transformations for $\mathrm{P_{II}}$

Bäcklund transformations for $\mathrm{P_{II}}$ are given in the following theorem.

Theorem 4.1. *Let $w \equiv w(z;\alpha)$ be a solution of $\mathrm{P_{II}}$, then the transformations*

$$\mathcal{S}: \qquad w(z;-\alpha) = -\,w, \tag{4.1}$$

$$\mathcal{T}_\pm: \qquad w(z;\alpha\pm 1) = -\,w - \frac{2\alpha\pm 1}{2w^2 \pm 2w' + z}, \tag{4.2}$$

give solutions of $\mathrm{P_{II}}$, provided that $\alpha \neq \mp\frac{1}{2}$ in (4.2).

Proof. See Gambier [68] and Lukashevich [155]. □

Gambier [68] also discovered the following special transformation of $\mathrm{P_{II}}$

$$W(\zeta;\tfrac{1}{2}\varepsilon) = \frac{2^{-1/3}\varepsilon}{w(z;0)}\frac{\mathrm{d}w}{\mathrm{d}z}(z;0), \tag{4.3a}$$

$$w^2(z;0) = 2^{-1/3}\left\{W^2(\zeta;\tfrac{1}{2}\varepsilon) - \varepsilon\frac{\mathrm{d}W}{\mathrm{d}\zeta}(\zeta;\tfrac{1}{2}\varepsilon) + \tfrac{1}{2}\zeta\right\}, \tag{4.3b}$$

where $\zeta = -2^{1/3}z$ and $\varepsilon = \pm 1$; see also [36]. Combined with the Bäcklund transformation (4.2), the transformation (4.3) provides a relation between two $\mathrm{P_{II}}$ equations whose parameters α are either integers or half odd-integers. Hence this yields a mapping between the rational solutions of $\mathrm{P_{II}}$, which arise when $\alpha = n$ for $n \in \mathbb{Z}$, see §5.1, and the one-parameter Airy function solutions, which arise when $\alpha = n + \frac{1}{2}$ for $n \in \mathbb{Z}$, see §7.1.

The classification of solutions of $\mathrm{P_{II}}$ are given in Theorem 5.1. Umemura [225] discusses geometrical aspects of the Bäcklund transformations of $\mathrm{P_{II}}$ in terms of affine Weyl groups; see also §4.6.

The solutions $w_\alpha = w(z;\alpha)$, $w_{\alpha\pm 1} = w(z;\alpha\pm 1)$ also satisfy the nonlinear three-point recurrence relation

$$\frac{\alpha+\frac{1}{2}}{w_{\alpha+1}+w_\alpha} + \frac{\alpha-\frac{1}{2}}{w_\alpha + w_{\alpha-1}} + 2w_\alpha^2 + z = 0, \tag{4.4}$$

a difference equation which is known as an alternative form of discrete $\mathrm{P_I}$ [56]; see also [39]. The difference equation (4.4) is obtained by eliminating w' between the transformations $\mathcal{T}_\pm$ given by (4.2). Note that for $\mathrm{P_{II}}$, the independent variable z varies and the parameter α is fixed, whilst for the discrete equation (4.4), z is a fixed parameter and α varies. This is analogous to the situation for classical special functions such as Bessel function $J_\nu(z)$ which satisfies both a differential equation

$$z^2 J_\nu''(z) + zJ_\nu'(z) + (z^2-\nu^2)J_\nu(z) = 0, \tag{4.5}$$

in which z varies and ν is fixed, and a difference equation

$$zJ_{\nu+1}(z) - 2\nu J_\nu(z) + zJ_{\nu-1}(z) = 0, \tag{4.6}$$

in which z is fixed and ν varies.

4.2 Bäcklund Transformations for $\mathrm{P_{III}}$

The discrete symmetries of $\mathrm{P_{III}}$ give rise to the following theorem.

Theorem 4.2. *Let* $w_j = w(z;\alpha_j,\beta_j,\gamma_j,\delta_j)$, $j=0,1,2$, *be solutions of* $\mathrm{P_{III}}$ *with*

$$(\alpha_1,\beta_1,\gamma_1,\delta_1) = (-\alpha_0,-\beta_0,\gamma_0,\delta_0), \tag{4.7a}$$

$$(\alpha_2,\beta_2,\gamma_2,\delta_2) = (-\beta_0,-\alpha_0,-\delta_0,-\gamma_0). \tag{4.7b}$$

Then

$$\mathcal{S}_1: \qquad w_1 = -w_0, \qquad\qquad \mathcal{S}_2: \qquad w_2 = 1/w_0. \tag{4.8}$$

Proof. See Fokas and Ablowitz [54]. □

For the generic case of $\mathrm{P_{III}}$ with $\gamma\delta \neq 0$ we may set $\gamma = 1$ and $\delta = -1$, without loss of generality, by rescaling the variables if necessary. The Bäcklund transformations are given by

Theorem 4.3. *Let* $w_j = w(z;\alpha_j,\beta_j,1,-1)$, $j=0,1,2,3,4$, *be solutions of* $\mathrm{P_{III}}$ *with*

$$\alpha_1 = \alpha_3 = \alpha_0 + 2, \qquad \alpha_2 = \alpha_4 = \alpha_0 - 2,$$

$$\beta_1 = \beta_2 = \beta_0 + 2, \qquad \beta_3 = \beta_4 = \beta_0 - 2.$$

Then

$$\mathcal{T}_1: \qquad w_1 = \frac{zw_0' + zw_0^2 - \beta w_0 - w_0 + z}{w_0(zw_0' + zw_0^2 + \alpha w_0 + w_0 + z)}, \tag{4.9a}$$

$$\mathcal{T}_2: \qquad w_2 = -\frac{zw_0' - zw_0^2 - \beta w_0 - w_0 + z}{w_0(zw_0' - zw_0^2 - \alpha w_0 + w_0 + z)}, \tag{4.9b}$$

$$\mathcal{T}_3: \qquad w_3 = -\frac{zw_0' + zw_0^2 + \beta w_0 - w_0 - z}{w_0(zw_0' + zw_0^2 + \alpha w_0 + w_0 - z)}, \tag{4.9c}$$

$$\mathcal{T}_4: \qquad w_4 = \frac{zw_0' - zw_0^2 + \beta w_0 - w_0 - z}{w_0(zw_0' - zw_0^2 - \alpha w_0 + w_0 - z)}. \tag{4.9d}$$

Proof. See Fokas and Ablowitz [54], Gromak [72]; also Gromak, Laine and Shimomura [83, §34], Milne, Clarkson and Bassom [169]. □

The Bäcklund transformations for the special cases of $\mathrm{P_{III}}$ when either (i), $\gamma = 0$ and $\alpha\delta \neq 0$, then set $\alpha = 1$ and $\delta = -1$, without loss of generality, or (ii), $\delta = 0$ and $\beta\gamma \neq 0$, then set $\beta = -1$ and $\gamma = 1$, without loss of generality, are given in the following theorem.

Theorem 4.4. *Let* $w_j = w(z;1,\beta_j,0,-1)$, $j = 0,5,6$, *be solutions of* $\mathrm{P_{III}}$ *with*

$$\beta_5 = \beta_0 + 2, \qquad \beta_6 = \beta_0 - 2.$$

Then

$$\mathcal{T}_5: \qquad w_5 = \frac{zw_0' + z - (\beta_0 + 1)w_0}{w_0^2}, \tag{4.10a}$$

$$\mathcal{T}_6: \qquad w_6 = -\frac{zw_0' - z + (\beta_0 - 1)w_0}{w_0^2}. \tag{4.10b}$$

Similar results hold for $\mathrm{P_{III}}$ *with* $\delta = 0$ *and* $\beta\gamma \neq 0$.

Proof. Gromak [71]; also Gromak, Laine and Shimomura [83, §33], Milne, Clarkson and Bassom [169]. □

Discrete Painlevé equations arising from the Bäcklund transformations of $\mathrm{P_{III}}$ given by (4.9) and (4.10) are discussed in [39, 56, 86].

4.3 Bäcklund Transformations for $\mathrm{P_{IV}}$

Bäcklund transformations for $\mathrm{P_{IV}}$ are given in the following theorem.

Theorem 4.5. *Let* $w_0 = w(z;\alpha_0,\beta_0)$ *and* $w_j^{\pm} = w(z;\alpha_j^{\pm},\beta_j^{\pm})$, $j = 1,2,3,4$ *be solutions of* $\mathrm{P_{IV}}$ *with*

$$\begin{aligned}
\alpha_1^{\pm} &= \tfrac{1}{4}\left(2 - 2\alpha_0 \pm 3\sqrt{-2\beta_0}\right), & \beta_1^{\pm} &= -\tfrac{1}{2}\left(1 + \alpha_0 \pm \tfrac{1}{2}\sqrt{-2\beta_0}\right)^2,\\
\alpha_2^{\pm} &= -\tfrac{1}{4}\left(2 + 2\alpha_0 \pm 3\sqrt{-2\beta_0}\right), & \beta_2^{\pm} &= -\tfrac{1}{2}\left(1 - \alpha_0 \pm \tfrac{1}{2}\sqrt{-2\beta_0}\right)^2,\\
\alpha_3^{\pm} &= \tfrac{3}{2} - \tfrac{1}{2}\alpha_0 \mp \tfrac{3}{4}\sqrt{-2\beta_0}, & \beta_3^{\pm} &= -\tfrac{1}{2}\left(1 - \alpha_0 \pm \tfrac{1}{2}\sqrt{-2\beta_0}\right)^2,\\
\alpha_4^{\pm} &= -\tfrac{3}{2} - \tfrac{1}{2}\alpha_0 \mp \tfrac{3}{4}\sqrt{-2\beta_0}, & \beta_4^{\pm} &= -\tfrac{1}{2}\left(-1 - \alpha_0 \pm \tfrac{1}{2}\sqrt{-2\beta_0}\right)^2.
\end{aligned}$$

Then

$$\mathcal{T}_1^{\pm}: \qquad w_1^{\pm} = \frac{w_0' - w_0^2 - 2zw_0 \mp \sqrt{-2\beta_0}}{2w_0}, \tag{4.11a}$$

$$\mathcal{T}_2^{\pm}: \qquad w_2^{\pm} = -\frac{w_0' + w_0^2 + 2zw_0 \mp \sqrt{-2\beta_0}}{2w_0}, \tag{4.11b}$$

$$\mathcal{T}_3^{\pm}: \qquad w_3^{\pm} = w_0 + \frac{2\left(1 - \alpha_0 \mp \tfrac{1}{2}\sqrt{-2\beta_0}\right)w_0}{w_0' \pm \sqrt{-2\beta_0} + 2zw_0 + w_0^2}, \tag{4.11c}$$

$$\mathcal{T}_4^{\pm}: \qquad w_4^{\pm} = w_0 + \frac{2\left(1 + \alpha_0 \pm \tfrac{1}{2}\sqrt{-2\beta_0}\right)w_0}{w_0' \mp \sqrt{-2\beta_0} - 2zw_0 - w_0^2}, \tag{4.11d}$$

valid when the denominators are non-zero, and where the upper signs or the lower signs are taken throughout each transformation.

Proof. See Lukashevich [152], Gromak [74, 76]; also Bassom, Clarkson and Hicks [17], Gromak, Laine and Shimomura [83, §25]. □

Discrete Painlevé equations arising from the Bäcklund transformations of $\mathrm{P_{IV}}$ given by (4.11) are discussed in [39, 56, 86].

4.4 Bäcklund Transformations for $\mathrm{P_V}$

The discrete symmetries of $\mathrm{P_V}$ give rise to the following theorem.

Theorem 4.6. *Let* $w_j(z_j) = w(z_j; \alpha_j, \beta_j, \gamma_j, \delta_j)$, $j = 0, 1, 2$ *be solutions of* $\mathrm{P_V}$ *with*

$$\begin{aligned} z_1 &= -z_0, \qquad & (\alpha_1, \beta_1, \gamma_1, \delta_1) &= (\alpha_0, \beta_0, -\gamma_0, \delta_0), \\ z_2 &= z_0, \qquad & (\alpha_2, \beta_2, \gamma_2, \delta_2) &= (-\beta_0, -\alpha_0, -\gamma_0, \delta_0). \end{aligned}$$

Then

$$\mathcal{S}_1: \qquad w_1(z_1) = w(z_0), \qquad\qquad \mathcal{S}_2: \qquad w_2(z_2) = 1/w(z_0). \tag{4.12}$$

The Bäcklund transformations for the generic case of $\mathrm{P_V}$ with $\delta \neq 0$, when we may set $\delta = -\frac{1}{2}$, without loss of generality, by rescaling the variables if necessary, are given by

Theorem 4.7. *Let* $w_0 = w(z; \alpha_0, \beta_0, \gamma_0, -\frac{1}{2})$ *and* $w_j = W(z; \alpha_j, \beta_j, \gamma_j, -\frac{1}{2})$, $j = 1, 2, 3,$ *be solutions of* $\mathrm{P_V}$, *where*

$$\alpha_j = \tfrac{1}{8}\left[\gamma_0 + \varepsilon_1\left(1 - \varepsilon_3\sqrt{-2\beta_0} - \varepsilon_2\sqrt{2\alpha_0}\right)\right]^2 \tag{4.13a}$$

$$\beta_j = -\tfrac{1}{8}\left[\gamma_0 - \varepsilon_1\left(1 - \varepsilon_3\sqrt{-2\beta_0} - \varepsilon_2\sqrt{2\alpha_0}\right)\right]^2 \tag{4.13b}$$

$$\gamma_j = \varepsilon_1\left(\varepsilon_3\sqrt{-2\beta_0} - \varepsilon_2\sqrt{2\alpha_0}\right) \tag{4.13c}$$

and $\varepsilon_j = \pm 1$, *independently. Also let*

$$\Phi = zw_0' - \varepsilon_2\sqrt{2\alpha_0}\, w_0^2 + \varepsilon_3\sqrt{-2\beta_0} + \left(\varepsilon_2\sqrt{2\alpha_0} - \varepsilon_3\sqrt{-2\beta_0} + \varepsilon_1 z\right) w_0, \tag{4.14}$$

and assume $\Phi \neq 0$. *Then*

$$\mathcal{T}_j: \qquad w_j = (\Phi - 2\varepsilon_1 z w_0)/\Phi, \tag{4.15}$$

provided that the numerator on the right-hand side does not vanish. For $j = 1, 2, 3,$ *and* $\varepsilon_j = \pm 1$, *independently, there are eight distinct transformations of type* $\mathcal{T}_j$.

Proof. See Gromak [73]; also Gromak and Filipuk [79, 80, 81], Gromak, Laine and Shimomura [83, §39]. □

4.5 Bäcklund Transformations for $\mathrm{P_{VI}}$

The discrete symmetries of $\mathrm{P_{VI}}$ give rise to the following theorem.

Theorem 4.8. *Let $w_j(z_j) = w_j(z_j;\alpha_j,\beta_j,\gamma_j,\delta_j)$, $j = 0,1,2,3$, be solutions of* $\mathrm{P_{VI}}$ *with*

$$z_1 = 1/z_0, \qquad z_2 = 1 - z_0, \qquad z_3 = 1/z_0,$$

and

$$\begin{aligned}(\alpha_1,\beta_1,\gamma_1,\delta_1) &= (\alpha_0,\beta_0,-\delta_0+\tfrac{1}{2},-\gamma_0+\tfrac{1}{2}),\\ (\alpha_2,\beta_2,\gamma_2,\delta_2) &= (\alpha_0,-\gamma_0,-\beta_0,\delta_0),\\ (\alpha_3,\beta_3,\gamma_3,\delta_3) &= (-\beta_0,-\alpha_0,\gamma_0,\delta_0).\end{aligned}$$

Then

$$\mathcal{S}_1: \qquad w_1(z_1) = w_0(z_0)/z_0, \tag{4.16a}$$
$$\mathcal{S}_2: \qquad w_2(z_2) = 1 - w_0(z_0), \tag{4.16b}$$
$$\mathcal{S}_3: \qquad w_3(z_3) = 1/w_0(z_0). \tag{4.16c}$$

Proof. See Okamoto [185]. □

The transformations $\mathcal{S}_j$, for $j = 1,2,3$, given by (4.16) generate a group of order 24; see Iwasaki, Kimura, Shimomura and Yoshida [109].

Theorem 4.9. *Let $w(z;\alpha,\beta,\gamma,\delta)$ and $W(z;A,B,C,D)$ be solutions of* $\mathrm{P_{VI}}$ *with*

$$(\alpha,\beta,\gamma,\delta) = (\tfrac{1}{2}(\theta_\infty-1)^2, -\tfrac{1}{2}\theta_0^2, \tfrac{1}{2}\theta_1^2, \tfrac{1}{2}(1-\theta_2^2)), \tag{4.17}$$
$$(A,B,C,D) = (\tfrac{1}{2}(\Theta_\infty-1)^2, -\tfrac{1}{2}\Theta_0^2, \tfrac{1}{2}\Theta_1^2, \tfrac{1}{2}(1-\Theta_2^2)), \tag{4.18}$$

and

$$\theta_j = \Theta_j - \tfrac{1}{2}\sigma, \tag{4.19}$$

for $j = 0,1,z,\infty$, where

$$\sigma = \theta_0+\theta_1+\theta_z+\theta_\infty-1 = 1-(\Theta_0+\Theta_1+\Theta_z+\Theta_\infty). \tag{4.20}$$

Then

$$\begin{aligned}\frac{\sigma}{w-W} &= \frac{z(z-1)W'}{W(W-1)(W-z)} + \frac{\Theta_0}{W} + \frac{\Theta_1}{W-1} + \frac{\Theta_2-1}{W-z}\\ &= \frac{z(z-1)w'}{w(w-1)(w-z)} + \frac{\theta_0}{w} + \frac{\theta_1}{w-1} + \frac{\theta_2-1}{w-z}.\end{aligned}$$

Proof. See Okamoto [185]; also Conte and Musette [43, 44], Gromak and Filipuk [82], Gromak, Laine and Shimomura [83, §42]. □

$\mathrm{P_{VI}}$ also has quadratic and quartic transformations. Let $w = w(z;\alpha,\beta,\gamma,\delta)$ be a solution of $\mathrm{P_{VI}}$. The quadratic transformation

$$u_1(\zeta_1) = \frac{(1-w)(w-z)}{(1+\sqrt{z})^2 w}, \qquad \zeta_1 = \left(\frac{1-\sqrt{z}}{1+\sqrt{z}}\right)^2, \tag{4.21}$$

transforms $\mathrm{P_{VI}}$ with $\alpha = -\beta$ and $\gamma = \frac{1}{2} - \delta$ to $\mathrm{P_{VI}}$ with $(\alpha_1,\beta_1,\gamma_1,\delta_1) = (4\alpha, -4\gamma, 0, \frac{1}{2})$. The quartic transformation

$$u_2(\zeta_2) = \frac{(w^2-z)^2}{4w(w-1)(w-z)}, \qquad \zeta_2 = z, \tag{4.22}$$

transforms $\mathrm{P_{VI}}$ with $\alpha = -\beta = \gamma = \frac{1}{2} - \delta$ to $\mathrm{P_{VI}}$ with $(\alpha_2,\beta_2,\gamma_2,\delta_2) = (16\alpha, 0, 0, \frac{1}{2})$. Also

$$u_3(\zeta_3) = \left(\frac{1-z^{1/4}}{1+z^{1/4}}\right)^2 \left(\frac{\sqrt{w}+z^{1/4}}{\sqrt{w}-z^{1/4}}\right)^2, \qquad \zeta_3 = \left(\frac{1-z^{1/4}}{1+z^{1/4}}\right)^4,$$

transforms $\mathrm{P_{VI}}$ with $\alpha = \beta = 0$ and $\gamma = \frac{1}{2} - \delta$ to $\mathrm{P_{VI}}$ with $\alpha_3 = \beta_3$ and $\gamma_3 = \frac{1}{2} - \delta_3$.

4.6 Affine Weyl Groups

The parameter space of $\mathrm{P_{II}}$–$\mathrm{P_{VI}}$ can be identified with the Cartan subalgebra of a simple Lie algebra and the corresponding affine Weyl groups $\widetilde{A}_1$, $\widetilde{C}_2$, $\widetilde{A}_2$, $\widetilde{A}_3$, $\widetilde{D}_4$, act on $\mathrm{P_{II}}$–$\mathrm{P_{VI}}$, respectively, as a group of Bäcklund transformations (cf. [174, 176, 184, 185, 186, 187, 225]). An affine Weyl group is essentially a group of translations and reflections on a lattice. For the Painlevé equations, this lattice is in the parameter space.

Example 4.1. The Bäcklund transformations of $\mathrm{P_{II}}$ given by $\mathcal{S}$ (4.1) and $\mathcal{T}_\pm$ (4.2) are affine transformations

$$\mathcal{S}(\alpha) = -\alpha, \qquad \mathcal{T}_\pm(\alpha) = \alpha \pm 1, \tag{4.23}$$

for $\alpha \in \mathbb{C}$. Consider the subgroup $\mathcal{G}$ of the affine transformation group on $\mathbb{C}$ generated by $\langle \mathcal{S}, \mathcal{T}_+, \mathcal{T}_- \rangle$. Then

$$\mathcal{S}^2 = \mathcal{I}, \qquad \mathcal{T}_+\mathcal{T}_- = \mathcal{T}_-\mathcal{T}_+ = \mathcal{I}, \qquad \mathcal{T}_+ = \mathcal{S}\mathcal{T}_-\mathcal{S}, \tag{4.24}$$

with $\mathcal{I}$ the identity transformation, and so $\langle \mathcal{S} \rangle \cong \mathbb{Z}/2\mathbb{Z}$, the Weyl group of the root system of type A_1, and $\langle \mathcal{T}_+, \mathcal{T}_- \rangle \cong \mathbb{Z}$. Therefore $\mathcal{G} \cong \mathbb{Z}/2\mathbb{Z} \ltimes \mathbb{Z}$, the Weyl group of the affine root system of type $\widetilde{A}_1$. This transformation group of the parameter space can be lifted to a transformation group on the set of solutions through the Bäcklund transformations for the system (3.8)

$$\mathcal{S}(q,p) = \left(-q, -p + 2q^2 + z\right), \tag{4.25}$$

$$\mathcal{T}_+(q,p) = \begin{cases} \left(-q - \dfrac{\alpha+\frac{1}{2}}{p}, -p + 2\left(q + \dfrac{\alpha+\frac{1}{2}}{p}\right)^2 + z\right), & \text{if } \alpha \neq -\frac{1}{2}, \\ \left(-q, -p + 2q^2 + z\right), & \text{if } \alpha = -\frac{1}{2}, \end{cases} \tag{4.26}$$

$$\mathcal{T}_-(q,p) = \begin{cases} \left(-q + \dfrac{\alpha-\frac{1}{2}}{p - 2q^2 - z}, -p + 2q^2 + z\right), & \text{if } \alpha \neq \frac{1}{2}, \\ \left(-q, -p + 2q^2 + z\right), & \text{if } \alpha = \frac{1}{2} \end{cases} \tag{4.27}$$

see [184, 225, 227] for further details.

5 Rational Solutions

The Painlevé equations $\mathrm{P_{II}}$–$\mathrm{P_{VI}}$ possess sets of rational solutions, often called hierarchies, for special values of the parameters. These hierarchies can be generated from "seed solutions", such as the simple solution $w = 0$ for $\mathrm{P_{II}}$, using the Bäcklund transformations discussed in §4 above and frequently are expressed in the form of determinants.

5.1 Rational Solutions of $\mathrm{P_{II}}$

According to Erugin [52] and Gromak [75], it was once a widely held view that all solutions of the Painlevé equations were new transcendents and there have been a few incorrect proofs that $\mathrm{P_{II}}$ admits no elementary solutions or just the three rational solutions $w = 0$, when $\alpha = 0$, and $w = \pm 1/z$, when $\alpha = \mp 1$.

Solutions of $\mathrm{P_{II}}$ are classified in the following theorem.

Theorem 5.1.

1. *For every $\alpha = n \in \mathbb{Z}$ there exists a unique rational solution of* $\mathrm{P_{II}}$.
2. *For every $\alpha = n + \frac{1}{2}$, with $n \in \mathbb{Z}$, there exists a unique one-parameter family of classical solutions of* $\mathrm{P_{II}}$*, each of which is rationally written in terms of Airy functions.*
3. *For all other values of α, the solution of* $\mathrm{P_{II}}$ *is nonclassical, i.e. transcendental.*

Proof. See Umemura and Watanabe [227]; also [11, 13, 53, 77]. □

5.2 The Yablonskii-Vorob'ev Polynomials

There are special polynomials associated with the rational solutions of $\mathrm{P_{II}}$ which are defined in the following theorem.

Theorem 5.2. *Rational solutions of* $\mathrm{P_{II}}$ *exist if and only if* $\alpha = n \in \mathbb{Z}$, *which are unique, and have the form*

$$w(z;n) = \frac{\mathrm{d}}{\mathrm{d}z}\left\{\ln\left[\frac{Q_{n-1}(z)}{Q_n(z)}\right]\right\}, \tag{5.1}$$

for $n \geq 1$, *where the polynomials* $Q_n(z)$ *satisfy the differential-difference equation*

$$Q_{n+1}Q_{n-1} = zQ_n^2 - 4\left[Q_nQ_n'' - (Q_n')^2\right], \tag{5.2}$$

with $Q_0(z) = 1$ *and* $Q_1(z) = z$. *The other rational solutions are given by*

$$w(z;0) = 0, \qquad w(z;-n) = -w(z;n). \tag{5.3}$$

Proof. See Vorob'ev [229] and Yablonskii [232]; also Clarkson and Mansfield [38], Fukutani, Okamoto and Umemura [66], Taneda [211], Umemura [224], Umemura and Watanabe [227]. □

The polynomials $Q_n(z)$ are monic polynomials of degree $\frac{1}{2}n(n+1)$ with integer coefficients, and are called the *Yablonskii-Vorob'ev polynomials*. The first few of these polynomials are given in Table 1, and the associated rational solutions $w(z;n)$ of $\mathrm{P_{II}}$ are given in Table 2.

Properties of the roots of the Yablonskii-Vorob'ev polynomials are discussed in the following theorems.

Theorem 5.3. *For every positive integer* n, *the polynomial* $Q_n(z)$ *has simple roots. Further the polynomials* $Q_n(z)$ *and* $Q_{n+1}(z)$ *do not have a common root.*

Table 1. Yablonskii-Vorob'ev polynomials

$Q_0(z) = 1$
$Q_1(z) = z$
$Q_2(z) = z^3 + 4$
$Q_3(z) = z^6 + 20z^3 - 80$
$Q_4(z) = (z^9 + 60z^6 + 11200)z$
$Q_5(z) = z^{15} + 140z^{12} + 2800z^9 + 78400z^6 - 313600z^3 - 6272000$
$Q_6(z) = z^{21} + 280z^{18} + 18480z^{15} + 627200z^{12} - 17248000z^9 + 1448832000z^6$ $+ 19317760000z^3 - 38635520000$
$Q_7(z) = (z^{27} + 504z^{24} + 75600z^{21} + 5174400z^{18} + 62092800z^{15} + 13039488000z^{12}$ $- 828731904000z^9 - 49723914240000z^6 - 3093932441600000)z$

Table 2. Rational solutions of $\mathrm{P_{II}}$

$$w(z;1) = -\frac{1}{z}$$

$$w(z;2) = \frac{1}{z} - \frac{3z^2}{z^3+4}$$

$$w(z;3) = \frac{3z^2}{z^3+4} - \frac{6z^2(z^3+10)}{z^6+20z^3-80}$$

$$w(z;4) = -\frac{1}{z} + \frac{6z^2(z^3+10)}{z^6+20z^3-80} - \frac{9z^5(z^3+40)}{z^9+60z^6+11200}$$

$$w(z;5) = \frac{1}{z} + \frac{9z^8+360z^5}{z^9+60z^6+11200} - \frac{15z^{14}+1680z^{11}+25200z^8+470400z^5-9408000z^2}{z^{15}+140z^{12}+2800z^9+78400z^6-3136000z^3-627200}$$

Proof. See Fukutani, Okamoto and Umemura [66], who also give a purely algebraic proof of Theorem 5.2. □

Theorem 5.4. *The polynomial* $Q_n(z)$ *is divisible by* z *if and only if* $n \equiv 1 \bmod 3$. *Further* $Q_n(z)$ *is a polynomial in* z^3 *if* $n \not\equiv 1 \bmod 3$ *and* $Q_n(z)/z$ *is a polynomial in* z^3 *if* $n \equiv 1 \bmod 3$.

Proof. See Taneda [211]. □

Remarks

1. The hierarchy of rational solutions for $\mathrm{P_{II}}$ given by (5.1) can also be derived using the Bäcklund transformation $\mathcal{T}_+$ (4.2) of $\mathrm{P_{II}}$, with "seed solution" $w_0 \equiv w(z;0) = 0$, i.e.,

$$w_{n+1} = -w_n - \frac{2n+1}{2w_n^2 + 2w_n' + z}, \tag{5.4}$$

where $w_m \equiv w(z;m)$, with $w_0 = 0$ and $w_{-m} = -w_m$, as is easily verified.

2. It is clear from the recurrence relation (5.2) that the $Q_n(z)$ are rational functions, though it is not obvious that in fact they are polynomials since one is dividing by $Q_{n-1}(z)$ at every iteration. Indeed it is somewhat remarkable that the $Q_n(z)$ defined by (5.2) are polynomials.
3. The recurrence relation (5.2) for the polynomials $Q_n(z)$ can be rewritten in the form

$$(2\mathrm{D}_z^2 - z)Q_n \bullet Q_n = -Q_{n+1}Q_{n-1}, \tag{5.5}$$

where D_z is the Hirota operator defined by

$$\mathrm{D}_z F(z) \bullet G(z) = \left[\left(\frac{\mathrm{d}}{\mathrm{d}z_1} - \frac{\mathrm{d}}{\mathrm{d}z_2}\right) F(z_1)G(z_2)\right]_{z_1=z_2=z}. \tag{5.6}$$

4. Letting $Q_n(z) = c_n \tau_n(z) \exp(z^3/24)$, with $c_n = (2\mathrm{i})^{n(n+1)}$, in (5.2) yields the Toda equation
$$\tau_n \tau_n'' - (\tau_n')^2 = \tau_{n+1}\tau_{n-1}. \tag{5.7}$$
5. The Yablonskii-Vorob'ev polynomials $Q_n(z)$ possess the discrete symmetry
$$Q_n(\omega z) = \omega^{n(n+1)/2} Q_n(z), \tag{5.8}$$
where $\omega^3 = 1$ and $\frac{1}{2}n(n+1)$ is the degree of $Q_n(z)$.
6. From Theorems 5.3 and 5.4, since each polynomial $Q_n(z)$ has only simple roots then it can be written as
$$Q_n(z) = \prod_{k=1}^{n(n+1)/2} (z - a_{n,k}), \tag{5.9}$$
where $a_{n,k}$, for $k = 1, \ldots, \frac{1}{2}n(n+1)$, are the roots and thus
$$\frac{Q_n'(z)}{Q_n(z)} = \sum_{k=1}^{n(n+1)/2} \frac{1}{z - a_{n,k}}. \tag{5.10}$$
Therefore the rational solution can be written as
$$\begin{aligned} w(z;n) &= \frac{Q_{n-1}'(z)}{Q_{n-1}(z)} - \frac{Q_n'(z)}{Q_n(z)} \\ &= \sum_{k=1}^{n(n-1)/2} \frac{1}{z - a_{n-1,k}} - \sum_{k=1}^{n(n+1)/2} \frac{1}{z - a_{n,k}}, \end{aligned} \tag{5.11}$$
and so $w(z;n)$ has n roots, $\frac{1}{2}n(n-1)$ with residue $+1$ and $\frac{1}{2}n(n+1)$ with residue -1; see also [78].
7. The roots $a_{n,k}$ of the polynomial $Q_n(z)$ satisfy
$$\sum_{k=1,k\neq j}^{n(n+1)/2} \frac{1}{(a_{n,j} - a_{n,k})^3} = 0, \qquad j = 1, 2, \ldots, \tfrac{1}{2}n(n+1), \tag{5.12a}$$
$$\sum_{j=1}^{n(n+1)/2} \sum_{k=j+1}^{n(n+1)/2} \frac{1}{(a_{n,j} - a_{n,k})^2} = 0. \tag{5.12b}$$
These follow from the study of rational solutions of the Korteweg-de Vries (KdV) equation
$$u_t + 6uu_x + u_{xxx} = 0, \tag{5.13}$$
and a related many-body problem by Airault, McKean and Moser [12]; see also [6, 10, 11, 29].

8. The Yablonskii-Vorob'ev polynomials also satisfy the bilinear equations

$$\mathrm{D}_z Q_{n+1} \bullet Q_{n-1} = (2n+1)Q_n^2, \tag{5.14a}$$
$$\mathrm{D}_z^2 Q_{n+1} \bullet Q_n = 0. \tag{5.14b}$$
$$(\mathrm{D}_z^3 - z\mathrm{D}_z + n + 1)Q_{n+1} \bullet Q_n = 0, \tag{5.14c}$$

   where D_z is the Hirota operator (5.6). Equations (5.14) can be derived by taking scaling reductions of bilinear equations in [6, 10] associated with rational solutions of the KdV equation (5.13); see also [11]. Equation (5.14a) is proved in [66, 211]; see also [130].
9. The Yablonskii-Vorob'ev polynomials are closely related with Schur functions [124, 226] and so it can be proved using (5.14) that the rational solution of $\mathrm{P_{II}}$ can be expressed using determinants as shown in §5.3.
10. Kametaka [126] has obtained a sharp estimate for the maximum modulus of the poles of the Yablonskii-Vorob'ev polynomials. It is shown that if $A_n = \max_{1 \le k \le n(n+1)/2}\{|a_{n,k}|\}$ then $n^{2/3} \le A_{n+2} \le 4n^{2/3}$, for $n \ge 0$. In [128] Kametaka studies the irreducibility of the Yablonskii-Vorob'ev polynomials.
11. Kaneko and Ochiai [130] derive formulae for the coefficients of the lowest degree term of the Yablonskii-Vorob'ev polynomials; the other coefficients remain to be determined, which is an interesting problem.

Clarkson and Mansfield [38] investigated the locations of the roots of the Yablonskii-Vorob'ev polynomials in the complex plane and showed that these roots have a very regular, approximately triangular structure. An earlier study of the distribution of the roots of the Yablonskii-Vorob'ev polynomials is given by Kametaka, Noda, Fukui and Hirano [129]; see also [109]. A plot of the roots of $Q_{25}(z)$ is given in Figure 1; see [38] for plots of other Yablonskii-Vorob'ev polynomials. From these plots we make the following observations

1. The roots of the Yablonskii-Vorob'ev polynomials form approximately equilateral triangles, in fact approximate "Pascal triangles". The values of the roots given in [38] show that they actually lie on curves rather than straight lines.
2. The plots are invariant under rotations through $\frac{2}{3}\pi$ and reflections in the real z-axis and the lines $\arg(z) = n\pi/3$, for $n = \pm 1, \pm 2$. This is because $\mathrm{P_{II}}$ admits the finite group of order 6 of scalings and reflections

$$w \to \varepsilon\mu^2 w, \qquad z \to \mu z, \qquad \alpha \to \varepsilon\alpha, \tag{5.15}$$

   where $\mu^3 = 1$ and $\varepsilon^2 = 1$.
3. The roots of $Q_n(z) = 0$ lie on circles with centre the origin. If we define

$$q_n(\zeta) = \begin{cases} Q_n(\zeta^{1/3}), & \text{if } n \not\equiv 1 \bmod 3, \\ Q_n(\zeta^{1/3})/\zeta^{1/3}, & \text{if } n \equiv 1 \bmod 3. \end{cases}$$

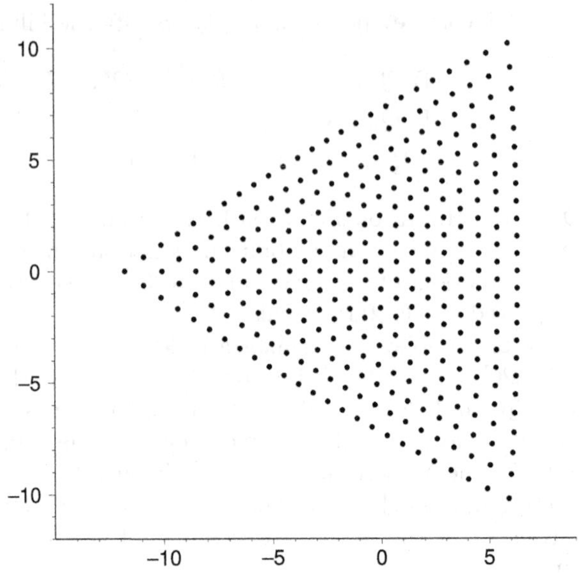

Fig. 1. Roots of the Yablonskii–Vorob'ev polynomial $Q_{25}(z)$

Then the radii of the circles are given by the third roots of the absolute values of the non-zero roots of $q_n(\zeta) = 0$, with three equally spaced roots of $Q_n(z) = 0$ on circles for the real roots of $q_n(\zeta) = 0$ and six roots, three complex conjugate pairs, of $Q_n(z) = 0$ on a circles for the complex roots of $q_n(\zeta) = 0$.

Using the Hamiltonian formalism for $\mathrm{P_{II}}$ described in §3, it can be shown that the Yablonskii-Vorob'ev polynomials $Q_n(z)$ satisfy a fourth order bilinear ordinary differential equation and a fourth order, second degree, hexa-linear (i.e., homogeneous of degree six) difference equation. It is shown in [33] that rational solutions of (3.10) have the form

$$\sigma_n = -\tfrac{1}{8}z^2 + \frac{\mathrm{d}}{\mathrm{d}z}\ln Q_n. \tag{5.16}$$

Differentiating (3.10) with respect to z yields

$$\sigma''' + 6(\sigma')^2 + 2z\sigma' - \sigma = 0, \tag{5.17}$$

and then substituting (5.16) into (5.17) yields the fourth order, bilinear equation

$$Q_nQ_n'''' - 4Q_n'Q_n''' + 3(Q_n'')^2 - z\left[Q_nQ_n'' - (Q_n')^2\right] - Q_nQ_n' = 0. \tag{5.18}$$

We remark that substituting (5.16) into (3.10) yields the third order, second degree, quad-linear (i.e., homogeneous of degree four) equation

$$\begin{aligned}Q_n^2 (Q_n''')^2 + Q_n''' \left[4 (Q_n')^3 - 6 Q_n Q_n' Q_n'' - \tfrac{1}{2} Q_n^3\right] + 4 Q_n (Q_n'')^3 \\ - (Q_n'')^2 \left[3 (Q_n')^2 + z Q_n^2\right] + \tfrac{1}{2} Q_n Q_n' Q_n'' (4 z Q_n' - Q_n) \\ - (Q_n')^3 (z Q_n' - Q_n) + \tfrac{1}{2} z Q_n^3 Q_n' - \tfrac{1}{4} n(n+1) Q_n^4 = 0. \quad (5.19)\end{aligned}$$

Additionally Q_n satisfies the fourth order, second degree, hexa-linear difference equation

$$\begin{aligned}16(2n+1)^4 Q_n^6 - 8(2n+1)^2 (Q_{n+2} Q_n^3 Q_{n-1}^2 + 2 Q_{n+1}^3 Q_{n-1}^3 + Q_{n-2} Q_n^3 Q_{n+1}^2 \\ - 4z Q_{n+1}^2 Q_n^2 Q_{n-1}^2) + (Q_{n+2} Q_{n-1}^2 - Q_{n+1}^2 Q_{n-2})^2 = 0, \quad (5.20)\end{aligned}$$

(see [33] for details). Hence the Yablonskii-Vorob'ev polynomials Q_n satisfy nonlinear ordinary differential equations (5.18) and (5.19), the difference equation (5.20) as well as the differential-difference equation (5.2); see [33] for further differential-difference equations satisfied by the Yablonskii-Vorob'ev polynomials. This is analogous to classical orthogonal polynomials, such as Hermite, Laguerre and Jacobi polynomials, which satisfy linear ordinary differential, difference and differential-difference equations (cf. [9, 15, 213]).

It seems reasonable to expect that the ordinary differential equations (5.18) and (5.19) will be useful for proving properties of the Yablonskii-Vorob'ev polynomials since there are more techniques for studying properties of solutions of ordinary differential equations than for difference equations or differential-difference equations. For example, suppose we seek a monic polynomial solution of (5.19) with $\alpha = n$ in the form

$$Q_n(z) = z^r + a_1 z^{r-1} + \cdots + a_{r-1} z + a_r.$$

Since (5.18) is homogeneous then we can assume, without loss of generality, that the coefficient of z^r is unity. Then it is easy to show that necessarily $r = \frac{1}{2} n(n+1)$, which is a simple proof of the degree of Q_n. Similarly it is straightforward to show using (5.18) that $a_{3j-1} = 0$ and $a_{3j-2} = 0$ and to derive recurrence relations for the coefficients a_{3j}.

An important, well-known property of classical orthogonal polynomials, such as the Hermite, Laguerre or Legendre polynomials whose roots all lie on the real line (cf. [9, 15, 213]), is that the roots of successive polynomials interlace. Thus for a set of orthogonal polynomials $\varphi_n(z)$, for $n = 0, 1, 2, \ldots$, if $z_{n,m}$ and $z_{n,m+1}$ are two successive roots of $\varphi_n(z)$, i.e., $\varphi_n(z_{n,m}) = 0$ and $\varphi_n(z_{n,m+1}) = 0$, then $\varphi_{n-1}(\zeta_{n-1}) = 0$ and $\varphi_{n+1}(\zeta_{n+1}) = 0$ for some ζ_{n-1} and ζ_{n+1} such that $z_{n,m} < \zeta_{n-1}, \zeta_{n+1} < z_{n,m+1}$. Further the derivatives $\varphi_n'(z)$ and $\varphi_{n+1}'(z)$ also have roots in the interval $(z_{n,m}, z_{n,m+1})$, that is $\varphi_n'(\xi_n) = 0$ and $\varphi_{n+1}'(\xi_{n+1}) = 0$ for some ξ_n and ξ_{n+1} such that $z_{n,m} < \xi_n, \xi_{n+1} < z_{n,m+1}$. An interesting question is whether there are analogous results for the Yablonskii-Vorob'ev polynomials $Q_n(z)$. Clearly there are notable differences since the Yablonskii-Vorob'ev polynomials $Q_n(z)$ are polynomials of degree $\frac{1}{2}n(n+1)$ with complex roots whereas classical orthogonal polynomials $\varphi_n(z)$, have n real roots. The triangular pattern of the roots of $Q_n(z) = 0$ is highly symmetric and structured, suggesting that they have interesting properties. The roots of $Q_{n+1}(z) = 0$ form a pattern of n^2 triangles formed by joining the nearest neighbours. Some preliminary investigations of the "interlacing of roots", the locations of the roots for two successive Yablonskii-Vorob'ev polynomials, $Q_n(z) = 0$ and $Q_{n+1}(z) = 0$, are given in [38]. This "interlacing of roots" clearly warrants further analytical study as does an investigation of the relative locations of the roots for Yablonskii-Vorob'ev polynomials $Q_n(z) = 0$ and its derivative $Q'_n(z) = 0$.

5.3 Determinantal Representation of Rational Solutions of $\mathrm{P_{II}}$

Kajiwara and Ohta [124], see also [122], proved Theorem 5.5 on the representation of rational solutions of $\mathrm{P_{II}}$ using determinants. Their results have been scaled so that they are for the standard version of $\mathrm{P_{II}}$, rather than the equation in [124] which is equivalent to $\mathrm{P_{II}}$ through a scaling of the variables. We remark that Flaschka and Newell [53], following the earlier work of Airault [11], expressed the rational solutions of $\mathrm{P_{II}}$ given by (5.1) as the logarithmic derivatives of determinants.

Theorem 5.5. *Let $p_k(z)$ be the polynomial defined by*

$$\sum_{k=0}^{\infty} p_k(z)\lambda^k = \exp\left(z\lambda - \tfrac{4}{3}\lambda^3\right), \tag{5.21}$$

and $\tau_n(z)$ be the $n \times n$ determinant

$$\tau_n(z) = \mathcal{W}\left(p_1(z), p_3(z), \ldots, p_{2n-1}(z)\right) \equiv \begin{vmatrix} p_1(z) & p_3(z) & \cdots & p_{2n-1}(z) \\ p_1'(z) & p_3'(z) & \cdots & p_{2n-1}'(z) \\ \vdots & \vdots & \ddots & \vdots \\ p_1^{(n-1)}(z) & p_3^{(n-1)}(z) & \cdots & p_{2n-1}^{(n-1)}(z) \end{vmatrix}, \tag{5.22}$$

for $n \geq 1$. Then

$$w(z;n) = \frac{\mathrm{d}}{\mathrm{d}z}\left\{\ln\left[\frac{\tau_{n-1}(z)}{\tau_n(z)}\right]\right\}, \tag{5.23}$$

satisfies $\mathrm{P_{II}}$ with $\alpha = n$, for $n \geq 1$.

We remark that Kajiwara and Ohta [124] wrote the determinantal representation of $\tau_n(z)$ in the form

$$\tau_n(z) = \begin{vmatrix} p_n(z) & p_{n+1}(z) & \cdots & p_{2n-1}(z) \\ p_{n-2}(z) & p_{n-1}(z) & \cdots & p_{2n-3}(z) \\ \vdots & \vdots & \ddots & \vdots \\ p_{-n+2}(z) & p_{-n+3}(z) & \cdots & p_1(z) \end{vmatrix}, \tag{5.24}$$

with $p_k(z) = 0$ for $k < 0$, which is equivalent to (5.22).

Table 3. The polynomials $p_n(z)$ given by (5.21)

$p_0(z) = 1$
$p_1(z) = z$
$p_2(z) = z^2/2$
$p_3(z) = (z^3 - 8)/3!$
$p_4(z) = (z^3 - 32)z/4!$
$p_5(z) = (z^3 - 80)z^2/5!$
$p_6(z) = (z^6 - 160z^3 + 640)/6!$
$p_7(z) = (z^6 - 280z^3 + 4480)z/7!$
$p_8(z) = (z^6 - 448z^3 + 17920)z^2/8!$
$p_9(z) = (z^9 - 672z^6 + 53760z^3 - 143360)/9!$
$p_{10}(z) = (z^9 - 960z^6 + 134400z^3 - 1433600)z/10!$

The first few of the polynomials $p_n(z)$ are given in Table 3. The polynomials $p_k(z)$ defined by (5.21) satisfy third-order linear recursion relations and differential equations as follows.

Lemma 5.1. *The polynomials $p_k(z)$ defined by (5.21) satisfy the third-order recursion relation*

$$(n+3)p_{n+3} = zp_{n+2} - 4p_n, \qquad n \geq 0, \tag{5.25}$$

where $p_0 = 1$, $p_1 = z$ and $p_2 = \frac{1}{2}z^2$ and the third-order differential equation

$$4p_n''' - zp_n' + np_n = 0, \qquad n \geq 1, \tag{5.26}$$

Proof. First note that $\Phi(z, \lambda) = \exp\left(z\lambda - \frac{4}{3}\lambda^3\right)$ satisfies

$$\Phi_\lambda = (z - 4\lambda^2)\Phi.$$

Then substituting $\Phi(z, \lambda) = \sum_{n=0}^{\infty} p_n \lambda^n$ into this yields

$$\sum_{n=1}^{\infty} np_{n-1}\lambda^n = (z - 4\lambda^2) \sum_{n=0}^{\infty} p_n \lambda^n,$$

which can be rewritten as

$$\sum_{n=0}^{\infty} [(n+3)p_{n+3} - zp_{n+2} + 4p_n]\lambda^n = 0,$$

and so we obtain (5.25) since the coefficient of every power of λ is zero; see also [130]. Also Φ satisfies

$$\Phi_z = \lambda\Phi.$$

Substituting $\Phi(z,\lambda) = \sum_{n=0}^{\infty} p_n\lambda^n$ into this yields

$$\sum_{n=1}^{\infty} \left(p_n' - p_{n-1}\right)\lambda^n = 0,$$

and so $p_n'(z) = p_{n-1}(z)$, for $n \geq 1$. Substituting this into (5.25) yields (5.26). $\square$

We remark that the general solution of the third-order equation

$$4p''' - zp' + np = 0, \tag{5.27}$$

is

$$p(z) = c_1\, {}_1F_2(-\tfrac{1}{3}n; \tfrac{1}{3}, \tfrac{2}{3}; \zeta) + c_2 z\, {}_1F_2(\tfrac{1}{3}(1-n); \tfrac{2}{3}, \tfrac{4}{3}; \zeta) \\ + c_3 z^2\, {}_1F_2(\tfrac{1}{3}(2-n); \tfrac{4}{3}, \tfrac{5}{3}; \zeta), \tag{5.28}$$

where ${}_1F_2(a; b_1, b_2; \zeta)$ is a generalized hypergeometric function, with $\zeta = z^3/36$, and c_1, c_2, c_3 arbitrary constants. The generalized hypergeometric function ${}_1F_2(a; b_1, b_2; \zeta)$, with a, b_1, b_2 arbitrary constants is defined by

$${}_1F_2(a; b_1, b_2; \zeta) = \sum_{k=0}^{\infty} \frac{(a)_k}{(b_1)_k (b_2)_k} \frac{\zeta^k}{k!},$$

where

$$(a)_k = a(a+1)\ldots(a+k-1) = \frac{\Gamma(a+k)}{\Gamma(a)},$$

with $\Gamma(a)$ the Gamma function (cf. [9]). If $n \in \mathbb{Z}^+$ then one of the functions ${}_1F_2(-\frac{1}{3}n; \frac{1}{3}, \frac{2}{3}; z^3/36)$, ${}_1F_2(\frac{1}{3}(1-n); \frac{2}{3}, \frac{4}{3}; z^3/36)$ or ${}_1F_2(\frac{1}{3}(2-n); \frac{4}{3}, \frac{5}{3}; z^3/36)$ is a polynomial in z, in fact a multiple of $p_n(z)$, since ${}_1F_2(-m; b_1, b_2; \zeta)$ is a polynomial for $m \in \mathbb{Z}^+$.

These "tau-functions" $\tau_n(z)$ are related to the Yablonskii-Vorob'ev polynomials $Q_n(z)$ through

$$\tau_n(z) = c_n Q_n(z), \qquad c_n = \prod_{j=1}^{n} (2j+1)^{j-n}, \tag{5.29}$$

and satisfy the differential-difference equation

$$(2n+1)\tau_{n-1}\tau_{n+1} + (2\mathrm{D}_z^2 - z)\tau_n \bullet \tau_n = 0, \tag{5.30}$$

where D_z is the Hirota operator (5.6).

An alternative representation of rational solutions of $\mathrm{P_{II}}$ in terms of determinants is given in the following theorem.

Theorem 5.6. *Let $q_k(z)$ be the polynomial defined by*

$$q_{n+1}(z) = 2\frac{\mathrm{d}q_n}{\mathrm{d}z} - \frac{1}{2}\sum_{k=0}^{n-1} q_k(z)q_{n-k-1}(z), \qquad n \geq 1, \tag{5.31}$$

with $q_0(z) = z$ and $q_1(z) = 2$, and $\sigma_n(z)$ be the $n \times n$ Hankel determinant

$$\sigma_n(z) = \begin{vmatrix} q_0(z) & q_1(z) & \cdots & q_{n-1}(z) \\ q_1(z) & q_2(z) & \cdots & q_n(z) \\ \vdots & \vdots & \ddots & \vdots \\ q_{n-1}(z) & q_n(z) & \cdots & q_{2n-2}(z) \end{vmatrix}, \qquad n \geq 0. \tag{5.32}$$

Then

$$w(z;n) = \frac{\mathrm{d}}{\mathrm{d}z}\left\{\ln\left[\frac{\sigma_{n-1}(z)}{\sigma_n(z)}\right]\right\} = \frac{\sigma_{n-1}'(z)}{\sigma_{n-1}(z)} - \frac{\sigma_n'(z)}{\sigma_n(z)}, \qquad n \geq 1, \tag{5.33}$$

satisfies $\mathrm{P_{II}}$ *with* $\alpha = n$.

Proof. See Kajiwara and Ohta [124]; also Kajiwara and Masuda [122]. □

The generating function associated with the coefficients $q_n(z)$ in Theorem 5.6 is discussed in the following theorem.

Theorem 5.7. *Let $\theta(z,t)$ be an entire function of two variables defined by*

$$\theta(z,t) = \exp(\tfrac{2}{3}t^3)\,\mathrm{Ai}(t^2 - \tfrac{1}{2}z), \tag{5.34}$$

where $\mathrm{Ai}(\xi)$ is the Airy function. Then there exists an asymptotic expansion

$$\frac{\partial}{\partial t}\ln\theta(z,t) \sim -\tfrac{1}{2}\sum_{n=0}^{\infty} q_n(z)(-2t)^n, \qquad t \to \infty, \tag{5.35}$$

in any proper subsector of the sector $|\arg t| < \frac{1}{2}\pi$.

Proof. See Imasaki, Kajiwara and Nakamura [108]; also Joshi, Kajiwara and Mazzocco [116] who show that the coefficients $q_n(z)$ are related to the asymptotic solution at infinity of the linear problem of which the $\mathrm{P_{II}}$ describes the isomonodromic deformations. □

Additionally Imasaki, Kajiwara and Nakamura [108] show that the generating function

$$Q(z,t) = \sum_{n=0}^{\infty} q_n(z)(-2t)^n, \tag{5.36}$$

satisfies the Riccati equation

$$tQ_t - \tfrac{1}{2}tQ^2 - (4t^3+1)Q = 4t^2(zt-1). \tag{5.37}$$

Making the transformation

$$Q(z,t) = -4t^2 - 4t^2 \frac{\mathrm{d}}{\mathrm{d}\zeta} \ln u(\zeta), \qquad \zeta = t^2 - \tfrac{1}{2}z, \tag{5.38}$$

in (5.37) yields the Airy equation

$$\frac{\mathrm{d}^2 u}{\mathrm{d}\zeta^2} - \zeta u = 0. \tag{5.39}$$

It is well-known that the Airy equation (5.39) has the formal solutions

$$U_{\pm}(\zeta) = \tfrac{1}{2} z^{-1/4} \pi^{-1/2} \exp(\pm\tfrac{2}{3} z^{3/2}) \sum_{n=0}^{\infty} \frac{(\frac{1}{6})_n (\frac{5}{6})_n}{n!} \left(\pm\tfrac{4}{3} z^{3/2}\right)^{-n}, \tag{5.40}$$

and that the Airy function $\mathrm{Ai}(\zeta)$ has the asymptotic representation $\mathrm{Ai}(z) \sim U_-(\zeta)$, as $\zeta \to \infty$, in any proper subsector of the sector $|\arg z| < \pi$.

Making the transformation

$$Q(z,t) = -4t^2 - 2t \frac{\partial}{\partial t} \ln P(z,t), \tag{5.41}$$

in (5.37) yields the linear equation

$$tP_{tt} - P_t + 2t^3(z - 2t^2)P = 0, \tag{5.42}$$

which has formal solution

$$P(z,t) = \frac{1}{2\sqrt{t\pi}} \exp\left(-\tfrac{2}{3}t^3 + \tfrac{1}{2}zt\right) \exp\left\{ -\sum_{n=1}^{\infty} \frac{q_{n+1}(z)}{4n} (-2t)^n \right\}, \tag{5.43}$$

and so $P(z,t) = U_-(\zeta)$.

5.4 Rational Solutions of $\mathrm{P_{III}}$

Consider the generic case of $\mathrm{P_{III}}$ when $\gamma\delta \neq 0$, then we set $\gamma = 1$ and $\delta = -1$, without loss of generality (by rescaling w and z if necessary), and so

$$w'' = \frac{(w')^2}{w} - \frac{w'}{z} + \frac{\alpha w^2 + \beta}{z} + w^3 - \frac{1}{w}. \tag{5.44}$$

Rational solutions of this equation are classified in the following theorem.

Theorem 5.8. *Equation (5.44) has rational solutions if and only if*

$$\varepsilon_1 \alpha + \varepsilon_2 \beta = 4n, \tag{5.45}$$

with $n \in \mathbb{Z}$ and $\varepsilon_1 = \pm 1$, $\varepsilon_2 = \pm 1$, independently. Generically (except when α and β are both integers), these rational solutions have the form

$$w(z) = P_{n^2}(z)/Q_{n^2}(z),$$

where $P_{n^2}(z)$ and $Q_{n^2}(z)$ are polynomials of degree n^2 with no common roots.

Proof. See Lukashevich [153]; also Milne, Clarkson and Bassom [169] and Murata [172]. □

We remark that rational solutions of (5.44) lie on the lines $\alpha + \varepsilon\beta = 4n$ in the α-β plane. For examples and plots see Milne, Clarkson and Bassom [169]. Determinantal representations, with entries expressed in terms of associated Laguerre polynomials $L_k^m(z)$, of rational solutions of $\mathrm{P_{III}}$ are given by Kajiwara and Masuda [123].

Umemura [226], see also [123, 233], derived special polynomials associated with rational solutions of $\mathrm{P_{III}}$; though these are actually polynomials in $1/z$ rather than polynomials in z. However it is straightforward to determine special polynomials associated with rational solutions of $\mathrm{P_{III}}$ that are polynomials in z which are given in the following theorem.

Theorem 5.9. *Suppose that $S_n(z;\mu)$ satisfies the recursion relation*

$$S_{n+1}S_{n-1} = -z\left[S_n\frac{\mathrm{d}^2S_n}{\mathrm{d}z^2} - \left(\frac{\mathrm{d}S_n}{\mathrm{d}z}\right)^2\right] - S_n\frac{\mathrm{d}S_n}{\mathrm{d}z} + (z+\mu)S_n^2, \tag{5.46}$$

with $S_{-1}(z;\mu) = S_0(z;\mu) = 1$. Then

$$\begin{aligned} w_n = w(z;\alpha_n,\beta_n,1,-1) &= 1 + \frac{\mathrm{d}}{\mathrm{d}z}\left\{\ln\left[\frac{S_{n-1}(z;\mu-1)}{S_n(z;\mu)}\right]\right\} \\ &\equiv \frac{S_n(z;\mu-1)\,S_{n-1}(z;\mu)}{S_n(z;\mu)\,S_{n-1}(z;\mu-1)}, \end{aligned} \tag{5.47}$$

satisfies $\mathrm{P_{III}}$ with $\alpha_n = 2n+2\mu-1$ and $\beta_n = 2n-2\mu+1$ and

$$\begin{aligned} \widehat{w}_n = w(z;\widehat{\alpha}_n,\widehat{\beta}_n,1,-1) &= 1 + \frac{\mathrm{d}}{\mathrm{d}z}\left\{\ln\left[\frac{S_{n-1}(z;\mu)}{S_n(z;\mu-1)}\right]\right\} \\ &\equiv \frac{S_n(z;\mu)\,S_{n-1}(z;\mu-1)}{S_n(z;\mu-1)\,S_{n-1}(z;\mu)}, \end{aligned} \tag{5.48}$$

satisfies $\mathrm{P_{III}}$ with $\widehat{\alpha}_n = -2n+2\mu-1$ and $\widehat{\beta}_n = -2n-2\mu+1$.

Proof. See Clarkson [31], which generalizes the work of Kajiwara and Masuda [123]. □

The rational solutions of $\mathrm{P_{III}}$ defined by (5.47) and (5.48) can be generalized using the Bäcklund transformation (4.8) to include all those described in Theorem 5.8 satisfying the condition $\alpha + \beta = 4n$. Rational solutions of $\mathrm{P_{III}}$ satisfying the condition $\alpha - \beta = 4n$ are obtained by letting $w \to \mathrm{i}w$ and $z \to \mathrm{i}z$ in (5.47) and (5.48), and then using the Bäcklund transformation (4.8). Plots of the roots of the polynomials $S_n(z;\mu)$ for various μ are given in [31]. Initially for μ sufficiently large and negative, the $\frac{1}{2}n(n+1)$ roots form an approximate triangle with n roots on each side. Then as μ increases, the roots in turn coalesce and eventually for μ sufficiently large and positive they form another approximate triangle, similar to the original triangle, though with its orientation reversed. It is straightforward to determine when the roots of $S_n(z;\mu)$ coalesce using discriminants of polynomials. Suppose that $f(z) = z^m + a_{m-1}z^{m-1} + \ldots + a_1 z + a_0$ is a monic polynomial of degree m with roots $\alpha_1, \alpha_2, \ldots, \alpha_m$, so $f(z) = \prod_{j=1}^{m}(z - \alpha_j)$. Then the *discriminant* of $f(z)$ is

$$\mathrm{Dis}(f) = \prod_{1\leq j<k\leq m} (\alpha_j - \alpha_k)^2.$$

Hence the polynomial f has a multiple root when $\mathrm{Disc}(f) = 0$. It is straightforward to show that

$$\begin{aligned}
\mathrm{Dis}(S_3(z;\mu)) &= 3^{12}5^5\mu^6(\mu^2-1)^2\\
\mathrm{Dis}(S_4(z;\mu)) &= 3^{27}5^{20}7^7\mu^{14}(\mu^2-1)^6(\mu^2-4)^2\\
\mathrm{Dis}(S_5(z;\mu)) &= 3^{66}5^{45}7^{28}\mu^{26}(\mu^2-1)^{14}(\mu^2-4)^6(\mu^2-9)^2
\end{aligned}$$

Thus $S_3(z;\mu)$ has multiple roots when $\mu = 0, \pm1$, $S_4(z;\mu)$ has multiple roots when $\mu = 0, \pm1, \pm2$, and $S_5(z;\mu)$ has multiple roots when $\mu = 0, \pm1, \pm2, \pm3$. Further the multiple roots occur at $z = 0$. Hence it is natural to conjecture that $S_n(z;\mu)$ has multiple roots at $z = 0$ when $\mu = 0, \pm1, \pm2, \ldots, \pm(n-2)$.

Using the Hamiltonian formalism for $\mathrm{P_{III}}$, see §3, it is shown in [31] that the polynomials $S_n(z;\mu)$ satisfy an fourth order bilinear equation and a sixth order, hexa-linear (homogeneous of degree six) difference equation.

5.5 Rational Solutions of $\mathrm{P_{IV}}$

Simple rational solutions of $\mathrm{P_{IV}}$ are

$$w_1(z,\pm2,-2) = \pm1/z, \quad w_2(z,0,-2) = -2z, \quad w_3(z,0,-\tfrac{2}{9}) = -\tfrac{2}{3}z. \tag{5.49}$$

There are three sets of rational solutions

$$w_1(z;\alpha_1,\beta_1) = P_{1,n-1}(z)/Q_{1,n}(z), \tag{5.50a}$$
$$w_2(z;\alpha_2,\beta_2) = -2z + P_{2,n-1}(z)/Q_{2,n}(z), \tag{5.50b}$$
$$w_3(z;\alpha_3,\beta_3) = -\tfrac{2}{3}z + P_{3,n-1}(z)/Q_{3,n}(z), \tag{5.50c}$$

where $P_{j,n-1}(z)$ and $Q_{j,n}(z)$ are polynomials of degrees $n-1$ and n, respectively, with no common roots, which have the solutions (5.49) as the simplest members. These sets are known as the "$-1/z$ hierarchy", the "$-2z$ hierarchy" and the "$-\frac{2}{3}z$ hierarchy", respectively (cf. [17]).

Theorem 5.10. *In general,* $\mathrm{P_{IV}}$ *has rational solutions if and only if either*

$$\alpha = m, \qquad \beta = -2(2n-m+1)^2, \tag{5.51}$$

or

$$\alpha = m, \qquad \beta = -2(2n-m+\tfrac{1}{3})^2, \tag{5.52}$$

with $m, n \in \mathbb{Z}$.

Proof. See Lukashevich [152], Gromak [76], Murata [171]; also Gromak, Laine and Shimomura [83, §26]. □

We remark that the rational solutions when the parameters satisfy (5.51) are special cases of special function solutions given in Theorem 7.5.

The "$-1/z$ hierarchy" and the "$-2z$ hierarchy" form the set of rational solutions of $\mathrm{P_{IV}}$ with parameters given by (5.51) and the "$-\frac{2}{3}z$ hierarchy" forms the set with parameters given by (5.52). The rational solutions of $\mathrm{P_{IV}}$ with parameters given by (5.51) lie at the vertices of the "Weyl chambers" and those with parameters given by (5.52) lie at the centres of the "Weyl chamber" [227]. These are summarized in Figure 2 which depicts the $(\alpha, \sqrt{-2\beta})$ plane, where α and β are the parameters in $\mathrm{P_{IV}}$. The dots at the vertices denote the rational solutions of $\mathrm{P_{IV}}$ with parameter values given by (5.51) and the dots in the triangles denote the rational solutions with parameter values given by (5.52). On the horizontal and diagonal lines, $\mathrm{P_{IV}}$ possesses special function solutions which are expressible in terms of Whittaker functions $M_{\kappa,\mu}(\xi)$ and $W_{\kappa,\mu}(\xi)$, or equivalently parabolic cylinder functions $D_\nu(\xi)$; see §7.3.

For examples and plots see Bassom, Clarkson and Hicks [17]; also Clarkson [32]. Determinantal representations, with entries expressed in terms of Hermite polynomials $H_n(z)$, of rational solutions of $\mathrm{P_{IV}}$ are given by Forrester and Witte [61], Kajiwara and Ohta [125] and Noumi and Yamada [176].

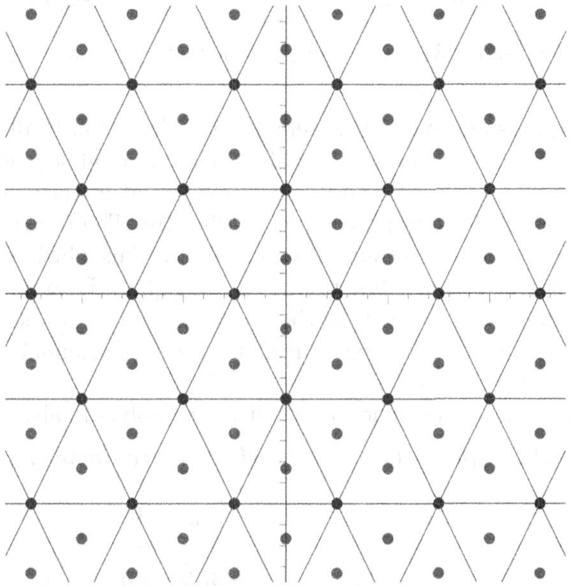

Fig. 2. Rational solutions of $\mathrm{P_{IV}}$ with parameters (5.51) at the vertices and (5.52) in the triangles. The lines denote special function solutions

In a comprehensive study of $\mathrm{P_{IV}}$, Okamoto [184] defined two sets of polynomials associated with rational solutions of $\mathrm{P_{IV}}$, analogous to the Yablonskii-Vorob'ev polynomials. Noumi and Yamada [176] generalized Okamoto's results and introduced the *generalized Hermite polynomials* $H_{m,n}(z)$, defined in Theorem 5.11, and the *generalized Okamoto polynomials* $Q_{m,n}(z)$, defined in Theorem 5.12; see also [32].

Theorem 5.11. *Suppose $H_{m,n}(z)$ satisfies the recurrence relations*

$$2mH_{m+1,n}H_{m-1,n} = H_{m,n}H''_{m,n} - \left(H'_{m,n}\right)^2 + 2mH^2_{m,n}, \tag{5.53a}$$

$$2nH_{m,n+1}H_{m,n-1} = -H_{m,n}H''_{m,n} + \left(H'_{m,n}\right)^2 + 2nH^2_{m,n}, \tag{5.53b}$$

with $H_{0,0} = H_{1,0} = H_{0,1} = 1$ and $H_{1,1} = 2z$, then

$$w^{(i)}_{m,n} = \frac{\mathrm{d}}{\mathrm{d}z}\left\{\ln\left(\frac{H_{m+1,n}}{H_{m,n}}\right)\right\}, \tag{5.54a}$$

$$w^{(ii)}_{m,n} = \frac{\mathrm{d}}{\mathrm{d}z}\left\{\ln\left(\frac{H_{m,n}}{H_{m,n+1}}\right)\right\}, \tag{5.54b}$$

$$w^{(iii)}_{m,n} = -2z + \frac{\mathrm{d}}{\mathrm{d}z}\left\{\ln\left(\frac{H_{m,n+1}}{H_{m+1,n}}\right)\right\}, \tag{5.54c}$$

where $w^{(j)}_{m,n} = w(z;\alpha^{(j)}_{m,n},\beta^{(j)}_{m,n})$, for $j = i, ii, iii$, are solutions of $\mathrm{P_{IV}}$, respectively, for the parameters $(\alpha^{(i)}_{m,n},\beta^{(i)}_{m,n}) = (2m+n+1, -2n^2)$, $(\alpha^{(ii)}_{m,n},\beta^{(ii)}_{m,n}) = (-(m+2n+1), -2m^2)$, and $(\alpha^{(iii)}_{m,n},\beta^{(iii)}_{m,n}) = (n-m, -2(m+n+1)^2)$.

Proof. See Clarkson [31, 121], which generalizes the work of Kajiwara and Masuda [123]. □

The rational solutions of $\mathrm{P_{IV}}$ defined by (5.54) include all solutions in the "$-1/z$" and "$-2z$" hierarchies, i.e. the set of rational solutions of $\mathrm{P_{IV}}$ with parameters given by (5.51), and can be expressed in terms of determinants whose entries are Hermite polynomials [125, 176]. These rational solutions of $\mathrm{P_{IV}}$ are special cases of the special function solutions which are expressible in terms of parabolic cylinder functions $D_\nu(\xi)$ (cf. [32]). Examples of generalized Hermite polynomials and plots of the locations of their roots are given in [32]. A plot of the roots of the generalized Hermite polynomial $H_{7,7}$ is given in Figure 3.

Next we consider the generalized Okamoto polynomials.

Theorem 5.12. *Suppose $Q_{m,n}(z)$ satisfies the recurrence relations*

$$\begin{aligned} Q_{m+1,n}Q_{m-1,n} &= \tfrac{9}{2}\left[Q_{m,n}Q''_{m,n} - \left(Q'_{m,n}\right)^2\right] \\ &\quad + \left[2z^2 + 3(2m+n-1)\right]Q^2_{m,n}, \end{aligned} \tag{5.55a}$$

$$\begin{aligned} Q_{m,n+1}Q_{m,n-1} &= \tfrac{9}{2}\left[Q_{m,n}Q''_{m,n} - \left(Q'_{m,n}\right)^2\right] \\ &\quad + \left[2z^2 + 3(1-m-2n)\right]Q^2_{m,n}, \end{aligned} \tag{5.55b}$$

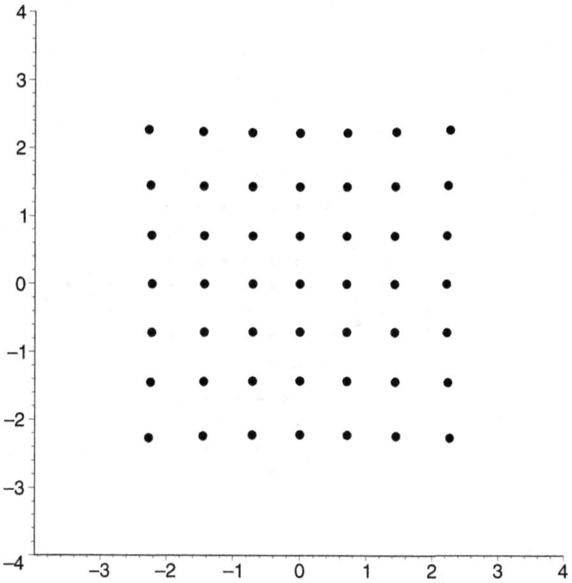

Fig. 3. Roots of the generalized Hermite polynomial $H_{7,7}(z)$

with $Q_{0,0} = Q_{1,0} = Q_{0,1} = 1$ *and* $Q_{1,1} = \sqrt{2}\,z$, *then*

$$\widetilde{w}^{(i)}_{m,n} = -\tfrac{2}{3}z + \frac{\mathrm{d}}{\mathrm{d}z}\left\{\ln\left(\frac{Q_{m+1,n}}{Q_{m,n}}\right)\right\}, \tag{5.56a}$$

$$\widetilde{w}^{(ii)}_{m,n} = -\tfrac{2}{3}z + \frac{\mathrm{d}}{\mathrm{d}z}\left\{\ln\left(\frac{Q_{m,n}}{Q_{m,n+1}}\right)\right\}, \tag{5.56b}$$

$$\widetilde{w}^{(iii)}_{m,n} = -\tfrac{2}{3}z + \frac{\mathrm{d}}{\mathrm{d}z}\left\{\ln\left(\frac{Q_{m,n+1}}{Q_{m+1,n}}\right)\right\}, \tag{5.56c}$$

where $\widetilde{w}^{(j)}_{m,n} = w(z;\widetilde{\alpha}^{(j)}_{m,n},\widetilde{\beta}^{(j)}_{m,n})$, *for* $j = i, ii, iii$, *are solutions of* $\mathrm{P_{IV}}$, *respectively, for the parameters* $(\alpha^{(i)}_{m,n},\beta^{(i)}_{m,n}) = (-(m+2n), -2(m-\frac{1}{3})^2)$, $(\alpha^{(ii)}_{m,n},\beta^{(ii)}_{m,n}) = (2m+n, -2(n-\frac{1}{3})^2)$, *and* $(\alpha^{(iii)}_{m,n},\beta^{(iii)}_{m,n}) = (n-m, -2(m+n+\frac{1}{3})^2)$.

Proof. See Noumi and Yamada [176]; also Clarkson [32]. □

The rational solutions of $\mathrm{P_{IV}}$ defined by (5.56) include all solutions in the "$-\frac{2}{3}z$" hierarchy, i.e., the set of rational solutions of $\mathrm{P_{IV}}$ with parameters given by (5.52), which also can be expressed in the form of determinants [125, 176]. Examples of generalized Okamoto polynomials and plots of the locations of their roots are given in [32]. A plot of the roots of the generalized Okamoto polynomial $Q_{7,7}$ is given in Figure 4.

Using the Hamiltonian formalism for $\mathrm{P_{IV}}$, see §3, it is shown in [35] that the generalized Hermite polynomials $H_{m,n}(z)$ and generalized Okamoto polynomials $Q_{m,n}(z)$, satisfy fourth order bilinear equations and homogeneous difference equations.

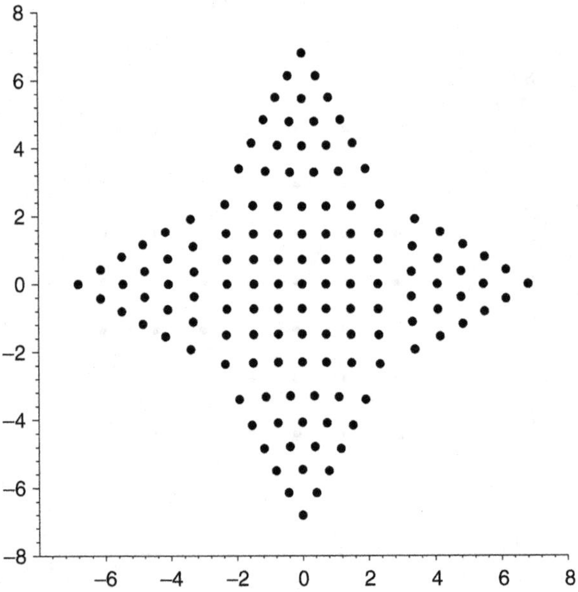

Fig. 4. Roots of the generalized Okamoto polynomial $Q_{7,7}(z)$

5.6 Rational Solutions of $\mathrm{P_V}$

Some simple rational solutions of $\mathrm{P_V}$ are

$$\begin{aligned} w(z; \tfrac{1}{2}, -\tfrac{1}{2}\mu^2, \kappa(2-\mu), -\tfrac{1}{2}\kappa^2) &= \kappa z + \mu, \\ w(z; \tfrac{1}{2}, \kappa^2\mu, 2\kappa\mu, \mu) &= \frac{\kappa}{z+\kappa}, \\ w(z; \tfrac{1}{8}, -\tfrac{1}{8}, -\kappa\mu, \mu) &= \frac{\kappa+z}{\kappa-z}, \end{aligned}$$

with κ and μ arbitrary constants.

Rational solutions of $\mathrm{P_V}$, with $\delta \neq 0$ when we set $\delta = -\frac{1}{2}$ without loss of generality, are classified in the following theorem.

Theorem 5.13. $\mathrm{P_V}$, *with* $\delta = -\frac{1}{2}$,

$$\begin{aligned} w'' = {} & \left(\frac{1}{2w} + \frac{1}{w-1}\right)(w')^2 - \frac{w'}{z} + \frac{(w-1)^2}{z^2}\left(\alpha w + \frac{\beta}{w}\right) \\ & + \frac{\gamma w}{z} - \frac{w(w+1)}{2(w-1)}, \end{aligned} \tag{5.57}$$

has a rational solution if and only if one of the following holds with $m, n \in \mathbb{Z}$ *and* $\varepsilon = \pm 1$.

(i) $\alpha = \frac{1}{2}(m+\varepsilon)^2$ *and* $\beta = -\frac{1}{2}n^2$, *where* $n > 0$, $m+n$ *is odd, and* $\alpha \neq 0$ *when* $|m| < n$,

(ii) $\alpha = \frac{1}{2}n^2$ *and* $\beta = -\frac{1}{2}(m+\varepsilon)^2$, *where* $n > 0$, $m+n$ *is odd, and* $\beta \neq 0$ *when* $|m| < n$,

(iii) $\alpha = \frac{1}{2}a^2$, $\beta = -\frac{1}{2}(a+n)^2$ *and* $\gamma = m$, *where* $m+n$ *is even and* a *arbitrary,*

(iv) $\alpha = \frac{1}{2}(b+n)^2$, $\beta = -\frac{1}{2}b^2$ *and* $\gamma = m$, *where* $m+n$ *is even and* b *arbitrary,*

(v) $\alpha = \frac{1}{8}(2m+1)^2$ *and* $\beta = -\frac{1}{8}(2n+1)^2$.

These rational solutions have the form

$$w(z) = \lambda z + \mu + P_{n-1}(z)/Q_n(z), \tag{5.58}$$

where λ *and* μ *are constants, and* $P_{n-1}(z)$ *and* $Q_n(z)$ *are polynomials of degree* $n-1$ *and* n, *respectively, with no common roots.*

Proof. See Kitaev, Law and McLeod [142]; also Gromak and Lukashevich [84], Gromak, Laine and Shimomura [83, §40]. □

Remarks

1. The rational solutions in cases (i) and (ii) are the special cases of the solutions of P_V expressible in terms of confluent hypergeometric functions ${}_1F_1(a;c;z)$, when the confluent hypergeometric function reduces to the associated Laguerre polynomial $L_k^m(\zeta)$ discussed in Theorem 7.6; see also Masuda [162].
2. The rational solutions in cases (i) and (ii) and those in cases (iii) and (iv) are related by the symmetry $\mathcal{S}_2$ (4.12).
3. Kitaev, Law & McLeod [142] did not explicitly give case (iv), though it is obvious by applying the symmetry $\mathcal{S}_2$ to case (iii).
4. Kitaev, Law & McLeod [142] also require that $\gamma \notin \mathbb{Z}$ in case (v), though this appears not to be necessary.
5. If $\delta = 0$ and $\gamma \neq 0$ in P_V then it is equivalent to the generic case if P_{III} [72]; see Theorem 6.3. Hence one obtains algebraic solutions of this special case of P_V in terms rational solutions of P_{III}, as shown in §6.2.

Determinantal representations, with entries expressed in terms of associated Laguerre polynomials $L_k^m(z)$, of rational solutions of P_V are given by Masuda, Ohta and Kajiwara [164]. Umemura [226] derived special polynomials associated with rational solutions of (5.57); see also [175, 233]. Masuda, Ohta and Kajiwara [164] generalized Umemura's result and derived special polynomials associated with all rational solutions in cases (iii), (iv), and (v) in Kitaev, Law and McLeod's classification described in Theorem 5.13 which are given in the following theorem.

Theorem 5.14. *Suppose that* $U_{m,n}(z;\mu)$ *satisfies the recursion relations*

$$U_{m+1,n}U_{m-1,n} = 8z\left[U_{m,n}U''_{m,n} - \left(U'_{m,n}\right)^2\right] + 8U_{m,n}U'_{m,n} + (z+2\mu-2-6m+2n)U^2_{m,n}, \quad (5.59a)$$

$$U_{m,n+1}U_{m,n-1} = 8z\left[U_{m,n}U''_{m,n} - \left(U'_{m,n}\right)^2\right] + 8U_{m,n}U'_{m,n} + (z-2\mu-2+2m-6n)U^2_{m,n}, \quad (5.59b)$$

with

$$U_{-1,-1}(z;\mu) = U_{-1,0}(z;\mu) = U_{0,-1}(z;\mu) = U_{0,0}(z;\mu) = 1. \quad (5.59c)$$

Then

$$w(z;\alpha^{(\mathrm{iii})}_{m,n},\beta^{(\mathrm{iii})}_{m,n},\gamma^{(\mathrm{iii})}_{m,n},\delta^{(\mathrm{iii})}_{m,n}) = -\frac{U_{m,n-1}(z;\mu)U_{m-1,n}(z;\mu)}{U_{m-1,n}(z;\mu-2)U_{m,n-1}(z;\mu+2)}, \quad (5.60a)$$

is a rational solution of $\mathrm{P_V}$ *for the parameters*

$$\left(\alpha^{(\mathrm{iii})}_{m,n},\beta^{(\mathrm{iii})}_{m,n},\gamma^{(\mathrm{iii})}_{m,n},\delta^{(\mathrm{iii})}_{m,n}\right) = \left(\tfrac{1}{8}\mu^2, -\tfrac{1}{8}(\mu-2m+2n)^2, -m-n, -\tfrac{1}{2}\right), \quad (5.60b)$$

and

$$w(z;\alpha^{(\mathrm{v})}_{m,n},\beta^{(\mathrm{v})}_{m,n},\gamma^{(\mathrm{v})}_{m,n},\delta^{(\mathrm{v})}_{m,n}) = -\frac{U_{m,n-1}(z;\mu+1)U_{m,n+1}(z;\mu-1)}{U_{m-1,n}(z;\mu-1)U_{m+1,n}(z;\mu+1)}, \quad (5.61a)$$

is a rational solution of $\mathrm{P_V}$ *for the parameters*

$$\left(\alpha^{(\mathrm{v})}_{m,n},\beta^{(\mathrm{v})}_{m,n},\gamma^{(\mathrm{v})}_{m,n},\delta^{(\mathrm{v})}_{m,n}\right) = \left(\tfrac{1}{8}(2m+1)^2, -\tfrac{1}{8}(2n+1)^2, m-n-\mu, -\tfrac{1}{2}\right). \quad (5.61b)$$

Proof. See Masuda, Ohta and Kajiwara [164]; also Clarkson [34]. □

Plots of the roots of these special polynomials are given by Clarkson [34]. These numerical simulations show that for μ sufficiently large and negative, the roots of $U_{m,n}(z;\mu)$ form two approximate triangles, one with $\frac{1}{2}m(m+1)$ roots and the other with $\frac{1}{2}n(n+1)$ roots, which are well separated. The discriminant of $U_{m,n}(z;\mu)$, i.e., $\mathrm{Dis}(U_{m,n}(z;\mu))$, is zero when $\mu = 2m-1, 2m-3, \ldots, 3, 1, -1, -3, \ldots, 3-2n, 1-2n$. As μ increases, some roots of $U_{m,n}(z;\mu)$ coalesce at $z=0$ when μ is a root of $\mathrm{Dis}(U_{m,n}(z;\mu)) = 0$. Finally for μ sufficiently large and positive, the roots of $U_{m,n}(z;\mu)$ again form two approximate triangles, one with $\frac{1}{2}m(m+1)$ roots and the other with $\frac{1}{2}n(n+1)$ roots, similar to those for μ negative, except that their positions are interchanged. The motion of the roots as μ varies is symmetric about $\mu = m-n$ and always there is symmetry about the real axis.

5.7 Rational Solutions of P_{VI}

Rational solutions of P_{VI} are summarized in the following theorem.

Theorem 5.15. *In the general case,* P_{VI} *has rational solutions if and only if*

$$a + b + c + d = 2n + 1, \tag{5.62}$$

where $n \in \mathbb{Z}$, $a = \varepsilon_1\sqrt{2\alpha}$, $b = \varepsilon_2\sqrt{-2\beta}$, $c = \varepsilon_3\sqrt{2\gamma}$ *and* $d = \varepsilon_4\sqrt{1-2\delta}$, *with* $\varepsilon_j = \pm 1$, $j = 1, 2, 3, 4$, *independently, and at least one of* a, b, c *or* d *is an integer.*

Proof. See Mazzocco [165]. These are special cases of the special function solutions which are expressed in terms of hypergeometric functions $F(a, b; c; z) \equiv {}_2F_1(a, b; c; z)$ when they reduce to the Jacobi polynomials $P_n^{(\alpha,\beta)}(z)$. □

6 Other Elementary Solutions

6.1 Elementary Solutions of P_{III}

It is straightforward to show that P_{III} possesses the algebraic solution $w = z^{1/3}$ provided that $\alpha + \delta = 0$ and $\beta = \gamma = 0$ and the solution $w = z^{-1/3}$ provided that $\alpha = \delta = 0$ and $\beta + \gamma = 0$.

Consider the special case of P_{III} when either (i), $\gamma = 0$ and $\alpha\delta \neq 0$, or (ii), $\delta = 0$ and $\beta\gamma \neq 0$. In case (i), we make the transformation

$$w(z) = (\tfrac{2}{3})^{1/2} u(\zeta), \qquad z = (\tfrac{2}{3})^{3/2} \zeta^3,$$

and set $\alpha = 1$, $\beta = 2\mu$ and $\delta = -1$, with μ an arbitrary constant, without loss of generality, which yields $P_{III}^{(7)}$

$$\frac{\mathrm{d}^2 u}{\mathrm{d}\zeta^2} = \frac{1}{u}\left(\frac{\mathrm{d}u}{\mathrm{d}\zeta}\right)^2 - \frac{1}{\zeta}\frac{\mathrm{d}u}{\mathrm{d}\zeta} + 4\zeta u^2 + 12\mu\zeta - \frac{4\zeta^4}{u}, \tag{6.1}$$

In case (ii), we make the transformation $w(z) = (\tfrac{3}{2})^{1/2}/u(\zeta)$, with $z = (\tfrac{2}{3})^{3/2}\zeta^3$ and set $\alpha = 2\mu$, $\beta = -1$ and $\gamma = 1$, with μ an arbitrary constant, without loss of generality, which again yields $P_{III}^{(7)}$.

Equation $P_{III}^{(7)}$ (6.1) is so named since it is of type D_7 in the terminology of Sakai [197]. Rational solutions of $P_{III}^{(7)}$ correspond to algebraic solutions of P_{III} with $\gamma = 0$ and $\alpha\delta \neq 0$, or $\delta = 0$ and $\beta\gamma \neq 0$. Lukashevich [151] and Gromak [71] obtained algebraic solutions of $P_{III}^{(7)}$, which are classified in the following theorem; see also [169].

Theorem 6.1. *Equation* $\mathrm{P}_{\mathrm{III}}^{(7)}$ *(6.1) has rational solutions if and only if* $\mu = n$, *with* $n \in \mathbb{Z}$, *which have the form*

$$u_n(\zeta) = P_{n^2+1}(\zeta)/Q_{n^2}(\zeta),$$

where $P_{n^2+1}(\zeta)$ *and* $Q_{n^2}(\zeta)$ *are monic polynomials of degrees* n^2+1 *and* n^2, *respectively, with integer coefficients and no common roots.*

A straightforward method for generating rational solutions of (6.1) is through the Bäcklund transformation

$$u_{\mu\pm1} = \frac{\zeta^3}{u_\mu^2} \pm \frac{\zeta}{2u_\mu^2}\frac{\mathrm{d}u_\mu}{\mathrm{d}\zeta} - \frac{3(2\mu\pm1)}{2u_\mu}, \tag{6.2}$$

where u_μ is the solution of (6.1) for parameter μ, using the "seed solution" $u_0(\zeta) = \zeta$ for $\mu = 0$, for details see Gromak, Laine and Shimomura [83, p164]; also [71, 169, 172].

Ohyama [181] derived special polynomials associated with the rational solutions of (6.1). These are essentially described in Theorem 6.2, though here the variables have been scaled and the expression of the rational solutions of (6.1) in terms of these special polynomials is explicitly given.

Theorem 6.2. *Suppose that* $R_n(\zeta)$ *satisfies the recursion relation*

$$2\zeta R_{n+1}R_{n-1} = -R_n\frac{\mathrm{d}^2R_n}{\mathrm{d}\zeta^2} + \left(\frac{\mathrm{d}R_n}{\mathrm{d}\zeta}\right)^2 - \frac{R_n}{\zeta}\frac{\mathrm{d}R_n}{\mathrm{d}\zeta} + 2(\zeta^2-n)R_n^2, \tag{6.3}$$

with $R_0(\zeta)=1$ *and* $R_1(\zeta)=\zeta^2$. *Then*

$$u_n(\zeta) = \frac{R_{n+1}(\zeta)\,R_{n-1}(\zeta)}{R_n^2(\zeta)} \equiv \frac{\zeta^2-n}{\zeta} - \frac{1}{2\zeta^2}\frac{\mathrm{d}}{\mathrm{d}\zeta}\left\{\zeta\frac{\mathrm{d}}{\mathrm{d}\zeta}\ln R_n(\zeta)\right\}, \tag{6.4}$$

satisfies (6.1) *with* $\mu = n$. *Additionally* $u_{-n}(\zeta) = -\mathrm{i}u_n(\mathrm{i}\zeta)$.

Plots of the locations of the roots of the polynomials $R_n(\zeta)$ are given in [31]. These plots show that the locations of the poles also have a very symmetric, regular structure and take the form of two "triangles" in a "bow-tie" shape. A plot of the roots of $R_{20}(\zeta)$ is given in Figure 5.

Other non-rational solutions of $\mathrm{P_{III}}$ include

$$w(z;0,-2\kappa,0,4\kappa\mu-\lambda^2) = z(\kappa\ln^2 z + \lambda\ln z + \mu),$$

$$w(z;-\nu^2\lambda,0,\nu^2(\lambda^2-4\kappa\mu),0) = \frac{z^{\nu-1}}{\kappa z^{2\nu}+\lambda z^\nu+\mu},$$

with κ, λ, μ and ν arbitrary constants [151, 153].

If $\beta=\delta=0$, then $\mathrm{P_{III}}$ has the first integral

$$z^2(w')^2 + 2zww' = (C + 2\alpha zw + \gamma z^2w^2)w^2,$$

where C is an arbitrary constant, which is solvable by quadrature. A similar result holds when $\alpha=\gamma=0$. Also if $\alpha=\beta=\gamma=\delta=0$, then $\mathrm{P_{III}}$ has general solution $w(z) = Cz^\mu$, with C and μ arbitrary constants.

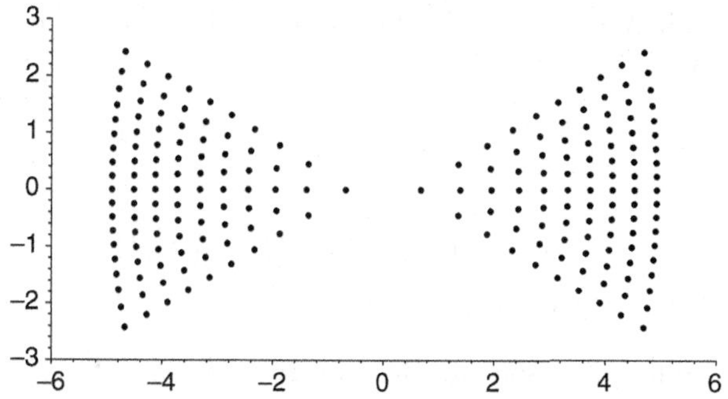

Fig. 5. Roots of the polynomial $R_{20}(\zeta)$

6.2 Elementary Solutions of $\mathrm{P_V}$

Elementary non-rational solutions of $\mathrm{P_V}$ include

$$w(z;\mu,-\tfrac{1}{8},-\mu\kappa^2,0)=1+\kappa\sqrt{z},\qquad w(z;0,0,\mu,-\tfrac{1}{2}\mu^2)=\kappa\exp(\mu z),$$

with κ and μ arbitrary constants [151].

If $\delta=0$ and $\gamma\neq 0$, then $\mathrm{P_V}$ is equivalent to the generic case if $\mathrm{P_{III}}$, which is summarized in the following theorem.

Theorem 6.3. *Suppose that* $v=v(\zeta;a,b,1,-1)$ *is a solution of* $\mathrm{P_{III}}$ *and*

$$\eta(\zeta)=\frac{\mathrm{d}v}{\mathrm{d}\zeta}-\varepsilon v^2+\frac{(1-\varepsilon a)v}{\zeta},\tag{6.5}$$

with $\varepsilon^2=1$. *Then*

$$w(z;\alpha,\beta,\gamma,\delta)=\frac{\eta(\zeta)-1}{\eta(\zeta)+1},\qquad z=\tfrac{1}{2}\zeta^2,\tag{6.6}$$

satisfies $\mathrm{P_V}$ *with*

$$(\alpha,\beta,\gamma,\delta)=((b-\varepsilon a+2)^2/32,-(b+\varepsilon a-2)^2/32,-\varepsilon,0).\tag{6.7}$$

Proof. See Airault [11], Gromak [72] and Lukashevich [154]. □

Making the change of variables $w(z)=u(\zeta)$, with $z=\frac{1}{2}\zeta^2$, in $\mathrm{P_V}$ with $\delta=0$ yields

$$\frac{\mathrm{d}^2u}{\mathrm{d}\zeta^2}=\left(\frac{1}{2u}+\frac{1}{u-1}\right)\left(\frac{\mathrm{d}u}{\mathrm{d}\zeta}\right)^2-\frac{1}{\zeta}\frac{\mathrm{d}u}{\mathrm{d}\zeta}+\frac{4(u-1)^2}{\zeta^2}\left(\alpha u+\frac{\beta}{u}\right)+2\gamma u.\tag{6.8}$$

Algebraic solutions of $\mathrm{P_V}$ with $\delta=0$ and $\gamma\neq 0$ are equivalent to rational solutions of (6.8) and so henceforth we shall only discuss rational solutions of (6.8). These are obtained by substituting the rational solutions of $\mathrm{P_{III}}$, which are classified in Theorem 5.8, into (6.5) and (6.6). Consequently we have the following classification of rational solutions for equation (6.8).

Theorem 6.4. *Necessary and sufficient conditions for the existence of rational solutions of (6.8) are either*

$$(\alpha, \beta, \gamma) = (\tfrac{1}{2}\mu^2, -\tfrac{1}{8}(2n-1)^2, -1), \tag{6.9}$$

or

$$(\alpha, \beta, \gamma) = (\tfrac{1}{8}(2n-1)^2, -\tfrac{1}{2}\mu^2, 1), \tag{6.10}$$

where $n \in \mathbb{Z}$ *and* μ *is arbitrary.*

Proof. See Milne, Clarkson and Bassom [169] and Murata [172]; also Gromak, Laine and Shimomura [83, §38]. □

We remark that the solutions of (6.8) satisfying (6.9) are related to those satisfying (6.10) by the transformation $\mathcal{S}_2$ given in (4.12). Thus we shall be concerned only with rational solutions of (6.8) satisfying (6.9). As shown above, there are special polynomials associated with the rational solutions of $\mathrm{P_{III}}$ given in Theorem 5.8. Therefore rational solutions of (6.8) are obtained by substituting the rational solutions of $\mathrm{P_{III}}$ given by (5.47) into (6.5,6.6). Hence, when $\varepsilon = 1$, rational solutions of (6.8) have the form

$$u_n(\zeta;\mu) = \frac{\zeta v_n'(\zeta;\mu) - \zeta v_n^2(\zeta;\mu) - 2(n+\mu)v_n(\zeta;\mu) - \zeta}{\zeta v_n'(\zeta;\mu) - \zeta v_n^2(\zeta;\mu) - 2(n+\mu)v_n(\zeta;\mu) + \zeta},$$

with $v_n(\zeta;\mu)$ given by (5.47). Consequently we obtain the following result.

Theorem 6.5. *Suppose that* $S_n(\zeta;\mu)$ *satisfies the recursion relation (5.46) with* $S_{-1}(\zeta;\mu) = S_0(\zeta;\mu) = 1$. *Then, for* $n \geq 1$, *the rational solution*

$$u_n(\zeta;\mu) = \frac{S_n(\zeta;\mu)S_{n-2}(\zeta;\mu)}{\mu S_{n-1}(\zeta;\mu+1)S_{n-1}(\zeta;\mu-1)}, \tag{6.11}$$

satisfies (6.8) with parameters given by (6.9).

It is straightforward to any specific value of n that (6.11) satisfies (6.8) with parameters given by (6.9). However, at present, Theorem 6.5 should be regarded as a conjecture rather than a theorem since there is no proof.

If $\gamma = \delta = 0$, then $\mathrm{P_V}$ has the first integral

$$z^2(w')^2 = (w-1)^2(2\alpha w^2 + Cw - 2\beta),$$

where C is an arbitrary constant, which is solvable by quadrature.

6.3 Elementary Solutions of $\mathrm{P_{VI}}$

An elementary algebraic solution of $\mathrm{P_{VI}}$ is

$$w(z; \tfrac{1}{2}\kappa^2, -\tfrac{1}{2}\kappa^2, \tfrac{1}{2}\mu^2, \tfrac{1}{2}(1-\mu^2)) = z^{1/2}, \tag{6.12}$$

with κ and μ arbitrary constants, which was given by Hitchin [98].

Dubrovin and Mazzocco [51] classify all algebraic solutions for the special case of $\mathrm{P_{VI}}$ with $\beta=\gamma=0$, $\delta=\frac{1}{2}$; this classification procedure can be generalized for any $\mathrm{P_{VI}}$ equation (see [165]). For further examples of algebraic solutions of $\mathrm{P_{VI}}$ see Andreev and Kitaev [14], Boalch [20, 21], Gromak, Laine and Shimomura [83, §48], Hitchin [99], Masuda [162], Mazzocco [166].

Let $\Lambda(u,z)$ be the elliptic function defined by

$$u=\int_0^{\Lambda}\frac{\mathrm{d}t}{\sqrt{t(t-1)(t-z)}}, \tag{6.13}$$

where the fundamental periods $2v_1$ and $2v_2$ are linearly independent solutions of the hypergeometric equation

$$z(1-z)v''+(1-2z)v'-\tfrac{1}{4}v=0, \tag{6.14}$$

Then $\mathrm{P_{VI}}$ with $\alpha=\beta=\gamma=0$ and $\delta=\frac{1}{2}$, i.e.,

$$\begin{aligned} w''=&\left(\frac{1}{w}+\frac{1}{w-1}+\frac{1}{w-z}\right)\frac{(w')^2}{2}-\left(\frac{1}{z}+\frac{1}{z-1}+\frac{1}{w-z}\right)w' \\ &+\frac{w(w-1)}{2z(z-1)(w-z)}, \end{aligned} \tag{6.15}$$

has general solution

$$w(z;0,0,0,\tfrac{1}{2})=\Lambda(C_1v_1+C_2v_2,z),$$

with C_1 and C_2 arbitrary constants, for details see Painlevé [189], Fuchs [64], Manin [159]. This solution is an essentially transcendental function of both constants of integration since $\mathrm{P_{VI}}$ with $\alpha=\beta=\gamma=0$ and $\delta=\frac{1}{2}$ does not admit a one-parameter algebraic first integral of the form $F(z,w,w',K)=0$, with K a constant.

7 Special Function Solutions

The Painlevé equations $\mathrm{P_{II}}$–$\mathrm{P_{VI}}$ possess hierarchies of solutions expressible in terms of classical special functions, for special values of the parameters through an associated Riccati equation,

$$w'=p_2(z)w^2+p_1(z)w+p_0(z), \tag{7.1}$$

where $p_2(z)$, $p_1(z)$ and $p_0(z)$ are rational functions. Hierarchies of solutions, which are often referred to as "one-parameter solutions" (since they have one arbitrary constant), are generated from "seed solutions" derived from the Riccati equation using the Bäcklund transformations given in §4. Furthermore, as for the rational solutions, these special function solutions are often expressed in the form of determinants.

We illustrate the procedure for deriving the Riccati equation associated with special function solutions of P_{II}, which are solved in terms of Airy functions, in next subsection. See, for example, Sachdev [196] for details of the derivation of the Riccati equations for P_{III}–P_{VI}; the results are summarized in Table 4.

Brief details of special function solutions for P_{III}–P_{VI} are given in §§7.2–7.5, respectively. Special function solutions of P_{III} are expressed in terms of Bessel functions $J_\nu(z)$ and $Y_\nu(z)$ [153, 160, 169, 172, 187, 228], of P_{IV} in terms of Weber-Hermite (parabolic cylinder) functions $D_\nu(z)$ [17, 76, 152, 171, 184, 227], of P_V in terms of Whittaker functions $M_{\kappa,\mu}(z)$ and $W_{\kappa,\mu}(z)$, or equivalently confluent hypergeometric functions ${}_1F_1(a;c;z)$ [154, 73, 186, 230], and of P_{VI} in terms of hypergeometric functions $F(a,b;c;z)$ [59, 156, 185]; see also [2, 74, 77, 107, 210]. Some classical orthogonal polynomials arise as particular cases of these special function solutions and thus yield rational solutions of the associated Painlevé equations. For P_{III} and P_V these are in terms of associated Laguerre polynomials $L_k^m(z)$ [26, 123, 164, 175], for P_{IV} in terms of Hermite polynomials $H_n(z)$ [17, 125, 171, 184], and for for P_{VI} in terms of Jacobi polynomials $P_n^{(\alpha,\beta)}(z)$ [162, 212]. In fact all rational solutions of P_{VI} arise as particular cases of the special solutions given in terms of hypergeometric functions [165].

Table 4. Special function solutions of P_{II}–P_{VI}

	$p_2(z)$	$p_1(z)$	$p_0(z)$	Condition on parameters	Special Function
P_{II}	ε_1	0	$\frac{1}{2}\varepsilon_1 z$	$\alpha = \frac{1}{2}\varepsilon_1$	Airy function $\mathrm{Ai}(z), \mathrm{Bi}(z)$
P_{III}	ε_1	$\dfrac{\varepsilon_1\alpha - 1}{z}$	ε_2	$\varepsilon_1\alpha + \varepsilon_2\beta = 2$, $\gamma = 1, \quad \delta = -1$	Bessel function $J_\nu(z), Y_\nu(z)$
P_{IV}	ε_1	$2\varepsilon_1 z$	$-2(1+\varepsilon_1\alpha)$	$\beta = -2(1+\varepsilon_1\alpha)^2$	Weber-Hermite (parabolic cylinder) function $D_\nu(z)$
P_V	$\dfrac{a}{z}$	$\varepsilon_3 + \dfrac{b-a}{z}$	$-\dfrac{b}{z}$	$a+b+\varepsilon_3\gamma = 1$, $\delta = -\frac{1}{2}$, $a = \varepsilon_1\sqrt{2\alpha}$, $b = \varepsilon_2\sqrt{-2\beta}$	Whittaker function $M_{\kappa,\mu}(z), W_{\kappa,\mu}(z)$
P_{VI}	$\dfrac{a}{z(z-1)}$	$b+c-\dfrac{a+c}{z}$	$-\dfrac{b}{z-1}$	$a+b+c+d=0$ $a = \varepsilon_1\sqrt{2\alpha}$, $b = \varepsilon_2\sqrt{-2\beta}$, $c = \varepsilon_3\sqrt{2\gamma}$, $d = \varepsilon_4\sqrt{1-2\delta}$	hypergeometric function $F(a,b;c;z)$

7.1 Special Function Solutions of $\mathrm{P_{II}}$

To obtain a special function solution of a Painlevé equation, one supposes that $w(z)$ satisfies the Riccati equation (7.1) for some functions $p_2(z)$, $p_1(z)$ and $p_0(z)$. Differentiating (7.1) yields

$$\begin{aligned} w'' &= p_2' w^2 + 2p_2 w w' + p_1' w + p_1 w' + p_0' \\ &= p_2' w^2 + p_1' w + p_1' + (2p_2 w + p_1)\left(p_2 w^2 + p_1 w + p_0\right) \\ &= 2p_2^2 w^3 + (p_2' + 3p_1 p_2)\, w^2 + \left(p_1' + 2p_0 p_2 + p_1^2\right) w + p_1' + p_1 p_0, \end{aligned} \tag{7.2}$$

and substituting this into $\mathrm{P_{II}}$ yields

$$\begin{aligned} 2p_2 w^3 + (p_2' + 3p_1 p_2)\, w^2 + \left(p_1' + 2p_0 p_2 + p_1^2\right) w + p_1' + p_1 p_0 & \\ = 2w^3 + zw + \alpha. & \end{aligned} \tag{7.3}$$

Equating coefficients of powers of w to zero, which are necessarily independent since w is any solution of $\mathrm{P_{II}}$, yields

$$\begin{aligned} w^3 &: \quad p_2^2(z) = 1, \\ w^2 &: \quad p_2'(z) + 3p_1(z)p_2(z) = 0, \\ w &: \quad p_1'(z) + 2p_0(z)p_2(z) + p_1^2(z) = z, \\ w^0 &: \quad p_0'(z) + p_1(z)p_0(z) = \alpha, \end{aligned}$$

and therefore

$$p_2(z) = \varepsilon, \qquad p_1(z) = 0, \qquad p_0(z) = \tfrac{1}{2}\varepsilon z, \qquad \alpha = \tfrac{1}{2}\varepsilon, \qquad \varepsilon^2 = 1. \tag{7.4}$$

Thus we obtain the Riccati equation

$$\varepsilon w' = w^2 + \tfrac{1}{2} z. \tag{7.5}$$

Any solution of this equation is also a solution of $\mathrm{P_{II}}$, provided that $\alpha = \frac{1}{2}\varepsilon$. Note that $\mathrm{P_{II}}$ can be written as

$$\varepsilon(\varepsilon w' - w^2 - \tfrac{1}{2}z)' = -2w(\varepsilon w' - w^2 - \tfrac{1}{2}z) + \alpha - \tfrac{1}{2}\varepsilon, \qquad \varepsilon^2 = 1.$$

Hence if $\alpha = \frac{1}{2}\varepsilon$, then special solutions of $\mathrm{P_{II}}$ can be obtained in terms of solutions of the Riccati equation (7.5). Setting $w = -\varepsilon\varphi'/\varphi$ in (7.5) yields

$$\varphi'' + \tfrac{1}{2} z\varphi = 0, \tag{7.6}$$

which is equivalent to the Airy equation and has general solution

$$\varphi(z) = C_1 \operatorname{Ai}(\xi) + C_2 \operatorname{Bi}(\xi), \qquad \xi = -2^{-1/3}\, z, \tag{7.7}$$

where $\operatorname{Ai}(\xi)$ and $\operatorname{Bi}(\xi)$ are the Airy functions and C_1 and C_2 are arbitrary constants.

Theorem 7.1. *The second Painlevé equation* $\mathrm{P_{II}}$ *has a one-parameter family of solutions expressible in terms of Airy functions given by (7.7) if and only if* $\alpha = n + \frac{1}{2}$, *with* $n \in \mathbb{Z}$.

Proof. See Gambier [68]. □

Theorem 7.2. *Let* $\varphi(z)$ *be the solution of (7.6) and* $\tau_n(z)$ *be the* $n \times n$ *determinant*

$$\tau_n(z) = \begin{vmatrix} \varphi(z) & \varphi'(z) & \cdots & \varphi^{(n-1)}(z) \\ \varphi'(z) & \varphi''(z) & \cdots & \varphi^{(n)}(z) \\ \vdots & \vdots & \ddots & \vdots \\ \varphi^{(n-1)}(z) & \varphi^{(n)}(z) & \cdots & \varphi^{(2n-2)}(z) \end{vmatrix}, \qquad n \geq 1, \tag{7.8}$$

where $\varphi^{(m)}(z) = \mathrm{d}^m\varphi/\mathrm{d}z^m$, *then*

$$w(z; n - \tfrac{1}{2}) = \frac{\mathrm{d}}{\mathrm{d}z}\left\{\ln\left[\frac{\tau_{n-1}(z)}{\tau_n(z)}\right]\right\} = \frac{\tau'_{n-1}(z)}{\tau_{n-1}(z)} - \frac{\tau'_n(z)}{\tau_n(z)}, \qquad n \geq 1, \tag{7.9}$$

satisfies $\mathrm{P_{II}}$ *with* $\alpha = n - \frac{1}{2}$.

Proof. Airault [11], Flaschka and Newell [53], Okamoto [184]. □

Theorem 7.3. *Suppose that* $\Psi_n(z)$ *satisfies the Toda equation*

$$\Psi_{n+1}\Psi_{n-1} = -\Psi_n \frac{\mathrm{d}^2\Psi_n}{\mathrm{d}z^2} + \left(\frac{\mathrm{d}\Psi_n}{\mathrm{d}z}\right)^2, \tag{7.10}$$

with $\Psi_0(z) = 1$ *and* $\Psi_1(z) = -\varphi(z)$ *where* $\varphi(z)$ *satisfies the Airy equation (7.6), then*

$$w(z; n - \tfrac{1}{2}) = \frac{\mathrm{d}}{\mathrm{d}z}\left\{\ln\left[\frac{\Psi_{n-1}(z)}{\Psi_n(z)}\right]\right\} = \frac{\Psi'_{n-1}(z)}{\Psi_{n-1}(z)} - \frac{\Psi'_n(z)}{\Psi_n(z)}, \qquad n \geq 1, \tag{7.11}$$

satisfies $\mathrm{P_{II}}$ *with* $\alpha = n - \frac{1}{2}$ *for* $n \in \mathbb{Z}^+$.

If we set $\Phi(z) \equiv \varphi'(z)/\varphi(z)$, where $\varphi(z)$ is given by (7.7), then the first few solutions in the Airy function solution hierarchy for $\mathrm{P_{II}}$ are given in Table 6.

Table 5. Solutions $\Psi_n(z)$ of the Toda equation (7.10)

$\Psi_1(z) = -\varphi,$
$\Psi_2(z) = (\varphi')^2 + \frac{1}{2}z\varphi^2,$
$\Psi_3(z) = (\varphi')^3 + \frac{1}{2}z\varphi'\varphi^2 - \frac{1}{4}\varphi^3$
$\Psi_4(z) = \frac{1}{2}z(\varphi')^4 + \varphi(\varphi')^3 + \frac{1}{2}z^2\varphi^2(\varphi')^2 + \frac{1}{2}z\varphi'\varphi^3 + \frac{1}{8}(z^3 - \frac{1}{2})\varphi^4$

Table 6. Airy function solutions of $\mathrm{P_{II}}$

$w(z;\tfrac{1}{2}) = -\Phi$
$w(z;\tfrac{3}{2}) = \Phi - \dfrac{1}{2\Phi^2+z}$
$w(z;\tfrac{5}{2}) = \dfrac{2z\Phi^2+\Phi+z^2}{4\Phi^3+2z\Phi-1} + \dfrac{1}{2\Phi^2+z}$
$w(z;\tfrac{7}{2}) = -\dfrac{3}{z} - \dfrac{2z\Phi^2+\Phi+z^2}{4\Phi^3+2z\Phi-1} + \dfrac{48\Phi^3+8z^2\Phi^2+28z\Phi+4z^3-9}{z(8z\Phi^4+16\Phi^3+8z^2\Phi^2+8z\Phi+2z^3-3)}$

These "Airy functions solutions", $w_n \equiv w(z;n/2)$ for $n = 1,3,5,\ldots$, of $\mathrm{P_{II}}$ are also the general solutions of a Fuchsian equation of the form

$$(w_n')^n + \sum_{j=0}^{n-1} P_j(w_n,z)(w_n')^j = 0, \tag{7.12}$$

where $P_j(w_n,z)$ are polynomials in w_n and z. For example, $w_3 \equiv w(z;\tfrac{3}{2})$ satisfies the first-order, third-degree equation

$$\begin{aligned}(w_3')^3 - \left(w_3^2+\tfrac{1}{2}z\right)(w_3')^2 - \left(w_3^4+zw_3^2+4w_3+\tfrac{1}{4}z^2\right)w_3' \\ + w_3^6+\tfrac{3}{2}zw_3^4+4w_3^3+\tfrac{3}{4}z^2w_3^2+2zw_3+\tfrac{1}{8}z^3+2=0. \end{aligned}\tag{7.13}$$

7.2 Special Function Solutions of $\mathrm{P_{III}}$

Special function solutions of $\mathrm{P_{III}}$ are expressed in terms of Bessel functions as given in the following theorem.

Theorem 7.4. *If $\gamma\delta \neq 0$ then we set $\gamma = 1$ and $\delta = -1$, without loss of generality. $\mathrm{P_{III}}$ with $\gamma = -\delta = 1$, has solutions expressible in terms of Bessel functions if and only if*

$$\varepsilon_1\alpha + \varepsilon_2\beta = 4n+2, \tag{7.14}$$

with $n \in \mathbb{Z}$ and $\varepsilon_1 = \pm 1$, $\varepsilon_2 = \pm 1$ independently.

Proof. See Gromak [74], Lukashevich [151, 153], Umemura and Watanabe [228]; also Gromak, Laine and Shimomura [83, §35]. □

In the case $\varepsilon_1\alpha + \varepsilon_2\beta = 2$, the Riccati equation is

$$zw' = \varepsilon_1 zw^2 + (\alpha\varepsilon_1 - 1)w + \varepsilon_2 z. \tag{7.15}$$

If $\alpha \neq \varepsilon_1$ then (7.15) has the solution

$$w(z) = -\varepsilon_1\,\varphi'(z)/\varphi(z), \tag{7.16}$$

where

$$\varphi(z) = z^{\nu}\left\{C_1 J_{\nu}(\zeta) + C_2 Y_{\nu}(\zeta)\right\}, \tag{7.17}$$

with $\zeta = \sqrt{\varepsilon_1\varepsilon_2}\, z$, $\nu = \frac{1}{2}\alpha\varepsilon_2$, and C_1, C_2 arbitrary constants. For examples and plots see Milne, Clarkson and Bassom [169]. Determinantal representations of special function solutions of $\mathrm{P_{III}}$ are given by Okamoto [187]; see also Forrester and Witte [62].

We remark that rational solutions of $\mathrm{P_{III}}$, with $\gamma = 1$ and $\delta = -1$, occur then $\varepsilon_1\alpha + \varepsilon_2\beta = 4$ and Bessel solutions of $\mathrm{P_{III}}$ occur then $\varepsilon_1\alpha + \varepsilon_2\beta = 4 + 2$. A question is what happens when these lines intersect? This is illustrated in the following example.

Example 7.1. If $\alpha + \beta = 4$ and $\alpha - \beta = 2$ then $\alpha = 3$ and $\beta = 1$. Hence the Riccati equation becomes

$$zw' = zw^2 + 2w - z,$$

then letting $w = -\varphi'/\varphi$ yields

$$z\varphi'' - 2\varphi' - z\varphi = 0,$$

which has general solution

$$\varphi(z) = C_1(z-1)\mathrm{e}^{z} + C_2(z+1)\mathrm{e}^{-z},$$

with C_1, C_2 arbitrary constants, and so

$$w(z) = -\frac{C_1 z\mathrm{e}^{z} - C_2 z\mathrm{e}^{-z}}{C_1(z-1)\mathrm{e}^{z} + C_2(z+1)\mathrm{e}^{-z}}.$$

Note that setting $C_1 = 0$ yields the rational solution $w = z/(z+1)$, whilst setting $C_2 = 0$ yields the rational solution $w = z/(1-z)$.

We remark that this example shows that rational solutions of $\mathrm{P_{III}}$ for specific parameters are not unique since, $\mathrm{P_{III}}$ with parameters $\alpha = 3$ and $\beta = 1$, $\gamma = 3$ and $\delta = -1$ has the rational solutions $z/(z+1)$ and $z/(1-z)$.

7.3 Special Function Solutions of $\mathrm{P_{IV}}$

Special function solutions of $\mathrm{P_{IV}}$ are expressed in in terms of parabolic cylinder functions as given in the following theorem.

Theorem 7.5. $\mathrm{P_{IV}}$ *has solutions expressible in terms of parabolic cylinder functions if and only if either*

$$\beta = -2(2n + 1 + \varepsilon\alpha)^2, \tag{7.18}$$

or

$$\beta = -2n^2, \tag{7.19}$$

with $n \in \mathbb{Z}$ and $\varepsilon = \pm 1$.

Proof. See Gambier [68], Gromak [76], Gromak and Lukashevich [84], Lukashevich [151, 152]; also Gromak, Laine and Shimomura [83, Chapter 6]. □

In the case when $n = 0$ in (7.18), then the associated Riccati equation is

$$w' = \varepsilon(w^2 + 2zw) - 2(1 + \varepsilon\alpha), \tag{7.20}$$

and yields the Weber-Hermite equation

$$\varphi'' - 2\varepsilon z\varphi' - 2(\alpha + \varepsilon)\varphi = 0, \tag{7.21}$$

which has solution

$$w(z) = -\varepsilon\varphi'(z)/\varphi(z), \tag{7.22}$$

where

$$\varphi(z) = \left\{C_1 D_\nu\left(\sqrt{2}\,z\right) + C_2 D_{-\nu}\left(\sqrt{2}\,z\right)\right\}\exp\left(\tfrac{1}{2}\varepsilon z^2\right) \tag{7.23}$$

with $\nu = -\frac{1}{2}(1 + \varepsilon + 2\alpha)$, C_1 and C_2 arbitrary constants, and $D_\nu\left(\sqrt{2}\,z\right)$ the parabolic cylinder function, provided that $\nu \notin \mathbb{Z}$.

When ν is a positive integer the parabolic cylinder functions reduce to Hermite polynomials times an exponential function; thus

$$w(z; -n, -2(n-1)^2) = -\frac{H'_{n-1}(z)}{H_{n-1}(z)}, \qquad n = 1, 2, 3, \ldots\,, \tag{7.24}$$

and

$$w(z; -n, -2(n+1)^2) = -2z + \frac{H'_n(z)}{H_n(z)}, \qquad n = 0, 1, 2, \ldots\,. \tag{7.25}$$

For examples and plots see Bassom, Clarkson and Hicks [17]. Determinantal representations of special function solutions of $\mathrm{P_{IV}}$ are given by Okamoto [184]; see also Forrester and Witte [61].

If $\alpha = -\frac{1}{2}\varepsilon$ then $\beta = -\frac{1}{2}$, with $\varepsilon = \pm 1$, and so the Weber-Hermite equation (7.21) has solution

$$\varphi(z) = \exp\left(\tfrac{1}{2}\varepsilon z^2\right)\left[C_1 D_{-1/2}\left(\sqrt{2}\,z\right) + C_2 D_{-1/2}\left(-\sqrt{2}\,z\right)\right], \tag{7.26}$$

and so

$$w(z; \tfrac{1}{2}, -\tfrac{1}{2}) = -\frac{\sqrt{2}\left[C_1 D_{1/2}\left(\sqrt{2}\,z\right) - C_2 D_{1/2}\left(-\sqrt{2}\,z\right)\right]}{C_1 D_{-1/2}\left(\sqrt{2}\,z\right) + C_2 D_{-1/2}\left(-\sqrt{2}\,z\right)}$$

$$w(z; -\tfrac{1}{2}, -\tfrac{1}{2}) = -2z + \frac{\sqrt{2}\left[C_1 D_{1/2}\left(\sqrt{2}\,z\right) - C_2 D_{1/2}\left(-\sqrt{2}\,z\right)\right]}{C_1 D_{-1/2}\left(\sqrt{2}\,z\right) + C_2 D_{-1/2}\left(-\sqrt{2}\,z\right)}.$$

Using these solutions as seed solutions one generates a hierarchy of solutions of $\mathrm{P_{IV}}$ expressed in terms of $D_{\pm 1/2}(\zeta)$. The special solutions of $\mathrm{P_{IV}}$ when $\alpha = \frac{1}{2}n$ and $\beta = -\frac{1}{2}n^2$ arise in quantum gravity; see Fokas, Its and Kitaev

[57, 58]. This hierarchy, the "half-integer" hierarchy, is discussed by Bassom, Clarkson and Hicks [17].

If $1+\varepsilon\alpha=0$, then (7.20) has solutions

$$w(z)=\begin{cases}\dfrac{2\exp(z^2)}{\sqrt{\pi}\,[C+\mathrm{i}\,\mathrm{erfc}(\mathrm{i}z)]}, & \text{if } \varepsilon=1,\\[2ex] \dfrac{2\exp(-z^2)}{\sqrt{\pi}\,[C+\mathrm{erfc}(z)]}, & \text{if } \varepsilon=-1,\end{cases} \tag{7.27}$$

where C is an arbitrary constant and $\mathrm{erfc}(z)$ is the complementary error function; see Gromak and Lukashevich [84], also Bassom, Clarkson and Hicks [17] who generate a hierarchy of solutions with (7.27) as the seed solution. These solutions have some rational solutions as special cases. Further a special case of this hierarchy occurs when $\alpha=2n+1$ and $\beta=0$ which gives *bound state solutions* that have exponential decay as $z\to\pm\infty$ and so are nonlinear analogues of bound states for the linear harmonic oscillator; for further details see [17, 18].

7.4 Special Function Solutions of $\mathrm{P_V}$

Special function solutions of $\mathrm{P_V}$ are expressed in in terms of Whittaker functions as given in the following theorem.

Theorem 7.6. *If $\delta\neq0$, then we may set $\delta=-\frac{1}{2}$. $\mathrm{P_V}$ then has solutions expressible in terms of Whittaker functions if and only if*

$$a+b+\varepsilon_3\gamma=2n+1, \tag{7.28}$$

or

$$(a-n)(b-n)=0, \tag{7.29}$$

where $n\in\mathbb{Z}$, $a=\varepsilon_1\sqrt{2\alpha}$ and $b=\varepsilon_2\sqrt{-2\beta}$, with $\varepsilon_j=\pm1$, $j=1,2,3$, independently.

Proof. See Gromak and Lukashevich [84], Lukashevich [154], Watanabe [230]; also Gromak, Laine and Shimomura [83, §40]. □

In the case when $n=0$ in (7.28), then the associated Riccati equation is

$$zw'=aw^2+(b-a+\varepsilon_3z)w-b. \tag{7.30}$$

If $a\neq0$, then (7.30) has the solution

$$w(z)=-\frac{z\varphi'(z)}{a\varphi(z)}, \tag{7.31}$$

where

$$\varphi(z) = \{C_1 M_{\kappa,\mu}(\zeta) + C_2 W_{\kappa,\mu}(\zeta)\}\, \zeta^{-(a-b+1)/2} \exp(\tfrac{1}{2}\zeta), \tag{7.32}$$

with $\zeta = \varepsilon_3 z$, $\kappa = \frac{1}{2}(a-b+1)$, $\mu = \frac{1}{2}(a+b)$, C_1 and C_2 arbitrary constants and $M_{\kappa,\mu}(\zeta)$, $W_{\kappa,\mu}(\zeta)$ Whittaker functions. Determinantal representations of special function solutions of $\mathrm{P_V}$ are given by Okamoto [186]; see also Forrester and Witte [62] and Masuda [163].

7.5 Special Function Solutions of $\mathrm{P_{VI}}$

Special function solutions of $\mathrm{P_{VI}}$ are expressed in in terms of hypergeometric functions as given in the following theorem.

Theorem 7.7. $\mathrm{P_{VI}}$ *has solutions expressible in terms of hypergeometric functions if and only if*

$$a + b + c + d = 2n + 1, \tag{7.33}$$

where $n \in \mathbb{Z}$, $a = \varepsilon_1\sqrt{2\alpha}$, $b = \varepsilon_1\sqrt{-2\beta}$, $c = \varepsilon_3\sqrt{2\gamma}$, *and* $d = \varepsilon_4\sqrt{1-2\delta}$, *with* $\varepsilon_j = \pm 1$, $j = 1, 2, 3, 4$, *independently.*

Proof. See Fokas and Yortsos [59], Lukashevich and Yablonskii [156], Okamoto [185]; also Gromak, Laine and Shimomura [83, §44]. □

If $n = 1$, then the associated Riccati equation is

$$w' = \frac{aw^2}{z(z-1)} + \frac{(b+c)z - a - c}{z(z-1)}\, w - \frac{b}{z-1}. \tag{7.34}$$

If $a \neq 0$ then (7.34) has the solution

$$w(z) = \frac{\zeta - 1}{a\varphi(\zeta)} \frac{\mathrm{d}\varphi}{\mathrm{d}\zeta}, \qquad \zeta = \frac{1}{1-z}, \tag{7.35}$$

where

$$\begin{aligned}\varphi(\zeta) &= C_1 F(b, -a; b+c; \zeta) \\ &\quad + C_2 \zeta^{-b+1-c} F(-a-b-c+1, -c+1; 2-b-c; \zeta),\end{aligned} \tag{7.36}$$

with C_1, C_2 arbitrary constants and $F(a, b; c; \zeta)$ is the hypergeometric function. Determinantal representations of special function solutions of $\mathrm{P_{VI}}$ are given by Okamoto [185]; see also Forrester and Witte [63] and Masuda [163].

There also is a family of solutions of $\mathrm{P_{VI}}$ with $\alpha = 2n^2$, for $n \in \mathbb{Z}$ with $n \neq 0$, $\beta = \gamma = 0$ and $\delta = \frac{1}{2}$, expressible in terms of solutions of the hypergeometric equation

$$z(1-z)v'' + (1-2z)v' - \tfrac{1}{4}v = 0, \tag{7.37}$$

for further details see Mazzocco [166]. For example, if $n = 1$, then the solution is

$$w(z; 2, 0, 0, \tfrac{1}{2}) = \frac{\left[v^2 + 4zvv' + 4z(z-1)(v')^2\right]^2}{8vv'[v + (2z-1)v'](v + 2zv')},$$

where $v(z)$ is any solution of (7.37).

8 Other Mathematical Properties

8.1 Hirota Bilinear Forms

Each of the Painlevé equations P_I–P_{VI} can be written in *Hirota bilinear form*, using the Hirota operator D_z defined by

$$D_z\,(F\bullet G)=\left(\frac{\mathrm{d}}{\mathrm{d}z_1}-\frac{\mathrm{d}}{\mathrm{d}z_2}\right)[F(z_1)\,G(z_2)]\bigg|_{z_1=z_2=z}=\frac{\mathrm{d}F}{\mathrm{d}z}G-F\frac{\mathrm{d}G}{\mathrm{d}z},\qquad(8.1)$$

see Hietarinta [93]. We illustrate this in the following example for P_{II}.

Example 8.1. Letting

$$w(z)=\frac{\mathrm{d}}{\mathrm{d}z}\left\{\ln\left[\frac{F(z)}{G(z)}\right]\right\}=\frac{F'(z)}{F(z)}-\frac{G'(z)}{G(z)},\qquad(8.2)$$

in P_{II} yields

$$\begin{aligned}F'''G-3F''G'+3F'G''-FG'''-z(F'G-FG')-\alpha FG,&\\=3\frac{F'G-FG'}{FG}\left(F''G-2F'G'+FG''\right),&\end{aligned}$$

which has the "decoupling"

$$F''G-2F'G'+FG''+\lambda(z)FG=0,\qquad(8.3a)$$
$$F'''G-3F''G'+3F'G''-FG'''=[z-3\lambda(z)](F'G-FG')+\alpha FG,\qquad(8.3b)$$

where $\lambda(z)$ is the separating function. Thus we obtain the bilinear representation of P_{II} given by

$$\left\{\mathrm{D}_z^2+\lambda(z)\right\}(F\bullet G)=0,\qquad(8.4a)$$
$$\left\{\mathrm{D}_z^3-[z-3\lambda(z)]\mathrm{D}_z-\alpha\right\}(F\bullet G)=0,\qquad(8.4b)$$

where D_z is the Hirota operator defined in (8.1) above. Fukutani [65] discusses P_{II} using this bilinear representation.

8.2 Coalescence Cascade

The Painlevé equations have the coalescence cascade

$$\begin{array}{ccccc}P_{VI}&\longrightarrow&P_{V}&\longrightarrow&P_{IV}&&\\&&\downarrow&&\downarrow&&\\&&P_{III}&\longrightarrow&P_{II}&\longrightarrow&P_{I}\end{array}\qquad(8.5)$$

see, for example, [101, 109] for further details. This is illustrated in the following example.

Example 8.2. If in P_{II} we make the transformation

$$w(z;\alpha) = \varepsilon W(\zeta) + \varepsilon^{-5}, \qquad z = \varepsilon^2\zeta - 6\varepsilon^{-10}, \qquad \alpha = 4\varepsilon^{-15}, \tag{8.6}$$

then

$$\frac{d^2W}{d\zeta^2} = 6W^2 + \zeta + \varepsilon^6(2W^3 + \zeta W), \tag{8.7}$$

and so in the limit as $\varepsilon \to 0$, $W(\zeta)$ satisfies P_I.

These other coalescences are given as follows:

1. If in P_{VI} we let

$$\begin{gathered} w(z;\alpha,\beta,\gamma,\delta) = y(x;a,b,c,d), \\ z = 1 + \varepsilon x, \qquad \gamma = \frac{c}{\varepsilon} - \frac{d}{\varepsilon^2}, \qquad \delta = \frac{d}{\varepsilon^2}, \end{gathered} \tag{8.8}$$

   then as $\varepsilon \to 0$, P_{VI} coalescences to P_V.
2. If in P_V we let

$$\begin{gathered} w(z;\alpha,\beta,\gamma,\delta) = 1 + \varepsilon y(x;a,b,c,d), \\ \alpha = \frac{a}{\varepsilon} + \frac{b}{\varepsilon^2}, \qquad \beta = -\frac{b}{\varepsilon^2}, \qquad \gamma = \varepsilon c, \qquad \delta = \varepsilon d, \end{gathered} \tag{8.9}$$

   then as $\varepsilon \to 0$, P_V coalescences to P_{III}.
3. If in P_V we let

$$\begin{gathered} w(z;\alpha,\beta,\gamma,\delta) = \tfrac{1}{2}\sqrt{2}\,\varepsilon y(x;a,b), \qquad z = 1 + \sqrt{2}\,\varepsilon x, \\ \alpha = \frac{1}{2\varepsilon^4}, \qquad \beta = \frac{b}{4}, \qquad \gamma = -\frac{1}{\varepsilon^4}, \qquad \delta = \frac{a}{\varepsilon^2} - \frac{1}{2\varepsilon^4}, \end{gathered} \tag{8.10}$$

   then as $\varepsilon \to 0$, P_V coalescences to P_{IV}.
4. If in P_{IV} we let

$$\begin{gathered} w(z;\alpha,\beta) = \tfrac{2^{2/3}}{\varepsilon} y(x;a) + \tfrac{1}{\varepsilon^3}, \\ z = \frac{\varepsilon x}{2^{2/3}} - \frac{1}{\varepsilon^3}, \qquad \alpha = -\frac{a}{2\varepsilon^6}, \qquad \beta = -\frac{1}{2\varepsilon^{12}}, \end{gathered} \tag{8.11}$$

   then as $\varepsilon \to 0$, P_{IV} coalescences to P_{II}.
5. If in P_{III} we let

$$\begin{gathered} w(z;\alpha,\beta,\gamma,\delta) = 1 + 2\varepsilon y(x;a), \qquad z = 1 + \varepsilon^2 x, \\ \alpha = -\frac{1}{2\varepsilon^6}, \qquad \beta = \frac{1}{2\varepsilon^6} + \frac{2a}{\varepsilon^3}, \qquad \gamma = -\delta = \frac{1}{4\varepsilon^6}, \end{gathered} \tag{8.12}$$

   then as $\varepsilon \to 0$, P_{III} coalescences to P_{II}.

8.3 The $\mathrm{P_{II}}$ Hierarchy

The Korteweg-de Vries (KdV) hierarchy can be written as

$$u_{t_{2n+1}} + \partial_x \mathcal{L}_{n+1}[u] = 0, \qquad n = 0, 1, 2, \ldots, \tag{8.13}$$

with $\partial_x = \partial/\partial x$, where the sequence $\mathcal{L}_n$ satisfies the Lenard recursion relation [148]

$$\partial_x \mathcal{L}_{n+1} = \left(\partial_x^3 + 4u\partial_x + 2u_x\right)\mathcal{L}_n.$$

Beginning with $\mathcal{L}_0[u] = \frac{1}{2}$, this then gives

$$\begin{aligned}
\mathcal{L}_1[u] &= u,\\
\mathcal{L}_2[u] &= u_{xx} + 3u^2,\\
\mathcal{L}_3[u] &= u_{xxxx} + 10uu_{xx} + 5u_x^2 + 10u^3,\\
\mathcal{L}_4[u] &= u_{xxxxxx} + 14uu_{xxxx} + 28u_xu_{xxx} + 21u_{xx}^2 + 70u^2u_{xx} + 70uu_x^2 + 35u^4.
\end{aligned}$$

and so on. The first four members of the KdV hierarchy are

$$\begin{aligned}
&u_{t_1} + u_x = 0,\\
&u_{t_2} + u_{xxx} + 6uu_x = 0,\\
&u_{t_3} + u_{xxxxx} + 10uu_{xxx} + 20u_xu_{xx} + 30u^2u_x = 0,\\
&u_{t_4} + u_{xxxxxxx} + 14uu_{xxxxx} + 42u_xu_{xxxx} + 70u_{xx}u_{xxx}\\
&\qquad + 70u^2u_{xxx} + 280uu_xu_{xx} + 70u_x^3 + 140u^3u_x = 0.
\end{aligned}$$

The modified KdV (mKdV) hierarchy is obtained from the KdV hierarchy via the Miura transformation $u = v_x - v^2$, and can be written as

$$v_{t_{2n+1}} + \partial_x\left(\partial_x + 2v\right)\mathcal{L}_n\left[v_x - v^2\right] = 0, \qquad n = 0, 1, 2, \ldots \tag{8.16}$$

The first three members of the mKdV hierarchy are

$$\begin{aligned}
&v_{t_1} + v_{xxx} - 6v^2v_x = 0,\\
&v_{t_2} + v_{xxxxx} - 10v^2v_{xxx} - 40v_xv_{xx} - 10v_x^3 + 30v^4v_x = 0,\\
&v_{t_3} + v_{xxxxxxx} - 14v^2v_{xxxxx} - 84vv_xv_{xxxx} - 140vv_{xx}v_{xxx} - 126v_x^2v_{xxx}\\
&\quad - 182v_xv_{xx}^2 + 70v^4v_{xxx} + 560v^3v_xv_{xx} + 420v^2v_x^3 - 140v^6v_x = 0.
\end{aligned}$$

A $\mathrm{P_{II}}$ hierarchy is obtained from equation (8.16) through the scaling reduction

$$v(x, t_{2n+1}) = w(z)/[(2n+1)t_{2n+1}]^{1/(2n+1)}, \quad z = x/[(2n+1)t_{2n+1}]^{1/(2n+1)},$$

which gives

$$\mathrm{P_{II}^{[n]}}[w, \alpha_n] \equiv \left(\frac{\mathrm{d}}{\mathrm{d}z} + 2w\right)\mathcal{L}_n[w' - w^2] - zw - \alpha_n = 0, \qquad n = 1, 2, 3, \ldots \tag{8.18}$$

with α_n an arbitrary constant, for $n = 1, 2, 3, \ldots$, where the case $n = 0$ has been excluded; see [11, 36, 53, 143] for further details. Since $\mathcal{L}_1[u] = u$ then the first member of this hierarchy is $\mathrm{P_{II}}$. Since $\mathcal{L}_2[u] = u_{xx} + 3u^2$, the second member of this hierarchy is the fourth order equation $\mathrm{P_{II}^{[2]}}$

$$w'''' = 10w^2 w'' + 10w\,(w')^2 - 6w^5 + zw + \alpha_2. \tag{8.19}$$

Since $\mathcal{L}_3[u] = u_{xxxx} + 10uu_{xx} + 5u_x^2 + 10u^3$, the third member of this hierarchy is the sixth order equation $\mathrm{P_{II}^{[3]}}$

$$\begin{aligned} w'''''' = {} & 14w^2 w'''' + 56ww'w''' + 42\,(w'')^2 - 70\left[w^4 - (w')^2\right] w'' \\ & - 140w^3\,(w')^2 + 20w^7 + zw + \alpha_3. \end{aligned} \tag{8.20}$$

Theorem 8.1. *Let $w \equiv w(z;\alpha)$ be a solution of $\mathrm{P_{II}^{[N]}}$, the Nth equation in the $\mathrm{P_{II}}$ hierarchy, then the transformations*

$$\mathcal{S}^{[N]}: \quad w(z; -\alpha) = -w, \tag{8.21}$$

$$\begin{aligned} \mathcal{T}_{\pm}^{[N]}: \quad & ww(z;\alpha_N \pm 1) = -w(z;\alpha_N) \\ & + \frac{2\alpha_N \pm 1}{2\mathcal{L}_N[\mp w'(z;\alpha_N) - w^2(z;\alpha_N)] - z}, \end{aligned} \tag{8.22}$$

give solutions of $\mathrm{P_{II}^{[N]}}$, provided that $\alpha \neq \mp\frac{1}{2}$ in (8.22).

Proof. See Airault [11] and Clarkson, Joshi and Pickering [36]. □

All rational solutions of $\mathrm{P_{II}^{[N]}}$ (8.18) can be generated using (8.22), (8.22) and the "seed solution" $w_0 = w(z;0) = 0$. Rational solutions for $\mathrm{P_{II}^{[N]}}$ are classified in the following theorem.

Theorem 8.2. *The equation $\mathrm{P_{II}^{[N]}}$ has a unique rational solution if and only if $\alpha_N = n \in \mathbb{Z}$, which has the form*

$$w(z) = P_{n^2-1}(z)/Q_{n^2}(z) \tag{8.23}$$

where $P_{n^2-1}(z)$ and $Q_{n^2}(z)$ are polynomials of degree $n^2 - 1$ and n^2 respectively, with no common roots.

Proof. See Gromak [78]; also Clarkson and Mansfield [38] who derive associated special polynomials and study their properties including the distribution of their roots in the complex plane. □

Other studies of equations in the $\mathrm{P_{II}}$ hierarchy include Gromak, Laine and Shimomura [83, §22], Gromak and Zenchenko [87], Joshi and Mazzocco [119], Kudryashov [144, 145], Kudryashov and Pickering [146], Kudryashov and Soukharev [147].

9 Asymptotic Expansions and Connection Formulae

9.1 First Painlevé Equation

The first type of asymptotic solution of $\mathrm{P_I}$ is given by

$$w(z) = z^{1/2}\left\{\wp(\tfrac{4}{5}z^{5/4} - a; -2, C) + o(1)\right\}, \qquad |z| \to \infty, \tag{9.1}$$

where a and C are arbitrary constants and $\wp(\zeta; g_2, g_3)$ is the Weierstrass elliptic function. There are five possible sectors of validity S_j, $j = 0, \pm1, \pm2$, given by

$$|\arg(z\mathrm{e}^{2j\mathrm{i}\pi/5})| \le \tfrac{1}{5}\pi - \delta(< \tfrac{1}{5}\pi), \tag{9.2}$$

and for given values of a, C, and j the solution (9.1) is unique.

A second type of asymptotic solution of $\mathrm{P_I}$ is given by

$$w(z) \sim \pm\frac{\mathrm{i}z^{1/2}}{\sqrt{6}}\sum_{n=0}^{\infty}\frac{a_n}{z^{5n/2}}, \qquad |z| \to \infty, \tag{9.3}$$

where $a_0 = 1$, and for $n \ge 0$

$$a_{n+1} = \frac{(1-25n^2)\mathrm{i}}{8\sqrt{6}}a_n - \tfrac{1}{2}\sum_{m=1}^{n} a_m a_{n+1-m}. \tag{9.4}$$

There are no free parameters in the coefficients of the divergent asymptotic series (9.3), but solutions having this expression have a free complex parameter hidden in exponentially small terms. There are five sectors of validity for (9.3), given by

$$|\arg(z\mathrm{e}^{(2j+1)\mathrm{i}\pi/5})| \le \tfrac{2}{5}\pi - \delta(< \tfrac{2}{5}\pi), \tag{9.5}$$

for $j = 0, \pm1, \pm2$. For any sector given by (9.5), there exists a solution of $\mathrm{P_I}$, a *tronquée solution*, whose asymptotic behavior as $|z| \to \infty$ in this sector is given by (9.3). There also exists a unique solution of $\mathrm{P_I}$, the *tritronquée solution*, whose asymptotic expansion is given by (9.3) in the sector

$$|\arg(z\mathrm{e}^{\mathrm{i}\pi})| \le \tfrac{4}{5}\pi - \delta(< \tfrac{4}{5}\pi). \tag{9.6}$$

There is a solution of $\mathrm{P_I}$ such that

$$w(z) = -\sqrt{\tfrac{1}{6}|z|} + d|z|^{-1/8}\sin\{\phi(z) - \theta_0\} + o(|z|^{-1/8}) \qquad z \to -\infty, \tag{9.7}$$

where

$$\phi(z) = (24)^{1/4}\left(\tfrac{4}{5}|z|^{5/4} + \tfrac{1}{8}d^2\ln|z|\right), \tag{9.8}$$

and d and θ_0 are constants.

There is also a solution of (1.1) such that

$$w(z) \sim \sqrt{\tfrac{1}{6}|z|} \qquad z \to -\infty. \tag{9.9}$$

This solution is not unique.

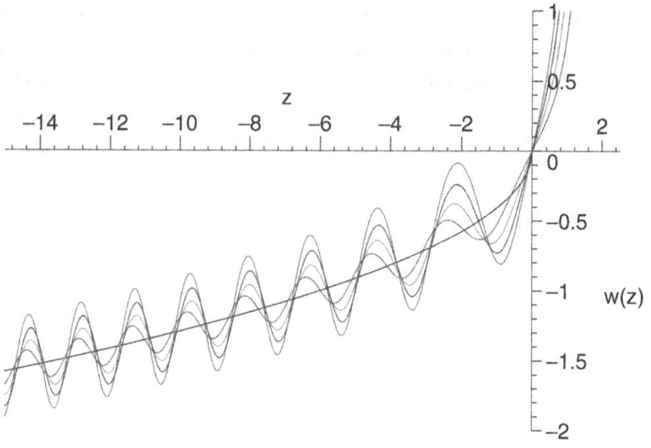

Fig. 6. $k = 0.5$ (red), 0.75 (green), 1 (blue), 1.25 (magenta)

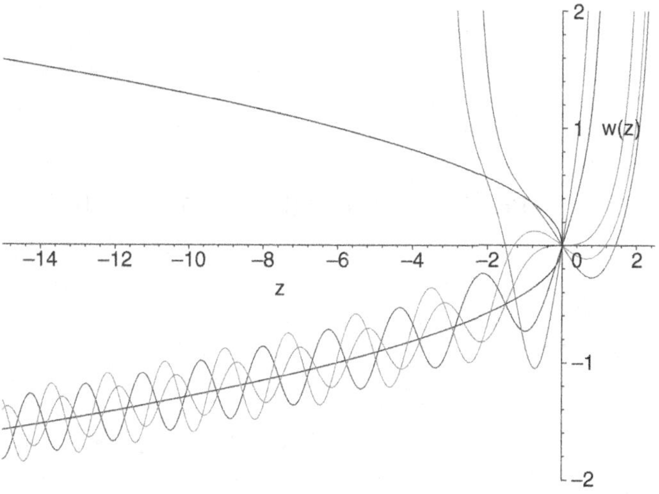

Fig. 7. $k = -0.5$ (red), -0.25 (green), 0 (gold), 1 (blue), 2 (magenta)

Next, for given initial conditions, $w(0) = 0$, $w'(0) = k$, with k real, $w(z)$ has at least one pole on the real axis. There are two special values of k, k_1 and k_2, with the properties $-0.451428 < k_1 < -0.451427$ and $1.851853 < k_2 < 1.851855$, such that: (a) if $k < k_1$, then $w(z) > 0$ for $z_0 < z < 0$, where z_0 is the first pole on negative real axis; (b) if $k_1 < k < k_2$, then $w(z)$ oscillates about and is asymptotic to $-\sqrt{\frac{1}{6}|z|}$; (c) if $k_2 < k$, then $w(z)$ changes sign once, from positive to negative as z passes from z_0 to 0. Plots of solutions of $\mathrm{P_I}$ with $w(0) = 0$ and $w'(0) = k$ for various values of k, and the parabola $6w^2 + z = 0$, which is shown in black, are given in Figures 6–9. The two graphs in Figure 8 are indistinguishable when z exceeds -5.2, approximately. The two graphs in Figure 9 are indistinguishable when z exceeds -4.8, approximately.

For further information about the asymptotic behaviour of solutions of P_I see Joshi and Kruskal [118], Kapaev [131, 135], Kapaev and Kitaev [137], Kitaev [141].

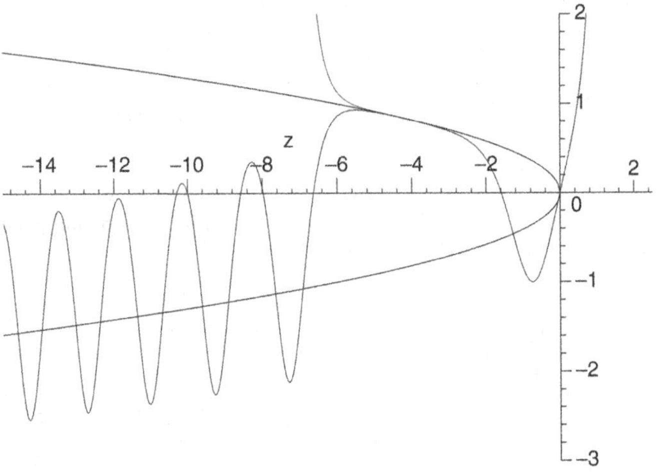

Fig. 8. $k = 1.851853$ (blue), 1.851855 (red)

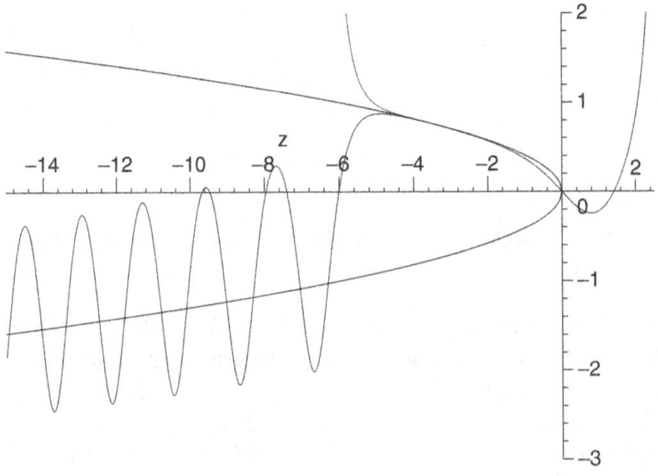

Fig. 9. $k = -0.451427$ (blue), -0.451428 (red)

9.2 Second Painlevé Equation

Generic $\mathrm{P_{II}}$ functions are meromorphic functions, which have an infinity of simple poles accumulating at the essential singularity at $z = \infty$. Making the transformation

$$w(z) = z^{1/2}u(\zeta), \qquad \zeta = \tfrac{2}{3}z^{3/2}, \tag{9.10}$$

in $\mathrm{P_{II}}$ gives

$$\frac{\mathrm{d}^2u}{\mathrm{d}\zeta^2} = 2u^3 + u - \frac{1}{\zeta}\frac{\mathrm{d}u}{\mathrm{d}\zeta} + \frac{u}{9\zeta^2} + \frac{2\alpha}{3\zeta}. \tag{9.11}$$

Thus, in three sectors of angle $\frac{2}{3}\pi$, the generic $\mathrm{P_{II}}$ function has the asymptotics

$$w(z) \sim z^{1/2}u(\zeta), \qquad \zeta = \tfrac{2}{3}z^{3/2}, \tag{9.12}$$

where $u(\zeta)$ satisfies the Jacobian elliptic function equation

$$\left(\frac{\mathrm{d}u}{\mathrm{d}\zeta}\right)^2 = u^4 + u^2 + K, \tag{9.13}$$

with K an arbitrary constant. The parameters in the elliptic function $u(\zeta)$ change across the Stokes lines at 0 and $\pm\frac{2}{3}\pi$ from the positive real axis of the complex z-plane.

There is a family of solutions of $\mathrm{P_{II}}$ with the asymptotic behaviour

$$w(z) \sim \pm\frac{iz^{1/2}}{\sqrt{2}}\sum_{n=0}^{\infty}\frac{a_n}{z^{3n/2}}, \qquad |z| \to \infty, \tag{9.14}$$

with $a_0 = 1$, $a_1 = \mp\frac{1}{2}\sqrt{2}\,\mathrm{i}\,\alpha$ and for $n \geq 0$

$$a_{n+2} = -\tfrac{1}{8}(9n^2-1)a_n - \tfrac{1}{2}\left\{\sum_{k=1}^{n+1}a_k a_{n+2-k} + \sum_{k=1}^{n+1}\sum_{m=0}^{k}a_m a_{k-m}a_{n+2-k}\right\}.$$

The first few coefficients in (9.14) are

$$a_2 = \tfrac{1}{8}(6\alpha^2+1), \qquad a_3 = \pm\sqrt{2}\,\mathrm{i}\,\frac{\alpha(16\alpha^2+11)}{16},$$
$$a_4 = -\frac{410\alpha^4+708\alpha^2+73}{128}, \qquad a_5 = \mp\sqrt{2}\,\mathrm{i}\,\frac{\alpha(768\alpha^4+2504\alpha^2+1021)}{128}.$$

Note that the asymptotic series (9.14) is a divergent series. There is a second family of solutions of $\mathrm{P_{II}}$ with the asymptotic behaviour

$$w(z) \sim -\frac{\alpha}{z}\sum_{n=0}^{\infty}\frac{b_n}{z^{3n}}, \qquad |z| \to \infty, \tag{9.15}$$

where $b_0 = 1$ and

$$b_{n+1} = (3n+2)(3n+1)b_n - 2\alpha^2 \sum_{k=0}^{n} \sum_{m=0}^{k} b_m b_{k-m} b_{n-k}, \qquad n \geq 0.$$

The first few coefficients in (9.15) are

$$\begin{aligned}
b_1 &= -2(\alpha^2 - 1),\\
b_2 &= 4(\alpha^2 - 1)(3\alpha^2 - 10),\\
b_3 &= -8(\alpha^2 - 1)(12\alpha^4 - 117\alpha^2 + 280),\\
b_4 &= 16(\alpha^2 - 1)(55\alpha^6 - 1091\alpha^4 + 7336\alpha^2 - 15400).
\end{aligned}$$

Again the asymptotic series (9.15) is a divergent series, unless $\alpha = \pm 1$.

As in the case of (9.3), there are no free parameters in (9.14) and (9.15), but solutions having these asymptotic expressions have a free complex parameter hidden in exponentially small terms. For further details see Its and Kapaev [106], Kapaev [136].

9.3 Connection Formulae for $\mathrm{P_{II}}$

Here we consider the special case of $\mathrm{P_{II}}$ with $\alpha = 0$, i.e.,

$$w'' = 2w^3 + zw, \tag{9.16}$$

with boundary condition

$$w(z) \to 0, \qquad z \to +\infty. \tag{9.17}$$

The "classic connection problem" for $\mathrm{P_{II}}$ concerning solutions of (9.16) satisfying (9.17) is given in the following theorem.

Theorem 9.1. *Any solution of (9.16), satisfying (9.17), is asymptotic to $k\,\mathrm{Ai}(x)$, for some k, with $\mathrm{Ai}(z)$ the Airy function. Conversely, for any k, there is a unique solution $w_k(z)$ of (9.16) which is asymptotic to $k\,\mathrm{Ai}(z)$ as $z \to +\infty$, for some k. If $|k| < 1$, then this solution exists for all real z as $z \to -\infty$, and as $z \to -\infty$*

$$w_k(z) = d|z|^{-1/4} \sin\left\{\tfrac{2}{3}|z|^{3/2} - \tfrac{3}{4}d^2 \log|z| - \theta_0\right\} + o(|z|^{-1/4}), \tag{9.18}$$

for some constants d and θ_0 which depend on k. If $|k| = 1$ then

$$w_k(z) \sim \mathrm{sgn}(k)\sqrt{-\tfrac{1}{2}z}, \qquad z \to -\infty. \tag{9.19}$$

If $|k| > 1$ then $w_k(z)$ has a pole at a finite z_0, dependent on k,

$$w_k(z) \sim \mathrm{sgn}(k)(z - z_0)^{-1} \qquad z \downarrow z_0. \tag{9.20}$$

Proof. See Hastings and McLeod [91]. □

The specific dependent of the constants d and θ_0 in (9.18) on the parameter k is given as follows.

Theorem 9.2. *The connection formulae d and θ_0 in the asymptotic expansion (9.18) are given by*

$$d^2(k) = -\pi^{-1}\ln(1-k^2), \tag{9.21}$$

$$\theta_0(k) = \tfrac{3}{2}d^2\ln 2 + \arg[\Gamma(1-\tfrac{1}{2}\mathrm{i}d^2)] - \tfrac{1}{4}\pi. \tag{9.22}$$

with $\Gamma(z)$ the Gamma function.

The amplitude connection formula (9.21) and the phase connection formula (9.22) were first conjectured, derived heuristically and subsequently verified numerically by Ablowitz and Segur [7, 198]. Some years later Clarkson and McLeod [40] gave a rigorous proof of (9.21), using the Gel'fand-Levitan-Marchenko integral equation (2.3) in §2.1. Suleimanov [209] derived (9.21) and (9.22) using the isomonodromy problem (2.12) in §2.2; see also [103, 104, 149]. Deift and Zhou [48, 49] rigorously proved that these connection formulae using a nonlinear version of the classical steepest descent method for oscillating Riemann-Hilbert problems. Subsequently Bassom, Clarkson, Law and McLeod [19] developed a uniform approximation method, which is rigorous, removes the need to match solutions and can leads to a simpler solution of this connection problem for the special case of $\mathrm{P_{II}}$ given by (9.16). As discussed in §10.2, the solution of (9.16) satisfying the boundary conditions (9.17) and (9.19) arises in a number of mathematical and physical problems.

Numerical studies of the solutions of (9.16) satisfying the boundary condition (9.17) are discussed by Miles [167, 168] and Rosales [193]. In Figure 10, $w_k(z)$ with $k = 0.5$ and $0.5\,\mathrm{Ai}(z)$ are plotted and the two graphs are indistinguishable when z exceeds -0.4, approximately. In Figures 11 and 12 $w_k(z)$ is plotted with $k = 1 \pm 0.001$ and $k = 1 \pm 10^{-10}$, respectively. These illustrate that when $0 < k < 1$ then $w_k(z)$ exists for all z whilst for $k > 1$ then $w_k(z)$ has a pole at a finite z_0, however the dependence of z_0 on k is currently unknown. In Figure 13, $w_k(z)$ with $k = 0.9$ and the asymptotic expansion (9.18) for $k = 0.9$ are plotted.

Letting $w \mapsto \mathrm{i}w$ in equation (9.16) gives

$$w'' + 2w^3 - zw = 0, \tag{9.23}$$

which is considered in [103, 104]. Any real solution of (9.23) for real $z \in \mathbb{R}$ is smooth for all z, and satisfies the asymptotic condition,

$$w(z) = d(-z)^{-1/4}\sin\left(\tfrac{2}{3}|z|^{3/2} + \tfrac{3}{4}d^2\ln|z| - \theta_0\right) + \mathcal{O}\left(|z|^{-5/4}\ln|z|\right), \qquad z \to -\infty, \tag{9.24}$$

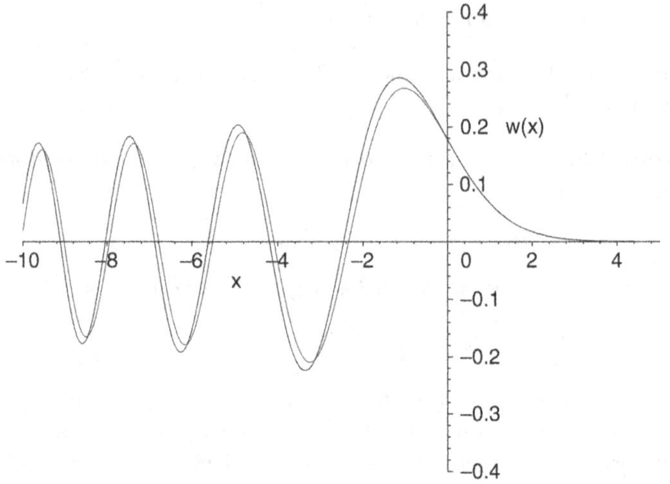

Fig. 10. $w_k(z)$ with $k = 0.5$ (blue) and $0.5\,\mathrm{Ai}(z)$ (red)

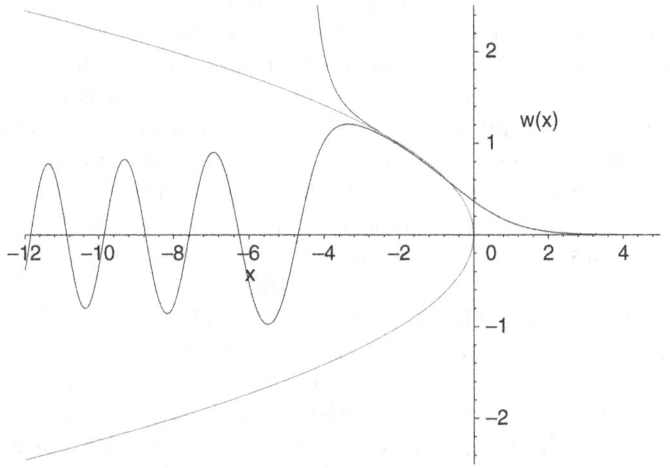

Fig. 11. $w_k(z)$ with $k = 0.999$ (blue), $k = 1.001$ (red)

with d and θ_0 arbitrary real constants, which determines the solution $w(z)$ uniquely. If the parameters d and θ_0 are related by

$$\theta_0 = \tfrac{3}{2}d^2 \ln 2 - \tfrac{1}{4}\pi - \arg\left[\Gamma\left(\tfrac{1}{2}\mathrm{i}d^2\right)\right] + n\pi, \qquad n \in \mathbb{Z}, \tag{9.25}$$

then the solution $w(z)$ decreases exponentially as $z \to +\infty$. More precisely,

$$w(z) \sim k\,\mathrm{Ai}(z), \qquad z \to +\infty, \tag{9.26}$$

where the connection formula for k is given by (cf., (9.18))

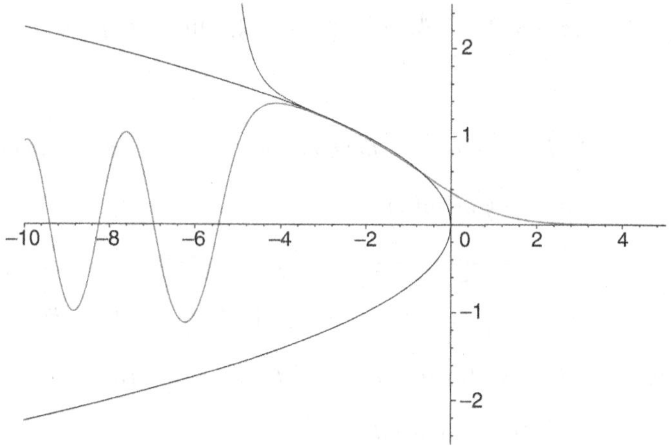

Fig. 12. $w_k(z)$ with $k = 1 - 10^{-10}$ (green), $k = 1 + 10^{-10}$ (blue)

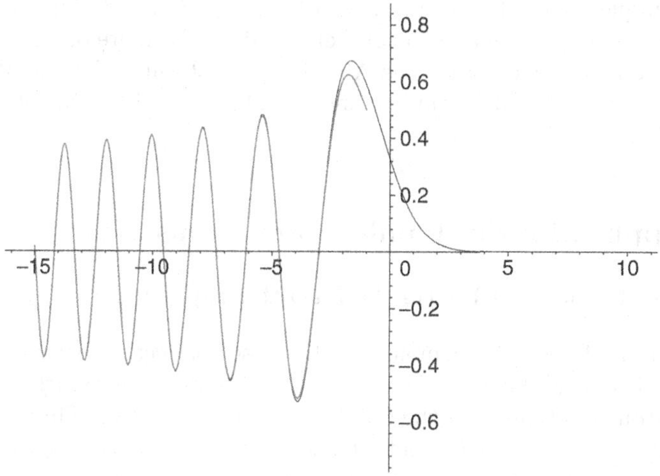

Fig. 13. $w_k(z)$ with $k = 0.9$ (blue) and the asymptotic expansion (9.18) for $k = 0.9$ (green)

$$d^2 = \pi^{-1} \ln(1 + k^2), \qquad \operatorname{sgn}(k) = (-1)^n. \tag{9.27}$$

For generic values of the parameters d and θ_0, i.e., if

$$\theta_0 - \tfrac{3}{2} d^2 \ln 2 + \tfrac{1}{4}\pi + \arg\left[\Gamma\left(\tfrac{1}{2} \mathrm{i} d^2\right)\right] \notin \pi\mathbb{Z}, \tag{9.28}$$

then the solution $w(z)$ oscillates and $w(z) \to \pm\sqrt{\tfrac{1}{2} z}$, as $z \to +\infty$. The corresponding asymptotics is given by

$$w(z)=\sigma\sqrt{\tfrac{1}{2}z}+\sigma(2z)^{-1/4}\rho\cos\left(\tfrac{2}{3}z^{3/2}-\tfrac{3}{2}\rho^2\ln z+\Theta\right) \\ +\,o\left(z^{-1/2}\right),\qquad z\to+\infty. \tag{9.29}$$

Theorem 9.3. *The connection formulae for σ, ρ and Θ are given by*

$$\sigma=-\operatorname{sgn}(\operatorname{Im}(\mu)), \tag{9.30}$$

$$\rho^2=\frac{1}{\sqrt{2}\,\pi}\ln\left(\frac{1+|\mu|^2}{2|\operatorname{Im}(\mu)|}\right),\qquad \rho>0, \tag{9.31}$$

$$\Theta=-\tfrac{3}{4}\pi-\tfrac{7}{2}\rho^2\ln 2+\arg[\Gamma(\mathrm{i}\rho^2)]+\arg(1+\mu^2), \tag{9.32}$$

where the complex number μ is given in terms of the parameters d and θ_0 by

$$|\mu|^2=\exp(\pi d^2)-1, \tag{9.33}$$

$$\arg(\mu)=\tfrac{3}{2}d^2\ln 2-\tfrac{1}{4}\pi-\arg\left[\Gamma\left(\tfrac{1}{2}\mathrm{i}d^2\right)\right]+\theta_0. \tag{9.34}$$

We remark that each of the equations (9.26) and (9.29) determines the solution $w(z)$ uniquely (see Its and Kapaev [104], Its, Fokas and Kapaev [103], Deift and Zhou [49]; also Its and Novokshënov [107]). For more on asymptotics and connection formulae for $\mathrm{P_{II}}$ see Abdullaev [1], Joshi and Kruskal [117, 118], Kapaev [132, 133, 134], Kapaev and Novokshënov [138], Novokshënov [177, 178, 179].

10 Applications of Painlevé Equations

10.1 Reductions of Partial Differential Equations

Ablowitz and Segur [7] demonstrated a close connection between completely integrable partial differential equations solvable by inverse scattering, so-called the soliton equations, and the Painlevé equations $\mathrm{P_I}$–$\mathrm{P_{VI}}$. This paper was the catalyst for much of the current interest in the Painlevé equations. The inverse scattering method was originally developed by Gardner, Greene, Kruskal and Miura [69] to solve the Cauchy problem for the Korteweg-de Vries (KdV) equation

$$u_t+6uu_x+u_{xxx}=0. \tag{10.1}$$

Example 10.1. The mKdV equation (2.2) is solvable by inverse scattering [4] and has the scaling reduction

$$u(x,t)=(3t)^{-1/3}w(z),\qquad z=x/(3t)^{1/3}, \tag{10.2}$$

where $w(z)$ satisfies $\mathrm{P_{II}}$ with α a constant of integration. Ablowitz and Segur [7] noted that since the Miura transformation $u=-v^2-v_x$, relates solutions of the mKdV equation (2.2) to solutions of the KdV equation (10.1) then solutions of the scaling reduction of the KdV equation given by

$$u(x,t) = (3t)^{-2/3}v(z), \qquad z = x/(3t)^{1/3}, \tag{10.3}$$

where $v(z)$ satisfies

$$v''' + 6vv' - (2v + zv') = 0, \tag{10.4}$$

whose solutions are also expressible in terms of solutions of $\mathrm{P_{II}}$. There exists a one-to-one correspondence between solutions of equation (10.4) and those of $\mathrm{P_{II}}$, given by

$$v = -w' - w^2, \qquad w = \frac{v' + \alpha}{2v - z} \tag{10.5}$$

see [54] for further details. The KdV equation (10.1) also has the similarity reduction

$$u(x,t) = w(z) - \lambda t, \qquad z = x + 3\lambda t^2, \tag{10.6}$$

with λ an arbitrary constant and where $w(z)$ satisfies

$$w'' + 3w^2 = \lambda z + A, \tag{10.7}$$

with A a constant of integration, which is equivalent to $\mathrm{P_I}$ through a scaling of the variables; see [54] for further details.

Example 10.2. The sine-Gordon equation

$$u_{xt} = \sin u, \tag{10.8}$$

is solvable by inverse scattering [4]. It has a scaling reduction

$$u(x,t) = v(z), \qquad z = xt. \tag{10.9}$$

After making the transformation $w = \exp(iv)$, to put the equation in rational form, then $w(z)$ satisfies

$$w'' = \frac{(w')^2}{w} - \frac{w'}{z} + \frac{w^2 - 1}{z}, \tag{10.10}$$

which is the special case of $\mathrm{P_{III}}$ with $\alpha = -\beta = 1$ and $\gamma = \delta = 0$; see [7] for further details.

Example 10.3. The Boussinesq equation

$$u_{tt} = u_{xx} - 6(u^2)_{xx} + u_{xxxx}, \tag{10.11}$$

is solvable by inverse scattering [3, 235]. It has the travelling wave solution

$$u(x,t) = v(\zeta), \qquad \zeta = x - ct, \tag{10.12}$$

with c an arbitrary constant and where $v(\zeta)$ satisfies

$$\frac{\mathrm{d}^2 v}{\mathrm{d}\zeta^2} = 6v^2 + (c^2 - 1)v + A\zeta + B, \tag{10.13}$$

with A and B constants of integration. Depending upon whether $A = 0$ or $A \neq 0$, then the solution $v(\zeta)$ of (10.13) is expressible in terms of either elliptic functions or solutions of $\mathrm{P_I}$, respectively [7]. The Boussinesq equation (10.11) also has the scaling reduction

$$u(x,t) = t^{-1}v(\zeta) + \tfrac{1}{6}, \qquad \zeta = x/t^{1/2}, \tag{10.14}$$

where $v(\zeta)$ satisfies

$$\frac{\mathrm{d}^4v}{\mathrm{d}\zeta^4} = 6v\frac{\mathrm{d}^2v}{\mathrm{d}\zeta^2} + 6\left(\frac{\mathrm{d}v}{\mathrm{d}\zeta}\right)^2 + \tfrac{1}{4}\zeta^2\frac{\mathrm{d}^2v}{\mathrm{d}\zeta^2} + \tfrac{7}{4}\zeta\frac{\mathrm{d}v}{\mathrm{d}\zeta} + 2v. \tag{10.15}$$

This equation is solvable in terms of solutions of $\mathrm{P_{IV}}$ since the solution of (10.15) is given by

$$v(\zeta) = \tfrac{1}{4}k^2\left[w' + w^2 + 2zw + \tfrac{2}{3}z^2 + \tfrac{2}{3}(\alpha - 1)\right], \qquad z = \tfrac{1}{2}k\,\zeta, \tag{10.16}$$

where $w(z)$ satisfies $\mathrm{P_{IV}}$ and $k^4 + 3 = 0$; see [37] for further details. Further the Boussinesq equation (10.11) has the "accelerating wave reduction"

$$u(x,t) = v(\zeta) - \tfrac{2}{3}\mu^2t^2, \qquad \zeta = x + \mu t^2, \tag{10.17}$$

with μ an arbitrary constant and where $v(\zeta)$ satisfies

$$\frac{\mathrm{d}^3v}{\mathrm{d}\zeta^3} = 6v\frac{\mathrm{d}v}{\mathrm{d}\zeta} - \frac{\mathrm{d}v}{\mathrm{d}\zeta} + 2\mu v - \tfrac{4}{3}\mu^2\zeta + A, \tag{10.18}$$

with A a constant of integration. This equation is equivalent to (10.4) since setting

$$v(\zeta) = -k^2v(z) + \tfrac{1}{6}(k^2z + 1), \qquad z = k\zeta,$$

with $k^3 = 2\mu$ and $A = \frac{1}{3}\mu$, in (10.18) yields (10.4) Thus (10.4) is solvable in terms of solutions of $\mathrm{P_{II}}$; see [37] for further details. We remark that Clarkson and Kruskal [37] show that there are six classes of symmetry reductions of the Boussinesq equation (10.11) which reduce it to ordinary differential equations that are solvable either in terms of solutions of $\mathrm{P_I}$, $\mathrm{P_{II}}$, $\mathrm{P_{IV}}$ or elliptic functions, depending upon the reduction.

Example 10.4. The nonlinear Schrödinger equation

$$\mathrm{i}u_t = u_{xx} - 2\sigma|u|^2u, \qquad \sigma = \pm 1, \tag{10.19}$$

was the second soliton equation to be solved by the inverse scattering method [234]. It has the scaling reduction

$$u(x,t) = t^{-1/2}U(\zeta)\exp\left(\tfrac{1}{2}\mathrm{i}\mu\ln t\right), \qquad \zeta = x/t^{1/2}, \tag{10.20}$$

with μ an arbitrary constant and $U(\zeta)$ satisfies

$$\frac{d^2U}{d\zeta^2} = \tfrac{1}{2}(i-\mu)U + \tfrac{1}{2}i\zeta\frac{dU}{d\zeta} + 2\sigma|U|^2U. \tag{10.21}$$

Setting $U(\zeta) = R(\zeta)\exp\{i\theta(\zeta)\}$ and equating real and imaginary parts yields

$$\frac{d^2R}{d\zeta^2} - R\left(\frac{d\theta}{d\zeta}\right)^2 = \tfrac{1}{2}R\zeta\frac{d\theta}{d\zeta} - \tfrac{1}{2}\mu R + 2\sigma R^3, \tag{10.22a}$$

$$2\frac{dR}{d\zeta}\frac{d\theta}{d\zeta} + R\frac{d^2\theta}{d\zeta^2} + \tfrac{1}{2}\zeta\frac{dR}{d\zeta} + \tfrac{1}{2}R = 0. \tag{10.22b}$$

Multiplying equation (10.22b) by R and integrating yields

$$\frac{d\theta}{d\zeta} = -\tfrac{1}{4}\zeta + \frac{C}{R^2(\zeta)} - \frac{1}{4R^2(\zeta)}\int^{\zeta} R^2(s)\,ds, \tag{10.23}$$

with C an arbitrary constant. Substituting this into equation (10.22a) and setting $W(\zeta) = \int^{\zeta} R^2(s)\,ds - 4C$ yields the third-order equation

$$2\frac{dW}{d\zeta}\frac{d^3W}{d\zeta^3} = \left(\frac{d^2W}{d\zeta^2}\right)^2 - (\tfrac{1}{4}\zeta^2 + 2\mu)\left(\frac{dW}{d\zeta}\right)^2 + \tfrac{1}{4}W^2 + 8\sigma\left(\frac{dW}{d\zeta}\right)^3, \tag{10.24}$$

which has first integral

$$\left(\frac{d^2W}{d\zeta^2}\right)^2 = -\tfrac{1}{4}\left(W - \zeta\frac{dW}{d\zeta}\right)^2 - 2\mu\left(\frac{dW}{d\zeta}\right)^2 + 4\sigma\left(\frac{dW}{d\zeta}\right)^3 + K\frac{dW}{d\zeta}, \tag{10.25}$$

with K an arbitrary constant. This is equation SD-I.c in the classification of second-order, second-degree ordinary differential equations with the Painlevé property due to Cosgrove and Scoufis [45], an equation first derived and solved by Chazy [27] and rederived by Bureau [25]. Equation (10.25) is solvable in terms of P_{IV} since the solution of (10.25) is given by

$$\begin{aligned} W(\zeta) = -\frac{\sigma k}{2}\Bigg\{&\frac{1}{4w}\left(\frac{dw}{dz}\right)^2 - \tfrac{1}{4}w^3 - zw^2 + (1+\alpha-z^2)w \\ &+ \frac{\beta}{2w} + \tfrac{2}{3}(\alpha+1)z + \tfrac{4}{3}i\mu z\Bigg\}, \qquad z = k\,\zeta, \end{aligned} \tag{10.26}$$

where $w(z)$ satisfies P_{IV} and $16k^4 + 1 = 0$, provided that

$$K = \tfrac{1}{9}\left[2\mu^2 + i\mu(\alpha+1) + (\alpha+1)^2\right], \qquad \beta = -\tfrac{2}{9}(\alpha + 2i\mu + 1)^2$$

see [24, 45, 67] for further details.

The nonlinear Schrödinger equation (10.19) also has the accelerating wave reduction

$$u(x,t) = U(\zeta)\exp\left\{\mathrm{i}\mu\left(\zeta t + \tfrac{1}{3}\mu t^3\right)\right\}, \qquad \zeta = x - \mu t^2, \tag{10.27}$$

with μ an arbitrary constant and $U(\zeta)$ satisfies

$$\frac{\mathrm{d}^2 U}{\mathrm{d}\zeta^2} = \mu\zeta U + 2\sigma|U|^2 U. \tag{10.28}$$

Setting $U(\zeta) = R(\zeta)\exp\{\mathrm{i}\theta(\zeta)\}$ and equating real and imaginary parts yields

$$\frac{\mathrm{d}^2 R}{\mathrm{d}\zeta^2} - R\left(\frac{\mathrm{d}\theta}{\mathrm{d}\zeta}\right)^2 = 2\sigma R^3 + \mu\zeta R, \tag{10.29a}$$

$$2\frac{\mathrm{d}R}{\mathrm{d}\zeta}\frac{\mathrm{d}\theta}{\mathrm{d}\zeta} + R\frac{\mathrm{d}^2\theta}{\mathrm{d}\zeta^2} = 0. \tag{10.29b}$$

Multiplying equation (10.29b) by R and integrating yields

$$\frac{\mathrm{d}\theta}{\mathrm{d}\zeta} = \frac{C}{R^2(\zeta)} \tag{10.30}$$

with C an arbitrary constant. Substituting this into equation (10.29a) and setting $R(\zeta) = W^{1/2}(\zeta)$ yields

$$\frac{\mathrm{d}^2 W}{\mathrm{d}\zeta^2} = \frac{1}{2W}\left(\frac{\mathrm{d}W}{\mathrm{d}\zeta}\right)^2 + 4W^2 + 2\mu\zeta W + \frac{2C^2}{W}. \tag{10.31}$$

After rescaling W and ζ, if necessary, this equation is equation XXXIV in Chapter 14 of Ince [101] which is solvable in terms of solutions of $\mathrm{P_{II}}$. The solution of (10.31) is

$$W(\zeta) = \tfrac{1}{2}\sigma k^2\left(\frac{\mathrm{d}w}{\mathrm{d}z} + w^2 + \tfrac{1}{2}z\right), \qquad z = -k\,\zeta, \tag{10.32}$$

where $w(z)$ satisfies $\mathrm{P_{II}}$, $k^3 = 2\mu$ and

$$\mu^2(\alpha + \tfrac{1}{2})^2 + 4C^2 = 0,$$

see [67] for further details.

10.2 Combinatorics

Let S_N be the group of permutations $\boldsymbol{\pi}$ of the numbers $1, 2, \ldots, N$. For $1 \le i_1 < \ldots < i_k \le N$, then $\boldsymbol{\pi}(i_1), \boldsymbol{\pi}(i_2), \ldots, \boldsymbol{\pi}(i_N)$ is an increasing subsequence of $\boldsymbol{\pi}$ of length k if

$$\boldsymbol{\pi}(i_1) < \boldsymbol{\pi}(i_2) < \cdots < \boldsymbol{\pi}(i_N).$$

Let $\ell_N(\boldsymbol{\pi})$ be the length of the longest increasing subsequence of $\boldsymbol{\pi}$ and define

$$q_N(n) \equiv \mathrm{Prob}(\ell_N(\boldsymbol{\pi}) \leq n). \tag{10.33}$$

The problem is to determine the asymptotics of $q_N(n)$ as $N \to \infty$, which Baik, Deift and Johansson [16] expressed in terms of solutions of $\mathrm{P_{II}}$ — see Theorem 10.1. Define the distribution function $F_2(s)$ by

$$F_2(s) = \exp\left\{ -\int_s^\infty (z-s)w^2(z)\,\mathrm{d}z \right\} \tag{10.34}$$

which is known as the Tracy-Widom distribution first introduced in [214], and $w(z)$ satisfies (9.16), the special case of $\mathrm{P_{II}}$ with $\alpha = 0$, and the boundary conditions

$$w(z) \sim \begin{cases} \mathrm{Ai}(z), & \text{as} \quad z \to \infty, \\ \sqrt{-\frac{1}{2}z}, & \text{as} \quad z \to -\infty. \end{cases} \tag{10.35}$$

where $\mathrm{Ai}(z)$ is the Airy function. Recall from Theorem 9.1 that Hastings and McLeod [91] proved there is a unique solution of (9.16) with boundary conditions (10.35), which is uniquely characterized by its asymptotics as $z \to \infty$. Baik, Deift and Johansson [16] proved the following theorem.

Theorem 10.1. *Let S_N be the group of all permutations of N numbers with uniform distribution and let $\ell_N(\boldsymbol{\pi})$ be the length of the longest increasing subsequence of $\boldsymbol{\pi} \in S_N$. Let χ be a random variable whose distribution function is the distribution function $F_2(t)$. Then, as $N \to \infty$,*

$$\chi_N := \frac{\ell_N(\boldsymbol{\pi}) - 2\sqrt{N}}{N^{1/6}} \to \chi,$$

in distribution, i.e.,

$$\lim_{N\to\infty} \mathrm{Prob}\left(\frac{\ell_N - 2\sqrt{N}}{N^{1/6}} \leq s \right) = F_2(s),$$

for all $s \in \mathbb{R}$.

The Tracy-Widom distribution function $F_2(s)$ given by (10.34) arose in random matrix theory were it gives the limiting distribution for the normalised largest eigenvalue in the Gaussian Unitary Ensemble (GUE) of $N \times N$ Hermitian matrices [214]. Specifically for the GUE

$$\lim_{N\to\infty} \mathrm{Prob}\left(\left(\lambda_{\max} - \sqrt{2N}\right) \sqrt{2}\, N^{1/6} \leq s \right) = F_2(s), \tag{10.36}$$

where $F_2(s)$ is given by (10.34). See, for example, [47, 61, 220, 221] and the references therein, for discussions of the application of Painlevé equations in combinatorics and random matrices.

We remark that work by Montgomery [170], followed by extensive numerical calculations by Odlyzko [180], on zeros of the Riemann zeta function have given convincing numerical evidence that the normalized consecutive spacings follow the GUE distribution; this is often referred to as the *Montgomery-Odlyzko law.* Rudnick and Sarnak [194, 195] have proved a restricted form of this hypothesis. These results of Rudnick and Sarnak [194, 195] are also compatible with the belief that the distribution of the spacings between zeros, not only of the Riemann zeta function, but also of quite general automorphic L-functions over $\mathbb{Q}$, are all given by this Montgomery-Odlyzko law. Subsequently Katz and Sarnak [139, 140] established the Montgomery-Odlyzko law for wide classes of zeta and L-functions over finite fields.

We further remark that the solution of $\mathrm{P_{II}}$ with $\alpha = 0$ (9.16) satisfying the boundary conditions (10.35) also arises in several other mathematical and physical applications including: spherical electric probe in a continuum plasma [22, 91], Görtler vortices in boundary layers [88, 89, 90], nonlinear optics [70], Bose-Einstein condensation [46, 157, 161, 222], superheating fields of superconductors [50, 92], universality of the edge scaling for nongaussian Wigner matrices [204], shape fluctuations in polynuclear growth models [114, 191, 192], and the distribution of eigenvalues for covariance matrices and Wishart distributions [115].

10.3 Orthogonal Polynomials

Suppose $\{p_n(x)\}_{n=0}^{\infty}$ are a set of orthonormal polynomials with respect to the weight function $w(x;z)$ on the interval (α, β), with $-\infty \le \alpha < \beta \le \infty$, i.e.,

$$\int_\alpha^\beta p_m(x)\, p_n(x)\, w(x;z)\, \mathrm{d}x = \delta_{nm}, \tag{10.37}$$

for $n, m = 1, 2, \ldots$, then the $p_n(x)$'s also satisfy the three-point recurrence relation

$$a_{n+1}(z)p_{n+1}(x) = [x - b_n(z)]p_n(x) - a_n(z)p_{n-1}(x), \tag{10.38}$$

for $n = 1, 2, \ldots$ (cf. [158]). If $w(-x) = w(x)$ for all real x, then $b_n \equiv 0$ and so (10.38) reduces to

$$a_{n+1}(z)p_{n+1}(x) = xp_n(x) - a_n(z)p_{n-1}(x). \tag{10.39}$$

Example 10.5. Consider the orthogonal polynomials $p_n(x)$ with respect to the weight function $w(x;z) = \exp(\frac{1}{3}x^3 + zx)$, then the orthogonal polynomials $p_n(x)$ satisfy

$$\int_{-\infty}^0 p_m(x)\, p_n(x)\, \exp(\tfrac{1}{3}x^3 + zx)\, \mathrm{d}x = \delta_{nm}. \tag{10.40}$$

Here a_n and b_n satisfy the difference equations

$$a_n^2 + a_{n+1}^2 + b_n^2 + z = 0, \qquad n + a_n^2(b_n + b_{n-1}) = 0, \tag{10.41}$$

and the differential-difference equations

$$a_n \frac{\mathrm{d}a_n}{\mathrm{d}z} = a_n^2 b_n + \tfrac{1}{2}n, \qquad \frac{\mathrm{d}b_n}{\mathrm{d}z} = -b_n^2 - 2a_n^2 - z. \tag{10.42}$$

Eliminating a_n in (10.41) yields

$$\frac{n+1}{b_{n+1}+b_n} + \frac{n}{b_n+b_{n-1}} = b_n^2 + z, \tag{10.43}$$

which is an alternative form of discrete P_I [56], whilst eliminating a_n in (10.42)

$$\frac{\mathrm{d}^2 b_n}{\mathrm{d}z^2} = 2b_n^3 + 2zb_n - 2n - 1, \tag{10.44}$$

which is equivalent to P_{II}.

Example 10.6. Consider the orthogonal polynomials $p_n(x)$ with respect to the weight function $w(x;z) = \exp(-\frac{1}{4}x^4 - zx^2)$, then the orthogonal polynomials $p_n(x)$ satisfy

$$\int_{-\infty}^{\infty} p_m(x)\, p_n(x)\, \exp(-\tfrac{1}{4}x^4 - zx^2)\, \mathrm{d}x = \delta_{nm}. \tag{10.45}$$

Here $b_n = 0$ and $u_n = a_n^2$ satisfy the differential-difference equation

$$\frac{\mathrm{d}u_n}{\mathrm{d}z} = u_n(u_{n-1} - u_{n+1}), \tag{10.46}$$

which is the Kac-van Moerbeke equation [120], and the difference equation

$$(u_{n+1} + u_n + u_{n-1})u_n = n - 2zu_n, \tag{10.47}$$

which is discrete P_I equation (dP_I) [57, 58]. From (10.46) and (10.47) we obtain

$$2u_{n+1} = \frac{n}{u_n} - \frac{1}{u_n}\frac{\mathrm{d}u_n}{\mathrm{d}z} - 2z - u_n, \tag{10.48a}$$

$$2u_{n-1} = \frac{n}{u_n} + \frac{1}{u_n}\frac{\mathrm{d}u_n}{\mathrm{d}z} - 2z - u_n. \tag{10.48b}$$

Letting $n \to n+1$ in (10.48b) and then eliminating u_{n+1} in (10.48a) yields P_{IV} with $(\alpha, \beta) = (-\frac{1}{2}n, -\frac{1}{2}n^2)$. We remark that Fokas, Its and Kitaev [57, 58] demonstrated a relationship between solutions of P_{IV} and dP_I (10.47) in the context of two-dimensional quantum gravity.

Recently a link between orthogonal polynomials with discontinuous weights and the Painlevé equations has been given by Chen and Pruessner [28].

11 Discussion

This paper gives an introduction to some of the fascinating properties which the Painlevé equations possess including connection formulae, Bäcklund transformations, associated discrete equations, and hierarchies of exact solutions. I feel that these properties show that the Painlevé equations may be thought of as nonlinear analogues of the classical special functions.

There are still several very important open problems relating to the following three major areas of modern theory of Painlevé equations.

(i) Asymptotics and connection formulae for the Painlevé equations using the isomonodromy method, for example the construction of uniform asymptotics around a nonlinear Stokes ray (cf., [106, 135, 136]).
(ii) Bäcklund transformations and exact solutions of Painlevé equations; a summary of many of the currently known results is given in [83].
(iii) The relationship between affine Weyl groups, Painlevé equations, Bäcklund transformations and discrete equations; see, for example, [173], for an introduction to this topic.

The ultimate objective is to provide a complete classification and unified structure for the exact solutions and Bäcklund transformations for the Painlevé equations (and the discrete Painlevé equations) — the presently known results are rather fragmentary and non-systematic.

Acknowledgements

I wish to thank Mark Ablowitz, Andrew Bassom, Chris Cosgrove, Galina Filipuk, Rod Halburd, Andrew Hicks, Andrew Hone, Alexander Its, Nalini Joshi, Martin Kruskal, Elizabeth Mansfield, Marta Mazzocco, Bryce McLeod, Alice Milne, Frank Nijhoff, Andrew Pickering, Craig Tracy and Helen Webster for their helpful comments and illuminating discussions.

References

1. A.S. Abdullaev, *On the theory of the second Painlevé equation,* Soviet Math. Dokl. **28** (1983), 726–729.
2. M.J. Ablowitz and P.A. Clarkson, "*Solitons, Nonlinear Evolution Equations and Inverse Scattering*", L.M.S. Lect. Notes Math., vol. **149**, Cambridge University Press, Cambridge (1991).
3. M.J. Ablowitz and R. Haberman, *Resonantly coupled nonlinear evolution equations,* Phys. Rev. Lett. **35** (1975), 1185–1188.
4. M.J. Ablowitz, D.J. Kaup, A.C. Newell and H. Segur, *The inverse scattering transform — Fourier analysis for nonlinear problems,* Stud. Appl. Math. **53** (1974), 249–315.

5. M.J. Ablowitz, A. Ramani and H. Segur, *A connection between nonlinear evolution equations and ordinary differential equations of Painlevé type. II,* J. Math. Phys. **21** (1980), 1006–1015.
6. M.J. Ablowitz and J. Satsuma, *Solitons and rational solutions of nonlinear evolution equations,* J. Math. Phys **19** (1978), 2180–2186.
7. M.J. Ablowitz and H. Segur, *Exact linearization of a Painlevé transcendent,* Phys. Rev. Lett. **38** (1977), 1103–1106.
8. M.J. Ablowitz and H. Segur, "*Solitons and the Inverse Scattering Transform*", SIAM, Philadelphia (1981).
9. M. Abramowitz and I.A. Stegun, "*Handbook of Mathematical Functions*", 10th edition, Dover, New York (1972).
10. M. Adler and J. Moser, *On a class of polynomials associated with the Korteweg-de Vries equation,* Comm. Math. Phys. **61** (1978), 1–30.
11. H. Airault, *Rational solutions of Painlevé equations,* Stud. Appl. Math. **61** (1979), 31–53.
12. H. Airault, H.P. McKean and J. Moser, *Rational and elliptic solutions of the KdV equation and related many-body problems,* Comm. Pure Appl. Math. **30** (1977), 95–148.
13. D.W. Albrecht, E.L. Mansfield and A.E. Milne, *Algorithms for special integrals of ordinary differential equations,* J. Phys. A: Math. Gen. **29** (1996), 973–991.
14. F.V. Andreev and A.V. Kitaev, *Transformations $RS_4^2(3)$ of the ranks ≤ 4 and algebraic solutions of the sixth Painlevé equation,* Comm. Math. Phys. **228** (2002), 151–176.
15. G. Andrews, R. Askey and R. Roy, "*Special Functions*", Cambridge University Press, Cambridge (1999).
16. J. Baik, P.A. Deift and K. Johansson, *On the distribution of the length of the longest increasing subsequence of random permutations,* J. Amer. Math. Soc. **12** (1999), 1119–1178.
17. A.P. Bassom, P.A. Clarkson and A.C. Hicks, *Bäcklund transformations and solution hierarchies for the fourth Painlevé equation,* Stud. Appl. Math. **95** (1995), 1–71.
18. A.P. Bassom, P.A. Clarkson, A.C. Hicks and J.B. McLeod, *Integral equations and exact solutions for the fourth Painlevé equation,* Proc. R. Soc. Lond. A **437** (1992), 1–24.
19. A.P. Bassom, P.A. Clarkson, C.K. Law and J.B. McLeod, *Application of uniform asymptotics to the second Painlevé transcendent,* Arch. Rat. Mech. Anal. **143** (1998), 241–271.
20. P. Boalch, *The fifty-two icosahedral solutions to Painlevé VI,* preprint `arXiv:math.AG/0406281` (2004).
21. P. Boalch, *From Klein to Painlevé via Fourier, Laplace and Jimbo,* Proc. London Math. Soc. **90** (2005), 167–208.
22. P.C.T. de Boer and L.S.S. Ludford, *Spherical electric probe in a continuum gas,* Plasma Phys. **17** (1975), 29–43.
23. A. Bobenko and U. Eitner, "*Painlevé equations in the Differential Geometry of Surfaces*", Lect. Notes Math., vol. **1753**, Springer-Verlag, Berlin (2000).
24. M. Boiti and F. Pempinelli, *Nonlinear Schrödinger equation, Bäcklund transformations and Painlevé transcendents,* Nuovo Cim. **59B** (1980), 40–58.
25. F. Bureau, *Équations différentielles du second ordre en Y et du second degré en $\ddot{Y}$ dont l'intégrale générale est à points critiques fixes,* Annali di Matematica **91** (1972), 163–281.

26. P.J. Charles, *Painlevé analysis and the study of continuous and discrete Painlevé equations*, PhD thesis, Institute of Mathematics & Statistics, University of Kent (2002).
27. J. Chazy, *Sur les équations différentielles du troisième ordre et d'ordre supérieur dont l'intégrale générale a ses points critiques fixes*, Acta Math. **34** (1911), 317–385.
28. Y. Chen and G. Pruessner, *Orthogonal polynomials with discontinuous weights*, J. Phys. A: Math. Gen. **38** (2005), L191–L198.
29. D.V. Choodnovsky and G.V. Choodnovsky, *Pole expansions of nonlinear partial differential equations*, Nuovo Cim. **40B** (1977), 339–353.
30. P.A. Clarkson, *Painlevé equations — nonlinear special functions*, J. Comp. Appl. Math. **153** (2003), 127–140.
31. P.A. Clarkson, *The third Painlevé equation and associated special polynomials*, J. Phys. A: Math. Gen. **36** (2003), 9507–9532.
32. P.A. Clarkson, *The fourth Painlevé equation and associated special polynomials*, J. Math. Phys. **44** (2003), 5350–5374.
33. P.A. Clarkson, *Remarks on the Yablonskii-Vorob'ev polynomials*, Phys. Lett. **A319** (2003), 137–144.
34. P.A. Clarkson, *Special polynomials associated with rational solutions of the fifth Painlevé equation*, J. Comp. Appl. Math. **178** (2005), 111–179.
35. P.A. Clarkson, *On rational solutions of the fourth Painlevé equation and its Hamiltonian*, in "*Group Theory and Numerical Analysis*" [Editors D. Gomez-Ullate, P. Winternitz *et al.*], CRM Proc. Lect. Notes Series, **39**, Amer. Math. Soc., Providence, RI, pp. 103–118 (2005).
36. P.A. Clarkson, N. Joshi and A. Pickering, *Bäcklund transformations for the second Painlevé hierarchy: a modified truncation approach*, Inverse Problems **15** (1999), 175–187.
37. P.A. Clarkson and M.D. Kruskal, *New similarity solutions of the Boussinesq equation*, J. Math. Phys. **30** (1989), 2201–2213.
38. P.A. Clarkson and E.L. Mansfield, *The second Painlevé equation, its hierarchy and associated special polynomials*, Nonlinearity **16** (2003), R1–R26.
39. P.A. Clarkson, E.L. Mansfield and H.N. Webster, *On the relation between the continuous and discrete Painlevé equations*, Theoret. and Math. Phys. **122** (2000), 1–16.
40. P.A. Clarkson and J.B. McLeod, *A connection formula for the second Painlevé transcendent*, Arch. Rat. Mech. Anal. **103** (1988), 97–138.
41. P.A. Clarkson and J.B. McLeod, *Integral equations and connection formulae or the Painlevé equations*, in "*Painlevé Transcendents, their Asymptotics and Physical Applications*" [P. Winternitz and D. Levi, Editors], NATO ASI Series B: Physics, vol. **278**, Plenum, New York, pp. 1–32 (1992).
42. R. Conte (Editor), "*The Painlevé Property, One Century Later*", CRM series in Mathematical Physics, Springer-Verlag, Berlin (1999).
43. R. Conte and M. Musette, *First-degree birational transformations of the Painlevé equations and their contiguity relations*, J. Phys. A: Math. Gen. **34** (2001), 10507–10522.
44. R. Conte and M. Musette, *New contiguity relation of the sixth Painlevé equation from a truncation*, Physica **D161** (2002), 129–141.
45. C.M. Cosgrove and G. Scoufis, *Painlevé classification of a class of differential equations of the second order and second degree*, Stud. Appl. Math. **88** (1993), 25–87.

46. F. Dalfovo, L. Pitaevskii and S. Stringari, *Order parameter at the boundary of a trapped Bose gas,* Phys. Rev. **A54** (1996), 4213–4217.
47. P. Deift, *Integrable systems and combinatorial theory,* A.M.S. Notices **47** (2000), 631–640.
48. P. Deift and X. Zhou, *A steepest descent method for oscillatory Riemann-Hilbert problems – asymptotics for the MKdV equation,* Ann. Math. **137** (1993), 295–368.
49. P. Deift and X. Zhou, *Asymptotics for the Painlevé II equation,* Comm. Math. Phys. **48** (1995), 277–337.
50. A. J. Dolgert, S.J. Di Bartolo and A.T. Dorsey, *Superheating fields of superconductors: asymptotic analysis and numerical results,* Phys. Rev. **B53** (1996), 5650–5660.
51. B. Dubrovin and M. Mazzocco, *Monodromy of certain Painlevé-VI transcendents and reflection groups,* Invent. Math. **141** (2000), 55–147.
52. N.P. Erugin, *The analytic theory and problems of the real theory of differential equations connected with the [Lyapunov] first method and with the method of analytic theory,* Differential Equations **3** (1967), 943–966.
53. H. Flaschka and A.C. Newell, *Monodromy- and spectrum preserving deformations. I,* Comm. Math. Phys. **76** (1980), 65–116.
54. A.S. Fokas and M.J. Ablowitz, *On a unified approach to transformations and elementary solutions of Painlevé equations,* J. Math. Phys. **23** (1982), 2033–2042.
55. A.S. Fokas and M.J. Ablowitz, *On the initial value problem of the second Painlevé transcendent,* Comm. Math. Phys. **91** (1983), 381–403.
56. A.S. Fokas, B. Grammaticos and A. Ramani, *From continuous to discrete Painlevé equations,* J. Math. Anal. Appl. **180** (1993), 342–360.
57. A.S. Fokas, A.R. Its and A.V. Kitaev, *Discrete Painlevé equations and their appearance in quantum-gravity,* Comm. Math. Phys. **142** (1991), 313–344.
58. A.S. Fokas, A.R. Its and A.V. Kitaev, *The isomonodromy approach to matrix models in 2D quantum-gravity,* Comm. Math. Phys. **147** (1992), 395–430.
59. A.S. Fokas and Y.C. Yortsos, *The transformation properties of the sixth Painlevé equation and one-parameter families of solutions,* Lett. Nuovo Cim. **30** (1981), 539–544.
60. A.S. Fokas and X. Zhou, *On the solvability of Painlevé–II and Painlevé–IV,* Comm. Math. Phys. **144** (1992), 601–622.
61. P.J. Forrester and N.S. Witte, *Application of the τ-function theory of Painlevé equations to Random Matrices: PIV, PII and the GUE,* Comm. Math. Phys. **219** (2001), 357–398.
62. P.J. Forrester and N.S. Witte, *Application of the τ-function theory of Painlevé equations to random matrices: PV, PIII, the LUE, JUE and CUE,* Comm. Pure Appl. Math. **55** (2002), 679–727.
63. P.J. Forrester and N.S. Witte, *Application of the τ-function theory of Painlevé equations to random matrices:* $\mathrm{P_{VI}}$, *the JUE, CyUE, cJUE and scaled limits,* Nagoya Math. J. **174** (2004), 29–114.
64. R. Fuchs, *Über lineare homogene Differentialgleichungen zweiter Ordnung mit drei im endlich gelegene wesentlich singulären Stellen,* Math. Ann. **63** (1907), 301–321.
65. S. Fukutani, *Hirota bilinear relations and Toda equation of the second Painlevé equation,* Funkcial. Ekvac. **43** (2000), 395–403.

66. S. Fukutani, K. Okamoto and H. Umemura, *Special polynomials and the Hirota bilinear relations of the second and fourth Painlevé equations*, Nagoya Math. J. **159** (2000), 179–200.
67. L. Gagnon, B. Grammaticos, A. Ramani and P. Winternitz, *Lie symmetries of a generalised nonlinear Schrödinger equation: III. Reductions to third-order ordinary differential equations*, J. Phys. A: Math. Gen. **22** (1989), 499–509.
68. B. Gambier, *Sur les équations différentielles du second ordre et du premier degré dont l'intégrale générale est à points critiques fixés*, Acta Math. **33** (1909), 1–55.
69. C.S. Gardner, J.M. Greene, M.D. Kruskal and R.M. Miura, *Method for solving the Korteweg-de Vries equation*, Phys. Rev. Lett. **19** (1967), 1095–1097.
70. J.A. Giannini and R.I. Joseph, *The role of the second Painlevé transcendent in nonlinear optics*, Phys. Lett. **A141** (1989), 417–419.
71. V.I. Gromak, *The solutions of Painlevé's third equation*, Differential Equations **9** (1973), 1599–1600.
72. V.I. Gromak, *Theory of Painlevé's equation*, Differential Equations **11** (1975), 285–287.
73. V.I. Gromak, *Solutions of Painlevé's fifth equation*, Differential Equations **12** (1976), 519–521.
74. V.I. Gromak, *One–parameter systems of solutions of Painlevé's equations*, Differential Equations **14** (1978), 1510–1513.
75. V.I. Gromak, *Solutions of the second Painlevé equation*, Differential Equations **18** (1982), 537–545.
76. V.I. Gromak, *Theory of the fourth Painlevé equation*, Differential Equations **23** (1987), 506–513.
77. V.I. Gromak, *Bäcklund transformations of Painlevé equations and their applications*, in "*The Painlevé Property, One Century Later*" [R. Conte, Editor], CRM series in Mathematical Physics, Springer-Verlag, Berlin, pp. 687–734 (1999).
78. V.I. Gromak, *Bäcklund transformations of the higher order Painlevé equations*, in "*Bäcklund and Darboux Transformations. The Geometry of Solitons*" [A. Coley, D. Levi, R. Milson, C. Rogers and P. Winternitz, Editors], CRM Proc. Lect. Notes Series, **29**, Amer. Math. Soc., Providence, pp. 3–28 (2000).
79. V.I. Gromak and G.V. Filipuk, *Functional relations between solutions of the fifth Painlevé equation*, Differential Equations **37** (2001), 614–620.
80. V.I. Gromak and G.V. Filipuk, *The Bäcklund transformations of the fifth Painlevé equation and their applications*, Math. Model. Anal. **6** (2001), 221–230.
81. V.I. Gromak and G.V. Filipuk, *On algebraic solutions of the fifth Painlevé equation*, Differential Equations **39** (2003), 322–330.
82. V.I. Gromak and G.V. Filipuk, *On the transformations of the sixth Painlevé equation*, J. Nonlinear Math. Phys. **10,** Supp. 2 (2003), 57–67.
83. V.I. Gromak, I. Laine and S. Shimomura, "*Painlevé Differential Equations in the Complex Plane*", Studies in Math., vol. **28**, de Gruyter, Berlin, New York (2002).
84. V.I. Gromak and N.A. Lukashevich, *Special classes of solutions of Painlevé's equations*, Differential Equations **18** (1982), 317–326.
85. V.I. Gromak and T.S. Stepanova, *On the general representation of the meromorphic solutions of higher analogues of the second Painlevé equation*, in

"*Mathematical Modelling and Complex Analysis*" [R. Ciegis, Editor], Viliniaus Gedimino Technikos Universitetas, pp. 61–65 (1997).

86. V.I. Gromak & V.V. Tsegel'nik, *Some properties of the Bäcklund transformations of the Painlevé equations*, Differential Equations **32** (1996), 1024–1029.
87. V.I. Gromak and A.S. Zenchenko, *On the theory of higher-order Painlevé equations*, Differential Equations **40** (2004), 625–633.
88. P. Hall, *On the nonlinear evolution of Görtler vortices in non-parallel boundary layers*, IMA J. Appl. Math. **29** (1982), 173–196.
89. P. Hall, *Taylor-Görtler vortices in fully-developed or boundary-layer flows — linear-theory*, J. Fluid Mech. **124** (1982), 475–494.
90. P. Hall and W.D. Lakin, *The fully nonlinear development of Görtler vortices in growing boundary layers*, Proc. R. Soc. Lond. A **415** (1988), 421–444.
91. S.P. Hastings and J.B. McLeod, *A boundary value problem associated with the second Painlevé transcendent and the Korteweg-de Vries equation*, Arch. Rat. Mech. Anal. **73** (1980), 31–51.
92. B. Helffer and F.B. Weissler, *On a family of solutions of the second Painlevé equation related to superconductivity*, Europ. J. Appl. Math. **9** (1998), 223–243.
93. J. Hietarinta, *Painlevé equations in terms of entire functions*, in "*The Painlevé Property, One Century Later*" [R. Conte, Editor], CRM Series in Mathematical Physics, Springer-Verlag, Berlin, pp. 661–686 (1999).
94. A. Hinkkanen and I. Laine, *Solutions of the first and second Painlevé equations are meromorphic*, J. Anal. Math. **79** (1999), 345–377.
95. A. Hinkkanen and I. Laine, *Solutions of a modified third Painlevé equation are meromorphic*, J. Anal. Math. **85** (2001), 323–337.
96. A. Hinkkanen and I. Laine, *Solutions of a modified fifth Painlevé equation are meromorphic*, Rep. Univ. Jyvaskyla Dep. Math. Stat. **83** (2001), 133–146.
97. A. Hinkkanen and I. Laine, *The meromorphic nature of the sixth Painlevé transcendents*, J. Anal. Math. **94** (2004), 319–342.
98. N.J. Hitchin, *Poncelet polygons and the Painlevé equations*, in "*Geometry and analysis*" [Edited by Ramanan], Tata Inst. Fund. Res., Bombay, 1992, pp. 151–185 (1995).
99. N.J. Hitchin, *A lecture on the octahedron*, Bull. London Math. Soc. **35** (2003), 577–600.
100. A.N.W. Hone, *Non-autonomous Hénon-Heiles systems*, Physica **D118** (1998), 1–16.
101. E.L. Ince, "*Ordinary Differential Equations*", Dover, New York (1956).
102. A.R. Its, *The Painlevé transcendents as nonlinear special functions*, in "*Painlevé Transcendents, their Asymptotics and Physical Applications*" [P. Winternitz and D. Levi, Editors], NATO ASI Series B: Physics, vol. **278**, Plenum, New York, pp. 49–59 (1992).
103. A.R. Its, A.S. Fokas and A.A. Kapaev, *On the asymptotic analysis of the Painlevé equations via the isomonodromy method*, Nonlinearity **7** (1994), 1291–1325.
104. A.R. Its and A.A. Kapaev, *The method of isomonodromy deformations and connection formulas for the second Painlevé transcendent*, Math. USSR Izvestiya **31** (1988), 193–207.
105. A.R. Its and A.A. Kapaev, *The nonlinear steepest descent approach to the asymptotics of the second Painlevé transcendent in the complex domain*, in "*MathPhys odyssey, 2001*" [M. Kashiwara and T. Miwa, Editors], Prog. Math. Phys., vol. **23**, Birkhäuser–Boston, Boston, MA, 273–311 (2002).

106. A.R. Its and A.A. Kapaev, *Quasilinear Stokes phenomenon for the second Painlevé transcendent,* Nonlinearity **16** (2003), 363–386.
107. A.R. Its and V.Yu. Novokshënov, "*The Isomonodromic Deformation Method in the Theory of Painlevé equations*", Lect. Notes Math., vol. **1191**, Springer-Verlag, Berlin (1986).
108. K. Iwasaki, K. Kajiwara and T. Nakamura, *Generating function associated with the rational solutions of the Painlevé II equation,* J. Phys. A: Math. Gen. **35** (2002), L207–L211.
109. K. Iwasaki, H. Kimura, S. Shimomura and M. Yoshida, "*From Gauss to Painlevé: a Modern Theory of Special Functions*", Aspects of Mathematics E, vol. **16**, Vieweg, Braunschweig, Germany (1991).
110. M. Jimbo and T. Miwa, *Monodromy preserving deformations of linear ordinary differential equations with rational coefficients. II,* Physica **D2** (1981), 407–448.
111. M. Jimbo and T. Miwa, *Monodromy preserving deformations of linear ordinary differential equations with rational coefficients. III,* Physica **D4** (1981), 26–46.
112. M. Jimbo, T. Miwa, Y. Mori and M. Sato, *Density matrix of an impenetrable Bose gas and the fifth Painlevé transcendent,* Physica **D1** (1980), 80–158.
113. M. Jimbo, T. Miwa and K. Ueno, *Monodromy preserving deformations of linear ordinary differential equations with rational coefficients. I,* Physica **D2** (1981), 306–352.
114. K. Johansson, *Shape fluctuations and random matrices,* Comm. Math. Phys. **209** (2000), 437–476.
115. I.M. Johnstone, *On the distribution of the largest principal component,* Ann. Stat. **29** (2001), 295–327.
116. N. Joshi, K. Kajiwara and M. Mazzocco, *Generating function associated with the determinant for the solutions of the Painlevé II equation,* Astérique **297** (2005), 67–78.
117. N. Joshi and M.D. Kruskal, *An asymptotic approach to the connection problem for the first and second Painlevé equations,* Phys. Lett. **A130** (1988), 129–137.
118. N. Joshi and M.D. Kruskal, *The Painlevé connection problem — an asymptotic approach,* Stud. Appl. Math. **86** (1992), 315–376.
119. N. Joshi and M. Mazzocco, *Existence and uniqueness of tri-tronquée solutions of the second Painlevé hierarchy,* Nonlinearity **16** (2003), 427–439.
120. M. Kac and P. van Moerbeke, *On an explicitly soluble system of nonlinear differential equations related to certain Toda lattices,* Adv. Math. **16** (1975), 160–169.
121. K. Kajiwara, *On a q-difference Painlevé III equation: II. Rational solutions,* J. Nonl. Math. Phys. **10** (2003), 282–303.
122. K. Kajiwara and T. Masuda, *A generalization of determinant formulae for the solutions of Painlevé II and XXXIV equations,* J. Phys. A: Math. Gen. **32** (1999), 3763–3778.
123. K. Kajiwara and T. Masuda, *On the Umemura polynomials for the Painlevé III equation,* Phys. Lett. **A260** (1999), 462–467.
124. K. Kajiwara and Y. Ohta, *Determinantal structure of the rational solutions for the Painlevé II equation,* J. Math. Phys. **37** (1996), 4393–4704.
125. K. Kajiwara and Y. Ohta, *Determinant structure of the rational solutions for the Painlevé IV equation,* J. Phys. A: Math. Gen. **31** (1998), 2431–2446.
126. Y. Kametaka, *On poles of the rational solution of the Toda equation of Painlevé*-II *type,* Proc. Japan Acad. Ser. A Math. Sci. **59** (1983), 358–360.

127. Y. Kametaka, *On rational similarity solutions of KdV an m-KdV equations,* Proc. Japan Acad. Ser. A Math. Sci. **59** (1983), 407–409.
128. Y. Kametaka, *On the irreducibility conjecture based on computer calculation for Yablonskiĭ–Vorobev polynomials which give a rational solution of the Toda equation of Painlevé-*II *type,* Japan J. Appl. Math. **2** (1985), 241–246.
129. Y. Kametaka, M. Noda, Y. Fukui and S. Hirano, *A numerical approach to Toda equation and Painlevé II equation,* Mem. Fac. Eng. Ehime Univ. **9** (1986), 1–24.
130. M. Kaneko and H. Ochiai, *On coefficients of Yablonskii-Vorob'ev polynomials,* J. Math. Soc. Japan **55** (2003), 985–993.
131. A.A. Kapaev, *Asymptotic behavior of the solutions of the Painlevé equation of the first kind,* Differential Equations **24** (1988), 1107–1115.
132. A.A. Kapaev, *Asymptotics of solutions for the second Painlevé equation,* Theoret. and Math. Phys. **77** (1989), 1227–1234.
133. A.A. Kapaev, *Irregular singular point of the second Painlevé function and the nonlinear Stokes phenomenon,* Zapiski Nauch. Seminarov LOMI **187** (1991), 139–170 [in Russian].
134. A.A. Kapaev, *Global asymptotics of the second Painlevé transcendent,* Phys. Lett. **A167** (1992), 356–362.
135. A.A. Kapaev, *Quasi-linear Stokes phenomenon for the Painlevé first equation,* J. Phys. A: Math. Gen. **37** (2004), 11149–11167.
136. A.A. Kapaev, *Quasi-linear Stokes phenomenon for the Hastings-McLeod solution of the second Painlevé equation,* preprint arXiv:nlin.SI/0411009 (2004).
137. A.A. Kapaev and A.V. Kitaev, *Connection formulas for the first Painlevé transcendent in the complex plane,* Lett. Math. Phys. **27** (1993), 243–252.
138. A.A. Kapaev and V.Yu. Novokshënov, *Two-parameter family of real solutions of the second Painlevé equation,* Sov. Phys. Dokl. **31** (1986), 719–721.
139. N.M. Katz and P. Sarnak, "*Random Matrices, Frobenius Eigenvalues, and Monodromy*", Amer. Math. Soc., Providence, RI (1999).
140. N.M. Katz and P. Sarnak, *Zeroes of zeta functions and symmetry,* Bull. Amer. Math. Soc. **36** (1999), 1–26.
141. A.V. Kitaev, *Elliptic asymptotics of the first and the second Painlevé transcendents,* Russian Math. Surveys **49** (1994), 81–150.
142. A.V. Kitaev, C.K. Law and J.B. McLeod, *Rational solutions of the fifth Painlevé equation,* Differential Integral Equations **7** (1994), 967–1000.
143. N.A. Kudryashov, *The first and second Painlevé equations of higher order and some relations between them,* Phys. Lett. **A224** (1997), 353–360.
144. N.A. Kudryashov, *Two hierarchies of ordinary differential equations and their properties,* Phys. Lett. **A252** (1999), 173–179.
145. N.A. Kudryashov, *Amalgamations of the Painlevé equations,* J. Math. Phys. **44** (2003), 6160–6178.
146. N.A. Kudryashov and A. Pickering, *Rational and special solutions of the* P_{II} *hierarchy,* in "*SIDE III — Symmetries and Integrability of Difference Equations*" [D. Levi and O. Ragnisco, Editors], CRM Proc. Lect. Notes Series, vol. **25**, Amer. Math. Soc., Providence, RI, pp. 245–253 (2000).
147. N.A. Kudryashov and M.B. Soukharev, *Uniformization and transcendence of solutions for the first and second Painlevé hierarchies,* Phys. Lett. **A237** (1998), 206–216.
148. P.D. Lax, *Almost periodic solutions of the KdV equation,* SIAM Rev. **18** (1976), 351–375.

149. G. Lebeau and P. Lochak, *On the second Painlevé equation — the connection forumla via a Riemann-Hilbert problem and other results*, J. Differential Equations **68** (1987), 344–372.
150. Y. Li and Y. He, *On analytic properties of higher analogs of the second Painlevé equation*, J. Math. Phys. **43** (1992), 1106–1115.
151. N.A. Lukashevich, *Elementary solutions of certain Painlevé equations*, Differential Equations **1** (1965), 561–564.
152. N.A. Lukashevich, *Theory of the fourth Painlevé equation*, Differential Equations **3** (1967), 395–399.
153. N.A. Lukashevich, *On the theory of the third Painlevé equation*, Differential Equations **3** (1967), 994–999.
154. N.A. Lukashevich, *Solutions of the fifth equation of Painlevé equation*, Differential Equations **4** (1968), 732–735.
155. N.A. Lukashevich, *The second Painlevé equation*, Differential Equations **6** (1971), 853–854.
156. N.A. Lukashevich and A.I. Yablonskii, *On a class of solutions of the sixth Painlevé equation*, Differential Equations **3** (1967), 264–266.
157. E. Lundh, C.J. Pethick and H. Smith, *Zero-temperature properties of a trapped Bose-condensed gas: Beyond the Thomas-Fermi approximation*, Phys, Rev. **A55** (1997), 2126–2131.
158. A.P. Magnus, *Painlevé type differential equations for the recurrence coefficients of semi-classical orthogonal polynomials*, J. Comput. Appl. Anal. **57** (1995), 215–237.
159. Yu.I. Manin, *Sixth Painlevé equation, universal elliptic curve, and mirror of* P^2, in "*Geometry of Differential equations*" [A. Khovanskii, A. Varchenko and V. Vassiliev, Editors], AMS Transl. ser. 2, vol. **186 (39)**, pp. 131–151 (1998).
160. E.L. Mansfield and H.N. Webster, *On one-parameter families of Painlevé III*, Stud. Appl. Math. **101** (1998), 321–341.
161. D. Margetis, *Asymptotic formula for the condensate wave function of a trapped Bose gas*, Phys. Rev. **A61** (2000), 055601.
162. T. Masuda, *On a class of algebraic solutions to Painlevé VI equation, its determinant formula and coalescence cascade*, Funkcial. Ekvac. **46** (2003), 121–171.
163. T. Masuda, *Classical transcendental solutions of the Painlevé equations and their degeneration*, Tohoku Math. J. **56** (2004), 467–490.
164. T. Masuda, Y. Ohta and K. Kajiwara, *A determinant formula for a class of rational solutions of Painlevé V equation*, Nagoya Math. J. **168** (2002), 1–25.
165. M. Mazzocco, *Rational solutions of the Painlevé VI equation*, J. Phys. A: Math. Gen. **34** (2001), 2281–2294.
166. M. Mazzocco, *Picard and Chazy solutions to the Painlevé VI equation*, Math. Ann. **321** (2001), 157–195.
167. J.W. Miles, *On the second Painlevé transcendent*, Proc. R. Soc. Lond. A **361** (1978), 277–291.
168. J.W. Miles, *The second Painlevé transcendent: a nonlinear Airy function*, Mechanics Today **5** (1980), 297–313.
169. A.E. Milne, P.A. Clarkson and A.P. Bassom, *Bäcklund transformations and solution hierarchies for the third Painlevé equation*, Stud. Appl. Math. **98** (1997), 139–194.
170. H.L. Montgomery, *The pair correlation of zeros of the zeta function*, in "*Analytic Number Theory*" [H.G. Diamond, Editor], Proc. Sympos. Pure Math, vol. **24**, Amer. Math. Soc., Providence, R.I., pp. 181–193 (1973).

171. Y. Murata, *Rational solutions of the second and the fourth Painlevé equations,* Funkcial. Ekvac. **28** (1985), 1–32.
172. Y. Murata, *Classical solutions of the third Painlevé equations,* Nagoya Math. J. **139** (1995), 37–65.
173. M. Noumi, "*Painlevé Equations through Symmetry*", Trans. Math. Mono., vol. **223**, AMS, Providence, RI (2004).
174. M. Noumi and Y. Yamada, *Affine Weyl groups, discrete dynamical systems and Painlevé equations,* Comm. Math. Phys. **199** (1998), 281–295.
175. M. Noumi and Y. Yamada, *Umemura polynomials for the Painlevé V equation,* Phys. Lett. **A247** (1998), 65–69.
176. M. Noumi and Y. Yamada, *Symmetries in the fourth Painlevé equation and Okamoto polynomials,* Nagoya Math. J. **153** (1999), 53–86.
177. V.Yu. Novokshënov, *The Boutroux ansatz for the second Painlevé equation in the complex domain,* Math. USSR Izvestiya **37** (1991), 587–609.
178. V.Yu. Novokshënov, *A modulated elliptic function as a solution of the second Painlevé equation in the complex plane,* Soviet Math. Dokl. **41** (1991), 252–255.
179. V.Yu. Novokshënov, *Nonlinear Stokes phenomenon for the second Painlevé equation,* Physica **D63** (1993), 1–7.
180. A.M. Odlyzko, *On the distribution of spacings between zeros of the zeta function,* Math. Comp. **48** (1987), 273–308.
181. Y. Ohyama, *On the third Painlevé equations of type* D_7, preprint, Department of Pure and Applied Mathematics, Graduate School of Information Science and Technology, Osaka University, `http://www.math.sci.osaka-u.ac.jp/~ohyama/ohyama-home.html` (2001).
182. K. Okamoto, *On the* τ*-function of the Painlevé equations,* Physica **2D** (1981), 525–535.
183. K. Okamoto, *Isomonodromic deformation and the Painlevé equations, and the Garnier system,* J. Fac. Sci. Univ. Tokyo, Sect. IA Math. **33** (1986), 575–618.
184. K. Okamoto, *Studies on the Painlevé equations III. Second and fourth Painlevé equations,* $\mathrm{P_{II}}$ *and* $\mathrm{P_{IV}}$, Math. Ann. **275** (1986), 221–255.
185. K. Okamoto, *Studies on the Painlevé equations I. Sixth Painlevé equation* $\mathrm{P_{VI}}$, Ann. Mat. Pura Appl. **146** (1987), 337–381.
186. K. Okamoto, *Studies on the Painlevé equations II. Fifth Painlevé equation* $\mathrm{P_{V}}$, Japan. J. Math. **13** (1987), 47–76.
187. K. Okamoto, *Studies on the Painlevé equations IV. Third Painlevé equation* $\mathrm{P_{III}}$, Funkcial. Ekvac. **30** (1987), 305–332.
188. K. Okamoto, *The Hamiltonians associated to the Painlevé equations,* in "The *Painlevé Property, One Century Later*" [R. Conte, Editor], CRM series in Mathematical Physics, Springer-Verlag, Berlin, pp. 735–787 (1999).
189. P. Painlevé, *Sur les équations différentielles du second ordre à points critiques fixés,* C.R. Acad. Sci. Paris **143** (1906), 1111–1117.
190. E. Picard, *Mémoire sur la théorie des fonctions algébrique de deux variables,* J. de Math. **5** (1889), 135–318.
191. M. Prähofer and H. Spohn, *Statistical self-similarity of one-dimensional growth processes,* Physica **A279** (2000), 342–352.
192. M. Prähofer and H. Spohn, *Universal distributions for growth processes in* $1+1$ *dimensions and random matrices,* Phys. Rev. Lett. **84** (2000), 4882–4885.
193. R.S. Rosales, *The similarity solutions for the Korteweg-de Vries equation and the related Painlevé transcendent,* Proc. R. Soc. Lond. A **361** (1978), 265–275.

194. Z. Rudnick and P. Sarnak, *The n-level correlations of zeros of the zeta function,* C. R. Acad. Sci. Paris Ser. I Math. **319** (1994), 1027–1032.
195. Z. Rudnick and P. Sarnak, *Zeros of principal L–functions and random matrix theory,* Duke Math. J. **81** (1996), 269–322.
196. P.L. Sachdev, "*A Compendium of Nonlinear Ordinary Differential Equations*", Wiley, New York (1997).
197. H. Sakai, *Rational surfaces associated with affine root systems and geometry of the Painlevé equations,* Comm. Math. Phys. **220** (2001), 165–229.
198. H. Segur and M.J. Ablowitz, *Asymptotic solutions of nonlinear evolution equations and a Painlevé transcendent,* Physica **3D** (1981), 165–184.
199. S. Shimomura, *Value distribution of Painlevé transcendents of the first and second kind,* J. Anal. Math. **79** (2000), 333–346.
200. S. Shimomura, *Value distribution of Painlevé transcendents of the fifth kind,* Results Math. **38** (2000), 348–361.
201. S. Shimomura, *The first, the second and the fourth Painlevé transcendents are of finite order,* Proc. Japan Acad. Ser. A Math. Sci. **77** (2001), 42–45.
202. S. Shimomura, *Growth of the first, the second and the fourth Painlevé transcendents,* Math. Proc. Camb. Phil. Soc **134** (2003), 259–269.
203. S. Shimomura, *Growth of modified Painlevé transcendents of the fifth and the third kind,* Forum Math. **16** (2004), 231–247.
204. A. Soshnikov, *Universality at the edge of the spectrum in Wigner random matrices,* Comm. Math. Phys. **207** (1999), 697–733.
205. N. Steinmetz, *On Painlevé's equations I, II and IV,* J. Anal. Math. **79** (2000), 363–377.
206. N. Steinmetz, *Value distribution of the Painlevé transcendents,* Israel J. Math. **128** (2001), 29–52.
207. N. Steinmetz, *Lower estimates for the orders of growth of the second and fourth Painlevé transcendents,* Portugaliae Math. **61** (2004), 369–374.
208. N. Steinmetz, *Global properties of the Painlevé transcendents: New results and open questions,* Ann. Acad. Sci. Fenn.-M. **30** (2005), 71–98.
209. B.I. Suleĭmanov, *The relation between asymptotic properties of the second Painlevé equation in different directions towards infinity,* Diff. Eqns. **23** (1987), 569–576.
210. T. Tamizhmani, B. Grammaticos, A. Ramani and K.M. Tamizhmani, *On a class of special solutions of the Painlevé equations,* Physica **A295** (2001), 359–370.
211. M. Taneda, *Remarks on the Yablonskii-Vorob'ev polynomials,* Nagoya Math. J. **159** (2000), 87–111.
212. M. Taneda, *Representation of Umemura polynomials for the sixth Painlevé equation by the generalized Jacobi polynomials,* in "Physics and Combinatorics" [A.N. Kirillov, A. Tsuchiya and H. Umemura, Editors], World Scientific, Singapore, 366–376 (2001).
213. N.M. Temme, "*Special Functions. An Introduction to the Classical Functions of Mathematical Physics*", Wiley, New York (1996).
214. C.A. Tracy and H. Widom, *Level-spacing distributions and the Airy kernel,* Comm. Math. Phys. **159** (1994), 151–174.
215. C.A. Tracy and H. Widom, *Level-spacing distributions and the Bessel kernel,* Comm. Math. Phys. **161** (1994), 289–310.
216. C.A. Tracy and H. Widom, *Fredholm determinants, differential equations and matrix models,* Comm. Math. Phys. **163** (1994), 33–72.

217. C.A. Tracy and H. Widom, *On orthogonal and symplectic matrix ensembles*, Comm. Math. Phys. **177** (1996), 727.
218. C.A. Tracy and H. Widom, *Fredholm determinants and the mKdV/sinh-Gordon hierarchies*, Comm. Math. Phys. **179** (1996), 1–9.
219. C.A. Tracy and H. Widom, *Proofs of two conjectures related to the thermodynamic Bethe ansatz*, Comm. Math. Phys. **179** (1996), 667.
220. C.A. Tracy and H. Widom, *Random unitary matrices, permutations and Painlevé*, Comm. Math. Phys. **207** (1999), 665–685.
221. C.A. Tracy and H. Widom, *Airy kernel and Painlevé II*, in "Isomonodromic Deformations and Applications in Physics" [J. Harnad and A.R. Its, Editors], CRM Proc. Lect. Notes Series, **31**, Amer. Math. Soc., Providence, pp. 85–96 (2002).
222. J.A. Tuszyński, J. Middleton, P.L. Christiansen and J.M. Dixon, *Exact eigenfunctions of the linear ramp potential in the Gross-Pitaevskii equation for the Bose-Einstein condensate*, Phys. Lett. **A291** (2001), 220–225.
223. H. Umemura, *On the second proof of the irreducibility of the first differential equation of Painlevé*, Nagoya Math. J. **117** (1990), 125–171.
224. H. Umemura, *Painlevé equations and classical functions*, Sugaku Expositions **11** (1998), 77–100.
225. H. Umemura, *On the transformation group of the second Painlevé equation*, Nagoya Math. J. **157** (2000), 15–46.
226. H. Umemura, *Painlevé equations in the past 100 Years*, Amer. Math. Soc. Translations **204** (2001), 81–110.
227. H. Umemura and H. Watanabe, *Solutions of the second and fourth Painlevé equations I*, Nagoya Math. J. **148** (1997), 151–198.
228. H. Umemura and H. Watanabe, *Solutions of the third Painlevé equation I*, Nagoya Math. J. **151** (1998), 1–24.
229. A.P. Vorob'ev, *On rational solutions of the second Painlevé equation*, Differential Equations **1** (1965), 58–59.
230. H. Watanabe, *Solutions of the fifth Painlevé equation I*, Hokkaido Math. J. **24** (1995), 231–267.
231. E.E. Whittaker and G.M. Watson, "*Modern Analysis*", 4th Edition, Cambridge University Press, Cambridge (1927).
232. A.I. Yablonskii, *On rational solutions of the second Painlevé equation*, Vesti Akad. Nauk. BSSR Ser. Fiz. Tkh. Nauk. **3** (1959), 30–35 [in Russian].
233. Y. Yamada, *Special polynomials and generalized Painlevé equations*, in "Combinatorial Methods in Representation Theory" [K. Koike, M. Kashiwara, S. Okada, I. Terada and H.F. Yamada, Editors], Adv. Stud. Pure Math., **28**, Kinokuniya, Tokyo, Japan, 391–400 (2000).
234. V.E. Zakharov and A.B. Shabat, *Exact theory of two-dimensional self-focusing and one-dimensional of waves in nonlinear media*, Sov. Phys. JETP **34** (1972), 62–69.
235. V.E. Zakharov and A.B. Shabat, *A scheme for integrating the nonlinear equations of mathematical physics by the method of the inverse scattering problem*, Func. Anal. Appl. **8** (1974), 226–235.

Index

Lecture Notes in Mathematics

For information about earlier volumes
please contact your bookseller or Springer
LNM Online archive: springerlink.com

Vol. 1681: G. J. Wirsching, The Dynamical System Generated by the 3n+1 Function (1998)
Vol. 1682: H.-D. Alber, Materials with Memory (1998)
Vol. 1683: A. Pomp, The Boundary-Domain Integral Method for Elliptic Systems (1998)
Vol. 1684: C. A. Berenstein, P. F. Ebenfelt, S. G. Gindikin, S. Helgason, A. E. Tumanov, Integral Geometry, Radon Transforms and Complex Analysis. Firenze, 1996. Editors: E. Casadio Tarabusi, M. A. Picardello, G. Zampieri (1998)
Vol. 1685: S. König, A. Zimmermann, Derived Equivalences for Group Rings (1998)
Vol. 1686: J. Azéma, M. Émery, M. Ledoux, M. Yor (Eds.), Séminaire de Probabilités XXXII (1998)
Vol. 1687: F. Bornemann, Homogenization in Time of Singularly Perturbed Mechanical Systems (1998)
Vol. 1688: S. Assing, W. Schmidt, Continuous Strong Markov Processes in Dimension One (1998)
Vol. 1689: W. Fulton, P. Pragacz, Schubert Varieties and Degeneracy Loci (1998)
Vol. 1690: M. T. Barlow, D. Nualart, Lectures on Probability Theory and Statistics. Editor: P. Bernard (1998)
Vol. 1691: R. Bezrukavnikov, M. Finkelberg, V. Schechtman, Factorizable Sheaves and Quantum Groups (1998)
Vol. 1692: T. M. W. Eyre, Quantum Stochastic Calculus and Representations of Lie Superalgebras (1998)
Vol. 1694: A. Braides, Approximation of Free-Discontinuity Problems (1998)
Vol. 1695: D. J. Hart el, Markov Set-Chains (1998)
Vol. 1696: E. Bouscaren (Ed.): Model Theory and Algebraic Geometry (1998)
Vol. 1697: B. Cockburn, C. Johnson, C.-W. Shu, E. Tadmor, Advanced Numerical Approximation of Nonlinear Hyperbolic Equations. Cetraro, Italy, 1997. Editor: A. Quarteroni (1998)
Vol. 1698: M. Bhattacharjee, D. Macpherson, R. G. Möller, P. Neumann, Notes on In nite Permutation Groups (1998)
Vol. 1699: A. Inoue,Tomita-Takesaki Theory in Algebras of Unbounded Operators (1998)
Vol. 1700: W. A. Woyczyński, Burgers-KPZ Turbulence (1998)
Vol. 1701: Ti-Jun Xiao, J. Liang, The Cauchy Problem of Higher Order Abstract Differential Equations (1998)
Vol. 1702: J. Ma, J. Yong, Forward-Backward Stochastic Differential Equations and Their Applications (1999)
Vol. 1703: R. M. Dudley, R. Norvaiša, Differentiability of Six Operators on Nonsmooth Functions and p-Variation (1999)
Vol. 1704: H. Tamanoi, Elliptic Genera and Vertex Operator Super-Algebras (1999)
Vol. 1705: I. Nikolaev, E. Zhuzhoma, Flows in 2-dimensional Manifolds (1999)
Vol. 1706: S. Yu. Pilyugin, Shadowing in Dynamical Systems (1999)
Vol. 1707: R. Pytlak, Numerical Methods for Optimal Control Problems with State Constraints (1999)
Vol. 1708: K. Zuo, Representations of Fundamental Groups of Algebraic Varieties (1999)
Vol. 1709: J. Azéma, M. Émery, M. Ledoux, M. Yor (Eds.), Séminaire de Probabilités XXXIII (1999)
Vol. 1710: M. Koecher, The Minnesota Notes on Jordan Algebras and Their Applications (1999)
Vol. 1711: W. Ricker, Operator Algebras Generated by Commuting Projections: A Vector Measure Approach (1999)
Vol. 1712: N. Schwartz, J. J. Madden, Semi-algebraic Function Rings and Re ectors of Partially Ordered Rings (1999)
Vol. 1713: F. Bethuel, G. Huisken, S. Müller, K. Steffen, Calculus of Variations and Geometric Evolution Problems. Cetraro, 1996. Editors: S. Hildebrandt, M. Struwe (1999)
Vol. 1714: O. Diekmann, R. Durrett, K. P. Hadeler, P. K. Maini, H. L. Smith, Mathematics Inspired by Biology. Martina Franca, 1997. Editors: V. Capasso, O. Diekmann (1999)
Vol. 1715: N. V. Krylov, M. Röckner, J. Zabczyk, Stochastic PDE's and Kolmogorov Equations in In nite Dimensions. Cetraro, 1998. Editor: G. Da Prato (1999)
Vol. 1716: J. Coates, R. Greenberg, K. A. Ribet, K. Rubin, Arithmetic Theory of Elliptic Curves. Cetraro, 1997. Editor: C. Viola (1999)
Vol. 1717: J. Bertoin, F. Martinelli, Y. Peres, Lectures on Probability Theory and Statistics. Saint-Flour, 1997. Editor: P. Bernard (1999)
Vol. 1718: A. Eberle, Uniqueness and Non-Uniqueness of Semigroups Generated by Singular Diffusion Operators (1999)
Vol. 1719: K. R. Meyer, Periodic Solutions of the N-Body Problem (1999)
Vol. 1720: D. Elworthy, Y. Le Jan, X-M. Li, On the Geometry of Diffusion Operators and Stochastic Flows (1999)
Vol. 1721: A. Iarrobino, V. Kanev, Power Sums, Gorenstein Algebras, and Determinantal Loci (1999)
Vol. 1722: R. McCutcheon, Elemental Methods in Ergodic Ramsey Theory (1999)
Vol. 1723: J. P. Croisille, C. Lebeau, Diffraction by an Immersed Elastic Wedge (1999)
Vol. 1724: V. N. Kolokoltsov, Semiclassical Analysis for Diffusions and Stochastic Processes (2000)
Vol. 1725: D. A. Wolf-Gladrow, Lattice-Gas Cellular Automata and Lattice Boltzmann Models (2000)
Vol. 1726: V. Marić, Regular Variation and Differential Equations (2000)
Vol. 1727: P. Kravanja M. Van Barel, Computing the Zeros of Analytic Functions (2000)
Vol. 1728: K. Gatermann Computer Algebra Methods for Equivariant Dynamical Systems (2000)
Vol. 1729: J. Azéma, M. Émery, M. Ledoux, M. Yor (Eds.)

Séminaire de Probabilités XXXIV (2000)
Vol. 1730: S. Graf, H. Luschgy, Foundations of Quantization for Probability Distributions (2000)
Vol. 1731: T. Hsu, Quilts: Central Extensions, Braid Actions, and Finite Groups (2000)
Vol. 1732: K. Keller, Invariant Factors, Julia Equivalences and the (Abstract) Mandelbrot Set (2000)
Vol. 1733: K. Ritter, Average-Case Analysis of Numerical Problems (2000)
Vol. 1734: M. Espedal, A. Fasano, A. Mikelić, Filtration in Porous Media and Industrial Applications. Cetraro 1998. Editor: A. Fasano. 2000.
Vol. 1735: D. Yafaev, Scattering Theory: Some Old and New Problems (2000)
Vol. 1736: B. O. Turesson, Nonlinear Potential Theory and Weighted Sobolev Spaces (2000)
Vol. 1737: S. Wakabayashi, Classical Microlocal Analysis in the Space of Hyperfunctions (2000)
Vol. 1738: M. Émery, A. Nemirovski, D. Voiculescu, Lectures on Probability Theory and Statistics (2000)
Vol. 1739: R. Burkard, P. Deu hard, A. Jameson, J.-L. Lions, G. Strang, Computational Mathematics Driven by Industrial Problems. Martina Franca, 1999. Editors: V. Capasso, H. Engl, J. Periaux (2000)
Vol. 1740: B. Kawohl, O. Pironneau, L. Tartar, J.-P. Zolesio, Optimal Shape Design. Tróia, Portugal 1999. Editors: A. Cellina, A. Ornelas (2000)
Vol. 1741: E. Lombardi, Oscillatory Integrals and Phenomena Beyond all Algebraic Orders (2000)
Vol. 1742: A. Unterberger, Quantization and Nonholomorphic Modular Forms (2000)
Vol. 1743: L. Habermann, Riemannian Metrics of Constant Mass and Moduli Spaces of Conformal Structures (2000)
Vol. 1744: M. Kunze, Non-Smooth Dynamical Systems (2000)
Vol. 1745: V. D. Milman, G. Schechtman (Eds.), Geometric Aspects of Functional Analysis. Israel Seminar 1999-2000 (2000)
Vol. 1746: A. Degtyarev, I. Itenberg, V. Kharlamov, Real Enriques Surfaces (2000)
Vol. 1747: L. W. Christensen, Gorenstein Dimensions (2000)
Vol. 1748: M. Ruzicka, Electrorheological Fluids: Modeling and Mathematical Theory (2001)
Vol. 1749: M. Fuchs, G. Seregin, Variational Methods for Problems from Plasticity Theory and for Generalized Newtonian Fluids (2001)
Vol. 1750: B. Conrad, Grothendieck Duality and Base Change (2001)
Vol. 1751: N. J. Cutland, Loeb Measures in Practice: Recent Advances (2001)
Vol. 1752: Y. V. Nesterenko, P. Philippon, Introduction to Algebraic Independence Theory (2001)
Vol. 1753: A. I. Bobenko, U. Eitner, Painlevé Equations in the Differential Geometry of Surfaces (2001)
Vol. 1754: W. Bertram, The Geometry of Jordan and Lie Structures (2001)
Vol. 1755: J. Azéma, M. Émery, M. Ledoux, M. Yor (Eds.), Séminaire de Probabilités XXXV (2001)
Vol. 1756: P. E. Zhidkov, Korteweg de Vries and Nonlinear Schrödinger Equations: Qualitative Theory (2001)
Vol. 1757: R. R. Phelps, Lectures on Choquet's Theorem (2001)
Vol. 1758: N. Monod, Continuous Bounded Cohomology of Locally Compact Groups (2001)
Vol. 1759: Y. Abe, K. Kopfermann, Toroidal Groups (2001)
Vol. 1760: D. Filipović, Consistency Problems for Heath-Jarrow-Morton Interest Rate Models (2001)
Vol. 1761: C. Adelmann, The Decomposition of Primes in Torsion Point Fields (2001)
Vol. 1762: S. Cerrai, Second Order PDE's in Finite and In nite Dimension (2001)
Vol. 1763: J.-L. Loday, A. Frabetti, F. Chapoton, F. Goichot, Dialgebras and Related Operads (2001)
Vol. 1764: A. Cannas da Silva, Lectures on Symplectic Geometry (2001)
Vol. 1765: T. Kerler, V. V. Lyubashenko, Non-Semisimple Topological Quantum Field Theories for 3-Manifolds with Corners (2001)
Vol. 1766: H. Hennion, L. Hervé, Limit Theorems for Markov Chains and Stochastic Properties of Dynamical Systems by Quasi-Compactness (2001)
Vol. 1767: J. Xiao, Holomorphic Q Classes (2001)
Vol. 1768: M.J. P aum, Analytic and Geometric Study of Strati ed Spaces (2001)
Vol. 1769: M. Alberich-Carramiñana, Geometry of the Plane Cremona Maps (2002)
Vol. 1770: H. Gluesing-Luerssen, Linear Delay-Differential Systems with Commensurate Delays: An Algebraic Approach (2002)
Vol. 1771: M. Émery, M. Yor (Eds.), Séminaire de Probabilités 1967-1980. A Selection in Martingale Theory (2002)
Vol. 1772: F. Burstall, D. Ferus, K. Leschke, F. Pedit, U. Pinkall, Conformal Geometry of Surfaces in S^4 (2002)
Vol. 1773: Z. Arad, M. Muzychuk, Standard Integral Table Algebras Generated by a Non-real Element of Small Degree (2002)
Vol. 1774: V. Runde, Lectures on Amenability (2002)
Vol. 1775: W. H. Meeks, A. Ros, H. Rosenberg, The Global Theory of Minimal Surfaces in Flat Spaces. Martina Franca 1999. Editor: G. P. Pirola (2002)
Vol. 1776: K. Behrend, C. Gomez, V. Tarasov, G. Tian, Quantum Comohology. Cetraro 1997. Editors: P. de Bartolomeis, B. Dubrovin, C. Reina (2002)
Vol. 1777: E. García-Río, D. N. Kupeli, R. Vázquez-Lorenzo, Osserman Manifolds in Semi-Riemannian Geometry (2002)
Vol. 1778: H. Kiechle, Theory of K-Loops (2002)
Vol. 1779: I. Chueshov, Monotone Random Systems (2002)
Vol. 1780: J. H. Bruinier, Borcherds Products on O(2,1) and Chern Classes of Heegner Divisors (2002)
Vol. 1781: E. Bolthausen, E. Perkins, A. van der Vaart, Lectures on Probability Theory and Statistics. Ecole d' Eté de Probabilités de Saint-Flour XXIX-1999. Editor: P. Bernard (2002)
Vol. 1782: C.-H. Chu, A. T.-M. Lau, Harmonic Functions on Groups and Fourier Algebras (2002)
Vol. 1783: L. Grüne, Asymptotic Behavior of Dynamical and Control Systems under Perturbation and Discretization (2002)
Vol. 1784: L.H. Eliasson, S. B. Kuksin, S. Marmi, J.-C. Yoccoz, Dynamical Systems and Small Divisors. Cetraro, Italy 1998. Editors: S. Marmi, J.-C. Yoccoz (2002)
Vol. 1785: J. Arias de Reyna, Pointwise Convergence of Fourier Series (2002)
Vol. 1786: S. D. Cutkosky, Monomialization of Morphisms from 3-Folds to Surfaces (2002)
Vol. 1787: S. Caenepeel, G. Militaru, S. Zhu, Frobenius and Separable Functors for Generalized Module Categories and Nonlinear Equations (2002)

Vol. 1788: A. Vasil'ev, Moduli of Families of Curves for Conformal and Quasiconformal Mappings (2002)
Vol. 1789: Y. Sommerhäuser, Yetter-Drinfel'd Hopf algebras over groups of prime order (2002)
Vol. 1790: X. Zhan, Matrix Inequalities (2002)
Vol. 1791: M. Knebusch, D. Zhang, Manis Valuations and Prüfer Extensions I: A new Chapter in Commutative Algebra (2002)
Vol. 1792: D. D. Ang, R. Goren o, V. K. Le, D. D. Trong, Moment Theory and Some Inverse Problems in Potential Theory and Heat Conduction (2002)
Vol. 1793: J. Cortés Monforte, Geometric, Control and Numerical Aspects of Nonholonomic Systems (2002)
Vol. 1794: N. Pytheas Fogg, Substitution in Dynamics, Arithmetics and Combinatorics. Editors: V. Berthé, S. Ferenczi, C. Mauduit, A. Siegel (2002)
Vol. 1795: H. Li, Filtered-Graded Transfer in Using Noncommutative Gröbner Bases (2002)
Vol. 1796: J.M. Melenk, hp-Finite Element Methods for Singular Perturbations (2002)
Vol. 1797: B. Schmidt, Characters and Cyclotomic Fields in Finite Geometry (2002)
Vol. 1798: W.M. Oliva, Geometric Mechanics (2002)
Vol. 1799: H. Pajot, Analytic Capacity, Recti ability, Menger Curvature and the Cauchy Integral (2002)
Vol. 1800: O. Gabber, L. Ramero, Almost Ring Theory (2003)
Vol. 1801: J. Azéma, M. Émery, M. Ledoux, M. Yor (Eds.), Séminaire de Probabilités XXXVI (2003)
Vol. 1802: V. Capasso, E. Merzbach, B.G. Ivanoff, M. Dozzi, R. Dalang, T. Mountford, Topics in Spatial Stochastic Processes. Martina Franca, Italy 2001. Editor: E. Merzbach (2003)
Vol. 1803: G. Dolzmann, Variational Methods for Crystalline Microstructure – Analysis and Computation (2003)
Vol. 1804: I. Cherednik, Ya. Markov, R. Howe, G. Lusztig, Iwahori-Hecke Algebras and their Representation Theory. Martina Franca, Italy 1999. Editors: V. Baldoni, D. Barbasch (2003)
Vol. 1805: F. Cao, Geometric Curve Evolution and Image Processing (2003)
Vol. 1806: H. Broer, I. Hoveijn. G. Lunther, G. Vegter, Bifurcations in Hamiltonian Systems. Computing Singularities by Gröbner Bases (2003)
Vol. 1807: V. D. Milman, G. Schechtman (Eds.), Geometric Aspects of Functional Analysis. Israel Seminar 2000-2002 (2003)
Vol. 1808: W. Schindler, Measures with Symmetry Properties (2003)
Vol. 1809: O. Steinbach, Stability Estimates for Hybrid Coupled Domain Decomposition Methods (2003)
Vol. 1810: J. Wengenroth, Derived Functors in Functional Analysis (2003)
Vol. 1811: J. Stevens, Deformations of Singularities (2003)
Vol. 1812: L. Ambrosio, K. Deckelnick, G. Dziuk, M. Mimura, V. A. Solonnikov, H. M. Soner, Mathematical Aspects of Evolving Interfaces. Madeira, Funchal, Portugal 2000. Editors: P. Colli, J. F. Rodrigues (2003)
Vol. 1813: L. Ambrosio, L. A. Caffarelli, Y. Brenier, G. Buttazzo, C. Villani, Optimal Transportation and its Applications. Martina Franca, Italy 2001. Editors: L. A. Caffarelli, S. Salsa (2003)
Vol. 1814: P. Bank, F. Baudoin, H. Föllmer, L.C.G. Rogers, M. Soner, N. Touzi, Paris-Princeton Lectures on Mathematical Finance 2002 (2003)
Vol. 1815: A. M. Vershik (Ed.), Asymptotic Combinatorics with Applications to Mathematical Physics. St. Petersburg, Russia 2001 (2003)
Vol. 1816: S. Albeverio, W. Schachermayer, M. Talagrand, Lectures on Probability Theory and Statistics. Ecole d'Eté de Probabilités de Saint-Flour XXX-2000. Editor: P. Bernard (2003)
Vol. 1817: E. Koelink, W. Van Assche(Eds.), Orthogonal Polynomials and Special Functions. Leuven 2002 (2003)
Vol. 1818: M. Bildhauer, Convex Variational Problems with Linear, nearly Linear and/or Anisotropic Growth Conditions (2003)
Vol. 1819: D. Masser, Yu. V. Nesterenko, H. P. Schlickewei, W. M. Schmidt, M. Waldschmidt, Diophantine Approximation. Cetraro, Italy 2000. Editors: F. Amoroso, U. Zannier (2003)
Vol. 1820: F. Hiai, H. Kosaki, Means of Hilbert Space Operators (2003)
Vol. 1821: S. Teufel, Adiabatic Perturbation Theory in Quantum Dynamics (2003)
Vol. 1822: S.-N. Chow, R. Conti, R. Johnson, J. Mallet-Paret, R. Nussbaum, Dynamical Systems. Cetraro, Italy 2000. Editors: J. W. Macki, P. Zecca (2003)
Vol. 1823: A. M. Anile, W. Allegretto, C. Ringhofer, Mathematical Problems in Semiconductor Physics. Cetraro, Italy 1998. Editor: A. M. Anile (2003)
Vol. 1824: J. A. Navarro González, J. B. Sancho de Salas, $\mathscr{C}^\infty$ – Differentiable Spaces (2003)
Vol. 1825: J. H. Bramble, A. Cohen, W. Dahmen, Multiscale Problems and Methods in Numerical Simulations, Martina Franca, Italy 2001. Editor: C. Canuto (2003)
Vol. 1826: K. Dohmen, Improved Bonferroni Inequalities via Abstract Tubes. Inequalities and Identities of Inclusion-Exclusion Type. VIII, 113 p, 2003.
Vol. 1827: K. M. Pilgrim, Combinations of Complex Dynamical Systems. IX, 118 p, 2003.
Vol. 1828: D. J. Green, Gröbner Bases and the Computation of Group Cohomology. XII, 138 p, 2003.
Vol. 1829: E. Altman, B. Gaujal, A. Hordijk, Discrete-Event Control of Stochastic Networks: Multimodularity and Regularity. XIV, 313 p, 2003.
Vol. 1830: M. I. Gil', Operator Functions and Localization of Spectra. XIV, 256 p, 2003.
Vol. 1831: A. Connes, J. Cuntz, E. Guentner, N. Higson, J. E. Kaminker, Noncommutative Geometry, Martina Franca, Italy 2002. Editors: S. Doplicher, L. Longo (2004)
Vol. 1832: J. Azéma, M. Émery, M. Ledoux, M. Yor (Eds.), Séminaire de Probabilités XXXVII (2003)
Vol. 1833: D.-Q. Jiang, M. Qian, M.-P. Qian, Mathematical Theory of Nonequilibrium Steady States. On the Frontier of Probability and Dynamical Systems. IX, 280 p, 2004.
Vol. 1834: Yo. Yomdin, G. Comte, Tame Geometry with Application in Smooth Analysis. VIII, 186 p, 2004.
Vol. 1835: O.T. Izhboldin, B. Kahn, N.A. Karpenko, A. Vishik, Geometric Methods in the Algebraic Theory of Quadratic Forms. Summer School, Lens, 2000. Editor: J.-P. Tignol (2004)
Vol. 1836: C. Nastasescu, F. Van Oystaeyen, Methods of Graded Rings. XIII, 304 p, 2004.
Vol. 1837: S. Tavaré, O. Zeitouni, Lectures on Probability Theory and Statistics. Ecole d'Eté de Probabilités de Saint-Flour XXXI-2001. Editor: J. Picard (2004)
Vol. 1838: A.J. Ganesh, N.W. O'Connell, D.J. Wischik, Big Queues. XII, 254 p, 2004.
Vol. 1839: R. Gohm, Noncommutative Stationary Processes. VIII, 170 p, 2004.

Vol. 1840: B. Tsirelson, W. Werner, Lectures on Probability Theory and Statistics. Ecole d'Eté de Probabilités de Saint-Flour XXXII-2002. Editor: J. Picard (2004)
Vol. 1841: W. Reichel, Uniqueness Theorems for Variational Problems by the Method of Transformation Groups (2004)
Vol. 1842: T. Johnsen, A.L. Knutsen, K3 Projective Models in Scrolls (2004)
Vol. 1843: B. Jefferies, Spectral Properties of Noncommuting Operators (2004)
Vol. 1844: K.F. Siburg, The Principle of Least Action in Geometry and Dynamics (2004)
Vol. 1845: Min Ho Lee, Mixed Automorphic Forms, Torus Bundles, and Jacobi Forms (2004)
Vol. 1846: H. Ammari, H. Kang, Reconstruction of Small Inhomogeneities from Boundary Measurements (2004)
Vol. 1847: T.R. Bielecki, T. Björk, M. Jeanblanc, M. Rutkowski, J.A. Scheinkman, W. Xiong, Paris-Princeton Lectures on Mathematical Finance 2003 (2004)
Vol. 1848: M. Abate, J. E. Fornaess, X. Huang, J. P. Rosay, A. Tumanov, Real Methods in Complex and CR Geometry, Martina Franca, Italy 2002. Editors: D. Zaitsev, G. Zampieri (2004)
Vol. 1849: Martin L. Brown, Heegner Modules and Elliptic Curves (2004)
Vol. 1850: V. D. Milman, G. Schechtman (Eds.), Geometric Aspects of Functional Analysis. Israel Seminar 2002-2003 (2004)
Vol. 1851: O. Catoni, Statistical Learning Theory and Stochastic Optimization (2004)
Vol. 1852: A.S. Kechris, B.D. Miller, Topics in Orbit Equivalence (2004)
Vol. 1853: Ch. Favre, M. Jonsson, The Valuative Tree (2004)
Vol. 1854: O. Saeki, Topology of Singular Fibers of Differential Maps (2004)
Vol. 1855: G. Da Prato, P.C. Kunstmann, I. Lasiecka, A. Lunardi, R. Schnaubelt, L. Weis, Functional Analytic Methods for Evolution Equations. Editors: M. Iannelli, R. Nagel, S. Piazzera (2004)
Vol. 1856: K. Back, T.R. Bielecki, C. Hipp, S. Peng, W. Schachermayer, Stochastic Methods in Finance, Bressanone/Brixen, Italy, 2003. Editors: M. Fritelli, W. Runggaldier (2004)
Vol. 1857: M. Émery, M. Ledoux, M. Yor (Eds.), Séminaire de Probabilités XXXVIII (2005)
Vol. 1858: A.S. Cherny, H.-J. Engelbert, Singular Stochastic Differential Equations (2005)
Vol. 1859: E. Letellier, Fourier Transforms of Invariant Functions on Finite Reductive Lie Algebras (2005)
Vol. 1860: A. Borisyuk, G.B. Ermentrout, A. Friedman, D. Terman, Tutorials in Mathematical Biosciences I. Mathematical Neurosciences (2005)
Vol. 1861: G. Benettin, J. Henrard, S. Kuksin, Hamiltonian Dynamics – Theory and Applications, Cetraro, Italy, 1999. Editor: A. Giorgilli (2005)
Vol. 1862: B. Helffer, F. Nier, Hypoelliptic Estimates and Spectral Theory for Fokker-Planck Operators and Witten Laplacians (2005)
Vol. 1863: H. Fürh, Abstract Harmonic Analysis of Continuous Wavelet Transforms (2005)
Vol. 1864: K. Efstathiou, Metamorphoses of Hamiltonian Systems with Symmetries (2005)
Vol. 1865: D. Applebaum, B.V. R. Bhat, J. Kustermans, J. M. Lindsay, Quantum Independent Increment Processes I. From Classical Probability to Quantum Stochastic Calculus. Editors: M. Schürmann, U. Franz (2005)
Vol. 1866: O.E. Barndorff-Nielsen, U. Franz, R. Gohm, B. Kümmerer, S. Thorbjønsen, Quantum Independent Increment Processes II. Structure of Quantum Lévy Processes, Classical Probability, and Physics. Editors: M. Schürmann, U. Franz, (2005)
Vol. 1867: J. Sneyd (Ed.), Tutorials in Mathematical Biosciences II. Mathematical Modeling of Calcium Dynamics and Signal Transduction. (2005)
Vol. 1868: J. Jorgenson, S. Lang, $Pos_n(R)$ and Eisenstein Sereies. (2005)
Vol. 1869: A. Dembo, T. Funaki, Lectures on Probability Theory and Statistics. Ecole d'Eté de Probabilités de Saint-Flour XXXIII-2003. Editor: J. Picard (2005)
Vol. 1870: V.I. Gurariy, W. Lusky, Geometry of Müntz Spaces and Related Questions. (2005)
Vol. 1871: P. Constantin, G. Gallavotti, A.V. Kazhikhov, Y. Meyer, S. Ukai, Mathematical Foundation of Turbulent Viscous Flows, Martina Franca, Italy, 2003. Editors: M. Cannone, T. Miyakawa (2006)
Vol. 1872: A. Friedman (Ed.), Tutorials in Mathematical Biosciences III. Cell Cycle, Proliferation, and Cancer (2006)
Vol. 1873: R. Mansuy, M. Yor, Random Times and Enlargements of Filtrations in a Brownian Setting (2006)
Vol. 1874: M. Yor, M. Émery (Eds.), In Memoriam Paul-André Meyer - Séminaire de Probabilités XXXIX (2006)
Vol. 1875: J. Pitman, Combinatorial Stochastic Processes. Ecole d'Eté de Probabilités de Saint-Flour XXXII-2002. Editor: J. Picard (2006)
Vol. 1876: H. Herrlich, Axiom of Choice (2006)
Vol. 1877: J. Steuding, Value Distributions of *L*-Functions (2006)
Vol. 1878: R. Cerf, The Wulff Crystal in Ising and Percolation Models, Ecole d'Eté de Probabilités de Saint-Flour XXXIV-2004. Editor: Jean Picard (2006)
Vol. 1879: G. Slade, The Lace Expansion and its Applications, Ecole d'Eté de Probabilités de Saint-Flour XXXIV-2004. Editor: Jean Picard (2006)
Vol. 1880: S. Attal, A. Joye, C.-A. Pillet, Open Quantum Systems I, The Hamiltonian Approach (2006)
Vol. 1881: S. Attal, A. Joye, C.-A. Pillet, Open Quantum Systems II, The Markovian Approach (2006)
Vol. 1882: S. Attal, A. Joye, C.-A. Pillet, Open Quantum Systems III, Recent Developments (2006)
Vol. 1883: W. Van Assche, F. Marcellàn (Eds.), Orthogonal Polynomials and Special Functions, Computation and Application (2006)
Vol. 1884: N. Hayashi, E.I. Kaikina, P.I. Naumkin, I.A. Shishmarev, Asymptotics for Dissipative Nonlinear Equations (2006)
Vol. 1885: A. Telcs, The Art of Random Walks (2006)

Recent Reprints and New Editions

Vol. 1471: M. Courtieu, A.A. Panchishkin, Non-Archimedean L-Functions and Arithmetical Siegel Modular Forms. – Second Edition (2003)
Vol. 1618: G. Pisier, Similarity Problems and Completely Bounded Maps. 1995 – Second, Expanded Edition (2001)
Vol. 1629: J.D. Moore, Lectures on Seiberg-Witten Invariants. 1997 – Second Edition (2001)
Vol. 1638: P. Vanhaecke, Integrable Systems in the realm of Algebraic Geometry. 1996 – Second Edition (2001)
Vol. 1702: J. Ma, J. Yong, Forward-Backward Stochastic Differential Equations and their Applications. 1999. – Corrected 3rd printing (2005)